ACTUARIAL STATISTICS

An Introduction Using R

SECOND EDITION

Shailaja Deshmukh
Vidyagouri Prayag

Universities Press

ACTUARIAL STATISTICS: AN INTRODUCTION USING R, SECOND EDITION

UNIVERSITIES PRESS (INDIA) PRIVATE LIMITED

Registered Office
3-6-747/1/A & 3-6-754/1, Himayatnagar, Hyderabad 500 029, Telangana, India
info@universitiespress.com; www.universitiespress.com

Distributed by
Orient Blackswan Private Limited

Registered Office
3-6-752 Himayatnagar, Hyderabad 500 029, Telangana, India

Other Offices
Bengaluru, Chennai, Guwahati, Hyderabad
Kolkata, Mumbai, New Delhi, Noida, Patna

© Universities Press (India) Private Limited 2009, 2025
First published 2009
Reprinted 2009, 2021, 2022, 2023
Second Edition 2025
Reprinted 2026

Cover and book design
© Universities Press (India) Private Limited 2025

ISBN: 978-93-93330-79-6

Typeset in NimbusRom 10 pt by 12 pt *by*
Mohan Boda, Hyderabad

Printed at
Kensington Printing & Packaging, Noida 201 301

Published by
Universities Press (India) Private Limited
3-6-747/1/A & 3-6-754/1, Himayatnagar, Hyderabad 500 029, Telangana, India

Dedicated to our beloved students

Preface to the Second Edition

In the first edition of the book, we introduced some basic concepts related to computation of premiums and reserves for some standard life insurance products. We adopted a rigorous mathematical approach towards the theoretical concepts. To absorb the concepts easily, these were simultaneously supported by numerical computations with R software.

In this revised second edition of the book, the theoretical development has not been changed much. However, we have extended the role of R software. Some readers may not be familiar with R software, which is heavily used in the book. Hence, we have added a separate section on a brief introduction to R in Chapter 2. R codes involved in all monetary computations and computational exercises are included in the last section of each of Chapters 3 to 9. In all, there are 41 codes. We hope that these will help the readers to understand the concepts with ease, to reveal the hidden aspects of the procedures and to fulfill the need for visual demonstration of the concepts. The codes are deliberately kept simple, so that readers can understand the underlying theory with minimal effort.

The book is useful for students appearing for the fellowship examination of the Institute of Actuaries of India (IAI) and other actuarial societies in the UK, USA and Canada. The structure of the examination by IAI is described in Chapter 1. The present book covers some part from CS1, CS2, CM1, CM2 and SP2 (Life Insurance). The extended use of R will be of immense use to students writing examinations of IAI, since one of the modules of the syllabus of this society is 'Actuarial Statistics (CS)'. It consists of papers, CS1(Actuarial statistics) and CS2 (Risk modelling and survival analysis). For each of these subjects, there are two examinations, one is paper-based and the other is computer-based using R.

In the first edition, there are many exercises in each chapter, some of these also involve computations. In the revised edition, in each chapter, computational exercises based on R software are included to provide hands-on experience to students. There are in total, 30 computational exercises and 143 conceptual exercises. R codes given in the last section of each chapter are useful to work out these computational exercises. From our teaching experience, we are convinced that these exercises are beneficial to students to understand and enjoy the underlying theory. These were often given as assignments to students.

Many inequalities among the present values of annuities certain, the actuarial present values of life annuities and the actuarial present values of benefit in some insurance product are derived and illustrated via computation with R. These are rarely seen in the actuarial literature. These inequalities facilitate the comparative study of different actuarial quantities and hence strengthen the understanding of the insurance products.

One more additional novel feature of the revised edition is the inclusion of multiple choice questions (MCQs). There are around 160 MCQs for which answers are provided. These MCQs are helpful for better assimilation of the concepts contained in the book and to review the knowledge. The MCQs have been very helpful while conducting online examinations during the Covid pandemic.

In order to get the feel of real life insurance policies, we have added in Chapter 1, two examples of policies available in the Indian market. One is from the Life Insurance Corporation, India and the other is from the State Bank of India Life Insurance company. Similarly, information on two annuity contracts issued by the HDFC Life Insurance company is given. After studying the theory and techniques through Chapters 3 to 7, we revisit these real life examples in Chapter 7. We compute monetary quantities such as premiums and survival benefits for these policies using reasonable assumptions about the mortality law and the interest pattern. In all these cases, our computed quantities are very close to those specified by the respective insurance companies. We feel that this exercise certainly throws light on the application of theory in actual practice and helps in developing better insight in Actuarial Statistics.

In the entire book we discuss how to compute premiums and reserves for some standard life insurance models. Section 3.4 in the first edition is more relevant to the non-life insurance. Since the focus of the book is on life insurance models, we decided to omit this section in the second edition.

In the second edition, we have made the presentation of the material more concise and elegant as compared to the first edition. All the typos listed in the errata of the first edition have been corrected in the revised version.

The book has evolved out of the instructional material prepared for teaching a course on 'Actuarial Statistics' for several years at Savitribai Phule Pune University, formerly known as University of Pune, Pune, at Modern College, Pune and MIT, Pune, India. To some extent, the topics coincide with what we cover in the course.

The main motive is to provide a fairly thorough treatment of basic techniques, theoretically and computationally using R, so that the book will be suitable for self-study, particularly in the present era of online education. The style of the book is purposely kept conversational so that the reader may feel the presence of a teacher.

We thank Prof. T V Ramanathan, Head of the Department of Statistics, Savitribai Phule Pune University, for providing the necessary facilities. We take this opportunity to acknowledge Mr Madhu Reddy, Director, Universities Press Pvt Ltd, for accepting the revised second edition. We express our gratitude to the Editor, Dr Gita S Dattatri, for her critical review of the manuscript and helpful comments.

We are deeply grateful to our family members for their constant support and encouragement.

Last but not the least, we owe profound thanks to all our students whom we have taught during the last several years and who have been the driving force to take up this immense task. Their reactions and doubts in the class and our urge to make the theory crystal clear to them, compelled us to pursue this activity and to prepare various illustrations and exercises in this book.

All mistakes and ambiguities in the book are exclusively our responsibility. We would love to be informed of and rectify any mistake that a reader comes across in the book. Suggestions and comments from colleagues and readers are most welcome.

<table>
<tr><td>Email addresses</td><td>Shailaja Deshmukh and Vidyagouri Prayag</td></tr>
<tr><td>Shailaja Deshmukh: deshshailaja@gmail.com</td><td>Pune, India</td></tr>
<tr><td>Vidyagouri Prayag: vidyaprayag@gmail.com</td><td>29 June 2024, National Statistics day</td></tr>
</table>

Preface to the First Edition

With the liberalization of the insurance industry in India, many foreign insurance companies have set up joint venture insurance businesses in India. Actuarial science is the basis of insurance and it is unthinkable to carry on the business without deploying actuarial techniques. Insurance Regulatory and Development Authority (IRDA) has specified actuarial valuation of life and non-life insurance as a requirement. As a result, there is a spurt in the demand of professionals called actuaries. An actuary is a business professional who manages financial risks related to insurance products. Statistics is the foundation of actuarial science and as such students of Statistics have an added advantage to exploit the emerging opportunities in the insurance business and to play a key role of actuaries.

Visualizing a widening gap between the demand and supply of actuaries, the Department of Statistics, University of Pune, started offering an elective course 'Actuarial Statistics' to M.Sc. students from year 2000. The credit for the initiation of this course goes to the vision of Prof. J V Deshpande, retired from the Department of Statistics, University of Pune, and the active support of Dr Chitra Lele, Executive Vice President, Sciformix Technologies Pvt Ltd Mumbai and also Country Ambassador, India, for Society of Actuaries, USA. It proved to be a fruitful course. Some of the students indeed got placements in the insurance sector. This course has now been introduced in many Universities in India. I have been teaching this course for some years and noticed that the instructional material for the course is not easily available and students do need a book for the course. The aim of this book is to cater to the need of students and to familiarize them with the application of various basic concepts and statistical techniques in the insurance industry. It covers the syllabus of the post-graduate course on Actuarial Statistics, as recommended by the UGC, India. It will be also helpful to students who appear for examinations conducted by actuarial societies worldwide.

The insurance business involves a variety of facets such as finance, financial economics, marketing, legal aspects and the actuarial calculations. Actuarial calculations mainly involve determination of premium rates and computation of reserves. This book discusses calculation of premium rates and reserves for some standard insurance models in the seventh and eighth chapters respectively. Tracing backwards, chapters five and six discuss two important components of the premium and reserve calculations—actuarial present values of benefit in life insurance and annuity. Chapter 4 is completely devoted to the mathematical preliminaries which work as basic building blocks for the material discussed in all the following chapters. This chapter introduces the concept of future life time random variable, its distribution function and survival function in the international actuarial notation. It also discusses the construction of a life table and analytical laws of mortality. The third chapter addresses the basic question of feasibility of the insurance business with the help of a novel mathematical tool of utility function.

The first chapter introduces the insurance business, its working, role of statistics in insurance and the insurance business in India. The second chapter gives an introduction to a variety of statistical tools used in the following chapters. This chapter will help the readers who do not have a statistics background but still wish to pursue a career in the insurance sector. Chapter nine generalizes the theory of the previous chapters to the insurance models involving multiple lives.

Major statistical techniques, which form the basis of all the chapters, include distribution theory, survival function, force of mortality or hazard rate and the life tables. Various topics in this book are discussed in the context of life-insurance. However, similar concepts and techniques, with appropriate modifications, are also applicable in disability insurance and pension funding. Key terms are provided at the beginning of each chapter and numerous illustrative examples are included as well. For better assimilation of the material, a variety of problems are given in exercises at the end of each chapter along with their answers.

Computations of various quantities, for example, premiums and reserves, carry a lot of significance in the insurance business. For example, it is necessary to prepare a ready reckoner for the premiums corresponding to various insurance and annuity products. Time to time valuation of funds is also necessary for any insurance company for management of the funds. Computations can be done by any software such as Excel, Minitab and Matlab. However, for a variety of reasons, the recent trend is to use R software for statistical computations and data analysis. It is a freely available software from the public domain. In all the Universities in India and abroad, the use of R software is increasing tremendously. A number of recent books incorporate R software in the book for statistical analysis. To be consistent with the recent trend and demand, R software is used in this book to compute various monetary functions involved in the insurance business. R commands are given for all the computations and the meaning of these is explained, so that a reader who is not familiar with R can also use it. All the tables and solutions to all illustrative examples and exercises included in the book are worked out using R. All the figures in the book are drawn using R software and these are helpful to grasp the basic concepts. The last section of chapters 4, 5, 6, 7, 8 and 9 provides the annotated R commands to compute various monetary functions for the specified mortality and interest pattern. The command driven R software brings out very clearly the successive stages in statistical computations.

The manuscript of the book has been used to teach the one semester elective course for postgraduate students at the Department of Statistics, University of Pune for many years. I hope that this book will be instructive and will induce interest among students about the actuarial profession. Feedback, in the form of suggestions and comments from colleagues and students, is most welcome.

I am grateful to the University authorities for granting me study leave for the period 2004–2005, during which I started the present work. I thank the Head of the Department, for providing all the facilities, all my colleagues and family members for the encouragement and appreciation received throughout this venture. I am indeed thankful to the M.Sc. students, who opted for this course in the last couple of years and compelled me to study rigorously and to collect and set a variety of problems.

June 2009   Shailaja Deshmukh
Pune

Contents

List of Figures

List of Tables

Chapter 1

Insurance Business

Key Terms: Actuary, Benefit, Claim, Corporate business, Financial loss, Insured, Insurer, IRDAI, Life tables, Loss rate, Policy, Premium, Pure risk, Randomness, Sharing losses and risk pooling, Speculative risk.

1.1 Introduction

Every individual has some plans in life and has some expectations about the fulfilment of these plans. Sometimes expectations are not realized, may be because the plans were too ambitious and beyond the capabilities or may be because events beyond our control interfere and the plans collapse. For example, the property or business may get destroyed or damaged by fire or natural calamities such as storms, tsunamis, cyclones, torrential downpours, floods or earthquakes. Prolonged illness or death of an earning member of a family or a key person in business may result in considerable financial loss. These are the situations in which insurance helps to compensate for the financial loss.

Insurance system is a mechanism to reduce the adverse financial impact of some types of random events, such as untimely death, accident, fire, theft or natural calamities. Depending on these types, insurance products are classified as life insurance products and non-life or general insurance products. Life insurance products include whole life insurance, term insurance, endowment insurance and their variants. Non-life or general insurance products include vehicle insurance, property insurance and medical insurance.

Life insurance and non-life products are designed to protect against serious financial consequences that result from random events intruding on the plans of individuals or businesses. Thus, insurance is essentially a risk-financing method. It provides a much needed cover to industry, thereby encouraging its growth. Society cannot ignore risks or negative surprises. These are ever present, but can live with them under the umbrella of insurance. Insurance has thus become an important aspect of modern society. Of course, there are certain basic limitations on the insurance protection, which are indicated by the words 'financial consequences' and 'random'. In this chapter, we elaborate on these important issues. For better understanding, it is essential to have a broad knowledge of the insurance business, its working and its role in the economy of any nation.

Sections 3, 4 and 5 give an overview of the insurance business. The word 'random' indicates the key role of statistics in the insurance business. Section 6 briefly describes the role of statistics in the insurance business. Section 7 presents a brief overview of the insurance business in India since its inception.

In this book, commonly used statistical methods in the insurance business are introduced and are illustrated for some standard life insurance products. The entire discussion is at the introductory level and can be treated as a first step towards studying the professional course—actuarial science. The next section briefly introduces what actuarial science is.

1.2   What is Actuarial Science?

There are several insurance companies which provide insurance cover to policy holders in case of any eventuality such as accidents, hospitalization, household hazards, thefts or death. There are finance institutes who look after investment schemes, employee benefits, retirement benefits and pension schemes. The policy holders are required to pay a fixed amount as instalments at regular intervals to get the benefit in the event of any untoward incident or upon the maturity of the policy. An important decision in any insurance contract is the amount of a periodic premium which has to be fixed at the time of issue of the policy. It is one of the major tasks of an actuary.

An actuary is a business professional who manages financial risks related to the insurance product, pension and other financial corporate planning. He uses the techniques of statistics and probability, risk theory, finance, law, marketing and management to quantify the future risk with respect to insurance, annuities and pension programmes. Based on the past data, he proposes a model to predict the outcome of future contingencies. It helps to design insurance products to lessen the financial severity of the events, by appropriately setting the periodic premiums. Thus, the actuary has to combine the skills of a statistician, economist and financier.

To develop such skills, an actuary needs training in actuarial science. Actuarial science is a discipline that comprises many subjects—mathematics, statistics, economics, finance, risk management and computer science. If one refers to the course structure of a degree in actuarial science offered by any University/Institute, or a list of papers in the examination by any of the actuarial societies across the world that one has to pass to be an actuary, then the list includes papers on mathematics, statistics, economics, finance and risk management. For example, the actuarial course offered by the Institute of Actuaries of India (IAI) consists of 7 core principle (CS, CM, CB) papers, 3 core practices (CP) papers, specialist principles (SP) papers and specialist advanced (SA) papers. A candidate who has passed the Higher Secondary or Diploma is eligible for pursuing the course for becoming a Fellow Actuary. However, before enrolling as a student, one has to clear the Actuarial Common Entrance Test (ACET) examination. It is an objective entrance examination based on mathematics, statistics, english, data interpretation and logical reasoning. The ACET is conducted twice a year. Once registered as a student, one has to clear all the 13 papers to attain the prestigious fellowship of the IAI. The structure of the actuarial course offered by the IAI is elaborated below.

(i) The syllabus is covered under the following four stages of subjects: core principles (7 subjects), core practices (3 subjects), specialist principles (7 optional subjects out of which one can select any 2), specialist advanced (5 optional subjects out of which one can select 1).

(ii) In the core principles stage there are 3 modules—actuarial statistics, actuarial mathematics and business.

(iii) Actuarial statistics (CS): The module consists of papers, CS1 (Actuarial statistics) and CS2 (Risk modeling and survival analysis). For each of these subjects, there are two examinations, paper based and computer based, using R and Excel.

(iv) Actuarial mathematics (CM): The module consists of papers, CM1 (Actuarial mathematics) and CM2 (Financial engineering and loss reserving). For each of these subjects, there are two examinations, paper based and computer based, using Excel.

(v) Business (CB): The module consists of papers, CB1 (Business finance), CB2 (Business economics) and CB3 (Business management).

(vi) In the core practices stage, there are three modules, CP1 (Actuarial practice, consisting of two papers), CP2 (Modelling practice, consisting of one paper) and CP3 (Communication practice, consisting of one paper).

(vii) Other papers are on health insurance, general insurance, and so on. More details can be obtained from the IAI web site https://actuariesindia.org.

Thus, the syllabus is based on the core foundations of statistics, probability, mathematics, economics, analytics, modelling, finance and investment. Its objectives are for creating and building up knowledge, applications and higher order skills for students under various scenarios and situations involving risk and uncertainty.

A student member who has passed a minimum of all core principle series (CS1, CS2, CM1, CM2, CB1, CB2, CB3) and all core application series (CP1 to CP3) papers, is eligible to become an Associate Member of IAI. A candidate who completes all examinations (currently 13 papers) and successfully completes the India Fellowship Seminar along with a minimum of 3 years of actuarial experience, is declared as a Fellow of IAI. The present book covers some part from CS1, CS2, CM1, CM2 and SP2 (Life Insurance). Actuarial societies in UK (www.actuaries.org.uk), USA and Canada(www.soa.org) also have similar papers in their examination scheme.

The actuarial profession was formally established in 1848, with the formation of Institute of Actuaries, London. IAI was formed in 1944 and in 1979 became a member of the International Actuarial Association, an umbrella organization to all actuarial bodies across the world. Its objectives include the advancement of the actuarial profession in India, providing opportunities for interaction among members of the profession, facilitating research, arranging lectures on relevant subjects and providing facilities and guidance to those studying for the professional Actuarial Examination. IAI was initially started as a non-examining body when actuaries used to get qualified from Institute of Actuaries or Faculty of Actuaries of UK. It started conducting entrance examinations in India for students of Institute of Actuaries, UK, in 1975. In 1989, it started conducting examinations for its Indian qualification up to associateship level, and in

1992, it started conducting fellowship level exams. The IAI has been following the UK pattern of examinations since November 2000 following the global standards set by the International Actuarial Association. To become an actuary, one must be a fellow of a recognized professional examining body like the IAI, Navi Mumbai or the Institute of Actuaries, London. In India, traditionally, actuaries were found only in the life insurance sector but with the opening up of the economy they find placements in non-life insurance companies, banks, stock exchanges, private and government agencies and this is one field where demand exceeds supply.

Historically, actuarial science became a formal mathematical discipline in the late 17th century with the increased demand for long-term insurance coverage such as life insurance and annuities. This kind of long term coverage required that money be set aside to pay future benefits, such as annuity payments for many years in the future and death benefits, whenever claim arises. This requires estimating future contingent events, such as the rates of mortality by age, as well as the development of mathematical techniques for discounting the value of funds set aside and invested. This led to the development of an important actuarial concept, referred to as the present value of a future sum. In traditional life insurance, actuarial science focuses on the analysis of mortality, the production of life tables, and the application of compound interest to determine periodic premiums in life insurance, annuities. Chapters $4, 5, 6$ and 7 of the book study in detail all these concepts and computations using R for some life insurance products.

The other sector of the insurance industry is non-life or general insurance. It covers short-term forms of insurance such as health insurance, vehicle insurance, travel insurance, property and casualty or liability insurance. In non-life insurance, the coverage is generally provided on a renewable annual period, such as a yearly contract to provide homeowners an insurance policy covering damage to a house and its contents for one year. Coverage can be cancelled at the end of the period by either party. In the property and casualty insurance fields, companies tend to specialize because of the complexity and diversity of risks such as fire and theft. Beyond these, the industry needs to provide catastrophe insurance for weather-related risks, earthquakes, terrorism and all its implications. In health insurance, actuarial science focuses on the analysis of rates of disability, morbidity, mortality, fertility and other contingencies. Actuarial science also aids in the design of benefit structures, reimbursement standards, and the effects of proposed government standards on the cost of health care.

In view of the short term nature of the non-life insurance products, the tools to determine the premiums are different than those for the life insurance products. The present book does not cover the discussion on pricing of the non-life insurance products.

In the reinsurance fields, actuarial science is used to design and price reinsurance and to establish reserve funds for known claims and future claims and catastrophes. Reinsurance can be used to spread the risk, to smooth earnings and cash flow, to reduce reserve requirements and improve the quality of surplus. In the pension industry, actuarial methods are used to measure the costs of alternative strategies with regard to the design, maintenance or redesign of pension plans.

In all of these ventures, actuarial science has to deal with data collection, estimation, forecasting and valuation tools to provide financial and underwriting data for management to assess marketing opportunities and the degree of risk. Historically, actuarial science used deterministic models in the construction of tables and premiums. The science has gone through revolutionary changes during the last 50 years due to the invention of high speed computers and the development of stochastic actuarial models with modern financial theory. Actuarial science needs to operate at two levels: (i) at the product level to facilitate correct equitable pricing and reserving and (ii) at the corporate level to assess the overall risk to the enterprise from catastrophic events in relation to its underwriting capacity or surplus.

The work of an actuary involves a lot of number crunching and the nature of work is quite tedious; nevertheless, it offers rewards in terms of intellectual challenge, status, job satisfaction and earnings. It is essential that the actuaries understand the entire operation of the insurance and employee benefit plans, because their evaluations often influence organizational policies and practices. In fact, actuarial calculations and judgment can commit organizations financially for many future years. In view of such long-range financial commitments, an actuary is frequently involved in many phases of an organization's business, such as general management, marketing, research, underwriting, investments, accounting, administration and long-range planning. Actuaries develop such broad-based knowledge and experience that they are often found in highly responsible management positions, frequently working as insurance company's vice-presidents and presidents or as senior partners in actuarial consulting firms. Actuaries are one of the highest paid professionals in the world. A degree in actuarial science is globally recognized and provides a zero unemployment profession. Any person with a high degree of aptitude for mathematics, statistics and finance can take up this course and become an actuary.

In summary, actuarial science is a study of mathematics, economics and business with the purpose of understanding the practicalities of business and ways of managing financial risks. While running its business, any insurance company has to face a variety of uncertainties, at the same time the company has to be financially sound to give the customers their dues. To manage these operational issues, the study of actuarial science comes in handy.

In this book we will discuss how statistical methods are useful in the determination of premiums and reserves for some standard insurance products. we begin with an overview of insurance business in the next three sections.

1.3   Insurance Companies as Business Organizations

A business is defined as an organization established for the purpose of producing goods and services that consumers need and selling those goods and services to gain profit. Each business organization is structured in one of the following three ways.

(i) **A sole proprietorship:** It is owned and operated by one individual. The owner reaps all profits and is personally responsible for all the debts of the business.

(ii) **A partnership:** It is a business owned by two or more individuals who are known as partners.

(iii) **A corporation:** A corporation is a group of individuals who obtain a charter giving them certain legal rights and privileges. A corporation can own property, can buy and sell goods and services, can sue or be sued. The corporation continues beyond the death of any or all of its owners and this characteristic property provides an element of stability and permanence that a sole proprietorship and partnership cannot guarantee.

The stability makes the corporation the ideal form of business organization for an insurance company. Life Insurance Corporation (LIC) is known as the biggest insurance company in India before the privatization of this sector.

The insurance business is highly specialized, technical corporate business with a huge capital, where customer trust is of prime importance. Insurance companies are financial intermediaries that function in the economy as part of the financial services industry and make a significant contribution to the economic growth of the nation. A financial intermediary is an organization that helps to channel funds through an economy by accepting the surplus money of savers and supplying that money to borrowers who pay to use the money; for example, banks, financial institutes and so on. Banking institutions were developed for the purpose of receiving, investing and dispensing the savings of individuals and corporations. Unlike insurance systems, saving institutions do not make payments based on the size of a financial loss occurring from an event outside the control of the person suffering the loss. However, with new financial reform, various banks have entered into the insurance business; for example, HDFC, ICICI and SBI.

Insurance companies play a crucial role in the commercial life of the nation and act as lubricants of economic activities. The following points elaborate on the role of insurance companies in the economy.

(i) They contribute to country's economic stability and general welfare by compensating individuals and businesses for potential financial losses that might otherwise ruin them.

(ii) Insurance systems increase total production by encouraging individuals and corporations to embark on ventures where the possibility of large losses would inhibit such projects in the absence of insurance. For example, development of marine insurance for reducing the financial impact of the perils of the sea. Mutually advantageous trading activity might be too hazardous for some potential trading partners without an insurance system to cover possible losses.

(iii) Insurance companies invest huge sums of money in stocks, bonds, mortgages (conveyance of property as security for debt until money is repaid), government securities, and other income-producing enterprises. Insurance companies also help to guarantee repayment of loans, the completion of commercial projects and public works. People get cover for the risks of starting a new business or acquiring property by buying insurance. Thus, the insurance industry helps in increasing the production of goods and services.

(iv) Finally, insurance companies pay large amounts in taxes and provide employment to many individuals.

The basic purpose of insurance is to provide protection against the risk of financial loss. In order to understand how insurance works, it is necessary to understand the concept of risk and which types of risks are insurable.

1.4   Concept of Risk

We all are surrounded by risks. We take risks when we travel, when we engage in recreational activities and hobbies and even when we breathe. Some risks are significant, some are not. Within the common meaning of risk, there are two distinct elements—the idea of loss or injury and that of uncertainty. Risk exists when there is uncertainty about the future. Both individuals and businesses experience two kinds of risk—speculative risk and pure risk.

Speculative Risk: It involves three possible outcomes: loss, gain and no change. For example, when we purchase shares of stock, we speculate that the value of the stock will rise and we will earn a profit on our investment. At the same time, we know that the value of the stock could fall and we could lose some or all the money we invested. Finally, we know that the value of the stock could remain the same, we might not lose money, but we might not make profit either.

Pure Risk: It involves no possibility of gain; either a loss occurs or no loss occurs. An example of pure risk is the possibility that a person becomes disabled. If he is unable to work, he will experience a financial loss. If, on the other hand, he never becomes disabled, then he will incur no loss from that risk. The possibility of financial loss without the possibility of gain is called pure risk and this is the only kind of risk that can be insured. The purpose of insurance is to compensate for financial loss, not to provide an opportunity for financial gain.

Insurance products are designed in accordance with some basic principles that define which risks are insurable. In order for a risk, a potential loss, to be considered insurable, it must have certain characteristics, as listed below.

Characteristics of Insurable Risks

(i) **The risk must be pure:** As mentioned above, insurance covers situations involving only pure risk, that is, situations in which only losses can occur. Such situations include death, fire, accidents and natural calamities like earthquakes or floods. Insurance does not cover gambling and other speculative risks, in which either losses or gains may result.

(ii) **The loss must occur by chance:** In order for a potential loss to be insurable, the element of chance must be present. The loss should be caused either by an unexpected event or by an event that is not intentionally caused by the person covered by the insurance. Further, the size, the frequency and time of the loss must also be random. For example, life insurance is designed to compensate for the loss to a family if a wage-earning parent dies, the time of death in this case is random. Fire insurance pays all or part of the loss if a home owner's house is destroyed by fire, which is again a random event. Vehicle insurance helps to cover the cost of damage resulting from an accident, the occurrence and the time of such an event being random variables. Property insurance covers the damage due to flood or fire or windstorm, which are random events. Health insurance helps to pay medical bills, beyond the budget, when an individual suffers from certain diseases, the occurrence and severity of which are random variables.

(iii) **The loss must be measurable in terms of money:** Insurance is restricted to reduce those consequences of random events that can be measured in monetary terms. Death, disability, retirement are various forms of income loss. Theft, natural calamities and adverse court judgments cause loss of wealth and are direct forms of income loss. Damage of the property reduces the value of that property. On the other hand, social and psychological impacts of natural calamities, pain and suffering caused by those, cannot be measured in monetary units and hence cannot be insured.

(iv) **The loss must be defined:** For most types of insurance, an insurable loss must be defined in terms of time and money. In other words, the insurance company must be able to determine when to pay policy benefits and how much the benefit should be. Death, illness, disability and old age are generally identifiable conditions.

(v) **The loss must be significant:** Insignificant losses, like the loss of a cycle, are not insured, the administrative expenses being comparatively high in such cases. Losses that cause financial hardship are considered to be insurable; for example, a person injured in an accident may lose a significant amount of income if he is unable to work.

(vi) **The loss rate must be predictable:** In order to provide a specific type of insurance coverage, an insurer must be able to predict the probable rate of loss. It will enable him to determine the proper premium amount. These predictions of future losses are based on the concept that the probability of random losses can be determined on the basis of past data. An important concept that helps to assure the accuracy of predictions about the probability of an event is the law of large numbers. It states that, the more times we observe a particular event, the more likely it is that the observed proportion of occurrence of events will approximate the true probability of the event. Insurance companies collect specific information from a large number of individuals in order to identify the pattern of losses. Using these statistical records, life insurance companies develop charts called life tables or mortality tables. These tables indicate with great accuracy the number of individuals in a large group who are likely to die at a specific age. Insurance companies also develop similar charts, called morbidity tables, which display the rates of morbidity or incidence of sickness and accidents, by age, occurring among a given group of people. By using accurate mortality and morbidity tables, life and health insurers can predict the probable loss rates and use these to establish premium rates that will be adequate to pay claims.

(vii) **The loss must not be catastrophic to the insurer:** Catastrophic losses are not insurable because the insurer could not responsibly promise to pay benefits for the loss. To prevent the possibility of catastrophic losses, insurers spread the risk they choose to insure. For example, a property insurer would be unwise to issue policies covering all homes in the vicinity of an epicenter of earthquake, because one event of earthquake could result in more claims at one time than the insurer could pay. Alternatively, an insurer can reduce the possibility that it will suffer catastrophic losses by transferring risks to another insurer, by reinsuring those risks. Reinsurance is insurance that one insurance company, known as the ceding company, purchases from another insurance company, known as the reinsurer.

The above seven characteristics are useful in identifying the general kinds of losses that are insurable and provide a helpful framework for the study of insurance principles and products.

Note that insurance does not directly reduce the probability of loss. The existence of the vehicle insurance will not alter the probability of an accident. However, a well-designed insurance system will often provide financial benefits if some random activities result in a loss.

In summary, an insurance system is a mechanism for reducing the adverse financial impact of random events, which prevent the fulfilment of reasonable expectations.

One may wonder how an insurance company can afford to be financially responsible for the economic risks of individuals who sign the contract with the insurance company. To clarify this aspect, we discuss in the next section, the working of the insurance business.

1.5   How does the Insurance Business Operate?

To understand the mechanism of the operation of any insurance company, we list below some of the common terms used in insurance business.

 (i) **Insured or Life Assured:** An insured is a person, or a group, or an organization, whose life or property is covered by an insurance company.

 (ii) **Insurer:** The insurance company is known as an insurer.

(iii) **Insurance Policy:** An insurance policy, usually referred to as a policy, is a legal financial agreement between the insurer and the insured. The insurer makes a contract with the insured through an insurance policy to cover up financial risk in the event of the unfortunate demise of the insured. Thus, a policy is a written evidence of the contract of insurance. The company promises to pay the insured a certain sum of money for the types of losses stated in the policy, in exchange of some amount from the insured. The insured is also known as a policy holder.

 (iv) **Beneficiary (Nominee):** It is a person, the insured mentions in the policy document at the time of buying the policy, to receive the benefits of the insurance policy, in the case of the policy holder's demise.

 (v) **Life Cover:** Life cover, simply known as cover, is the amount that the insurer will pay to the beneficiary in case of the unfortunate event of death of the policy holder.

 (vi) **Insurance Claim or Claim:** Claim is a formal request by the insured to the insurer for reimbursement against losses covered under policy holder's insurance policy.

(vii) **Sum Assured or Benefit or Death Benefit:** A sum assured is an amount that is paid to the beneficiary of the plan, in the unfortunate event of the policyholder's demise. The insurance company pays this money as per the sum chosen by the insured at the time of purchasing the policy.

(viii) **Policy Term:** Policy term is the duration for which the life cover continues under that policy. It is a duration for which the policy is in force or the policy is active. It may be fixed or random, depending on the death or survival of the policy holder.

(ix) **Premium:** Insurance is an agreement between the insured and the insurer. The insured has to pay certain amount to the insurance company for the financial cover for losses as designated in the policy. The amount the insured pays to the insurer for receiving the benefits of the insurance policy is known as the premium. These payments can be made on a regular basis throughout the policy duration, or for a limited number of years or just once, as per the options available under the policy.

(x) **Proposer:** It is a person who pays the premiums of the policy. If the insured has bought the policy for himself, then he is both the insured and the proposer. However, if a person purchases an insurance policy for a family member, then the purchaser is the proposer and the family member is the insured.

(xi) **Premium Paying Terms:** The period for which the premiums are paid is known as the premium payment term. Depending on the nature of the policy, it may be fixed or random, again depending on the death or survival of the policy holder.

(xii) **Underwriting:** Underwriting is a process through which an individual or institution takes on financial risk for a fee. The insurer evaluates the risk to determine if the insurance company can afford to insure it. It thus involves pricing the risk by analyzing how likely it is that the insured would make a costly claim and the insurer would have to face the loss by issuing the policy.

(xiii) **Annuity:** In any insurance product, to get the benefit, the policy holder has to pay one-time or periodical premiums to the insurance company, as decided at the time of signing the contract. A series of periodic payments or a special type of cash flow, where payments occur at regular intervals, is known as an annuity. Thus, the payment of premiums by the insured to the insurer is an annuity.

(xiv) **Annuity Contract:** In the financial services industry, the term annuity is also used as a financial contract. It is a saving plan sold by the insurance companies to provide regular income to the insured. For example, an individual after retirement invests in the annuity plan to get the regular, either monthly or annual, income from the insurer. Thus, in such a contract, the insurer promises to make a series of periodic payments to the insured, in exchange of a single premium or a series of premiums from the insured. Annuity contract is often described as the flip side of life insurance, because the annuity contract provides benefit till the individual survives, whereas life insurance provides benefit after the death. Two such annuity contracts are presented below.

(xv) **Survival Benefit:** Survival benefit is the amount that the insured receives from the insurer in the annuity contract. In some life insurance products, after the premium paying term is over, the insured is offered the survival benefit. One such policy is described below.

(xvi) **Non-participating Policy:** An insurance company pays dividends if the money collected via premiums exceeds the amount needed to pay benefits and administrative cost. Dividends also may include a share of the profits the company earns on investments made with premium funds. Dividends may be paid on many types of insurance. But they are most commonly paid on life insurance. Policies that do not pay dividends are known as non-participating policies. An owner of a participating policy may receive the dividends in cash or let them accumulate with the insurance company, which pays interest on the amount. A policyholder may also use the dividends to pay the premiums on the policy or to buy additional insurance.

The life insurance products are broadly classified into three categories such as, whole life insurance, term insurance and endowment insurance. In all these products, payment of benefit depends on death or survival of the policy holder. For example, a whole life insurance offers a lifetime coverage and pays a fixed sum to the beneficiary of the insured, when the insured dies. In contrast, a term life insurance provides coverage for a specified period of time, the policy term, as decided in the contract. The policy benefit is payable only if the insured dies during the policy term. If the policy holder lives for the term and then dies any time after that, then the beneficiary does not get any benefit. The third type is an endowment insurance. In this product also a policy term is fixed. The benefit is paid to the insured, if death occurs before the term ends. If the insured survives to the end of the period of the policy, then the benefit is paid at the maturity.

Many products in the market are slight versions or combinations of these insurance products and annuity contracts. In order to get the feel of real life insurance policies, we give here two examples of policies available in the Indian market. One is from LIC, and the other from State Bank of India Life Insurance company. Two illustrations of annuity contract from HDFC Life are also discussed.

LIC's Jeevan Umang Plan: LIC's Jeevan Umang plan offers a combination of deferred annuity product and whole life insurance. This plan provides for annual survival benefits from the end of the premium paying term till maturity and a lump sum payment at the time of maturity or on death of the policyholder during the policy term. In addition, this plan also takes care of liquidity needs through loan facility. In this policy, the age at maturity is considered to be 100 years and the policy term is $100 - x$, where x denotes the age of the policy holder at the time of purchasing the policy. The premium paying term is decided in such a way that the policyholder will be paying premiums till the age of 70. A policy holder can choose any other option for premium payments. Thus, the premium paying term may be fixed, say 20 years, provided the policy is in force till that period. The modes of premium payment allowable are yearly, half-yearly, quarterly or monthly. The minimum age at entry is 90 days, while the maximum entry age is 55 years. For a person of 50 years of age, the premium paying term is 20 years meaning thereby that he is supposed to pay the premiums for 20 years, till he reaches the age of 70 years. The insured is covered for the rest of his life, hence the policy is called as whole life insurance. Minimum basic sum assured is Rs 2 lakhs while maximum is Rs 20 lakhs. For the life assured surviving to the end of the premium paying term, a survival benefit equal to 8% of the basic sum assured is payable each year. The first survival benefit payment is payable at the end of the premium paying term and thereafter on completion of each subsequent year till the life assured survives or till the policy anniversary prior to the date of maturity, whichever is earlier. We take a hypothetical case study for illustration.

Case Study 1: Mrs Vidula is a lady who was born on 12 December 1977. Her age as on 13 September 2022 is therefore 45 years. She chooses the Jeevan Umang plan at age 45. The cover she chooses is Rs 5 lakhs. Mrs Vidula thinks that she will be able to pay the premium till the age of 60 years as she would be earning till that time. Hence, she chooses the premium paying term as 15 years. From the calculation on the LIC website, Mrs Vidula will have to pay an annual premium of Rs 38772 for 15 years, including taxes, provided she survives till 60 (Source: https://licindia.in).

SBI Life—Saral Jeevan Bima: It is a term insurance product. A person of age between 18 and 60 years can avail this policy. The policy term can be from 5 to 40 years. The minimum basic sum assured is Rs 5 lakhs while the maximum is Rs 25 lakhs. Premium options are regular premium, that is, policy term and premium paying term are the same, limited premiums, 5 or 10 years, or a single premium. Premiums could be paid annually, half-yearly or monthly. On death of the life assured, during the policy term, the beneficiary receives the sum assured. No benefit is paid after the policy term is over, if death does not occur during the policy term. The following case study clarifies the product.

Case Study 2: Mr Ketan, a healthy 50 year old person, opts for a 10 year term Saral Jeevan Bima policy for a sum assured of Rs 20 lakhs. From the SBI life website, Mr Ketan is required to pay a premium of Rs 22140 for 10 years or till the policy is active. In case of the unfortunate event of death of Mr Ketan during the term of 10 years, a sum assured equal to Rs 20 lakhs will be paid to his beneficiary. If death does not occur during the term of 10 years, there is no benefit payment. For the sake of understanding, suppose Mr Ketan dies during the 5th year of the policy term. In that case, he would have paid 5 annual premium installments amounting to Rs $22140 \times 5 = 110700$. He dies at the age of 55. His sum assured is Rs 20 lakhs. This amount would be paid to the nominee. Note that the annual premium in this term insurance seems to be much smaller than that in the case study of Jeevan Umang insurance. This is because, the latter is term insurance and the insurance company has no liability to pay the assured benefit after the policy term is over. (Source: https://www.sbilife.co.in/en/individual-life-insurance/traditional/saral-jeevan-bima)

In the above mentioned two illustrations, readers can refer to their respective websites for further details. In the present book we develop the theory to decide some component of the basic premium in some such above mentioned cases.

We now briefly provide information on two annuity contracts issued by HDFC Life Insurance company.

Immediate Annuity Policy: It is a whole life annuity, purchased with a single premium, to be paid at the time of issue of the contract. Thus, the single premium is a purchase price of the annuity. In annuity contracts, a policy holder is referred to as an annuitant. In this contract, an annuitant receives the benefit annually till he survives. In the event of the death of the annuitant, the annuity payments will cease and the company will pay to the nominee, mentioned in the contract, the purchase price of the annuity. In immediate annuity policy, the benefit payments are made annually at the anniversary of the policy, the first to be given at the first anniversary.

Case Study 3: Mr Bapat of age 57, purchased an immediate annuity policy on 21 September 2011. The purchase price was Rs $10, 58, 652$. An annual benefit of Rs 68503 is payable annually, where the first payment was made on 21 September 2012.

HDFC Life Pension Guaranteed Policy: This is also a single premium immediate annuity contract, where the benefit payments are made quarterly. In the immediate policy, the first payment is made at the end of the first quarter. As in the previous annuity contract, an annuitant receives the benefit till he survives. In the event of death of the annuitant, annuity payments will cease and the company will pay to the nominee, mentioned in the contract, the purchase price of the annuity. It is a non-participating, non-linked annuity policy. The non-linked policy means the amount of single premium is not invested in the share market.

Case Study 4: Mr Singh of age 65, purchased the policy with a single premium of Rs 4911591 on 29 September 2020. The quarterly benefit payment is Rs $67, 708.43$. The annuity payment due dates are 27th December, 27th March and 27th September, the first payment being paid on 27 December 2020.

Note that in both the annuity contracts, the period of benefit payments is a random variable, it is the life length of the annuitant after signing the contract. In the light of such randomness, it is of interest to see how the amount of benefit payments is fixed at the time of policy issue.

It is clear from the above illustrations that many insurance products are combinations of three basic life insurance products and some annuity products. In addition to these, many insurance products offer riders such as annual survival benefit to the policy holder, savings such as term deposits, loan facility, tax benefits, and so on. Some offer facilities like option of limited premium paying term, waiting period, deferment of benefit and loyalty benefit. These options are based on the requirement of different sectors of the society based on age, occupation, sex, health status and location. Also, several other aspects such as legal and taxes play a role in framing the policies. As a result, the insurance products available in the market become complex and intricate and that is the reason why life insurance companies employ agents to explain the products to common people. Nonetheless, the underlying principle of computation of premiums for any life insurance remains the same (which we are going to learn in this book).

To enrich our students' understanding regarding how the theoretical techniques they have learnt throughout the book, get applied in real insurance products, we revisit these case studies in Chapter 7. We illustrate how the premiums in first two case studies and benefit amount in the next two case studies are determined under the assumption of suitable mortality pattern and rate of interest.

In general, insurance works on the principle of sharing losses and risk pooling. With risk pooling, individuals who face the uncertainty of a particular economic loss, for example, the loss of income because of disability, transfer this risk to an insurance company. Insurance companies know that not everyone, who is issued a policy to cover the risk of economic loss caused by disability, will suffer a disability. In reality, only a small percentage of the individuals, who purchase this type of insurance, will actually become disabled at some time during the period of insurance coverage. By collecting premiums from all individuals and businesses who wish to transfer the financial risk of disability, insurers spread the cost of the few losses that are expected to occur, among all the insured persons. Insurance then provides protection against the risk of economic loss by applying the following simple principle. If the economic losses which actually result from a given peril, such as disability, can be shared by large number of people who are all subject to risk of such losses and if the probability of loss is relatively small for each person, then the cost to each person will be relatively small.

The insurance company invests the money collected via premiums in stocks, bonds, mortgages, government securities and other income generating enterprises. The company pays benefits from the premiums it collects and from the investment income the premiums earn. Insurance works because policyholders are willing to trade a small, certain loss, namely the premiums, for the guarantee that they will be indemnified in the case of a larger random loss. In Chapter 3, we prove mathematically how the two parties come together and agree to sign a contract.

Insurance companies base premiums on the interest they expect to earn from investments made with the premiums and also on the costs of doing business and making a profit. The amount of the premiums may also be affected by a person's insurability; that is the risk the insurance company takes in providing coverage for that person. A person who has high blood pressure, diabetes, or some other medical problem may be charged higher premiums. A person whose leisure activities are considered dangerous may also need to pay higher premiums.

Although a policyholder may never have to claim benefits, the investment in terms of premiums is never wasted, insurance gives policyholders a feeling of security, for example, travel insurance while travelling abroad. They know they will be indemnified if a loss occurs. They can therefore own property, drive a car, operate a business, and engage in many other activities under the assurance of cover to possible financial loss.

The role of statistics in the insurance business is evident in the discussion on the characteristics of insurable risks. We now discuss it in more detail.

1.6   Role of Statistics in Insurance

Statistics is the foundation upon which important calculations of an insurance business are built. Many a time, one has to face uncertainty, which is one facet of risk. Insurance business helps to deal with uncertainty using many tools from probability and statistics. The role of statistics in the insurance business is clear once we take a brief tour of the chapters of this book. We provide below a quick overview of the contents of the book.

As noted in the previous section, in any insurance product, the benefit is paid to the beneficiary after the death of the insured. Time of death being an uncertain event, the time at which the benefit is to be paid is a random variable. As defined in Section 1.5, in all the insurance products, the period from the policy issue to the payment of claims is known as the policy term. During this period we say that the policy is in force or the policy is active. Since the life length of an individual is a random variable, the policy term is also a random variable. For example, if the whole life policy is issued at age 25, then the time period of interest to the insurer, is the life length of the insured beyond 25. It is known as the future lifetime beyond 25. The distribution of such a future life time random variable can be obtained from the distribution of life length random variable of the individual, conditional on the event that the individual has survived up to age 25. Hence, the distribution theory from statistics comes into the picture. Chapter 4 is devoted to the theory related to the distribution of future life time random variable. In all the subsequent chapters, distribution of the future life time random variable is a basic building block.

In any insurance product, the benefit is paid to the beneficiary either at the moment of death or at the end of policy year. Although it is said that the benefit is paid at the moment of death, it is not to be taken in the strict sense of the words. It may take a couple of days to report a death to the insurance company. It takes some time to settle the policy, but such a delay is usually small compared to the time for which the policy is in force and does not have a significant effect on the finance of the insurance companies. So it can be ignored and in all the derivations one can treat the death benefit as being paid at the moment of death. Such treatment simplifies the mathematics to a great extent without losing much accuracy. With this assumption, the time period for which the policy is in force is a continuous random variable. We study its distribution in Section 4.2.

In some products, the benefit is not paid as soon as the claim is made, but is paid at the end of policy year, also referred to as the end of year of death. For example, suppose a policy is written on 2 November 1960. The policy year is taken as 2 November 1960 to 1 November 1961, and 2 November is named as the policy anniversary. If the insured dies between 2 November 2026 and 1 November 2027, the benefit will be paid on 1 November 2027, since the policy year ends on 1 November 2027. Hence, the time period for which the policy is in force is 2 November 1960 to 1 November 2027, that is 67 years. Thus, when the benefit is paid at the end of year of death, then also the period for which the policy is in force is a random variable. It is related to a curtate future life time and we study it in Section 4.3.

In any insurance product, to get the benefit, the policy holder has to pay a one-time or periodic premiums to the insurance company, as decided at the time of signing the contract. Thus, the payment of premiums by the insured to the insurer is an annuity. In whole life annuity, described in Section 1.5, the payment of periodical benefit by the insurer to the insured is also an annuity. The number of payments of premiums or the number of benefit payments in life annuity products, is in many cases governed by the death or survival of the insured. Thus, the period of annuity payments in most of the cases is related to the future life time random variable.

Insurance companies invest the money collected via premiums in interest earning enterprises. In view of the long-term nature of these insurances, the amount of investment earnings, up to the time of payment, is also uncertain. Such uncertainty is due to two reasons—(i) random period for which policy is in force and (ii) variation in the rate of interest depending on the economic and financial status of the country. In the present book we assume that the rate of interest is non-random and it remains the same for the period of interest. Thus, the uncertainty in investment earnings is only due to the random period of investment, which is dictated by the future lifetime of the insured. Interested readers may refer to the Chapter 6 on 'Stochastic Interest Rate' of the book by Deshmukh [4], in which the rate of interest is considered to be a random variable.

Determining the premiums optimally is an important task of any insurance company. It depends mainly on the insurance product which involves three aspects—the amount of benefit, the time at which the benefit is to be paid and the mode of premium payments. The most important issue is that the company has to decide the premium amount at the time of issue of the policy. To decide the premiums, it is necessary to take into account the time value of the money involved in the premium input and the benefit output. This concept is discussed in Section 5.2.

Suppose the whole life insurance contract with a benefit value of 1 lakh is signed by an insured with a single premium to be paid at the time of issue of the policy. The prominent question is to decide the amount of the single premium, also known as the one-time purchase value of the policy. The insurer will invest the premium amount to earn interest on it and will pay the benefit from the accumulated value of the premium amount over the period for which the policy is in force. In this approach, expenses incurred and profit margin of the company are ignored. Hence, it is known as a net single premium. Thus, the net single premium amount should be such that it will accumulate to 1 lakh, with the specific rate of interest, over a specific period, which is of course random in any life insurance product. Similar arguments are used to decide periodical premiums. Chapter 7 is devoted to the determination of the different types of premiums for some standard insurance products. The main guiding principle in the determination of premiums is the equivalence principle. It is given by,

$$\text{Expected Present Value of Outflow} \ = \ \text{Expected Present Value of Inflow}$$

Inflow to the company is via premiums and interest and outflow of the company is via payment of claims and administrative expenses. In insurance literature, the expected present value is known as the actuarial present value.

Chapter 5 discusses how to decide the expected present value of the outflow, that is, benefit to be paid at the moment of death or at the end of year of death, for three basic insurance products. Chapter 6 is concerned with the expected present value of inflow via various types of modes of premium payments. These two chapters together help to decide in Chapter 7, the premiums to be charged for a variety of insurance products.

Two important calculations in the insurance industry are premium calculation and the reserve calculation. Chapter 8 is concerned with the computation of reserves, also known as policy fund values. It is a technical term. As time passes by, the company receives the premiums and pays out the benefits. If no claim has been made, the present value of the benefits will increase as the time for payment becomes nearer and the value of the premiums still to be received will reduce as the annuity value with which they are valued will decrease. The insurer wants to be confident that funds at hand, together with future premiums and interest earnings are sufficient to pay the promised benefits. Suppose a policy is issued to an individual of age x. After t units of time, if the policy is in force, some premiums are yet to be received. However, if at all the claim arises at age $x + t$, the insurer has to pay the benefit amount and hence must have sufficient funds at hand to pay the claim. This fund is known as reserve or policy value. Thus, reserve is liability representing the amount of money an insurer estimates for the fulfilment of future obligations.

Chapter 9 generalizes the theory of the previous chapters to the insurance models involving multiple lives.

Life insurance premiums are based mainly on statistical tables called mortality tables or life tables. These tables state the proportion of individuals of a given age who are expected to die during a year. These are obtained from the vital statistics registry, which keeps the record of time of deaths along with the cause of deaths. Chapter 4 discusses thoroughly the construction of life tables. One needs some assumptions in these derivations, which are based on the uniform distribution. Estimates are usually reliable when these are obtained from large data, as conveyed by the laws of large numbers. According to the laws of large numbers, when there is a bigger base for estimation, the estimates of probabilities of survival or death in a certain interval become more predictable.

The third chapter addresses the basic question of feasibility of insurance business, wherein the two parties agree to come together to sign a contract. It is explained with the help of a novel mathematical tool of utility function. For students with a statistics background, it is easy to understand the statistics involved in the insurance business. There may be many readers who do not have sufficient statistics background but still wish to pursue a career in the insurance sector. To help this group of readers, the second chapter gives an introduction to a variety of statistical tools used in subsequent chapters.

As the title of the book indicates, R software is heavily used throughout the book to illustrate the concepts with numerical examples. Some readers may not be familiar with R software. Hence, we have added a separate section on a brief introduction to R in Chapter 2.

Insurance works well only when the possible losses to the insured person can be estimated. Insurance companies take advantage of the laws of probability. Laws of probability are based on the laws of large numbers. For example, as the number of car insurance policyholders increases, the insurance company can use this law to predict with greater accuracy, the number of policyholders who will be involved in an accident. These laws enable an actuary, to determine the likelihood that an event will occur. Actuary uses this knowledge in the calculation of premiums and reserves. Actuary uses a variety of techniques from statistical inference and stochastic processes to estimate the rate of occurrence of events of interest. For example, the number of car accidents in a given region and in a given period can be modelled by the Poisson process. With the data on the number of car accidents in a given period, one can estimate the rate of occurrences of events with maximum likelihood estimation or with the method of moments.

The actuary has to evaluate the financial risk a company takes, when it sells an insurance policy or offers a pension program. In performing these duties, actuaries have many responsibilities. First, the actuary must make sure that there is enough cash on hand to pay benefits when people make claims on the insurance policies or draw income from their pension plans. Secondly, the actuary must also see to it that the price charged to participants' insurance or pension plan is fair. In the following chapters we will discuss how statistical methods help to achieve this goal.

Most insurance systems operate under dynamic conditions, so it is an important task of the actuary to have a plan for collecting and analyzing insurance operating data so that the insurance system can adapt. Adaptation in this case may mean changing premiums, paying an experience-based dividend or premium refund or modifying future policies. In this setup also, statistical tools such as time series analysis and regression analysis come handy. With the liberalization of the insurance industry in India, there is a vast scope for statistics students to exploit the emerging opportunities in the insurance business.

The last three decades have witnessed noticeable changes in India's financial scenario. There were reforms in the banking sector and in stock markets. Towards the end of 1999, India took the bold step of opening the insurance sector to private companies. Since this sector was liberalized, many new players have entered the Indian insurance market, such as, Bajaj Allianz Life Insurance Company, Aviva Life Insurance Company, Birla Sun Life Insurance Company, HDFC Standard Life Insurance Company, ICICI Prudential Life Insurance, SBI Life Insurance Company, OM Kotak Mandindra Life Insurance Company and many others. These companies are joint venture businesses, in which outside partners are world renowned names in the insurance business. As a consequence of such new emerging insurance companies, there is lot of scope for students of statistics in this field to play the key role of actuaries.

In the following section, a brief overview of the growth of the insurance business in India, since its inception up to 2021, is presented (Palande et al [12], (www.investindia.gov.in)). Insurance in India refers to the market for insurance in India which covers both the public and private sector organizations. It is listed in the constitution of India in the seventh schedule as a Union list subject, meaning it can be legislated by the Central government only.

1.7   Insurance Business in India

The formal insurance business, as we know it today in both the life as well as the non-life (general) sector, was introduced in India by the British in the beginning of the 19th century. Over a period of time the business spread, though not adequately. By 1956, as many as 154 Indian insurers, 16 non-Indian insurers and 75 provident societies (in all 245 entities) had entered the life insurance business in India. However, the geographical spread and the number of lives covered were rather small. The business was mainly confined to cities. During this period, a number of malpractices occurred in the industry, causing losses to the unsuspecting public. There were also some instances of mismanagement and misutilization of the funds collected. To safeguard the industry and the public, the then Government of India opted for nationalization of the insurance industry. The bill to provide for nationalization of the life insurance business was introduced in the Lok Sabha in February 1956 and the same became an Act on 1 July 1956. The well-known economist Dr C D Deshmukh, who was the finance minister at that time, is known as the architect of nationalization of the insurance business.

To take care of the life insurance business, the LIC was formed on 1 September 1956 by an Act of Parliament, with a capital contribution of Rs 50 million from the Government of India. LIC was set up as an autonomous corporation. At the time of its inception, the LIC had five zonal offices, 33 divisional offices and 212 branches. After the initial difficult period, over the years LIC made commendable progress. It has grown from a level of Rs 137.5 million sum assured under 5.4 million policies in 1957 to Rs 55699.7 million under 20.304 million policies as on 31 March 2021. The valuation surplus and consequently, the bonus to policyholders and the central government's share steadily increased over the years. Through its vast network of 2048 branches, 113 divisions and eight zonal offices spread over the country, its huge marketing force and number of full-time and part-time agents, the LIC reached various corners of the country and provided sales and services to the Indian public. It also reached illiterate people living in interior rural areas. Thus LIC had virtually a monopoly over the life insurance business. But, now there are 23 private life insurance companies in India. As a result, LIC's market share is slowly slipping to private giants like HDFC Life, ICICI Prudential Life Insurance, General Insurance Corporation of India and Exide Life Insurance. During the year 2020, life insurance issued 288.47 lakh new individual policies out of which LIC issued 75.9% of policies and the private life insurers issued 24.1% of the policies. (www.investindia.gov.in)

The management of non-life insurance was taken over by the central government on 13 May 1971 as a prelude to nationalization, because general insurance had become by and large city-oriented, catering mainly to the needs of organized trade and industry. The business was nationalized with effect from 1 January 1973 by the General Insurance Business Act, 1972. The General Insurance Corporation (GIC) was incorporated as a holding company in 1972 to look after the non-life insurance. 107 insurers were amalgamated and grouped into four companies, namely the National Insurance Company Ltd, the New India Assurance Company Ltd, the Oriental Insurance Company Ltd and the United India Insurance Company Ltd. The general insurance business grew in spread and volume after nationalization but still continued to show a bias towards city-based trade and industry for a number of years.

On the whole, the insurance industry in India experienced substantial growth in terms of absolute numbers. However, it was unable to tap the full potential of the market. In fact, this was one of the important considerations which motivated the Malhotra Committee to recommend, and the government to favor liberalization and introduction of competition. There are several other reasons and certain historical developments, nationally and globally, which persuaded the government of India to take steps to open this sector. De-regulation, globalization and privatization are the routes that were found to have been successful in many parts of the world. Accordingly, India too has opted for these routes. Some of the important factors leading to open up the insurance sector are as follows.

(i) **Global context:** Rapid developments in telecommunication and information technology have made the world financial markets highly dynamic and increasingly integrated. Naturally, the integration of the Indian insurance industry with world economy became inevitable. India has signed the General Agreement on Tariff and Trade (GATT) and has become a member of the World Trade Organization (WTO). There were international compulsions and pressures, so the structural changes were introduced in the insurance industry hoping that insurance industry could attain its full growth potential, could compete and could get benefit from the world economy.

(ii) **India's initiatives in other sectors:** Around 1993, the government of India decided to undertake structural changes in the financial sector with the aim to bring the Indian financial system to international standards in terms of its financial viability, competence, technology, regulation and credibility. It was noticed that the reforms in the banking sector and the capital market have benefited the economy as well as these sectors. As a result, it was decided to liberalize the insurance sector.

(iii) **Weaknesses:** Certain weaknesses as noted by the Malhotra committee that did surface in the nationalized insurance industry are as follows: poor customer service, vast marketing and services network inadequately responsive to customer needs, expensive insurance covers, excessive staff, poor work culture, lack of productivity and discipline, limited spread of rural and welfare oriented insurance, underdeveloped technology, poor investment skills as a result of excessive government-directed investment of funds, low returns from life insurance as compared to other savings instruments, under performance due to a variety of constraints put by the government and lack of competition due to government monopolies. To overcome these weaknesses, it was decided to privatize the insurance sector.

(iv) **Need for wider insurance cover:** Insurance cover was available to just about 22% of the insurable population in case of life insurance and about 22% of the potential general insurance around the year 2000. In a monopolistic market, there was no incentive for expanding the market share and as a consequence the coverage remained low and the customers had to be satisfied with a limited choice.

The Malhotra committee found that a majority of those who met the committee or responded to its questionnaire were in favor of liberalization. A large number of corporate sectors as well as individuals responded positively to the idea of allowing private entrants into the insurance industry. The general hope was that privatization would bring more efficient and better service, quicker evaluation and settlement of claims, less paper work and lower premiums. In general, the climate was favourable to liberalization of the insurance industry. There were some reservations and fears towards the change. One important apprehension was that like in the pre-nationalization days, there could be some misuse or interlocking of funds of the insurers. To take care of this fear, the Insurance Regulatory and Development Authority (IRDA) was set up and the Insurance Regulatory and Development act was passed in 1999. This organization is now called the Insurance Regulatory and Development Authority of India (IRDAI). The statement of objects and reasons of the act reveals that the Government intends to exercise adequate care with respect to the functioning of the industry.

The IRDAI is expected to regulate the entry and exit of insurance entities, the investment of their funds, the maintenance of accounts by them, rates and conditions of general insurance companies. The government expects the IRDAI to protect the interest of the policy holders and to promote efficiency in the conduct of insurance business. The Act specifies the duties, powers and functions of the IRDAI, which is expected to emerge as an effective watch-dog and regulator of the sector. Experience revealed that the IRDAI has been quite active on all these fronts.

With the liberalization, the insurance industry has had to face many challenges. Human resources constitute the most vital segment of any organization and great care is needed in recruitment, training and retention of talent. Of particular relevance would be training in actuarial science, management, marketing and technical subjects. Actuarial science is the very basis of insurance and it is unthinkable to carry on the insurance business without deploying actuarial and forecasting techniques on an increasing scale. In the words of the Malhotra committee: 'The actuary plays an essential role in the life insurance business, particularly in product development, determination of premium rates, study of mortality experience and construction of mortality tables, laying down underwriting standards, valuation of liabilities and distribution of surplus. The actuary can also be useful in investment management. In general insurance, actuaries should be consulted in matters related to rating and technical reserves'. IRDAI has specified actuarial valuation for health insurance as a requirement. Therefore, the general insurance companies too are looking out for availability of actuaries to be appointed by them.

On one hand, the supply of actuaries is already very limited and on the other, there is a need for producing actuaries in larger numbers in view of the increasing demand for them. As of now, the only way to acquire qualification as a full-fledged actuary is through the examinations conducted by the Institute of Actuaries, London. Since 1989, the Institute of Actuaries of India (IAI) has been conducting examinations for an Indian qualification in Actuarial Science and a person qualifying as a fellow of the Institute of Actuaries of India through these examinations is recognized as an Actuary under the Insurance Rules, 1939. There are some universities and private institutions offering courses in Actuarial Science. It is necessary to further strengthen the facilities and the efforts of the Actuarial Society. The industry itself could take an active lead in this regard, since the demand is confined not only to India, but may also come from other developing countries.

The non-life insurance sector also has tremendous scope of expansion, mainly in health insurance. The COVID pandemic has emphasized the importance of health care on the economy and health insurance is expected to play a central role in the effort to strengthen the health ecosystem. The following are some important growth drivers for the insurance business in India.

(i) **Favourite demographics:** Around 68% of India's population is young and 55% of its population is in the age group of $20 - 59$ (working population) in the year 2020 and it is estimated to rise up to 56% of the total population by 2025. These point towards a young insurable population in India.

(ii) **Wide middle class expansion:** By 2030 India will add 1.40 million middle income and 21 million high-income households, which will drive the demand and growth of the Indian insurance sector.

(iii) **Digital mode:** Customers are now starting to prefer digital modes for their insurance needs.

(iv) **COVID pandemic:** The COVID pandemic increased the insurance penetration rate and triggered awareness of insurance and demand for protection products especially health insurance.

(v) **Government initiatives:** Government initiatives such as PM-JAY, PMFBY and PMJJBY are increasing insurance penetration. PM-JAY is fully funded by a Government scheme called Ayushman Bharat. It is the largest health insurance scheme in the world. It covers 107 million vulnerable families meaning thereby about 500 million beneficiaries. Other schemes are Pradhan Mantri Jeevan Jyoti Bima Yojana (PMJJBY), Pradhan Mantri Jan Dhan Yojana (PMJDY)(Life cover), Pradhan Mantri Suraksha Bima Yojana (PMSBY), Pradhan Mantri Fasal Bima Yojana (PMFBY), Pradhan Mantri Vaya Vandana Yojana (PMVVY) and Restructured Weather Based Crop Insurance Scheme (RWBCIS). PMVVY is a 10 year endowment policy for senior citizens which assures pension after the maturity period as well as the life insurance cover.

To meet the increasing demand of actuaries in the growing insurance business, the students of statistics do have a lot of potential. With their knowledge of probability theory, distribution theory and mathematical statistics, they can easily master the tools of actuarial science. The aim of the book is essentially to introduce to the students of statistics, some basic tools of actuarial science in the life insurance sector.

Chapter 2

Introduction to Statistics and R Software

Key Terms : Bernoulli distribution, Binomial distribution, Chi-square distribution, Continuous random variable, Discrete random variable, Discrete uniform distribution, Distribution function, Exponential distribution, Gamma distribution, Geometric distribution, Mean, Moments, Moment generating function, Negative Binomial distribution, Normal distribution, Poisson distribution, Probability, Random variable, Survival function, Variance.

2.1   Introduction

Statistics as a subject comprises the art and science of collection, interpretation and analysis of data and the ability to draw logical generalities that relate to the phenomenon under investigation. In view of such essential stages of scientific method, statistics extensively enters the domain of all scientific investigations. As discussed in Chapter 1, statistics is the foundation of actuarial science and hence plays a major role in the insurance sector. This chapter is devoted to a brief introduction to statistics in Sections 2 to 5. We will restrict the content to those concepts and methods which are needed in the rest of the chapters. Interested readers may refer to Rohatgi and Saleh [16] and Ross [17] among many other books of statistics for further reading.

The novelty of this book is the use of R software (R [15]) to compute various monetary functions involved in a variety of insurance products and annuity contracts. The main computations are related to finding premiums and reserves. The use of software is essential in all these computations. There is a variety of software available for the computation such as MS Excel, Minitab, Matlab and SAS. In the last two decades, R software is strongly advocated and a large proportion of the world's leading statisticians uses it for statistical analysis. It is a high-level language and an environment for data analysis and graphics, created by Ross Ihaka and Robert Gentleman in 1996. It is both a software and a programming language considered as a dialect of the S language developed by AT and T Bell Laboratories. The current R software is the result of a collaborative effort with contributions from all over the world. It has become very popular in academics and also in the corporate world for a variety of reasons such as its good computing performance, excellent built-in help system, flexibility in graphical environment, vast coverage, availability of new, cutting edge applications in many fields and scripting and interfacing facilities. The most important advantage is that in spite of being the finest integrated software, it is freely available software. Keeping up with the recent trend of using R software for statistical computations, we too have used it extensively in the present book for illustrating the computations of a variety of monetary functions needed for determination of premiums and reserve. The last section of Chapters 3 to 8 presents R codes used for the computation in the respective chapters.

Some readers may be familiar with R software as it has been introduced in the curriculum of many undergraduate and post-graduate statistics programs. In the last section, we give a brief introduction to R, which will be useful to beginners. We have also tried to make the codes given in subsequent chapters self-explanatory.

2.2   Introduction to Statistics

Any realistic model of a real-world phenomenon has to take into account the possibility of randomness. This is usually achieved by allowing the model to be probabilistic in nature. The theory of probability is the foundation of statistics as a science. In this section we give a very brief introduction to theory of probability. A starting point in the development of theory of probability is the concept of a random experiment. An experiment in which a set of possible outcomes of the experiment is known, but an outcome of a given performance of the experiment is not known in advance, is known as a random experiment. As an illustration, the annual premium of vehicle insurance is based on the estimate of the claim the vehicle owner will make in a year. The vehicle may or may not meet with an accident. It is a random experiment with two possible outcomes—an accident occurs in a given year or an accident does not occur in a given year. In case there is no accident, there is no claim. If an accident occurs, it is known beforehand that the damage will be between two limits say, a and b. The actual value of the damage and hence of the claim is not known in advance. In the determination of the premiums, the insurer has to take into account the output of many such repeated cases of accidents and claims to model the distribution of the number of occurrences of accidents and of the claim amount.

In probability theory, we model the uncertainty of a random experiment. With each such experiment, we associate a set Ω, which is a set of all possible outcomes of the experiment. It is known as the sample space associated with the experiment. The elements of Ω are called sample points and a subset of Ω is known as an event. We say that an event A occurs if the outcome of the experiment is an element in set A. We assign a real number $P(A)$ to every event A, called the probability of A. To qualify as a probability, the set function P must satisfy the following three axioms.

Axiom 1 : $P(A) \geq 0 \ \forall \ A$

Axiom 2 : $P(\Omega) = 1$

Axiom 3 : If $A_1, A_2, \cdots$, are disjoint, then

$$P\left(\bigcup_{i=1}^{\infty} A_i\right) = \sum_{i=1}^{\infty} P(A_i)$$

Two common interpretations of $P(A)$ are in terms of relative frequency and degree of belief. In the relative frequency interpretation, $P(A)$ is the long run proportion of times that A will occur in repeated trials of the experiment. For example, suppose the probability of getting a head is 1/2, when a coin is tossed. It means that if we flip the coin a number of times, then the proportion of times we get heads is near to 1/2 and it tends to 1/2, as the number of tosses increases. The statements, 'the probability of an accident, with a major damage is 0.01' is interpreted as follows. Over a sufficiently long period of time, out of 100 accidents only 1 causes severe damage. The degree of belief interpretation is such that $P(A)$ measures an observer's strength of belief that A is true. In the statement 'the probability that Smita will pass the examination is 90%', the interpretation of probability is the degree of belief.

A number of properties of P can be derived from the axioms. Some of these properties are stated below.

(i) $0 \leq P(A) \leq 1$.

(ii) $P(\emptyset) = 0$, where $\emptyset$ denotes the empty set.

(iii) $A \subset B \Rightarrow P(A) \leq P(B)$.

(iv) $P(A^c) = 1 - P(A)$, where A^c denotes the complement of A.

(v) $P(A \cup B) = P(A) + P(B) - P(A \cap B)$.

If $A \cap B = \emptyset$, that is, if A and B are disjoint sets, we say that A and B are mutually exclusive events. In this case, the property (v) reduces to $P(A \cup B) = P(A) + P(B)$. It is usually referred to as the law of addition. There are many situations, where we have partial knowledge concerning the outcome of a random experiment. This partial knowledge enables us to re-evaluate the probabilities with which the event of interest will occur. For example, consider the following cooked up data, specifying probability of death before 65.

	Probability
Men	0.26
Women	0.20
Both men and women	0.23

Table 2.1 Probability of death before 65

If an actuary does not know the sex of a particular applicant, the premium will be based on a probability 0.23 of death before 65. However, if the actuary knows that the applicant is a woman, the premium will be lowered as the chance of death before 65 in women population is 0.20. The process of re-evaluating the probabilities in the light of additional information about the outcome of the experiment is made precise by the concept of conditional probability. Suppose A and B are events corresponding to a sample space Ω. Suppose $P(B) > 0$. Then

$$P(A|B) = P(A \cap B)/P(B)$$

is the conditional probability of A given that B has occurred. If $P(B) = 0$, $P(A|B)$ is not defined. From the expression of the conditional probability $P(A|B)$, it immediately follows that

$$P(A \cap B) = P(A|B)P(B)$$

It is known as the multiplication law. In some cases, the information provided by B in the conditional probability $P(A|B)$ does not affect the probability of event A, that is $P(A|B) = P(A)$. In this setup $P(A \cap B) = P(A)P(B)$ and we say that two events A and B are independent of each other. If A and B are independent, it follows that A and B^c, A^c and B, A^c and B^c are also independent events.

It is to be noted that, independence of events is not to be confused with mutually exclusive events. If two events, each with non-zero probability are mutually exclusive, they are obviously dependent, since the occurrence of one will automatically preclude the occurrence of the other. Similarly, if A and B are independent and $P(A) > 0$, $P(B) > 0$, then A and B cannot be mutually exclusive. The concept of independence of two events can be extended to more than two events.

Statistics is concerned with data and the link between sample spaces and events to data is provided by the concept of a random variable. A random variable X is a mapping, denoted as $X : \Omega \to \mathbb{R}$, that assigns a real number $X(\omega)$ to each outcome $\omega \in \Omega$. The simplest but the most frequently used random variable is the indicator random variable. It is defined as follows. Suppose $A \subset \Omega$ is an event. Then the function $I_A(\omega)$ defined as

$$I_A(\omega) \quad = \quad \begin{cases} 1, & \text{if} \quad \omega \in A \\ 0, & \text{if} \quad \omega \notin A \end{cases}$$

is known as an indicator random variable, as its value indicates whether A has occurred or not. For example, if an accident occurs in a given period, we say that the random variable takes a value 1, otherwise its value is 0. Thus, occurrence of an accident in a given period is an indicator random variable. In actuarial science, commonly encountered random variables are, the number of claims in a specific period, remaining life or future life of an individual or item, claim amount, total claim amount, time until termination of a joint life status or last survivor status. Sometimes, the rate of interest is also taken as a random variable as it varies in many dimensions from time to time, from place to place, by degree of security risk and by time to maturity.

The set of possible values of a random variable X is the sample space associated with X, referred to as support of X. Random variables are broadly classified into two classes, discrete and continuous depending upon the possible values they take. A random variable X is said to be a discrete random variable if the set of possible values of X is countable, that is, either finite or countably infinite (that is, it can be put into a one-to-one correspondence with the set of positive integers). Number of claims random variable is a discrete random variable, with possible values $0, 1, 2, \cdots ,$. Indicator random variable is a discrete random variable with possible values 0 and 1. A random variable X is said to be a continuous random variable if the set of possible values of X is an interval, finite or infinite. For example, age at the death of an individual can be any number between 0 to 110 years, say, that is, the age can be 75.68 years. Thus, life length of a human being is a continuous random variables. Height, weight, claim amounts, total claim amounts in a specific period, future life or remaining life of an individual or item are illustrations of continuous random variables.

Using the probability function P defined on the sample space, we define a cumulative distribution function, or simply referred to as a distribution function, of X as follows. The distribution function F of a random variable is a function from real line $\mathbb{R}$ to $[0, 1]$, defined by

$$F_X(x) \ = \ P[X \leq x] \ = \ P\{\omega | X(\omega) \leq x\}, \quad x \in \mathbb{R}$$

The distribution function completely determines the probability distribution of a random variable. The distribution function satisfies the following properties.

(i) $0 \leq F(x) \leq 1, \quad \forall\, x \in \mathbb{R}$.

(ii) F is non-decreasing on $\mathbb{R}$.

(iii) F is right continuous on $\mathbb{R}$, that is $\lim_{h \to 0} F(x+h) = F(x)\ \forall\ x \in \mathbb{R}$.

(iv) $\lim_{x \to -\infty} F(x) = 0$ and $\lim_{x \to \infty} F(x) = 1$.

Any function, defined on $\mathbb{R}$, satisfying the above four conditions, qualifies to be a distribution function of some random variable. For any two real numbers a and b,

$$P[a < X \leq b] = F(b) - F(a)$$

A more frequently used function in actuarial science is a survival function $S(x)$ and it is defined as,

$$S(x) \ = \ 1 - F(x) \ = \ P[X > x]$$

Properties of this function are discussed in Section 4.2. From the distribution function of X, we can find a variety of functions associated with the probability distribution of X.

Suppose $S = \{x_1, x_2, \cdots\}$ is a set of possible values of a discrete random variable X. We then define the probability mass function of X as

$$p_i = P[X = x_i] = P\{w | X(w) = x_i\}$$

It immediately follows that $0 \leq p_i \leq 1$ for all i and $\sum p_i = 1$, summation being over all possible values $x_1, x_2, \cdots$ in S. The connection between the distribution function and the probability mass function of a discrete random variable X is given by,

$$F(x) = P[X \leq x] = \sum_{x_i \leq x} p_i$$

$F(x)$ is a step function, with jumps at $x_1, x_2, \cdots$. with jump sizes $p_1, p_2, \cdots$, respectively. Suppose X is a random variable with the probability mass function as specified in Table 2.2.

X	8	9	10	11	12
Probability	0.1	0.3	0.3	0.2	0.1

Table 2.2 Probability mass function

Then the distribution function of X is given by,

$$F_X(x) \;=\; \begin{cases} 0 & \text{if} & x < 8 \\ 0.1 & \text{if} & 8 \le x < 9 \\ 0.4 & \text{if} & 9 \le x < 10 \\ 0.7 & \text{if} & 10 \le x < 11 \\ 0.9 & \text{if} & 11 \le x < 12 \\ 1 & \text{if} & x \ge 12 \end{cases}$$

In the case of a continuous random variable, its probability distribution is specified by a probability density function f. It is a non-negative function such that

$$\int_S f(x)\, dx \;=\; 1$$

where S is the support of X, which may be a finite or infinite interval. For a continuous random variable X, its distribution function F is a continuous function and further, if it is a differentiable function, then it is related to the probability density function as follows. For every real number x, we have

$$F(x) = \int_{-\infty}^{x} f(t)dt \quad \Longleftrightarrow \quad f(x) \;=\; \frac{dF(x)}{dx}$$

For a continuous random variable X, $P[X = x] = 0$, $\forall\, x \in \mathbb{R}$. Hence,

$$P[a \le X \le b] = \int_a^b f(x)dx = F(b) - F(a) \;\; \& \;\; S(x) = P[X \ge x]$$

Distribution of a discrete random variable X is completely specified by the probability mass function, while distribution of a continuous random variable X is completely specified by the probability density function. As an illustration, suppose X denotes the life time in hours of a 40 watt bulb. The possible value of X is any non-negative number. Then under certain conditions, the probability density function of X is given by,

$$f_X(x) \;=\; \begin{cases} (1/100)e^{-x/100}, & \text{if} & x \ge 0 \\ 0 & & \text{otherwise} \end{cases}$$

Distribution with this probability density function is known as an exponential distribution. It is extensively used to model the life length of electronic items. 100 is the value of the parameter of the distribution and in this case, it indicates the average life of the bulb. We will study some more discrete and continuous distributions in the following sections. The study of the probability distribution of a random variable is essentially the study of some parameters associated with the distribution. These parameters are the functions of moments of the distribution. Moments of the distribution are defined as follows. The raw moment of order $r (r > 0)$ of a random variable X, denoted by $E(X^r)$ or μ'_r is defined as,

$$\mu'_r = E(X^r) \quad = \quad \begin{cases} \sum\limits_{x} x^r p_x & \text{if} \quad X \quad \text{is discrete} \\ \int x^r f(x) dx & \text{if} \quad X \quad \text{is continuous} \end{cases}$$

provided the sum and integral exist, these being taken over all possible values of X. With $r = 1$, $E(X) = \mu'_1$ is the first moment of X and is known as the mean or average of X. It is generally denoted by μ. It provides a measure of central tendency of a distribution. Such a one-number summary of the distribution is quite often used in actuarial science. If X denotes the financial loss as a result of some random phenomena such as fire, accident or flood, then $E(X)$ specifies the average loss which is the centre of the distribution of X. It is used to determine the premiums. Under some conditions $E(X)$ can be obtained from the survival function of X as follows. If X is a discrete random variable with support as a set of non-negative integers, then $E(X) = \sum_{x=1}^{\infty} P[X \geq x]$. If X is a continuous random variable with support as a set of non-negative real numbers, then $E(X) = \int_0^{\infty} (1 - F(x)) dx$. This approach is often used in actuarial statistics. Two more measures of central tendency are median and mode. Median M is such that $F(M) = 0.5$. Mode M_0 is such that at M_0 the probability mass function or the probability density function, as the case may be, is maximum.

Given n observations on a random variable X, the median M is such that 50% of the observations on X will be below M and 50% will be above M. Similarly mode is that observation which occurs with maximum frequency. There are some more measures of location or central tendency. The median of a set of measurements is the value that divides the distribution into two halves, each containing 50% of the observations. In the same way, quartiles Q_1, Q_2 and Q_3 are the three values that divide the distribution into four equal parts. Thus Q_1, the lower quartile, is such that 25% of the observations are less than or equal to the Q_1. Q_2 is the median. Q_3, the upper quartile, is such that 75% of the observations are less than or equal to Q_3. The deciles are 9 values that divide the distribution into 10 equal parts. Thus D_1, the first decile, is such that 10% of the observations are less than or equal to D_1. D_2 is such that 20% of the observations are less than or equal to D_2. Other deciles are similarly defined. The percentiles are 99 values that divide the distribution into 100 equal parts. Thus P_1, the first percentile, is such that 1% of the observations are less than or equal to P_1. P_{85} is such that 85% of the observations are less than or equal to P_{85}, similarly P_{95} is such that 95% of the observations are less than or equal to P_{95}.

To measure the spread of the distribution around its centre another important parameter of the distribution is variance of X, denoted by $Var(X)$. It is defined as

$$\sigma_X^2 = Var(X) \quad = \quad E(X^2) - (E(X))^2 = E[X - E(X)]^2$$

The variance of the probability distribution is an indicator of how widely the variable is scattered around its centre. By definition, variance is expressed in square units. Hence, by taking the positive square root of variance, we define a measure, known as the standard deviation σ_X of X. Thus, $\sigma_X = (Var(X))^{1/2}$, with units same as the units of mean. Further, $100\sigma_X/E(X)$ is known as the coefficient of variation of X, as it specifies the dispersion of X per unit mean of the distribution, usually expressed in percentage. It is a unit-less number and hence is used to compare the variability or spread of two random variables of different types, such as height and weight. The following example illustrates the computation of the mean and variance for a discrete distribution. Suppose the random variable N denotes the number of claims in a week, with the following probability mass function.

N	0	1	2	5
Probability	0.5	0.2	0.2	0.1

Table 2.3 Probability mass function of N

We find the expected number $E(N)$ of claims and variance of N.

$$\begin{aligned}
E(N) &= 0 \times 0.5 + 1 \times 0.2 + 2 \times 0.2 + 5 \times 0.1 = 0.2 + 0.4 + 0.5 = 1.1 \\
E(N^2) &= 1^2 \times 0.2 + 2^2 \times 0.2 + 5^2 \times 0.1 = 0.2 + 0.8 + 2.5 = 3.5 \\
\Rightarrow \quad Var(N) &= E(N^2) - (E(N))^2 = 3.5 - (1.1)^2 = 2.29 \\
\text{and} \quad \sigma_N &= (2.29)^{1/2} \approx 1.5133
\end{aligned}$$

Note that $E(N) = 1.1$ indicates that on the average, there will be 1.1 claims per week.

Expectation of a function g of a random variable X is defined as,

$$E(g(X)) = \begin{cases} \sum g(x)p_x, & \text{if} \quad X \text{ is discrete} \\ \int g(x)f(x)dx, & \text{if} \quad X \text{ is continuous} \end{cases}$$

provided these exist. Note that to find $E(g(X))$, it is not necessary to obtain the distribution of $g(X)$. This result is heavily used in Chapter 4 to find the actuarial present values of the benefits corresponding to a variety of insurance products. There are certain moment inequalities. We refer to the book by Rohatgi and Saleh [16] for details. An important inequality which is used in Chapter 3 to establish the feasibility of insurance business is Jensen's inequality. It is stated below.

$$E(g(X)) \geq g(E(X)) \quad \text{if } g \text{ is convex} \quad \text{and} \quad E(g(X)) \leq g(E(X)) \quad \text{if } g \text{ is concave.}$$

A necessary and sufficient condition for a differential function g to be convex (concave) is the second derivative of g is positive (negative) for all possible values in the domain of g.

The third and fourth order moments of X are useful to study the shape of the distribution. All these and higher order moments can be obtained from a function known as moment generating function. It is defined as follows. A moment generating function $M_X(t)$ of a random variable X is defined as

$$M_X(t) = E(e^{tX})$$

provided the expectation exists for some t. If $M_X(t)$ exists in some neighbourhood of the origin, then r^{th} derivative of $M_X(t)$ at $t = 0$ gives the rth raw moment $\mu'_r = E(X^r)$. In this sense, the function $M_X(t)$ generates the moments. Under certain conditions, the moment generating function determines the distribution uniquely. In the following section, we discuss some discrete distributions commonly encountered in actuarial science.

2.3   Some Important Discrete Distributions

In the preceding section, we studied probability distributions in general. In this section, we study some discrete probability distributions and investigate their basic properties. The results of this section will be of considerable importance in theoretical as well as in practical applications. The probability distribution of a discrete random variable is completely specified by its probability mass function given by $p_x = P[X = x]$ for $x \in S$, where S is the set of possible values of X. Table 2.4 presents the summary of some standard discrete distributions by specifying its probability mass function, mean, variance and the moment generating function.

Distribution	Probability mass function	$E(X)$	$Var(X)$	$M_X(t)$
Degenerate	$P[X = C] = 1$	C	0	e^{tC}
Discrete uniform	$p_x = \frac{1}{N},$ $x = 1, 2, \cdots, N$	$\frac{N+1}{2}$	$\frac{N^2-1}{12}$	$\frac{e^t}{1-e^t}\frac{1-e^{tN}}{N}$
Bernoulli	$p_x = p^x q^{1-x}$ $q = 1 - p, x = 0, 1$	p	$p(1-p)$	$q + pe^t$
Binomial	$p_x = \binom{n}{x}p^x q^{n-x},$ $x = 0, 1, \cdots, n$	np	npq	$(q + pe^t)^n$
Geometric	$p_x = pq^x,$ $x = 0, 1, 2, \cdots,$	$\frac{q}{p}$	$\frac{q}{p^2}$	$p(1 - qe^t)^{-1}$
Negative binomial	$p_x = \binom{x+k-1}{x}p^k q^x,$ $x = 0, 1, 2, \cdots,$	$\frac{kq}{p}$	$\frac{kq}{p^2}$	$p^k(1 - qe^t)^{-k}$
Poisson	$p_x = \frac{e^{-\lambda}\lambda^x}{x!},$ $x = 0, 1, 2, \cdots,$	λ	λ	$e^{\lambda(e^t-1)}$

Table 2.4 Some standard discrete distributions

The simplest discrete distribution is the one in which a random variable X takes only one value, say 'C', with probability 1. The entire distribution is concentrated at a single point, thus there is no spread at all, which is indicated by zero variance. This distribution is referred to as a degenerate distribution or the one point distribution.

In the discrete uniform distribution, possible values of X are $\{1, 2, \cdots, N\}$ each with a probability $1/N$. Since the probability remains the same for all possible values of X, it is referred to as the discrete uniform distribution. N is the parameter of the distribution. We use this distribution in Section 6.5 to obtain the expression for the actuarial present value in m-thly annuities.

Bernoulli distribution arises in a Bernoulli process, named in honor of the Swiss mathematician James Bernoulli (1654–1705), who made significant contributions in the field of probability. When a random process or experiment, called a trial, results in only one of two mutually exclusive outcomes such as dead or alive, male or female, at least one claim or no claim in a specific period, the trial is known as a Bernoulli trial. One of the outcomes, usually the outcome of interest, is termed as 'success'. A random variable X is assigned a value 1 if success occurs otherwise X is assigned the value 0 and X is said to follow a Bernoulli distribution. The parameter p is known as the probability of success.

Suppose Bernoulli trials are repeated n times under the following two conditions—(i) the trials are independent, that is, the outcome of any particular trial is not affected by the outcome of any other trial and (ii) the probability of success p, remains constant from trial to trial. Suppose X denotes the number of successes in n trials. Then distribution of X is the binomial distribution and is denoted by $B(n, p)$. Its probability mass function is given by,

$$p_x = P[X = x] = \binom{n}{x} p^x q^{n-x}, \qquad x = 0, 1, \cdots, n$$

n and p are the parameters of the distribution with $0 < p < 1$ and n being a positive integer. Figure 2.1 shows the graph of probability mass function of binomial distribution for $n = 5$ and $p = 0.1$ to 0.9.

Note that the nature of probability mass function changes as the value of p changes. For $p = 0.5$, it is symmetric, while for $p > 0.5$, the probabilities in the upper tail are high as compared to the lower tail. In this case, the distribution is called negatively skewed. For $p < 0.5$, the picture is reverse to that for $p > 0.5$, giving rise to a positively skewed distribution. With $n = 1$, binomial $B(n, p)$ distribution is the Bernoulli distribution, given by the probability mass function $p_x = p^x q^{1-x}$, $x = 0, 1$.

Suppose a sequence of Bernoulli trials is repeated under the same two conditions, mentioned for binomial distribution, till the first success occurs. Then number X of failures before the first success is a random variable and is modelled by the geometric distribution with probability mass function given by

$$p_x = P[X = x] = p(1 - p)^x, \quad x = 0, 1, \cdots$$

The probability mass function of this distribution is in a geometric progression with a common factor $(1-p)$, hence, the name geometric distribution. In some cases, the distribution of number of Bernoulli trials needed to get the first success is termed as the geometric distribution. Its probability mass function is given by $p_x = p(1 - p)^{x-1}$, $x = 1, 2, \cdots$.

Negative binomial distribution is an extension of the geometric distribution in the sense that the number X of failures before a fixed number k of successes is noted. The distribution of X, is then known as the negative binomial distribution. With $k = 1$, negative binomial distribution reduces to the geometric distribution.

Poisson distribution is the most frequently encountered distribution to model certain phenomena in biology, medicine and also in insurance. It is named after the French Mathematician, Simeon Denis Poisson (1781–1840). Suppose X denotes the number of occurrences of a random event in a fixed interval of time or space. As in the case of binomial and

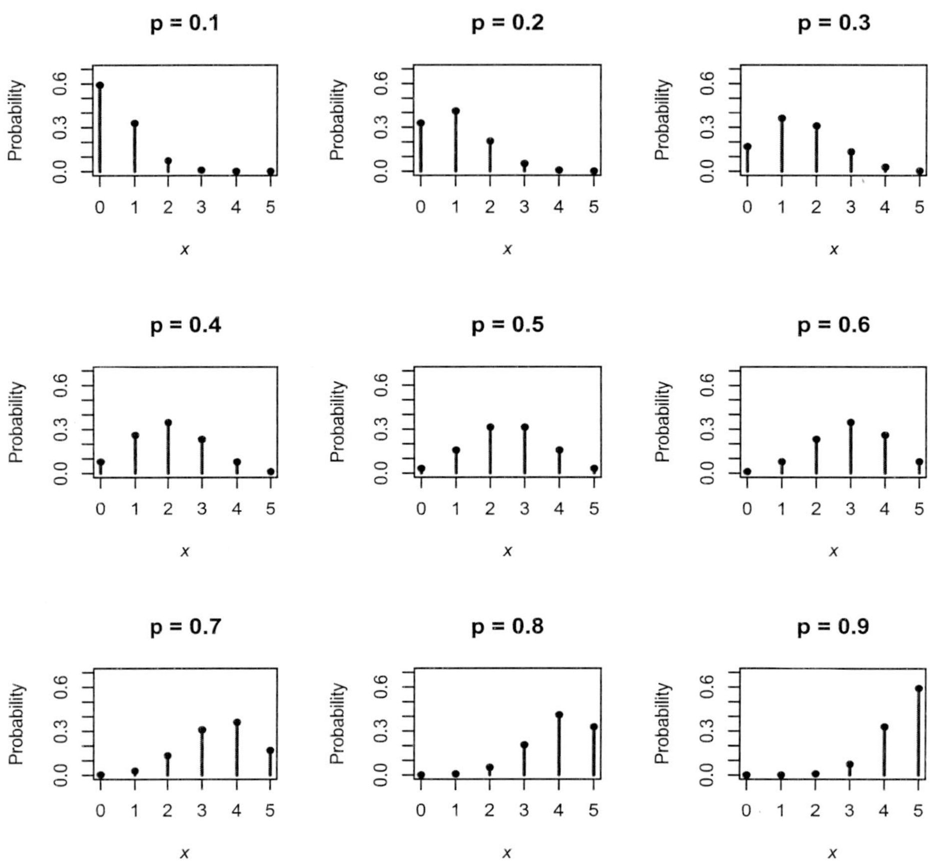

Figure 2.1 Binomial distribution: $B(5, p)$

negative binomial distribution, there are certain underlying assumptions regarding occurrence of events. These are listed below.

(i) The probability of a single occurrence of the event in a given interval is proportional to the length of the interval.

(ii) In any infinitesimally small portion of the interval, the probability of more than one occurrence of the event is negligible.

(iii) Theoretically, an infinite number of occurrences of the event are possible in a given interval.

(iv) The occurrence of an event in an interval has no effect on the probability of the occurrence of the event in any other non-overlapping interval.

Under these conditions, the distribution of X is a Poisson distribution with its probability mass function given by,

$$p_x = P[X = x] = e^{-\lambda}\lambda^x/x! , \quad x = 0, 1, 2, \cdots$$

λ, a positive real number, is the parameter of the distribution. The symbol e is the constant 2.7183 (approximately up to four decimals). In insurance, the number of claims in a specific period is usually modelled by the Poisson distribution. Figure 2.2 shows the graph of probability mass function of Poisson distribution for four values of λ.

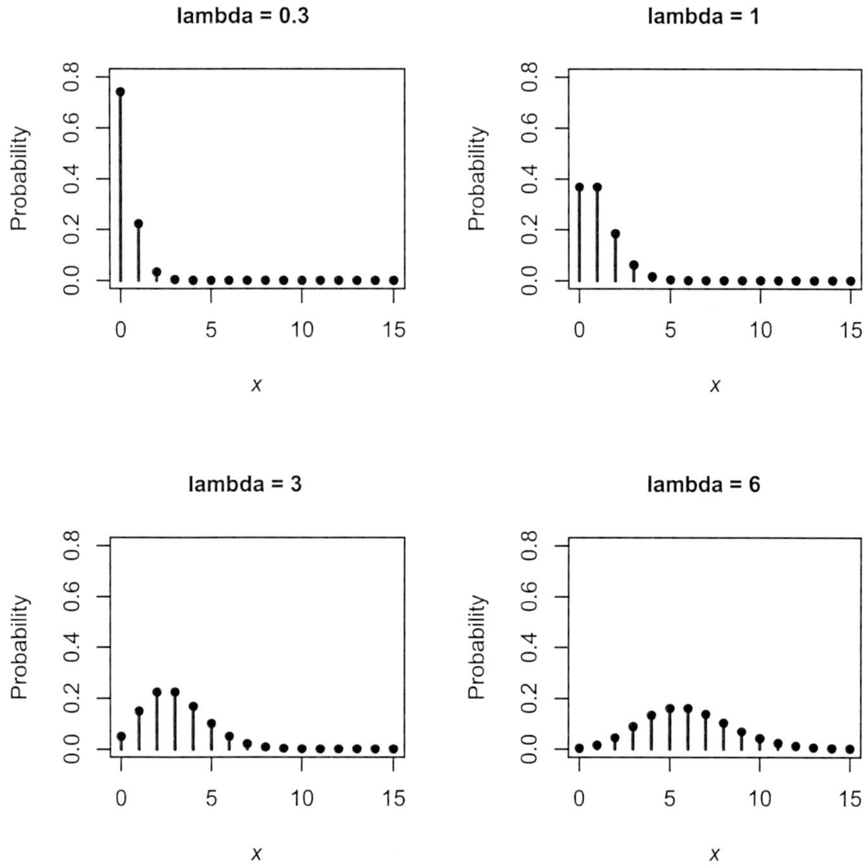

Figure 2.2 Poisson distribution with mean λ

The nature of probability mass function changes as the value of λ changes. For $\lambda = 0.3$, the probability at 0 is very high and for high x values, it is close to 0. For $\lambda = 1$, probabilities for 0 to 5 are positive and for remaining $x > 5$, these are almost 0. For $\lambda = 3$, probabilities for 0 to 8 are positive and for $x > 8$, these are almost 0. For $\lambda = 6$, the span of values of x for which probability is positive, increases and the graph of probability mass function is more or less symmetric.

When one of these discrete probability distributions is chosen to model the random phenomena under observation, the choice is usually made either for theoretical reasons or for empirical ones, that is, either because the probability distribution appears to be a logical consequence of certain properties of the random phenomena which are known to the experimenter, or because the probabilities derived from the distribution agree with the relative frequencies obtained from repeated observations of the phenomena. Histogram and frequency polygon corresponding to observed data help to guess the underlying model.

In the next section, we discuss some standard continuous distributions. As in the case of the discrete probability distributions discussed in this section, each continuous distribution is specified by the probability density function, presented in a parametric form, that is, the general shape of the distribution is given in a mathematical form with certain parameters. When these parameters are specified, the distribution gets completely determined. The conclusion, that a given random variable has a certain parametric distribution is a considerable achievement. When such a parametric form is found, not only the number of possible distributions that we must consider is greatly reduced, but insights beyond the range of past experience, that is, extrapolations and interpolations, are possible.

2.4 Some Important Continuous Distributions

As noted in Section 2, the set S of possible values of a continuous random variable X is an interval, finite or infinite. The distribution of X is completely specified by the probability density function $f(x), x \in S$ of X. Table 2.5 presents the probability density function, mean, variance and the moment generating function of some commonly encountered continuous distributions.

Distribution	$f(x)$	$E(X)$	$Var(X)$	$M_X(t)$
Uniform	$1/\theta,$ $0 < x < \theta$	$\theta/2$	$\theta^2/12$	$\frac{e^{t\theta}-1}{t\theta}$
Exponential	$\lambda e^{-\lambda x},$ $x > 0$	λ^{-1}	λ^{-2}	$\left(1 - \frac{t}{\lambda}\right)^{-1}$
Gamma	$\frac{\alpha^\lambda}{\Gamma(\lambda)} e^{-\lambda x} x^{\lambda-1},$ $x > 0$	λ/α	λ/α^2	$\left(1 - \frac{t}{\alpha}\right)^{-\lambda}$
Chi-square	$\frac{1}{2^p \Gamma(\frac{p}{2})} e^{-x/2} x^{\frac{p}{2}-1},$ $x > 0$	p	$2p$	$(1 - 2t)^{-\frac{p}{2}}$
Normal	$\frac{1}{\sqrt{2\pi}\sigma} e^{-\frac{1}{2\sigma^2}(x-\mu)^2},$ $-\infty < x < \infty$	μ	σ^2	$e^{\mu t + \frac{1}{2}\sigma^2 t^2}$

Table 2.5 Some standard continuous distributions

The uniform distribution is also labelled as the rectangular distribution as the graph of its probability density function comes out to be a rectangle. This is the simplest continuous distribution, but does arise very frequently in application. In actuarial science, the life length of a human being was modelled by the uniform distribution on $(0, 100)$ by De-Moivre, long back in 1827. However, it is observed that uniform distribution is not at all a good model for life length of a human being. In the uniform distribution, all intervals with the same length lying within the set of possible values have the same probability, a feature similar to that of discrete uniform distribution which assigns equal probability to all the possible values.

In the class of continuous distributions, exponential distribution is the unique one to have the property of constant failure rate which is also termed as the 'loss of memory' property. The exponential distribution arises quite frequently in applications concerned with a series of point events such as accidents, radioactive emissions, breakdowns of machines, and so on, which occur completely randomly in time or space. In such a series of point events, the length of time between successive occurrences is modelled by the exponential distribution in view of the following observations. If we make repeated measurements of the duration of time between successive occurrences, then the histogram obtained from such data often starts on the left with very tall rectangles corresponding to large probabilities assigned to small values of the random variable. As one moves to the right, the heights of the rectangles decrease rapidly. The steepness of the decrease in height of the graphs of the probability density function of exponential distribution depends on the value of λ; the larger the value of λ, the steeper the decrease in height of the graph of the probability density function.

Gamma distribution derives its name from the gamma function from mathematics, which is defined as,

$$\Gamma\alpha = \int_0^\infty y^{\alpha-1}e^{-\alpha y}dy$$

From Table 2.5, note that with $\lambda = 1$, gamma distribution reduces to the exponential distribution.

Normal distribution is the most widely studied and frequently used distribution in statistics. The form of the normal distribution was discovered early in the history of probability theory by De-Moivre, Laplace and Gauss. One of the very early applications of the normal distribution occurred in astronomy, where Laplace and Gauss used it to describe errors of measurement in the observation of the motions of planetoids. This distribution is frequently called the Gaussian distribution in recognition of the contributions of Carl Friedrish Gauss (1777–1855). The parameter μ is the center or the mean of the distribution and σ^2 is the measure of spread of the distribution. Normal distribution is completely specified by its mean and variance and hence is denoted by $N(\mu, \sigma^2)$. The probability density function of the distribution is symmetric around the mean, which is the same as the median and the mode (Figure 2.3).

Normal distribution plays an important role in probability and statistics. Many phenomena in nature have approximately normal distribution, justification of which follows from the well known Central Limit Theorem. We discuss it briefly in the next section. A random variable is said to have the standard normal distribution when the mean μ is zero and the variance σ^2 is one. The probability density function and the distribution function of the standard normal distribution are denoted by ϕ and Φ respectively. It can be shown that if X has the standard normal distribution, then $Y = \mu + \sigma X$ follows the normal distribution with parameters μ and

σ^2. Thus, the probabilities related to Y can be calculated from the corresponding probabilities for X. All statistical computing packages and most statistics textbooks provide the values of the distribution function, the survival function and the probability density function of the standard normal distribution.

Figure 2.3 shows the graph of the probability density function of exponential distribution, gamma distribution for certain values of parameters and for $N(5,1)$ and $N(0,1)$. For exponential distribution there is a high concentration of probability at the lower tail. For gamma distribution also there is a high concentration of probability at the lower tail, but less as compared to the exponential distribution. The nature of the graph of course changes as the values of the parameters change. The probability density function of normal distribution is symmetric around its mean and 99.73% values lie in the range $(\mu - 3\sigma, \mu + 3\sigma)$.

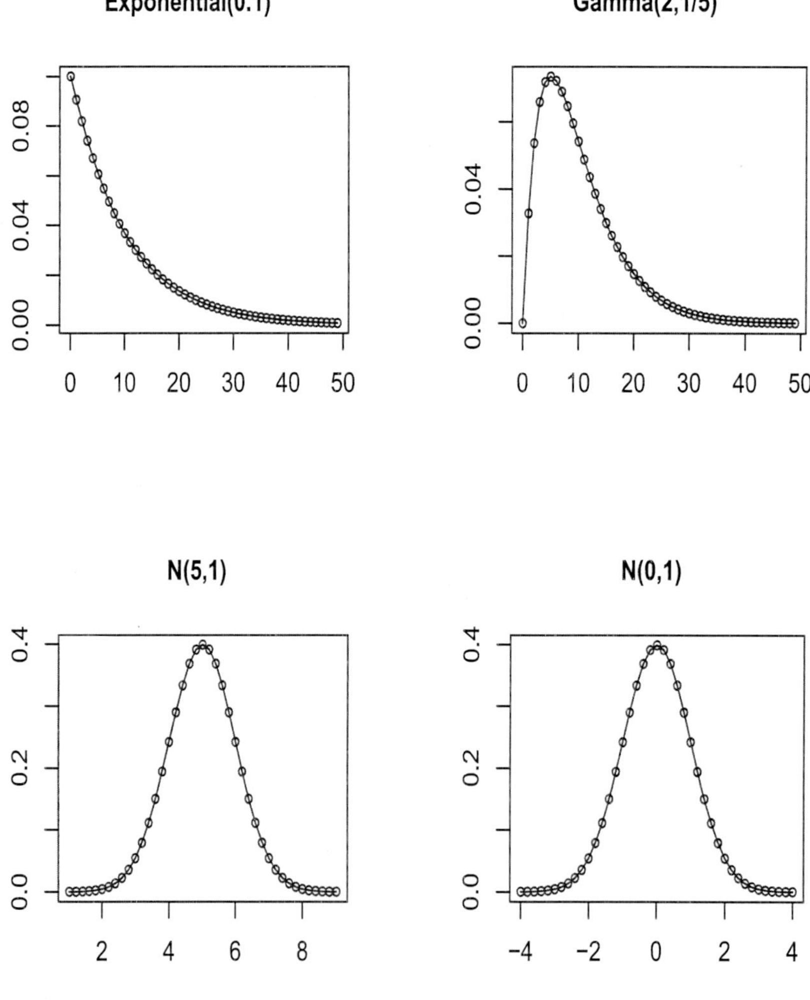

Figure 2.3 Probability density functions

The exponential and gamma distributions are two large classes of continuous distributions which describe the variation of non-negative random variables, such as lifetimes of mechanical, electrical, biological or social systems, waiting time for service, duration of epidemics, and so on. Another parametric family of distributions, which has been used successfully to model the variability of non-negative quantitative random phenomena is the family of Weibull distributions, named after the Swedish physicist, Waloddi Weibull, who in 1934 suggested this family. Functions related to this distribution are discussed in Section 4.2.

For more details on the distribution theory of discrete and continuous distributions, interested readers may refer to the books Johnson et al [9], Johnson et al [10] and Johnson et al [11].

2.5   Multivariate Distributions

In most of the investigations, data are collected on more than one inter-related characteristics of interest. While deciding the premiums in the insurance business, data on various variables describing the medical background of the individual and also on age, sex, hobbies, nature of job are collected as the premium calculations do vary as such related random variables vary. The methods to analyze the data including simultaneous measurements on many variables constitute an important branch of statistics, known as multivariate analysis. Here one needs to define the distribution of p random variables. $X_1, \cdots, X_p$, are referred to as the multivariate distribution, or the joint distribution of p random variables $X_1, X_2, \cdots, X_p$. To introduce the ideas involved in the notion of multivariate distributions, we first discuss the joint distribution of two random variables. Generalization to more than two random variables is straightforward.

Suppose we are interested in the study of two random variables, associated in some sense. In such a random experiment, if we concentrate upon the distribution of each of these variables separately, we may ignore any information about the association between the two. More precisely, we ignore the frequency with which certain values of any one of these random variables, say X, can be expected to occur in connection with certain values of another random variable, say Y. To incorporate such information, we study the bivariate distribution of X and Y. Two random variables X and Y are usually denoted in a vector notation as (X, Y). Suppose X and Y are both discrete random variables. Then the joint probability mass function $p(x, y)$ of (X, Y) is defined for all possible values (x, y) of (X, Y) as $p(x, y) = P[X = x, Y = y]$. It is clear that $p(x, y) \geq 0$ and addition of $p(x, y)$ for all possible values of X and Y is 1. The bivariate probability mass function of (X, Y) is many times presented in a bivariate table as in Table 2.6.

$X \setminus Y$	0	1	Total
0	1/9	2/9	3/9
1	2/9	4/9	6/9
Total	3/9	6/9	1

Table 2.6 Bivariate distribution $p(x, y)$ of X and Y

In the above illustration, possible values for both X and Y are 0 and 1 and $P[X = 0, Y = 1] = 2/9$. The addition of 4 elements in the table is 1. The last row and the last column of the table give the marginal totals. These essentially represent what are known as the marginal distributions. The joint mass function not only specifies all the relevant statistical information when X and Y are studied simultaneously, but also the information about X and Y when studied individually. From $p(x, y)$ we can obtain $p_1(x) = P[X = x]$ and $p_2(y) = P[Y = y]$ for all possible values of x and y. Thus, we have,

$$p_1(x) \;=\; P[X = x] = \sum_y p(x, y) \;\text{ and }\; p_2(y) = P[Y = y] = \sum_x p(x, y)$$

The collections $\{x, p_1(x)\}$ and $\{y, p_2(y)\}$ for possible values of x and y, represent the marginal distributions of X and Y respectively. In the above illustration, $P[Y = 0] = P[X = 0, Y = 0] + P[X = 1, Y = 0] = 1/9 + 2/9 = 3/9$. With the marginal distributions, we obtain $E(X)$, $E(Y)$, $Var(X)$ and $Var(Y)$ as in the univariate set up. In the bivariate set up one more measure, namely, covariance between X and Y denoted by, $Cov(X, Y)$ is defined as follows.

$$Cov(X, Y) \;=\; E\{(X - E(X))(Y - E(Y))\} = E(XY) - E(X) \times E(Y)$$

In the discrete set up, $E(XY)$ is obtained as

$$E(XY) = \sum \sum xy\, p(x, y)$$

the sum being taken over all possible values of x and y. $Cov\ (X, Y)$ is the measure of simultaneous variation of X and Y. If X and Y both increase (or decrease) together then $E[(X - E(X))(Y - E(Y))]$ is positive, whereas if X decreases while Y increases and conversely, then the covariance is negative. Thus, the covariance is a measure of association or joint variation between X and Y. One more measure of linear association between X and Y is given by the correlation coefficient $r_{xy} \equiv r$. It is defined as $r_{xy} \equiv r = Cov(X, Y)/(\sigma_X \times \sigma_Y)$. It is to be noted that, r lies between -1 and 1. X and Y are said to be uncorrelated if $r = 0$, that is, if $Cov(X, Y) = 0$. If $r = 1$, X and Y have a perfect positive correlation while $r = -1$ indicates a perfect negative correlation between the two variables. When $r > 0$, X and Y are said to be positively correlated while when $r < 0$, they are said to be negatively correlated. For the bivariate distribution displayed in Table 2.6,

$$
\begin{aligned}
E(X) &\;=\; E(Y) \;=\; 0 \times (3/9) + 1 \times (6/9) \;=\; 6/9 \\
E(X^2) &\;=\; E(Y^2) \;=\; 1 \times (6/9) \;=\; 6/9 \\
Var(X) &\;=\; Var(Y) \;=\; 6/9 - (6/9)^2 \;=\; 2/9 \\
E(XY) &\;=\; 0 \times (1/9) + 0 \times (2/9) + 0 \times (2/9) + 1 \times (4/9) \;=\; 4/9 \\
Cov(X, Y) &\;=\; 4/9 - (6/9)^2 \;=\; 0 \;\text{ and }\; r_{xy} \;=\; 0
\end{aligned}
$$

Thus, X and Y are uncorrelated random variables.

In the bivariate set up along with the marginal distributions, we discuss conditional distribution, which provides some relevant information. The concept of a conditional distribution is parallel to the concept of conditional probability $P(A|B)$ introduced in Section

2. Thus, when the realized value of $X = x$ is known, one would like to know what is the chance that another random variable Y takes the value y. Thus, we define, for all possible values y of Y,

$$p_{y|x} = P[Y = y|X = x] = P[X = x, Y = y]/P[X = x]$$

It is clear that the conditional probability is defined only if $P[X = x] > 0$. Specification of $p_{y|x}$ for all possible values of Y is referred to as the conditional distribution of Y given $X = x$. On similar lines, the conditional distribution of X given $Y = y$ is defined. From the conditional distribution, one can easily obtain the conditional mean and the conditional variance. Thus,

$$E(X|Y = y) = \sum_x x \; P[X = x|Y = y]$$

It is to be noted that $E(X|Y = y)$ is a function of y. Its values vary as the values of y vary. Thus, $E(X|Y)$ is a random variable whose value is $E(X|Y = y)$ when $Y = y$. In some cases, $E(X|Y = y)$ may not change as the values of y change. For example, for the bivariate distribution displayed in Table 2.6

$$P[X = x|Y = 0] = 1/3, \text{ if } \quad x = 0 \text{ and } 2/3, \text{ if } x = 1$$

$$P[X = x|Y = 1] = 1/3, \text{ if } x = 0 \text{ and } 2/3, \text{ if } x = 1$$

Hence, the conditional distribution of X given $Y = 0$ and given $Y = 1$ is the same as the marginal distribution of X. Thus, the particular realization of Y does not alter the distribution of X. In this case, we say that X and Y are independent random variables. More precisely, two random variables X and Y are said to be independent if for every x and y,

$$P[X = x, Y = y] = P[X = x]P[Y = y]$$

otherwise, we say that X and Y are dependent random variables.

We now discuss all these concepts when both the random variables X and Y are continuous. In this set up, the joint distribution of (X, Y) is specified by the joint probability density function $f(x, y)$. The joint probability density function $f(x, y)$ satisfies the following properties.

$$f(x, y) \geq 0 \; \forall \; x, y \qquad \text{and} \qquad \int_{-\infty}^{\infty} \int_{-\infty}^{\infty} f(x, y) \; dx \; dy = 1$$

The joint probability $P[X \in (a, b), Y \in (c, d)]$ is given by,

$$P[X \in (a, b), Y \in (c, d)] = \int_a^b \int_c^d f(x, y)dx \; dy$$

From the joint probability density function, we get the marginal probability density functions to specify the marginal distributions. Thus,

$$f_X(x) = \int_{-\infty}^{\infty} f(x, y)dy \quad \text{and} \quad f_Y(y) = \int_{-\infty}^{\infty} f(x, y)dx$$

are the marginal probability density functions of X and Y respectively. The conditional probability density function of X given $Y = y$ is defined as, $f_{X|Y=y}(x) = f(x,y)/f_Y(y)$, provided $f_Y(y)$ is not zero. The conditional distribution of Y given $X = x$ is defined by the conditional probability density function of Y given X. It is given by, $f_{Y|X=x}(y) = f(x,y)/f_X(x)$. The conditional mean of Y given $X = x$ is defined as $E(Y|X = x) = \int y f_{Y|X=x}(y)dy$. If the conditional probability density function of $Y|X = x$ for every x is the same as the marginal probability density function of Y then X and Y are said to be independent. More precisely X and Y are said to independent if $f(x,y) = f_X(x)\, f_Y(y)$ for all x and y, that is, if the joint probability density function is the product of marginal probability density functions. As an illustration, suppose the joint distribution of (X,Y) is specified by the probability density function

$$f(x,y) = \begin{cases} x+y, & \text{if} \quad 0 \le x \le 1, \ 0 \le y \le 1 \\ 0, & \text{otherwise} \end{cases}$$

Note that $f(x,y) \ge 0 \ \forall \ x$ and y and

$$\int_0^1 \int_0^1 (x+y)dx \, dy \ = \ \int_0^1 \left\{ \int_0^1 (x+y)dx \right\} dy$$
$$= \ \int_0^1 [x^2/2 + yx]_0^1 dy = \int_0^1 (1/2 + y)dy$$
$$= \ [y/2 + y^2/2]_0^1 = 1/2 + 1/2 = 1$$

Thus, $f(x,y)$ is indeed the joint probability density function of (X,Y). The marginal probability density function of X is

$$f_X(x) \ = \ \int_0^1 f(x,y)dy = \int_0^1 (x+y)dy = x + 1/2, \ \ 0 \le x \le 1$$

and that of Y is

$$f_Y(y) \ = \ \int_0^1 (x+y)dx \ = \ 1/2 + y, \ \ 0 \le y \le 1$$

The conditional probability density function $g(x)$ of X given $Y = y$ is

$$g(x) = f(x,y)/f_Y(y) = (x+y)/(y+1/2), \ \ 0 \le x \le 1$$

and the conditional probability density function $h(y)$ of Y given $X = x$ is

$$h(y) = f(x,y)/f_X(x) = (x+y)/(x+1/2), \ \ 0 \le y \le 1$$

It is to be noted that

$$f(x,y) \ne f_X(x)f_Y(y)$$

Thus, X and Y are not independent random variables. In other words, X and Y are dependent random variables.

When two characteristics are measured in a random experiment, it may happen that one is continuous and the other is discrete. In some insurance products the benefit to be paid depends on the cause of the death, thus the data on two variables are collected—one is the cause of death and the other is the time from policy issue to the claim of the policy. In this situation, the cause of death is modelled by a discrete random variable, while duration between policy issue and the claim of the policy is a continuous random variable. It is important to note that when one is discrete and the other is continuous, it is not possible to write down either the joint probability mass function or the joint probability density function. Such a situation arises while defining the actuarial present value of unit benefit to be payable at the moment of death in a n-year endowment insurance. The joint probabilities, in such a set are obtained from the joint distribution function. It is denoted by $F(x, y)$ and is defined for all x and y in R as,

$$F(x, y) = P[X \le x, Y \le y]$$

From the joint distribution function, one can obtain the marginal distribution functions. Thus the marginal distribution function of X is given by $F(x, \infty)$ and that of Y is given by $F(\infty, x)$. From marginal distribution functions, one can obtain marginal probability density function and probability mass function.

There is an interesting and frequently used relationship between the mean of marginal distribution and the conditional distribution and also between variance of marginal distribution and the conditional distribution. It is stated below.

$$E(X) = E(E(X|Y)) \quad \text{and} \quad Var(X) = E(Var(X|Y)) + Var(E(X|Y))$$

$E(X) = E(E(X|Y))$ is known as a rule of iterated expectation.

Sample mean $\overline{X}_n = (X_1 + \cdots + X_n)/n$ or sample total $S_n = X_1 + \cdots + X_n$ are frequently used to summarize the results of a random experiment. We briefly discuss below two important results from probability theory related to sample mean $\overline{X}_n$, which are heavily used in insurance business.

Weak Law of Large Numbers: If $X_1, \cdots, X_n$ are independent and identically distributed random variables with a common mean μ, then the probability that $\overline{X}_n$ converges to μ, approaches 1, as n increases. This result is known as weak law of large numbers (WLLN). In words, WLLN asserts that as n increases, the sample mean is a better approximation for the population mean. WLLN applied to Bernoulli random variables leads to the conclusion that as n increases, the relative frequency, that is, proportion of success approaches to the true probability of success, which is unknown in general. Thus, WLLN assures the better prediction of rates of occurrence of events leading to financial loss, if the customer base of the insurance company is large.

The determination of the distribution of S_n, when n is small may be a tedious or difficult task. For moderately large values of n, however, we can use the famous Central Limit Theorem (CLT) to find an approximation to the distribution of S_n. We discuss one such application in Chapter 5. A formal statement of CLT, in a particular setup is as follows.

Central Limit Theorem: If $X_1, X_2, \cdots, X_n$ are independent and identically distributed random variables with a common mean μ and positive, finite variance σ^2, then the distribution of $Z_n = (\sum X_i - n\mu)/\sqrt{n}\sigma$ can be approximated by the standard normal distribution for a moderately large n.

The remarkable aspect of the CLT is that Z_n has approximately a normal distribution regardless of the common distribution of independent random variable $X_1, \cdots, X_n$. The only requirement is that σ^2 must be positive and finite, positive variance excludes the case that the common distribution is degenerate.

In view of the CLT, in many applications, a random variable is often assumed to have a normal distribution, since quite frequently the random variable of interest may be conceptualized as being composed of the sum of many random variables. Figures 2.4 and 2.5 illustrate CLT graphically.

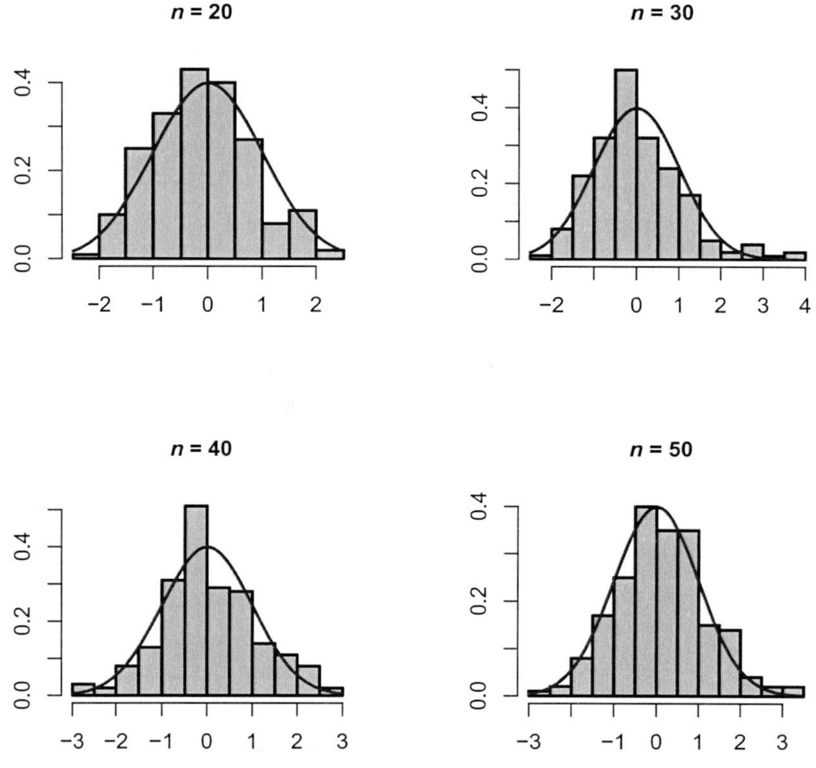

Figure 2.4 Central Limit Theorem for exponential distribution

We have generated 200 samples, each of size $n = 20$ from the exponential distribution with mean 0.5. Then we compute 200 values of Z_n and draw the histogram of these values with relative frequencies on y axis. Curve of the probability density function of the standard normal distribution is imposed on it. The same procedure is repeated for $n = 30, 40$ and 50. Figure 2.4 displays four histograms corresponding to four sample sizes. It is to be noted that as the sample size increases, the histogram becomes more and more bell shaped and the curve of probability density function of standard normal distribution is a better fit to the histogram. In

Figure 2.5, a similar procedure is adopted when the distribution of $X_1, \cdots, X_n$ is Poisson with mean four. In this figure also we observe that as the sample size increases, the histogram becomes more and more bell shaped and the curve of probability density function of standard normal distribution is better fit to the histogram.

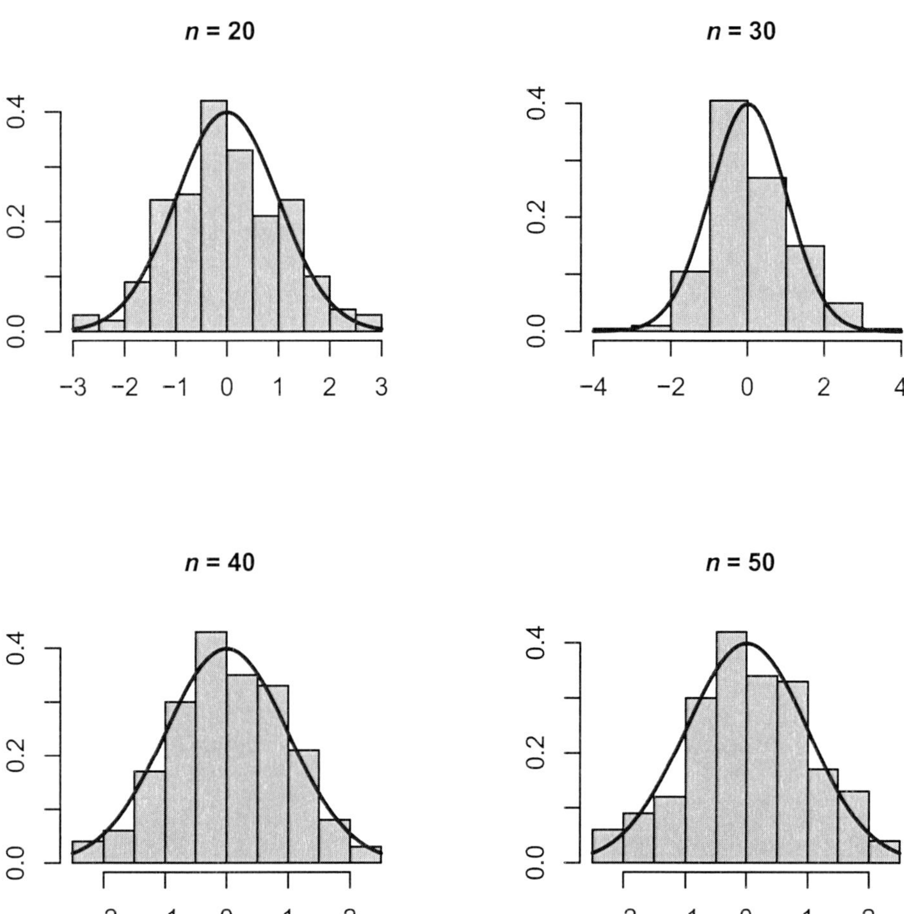

Figure 2.5 Central Limit Theorem for Poisson distribution

Note that in Figure 2.4, the underlying distribution is continuous while in Figure 2.5, it is discrete. Both the distributions are not symmetric, but still the approximate distribution of the standardized sample mean is symmetric normal distribution, for a moderately large sample size.

In computation of monetary functions discussed in this book, we need some built in functions given in R. Hence, we give below a very brief introduction to R software and some useful functions.

2.6   Introduction to R Software and Language

The most appealing feature of R software is that it is freely downloadable software. It can be downloaded from the site called CRAN (Comprehensive R Archive Network) with address `http://cran.r-project.org/`. From this site one needs to 'Download and Install R'. When R is installed properly, there is an image of the R icon on your desktop/laptop. To start R, one has to click on the R icon. The session is ended by typing q().

R is an interpreter based language, meaning that all commands typed on the key board are directly executed without building in the complete program like other programming languages such as C, Pascal, and so on. Furthermore, R syntax is very simple and intuitive. Variables, data, functions and results are stored in the active memory of R in the form of objects. The data analysis in R proceeds as an interactive dialogue with the interpreter. As soon as we type command at the prompt(>) and press the enter key, the interpreter responds by executing the command. R language includes the usual arithmetic operations, such as, + for addition, − for subtraction, * for multiplication, / for division and ˆ for exponentiation, with the usual hierarchy.

Like any other programming language, R contains data structures. Vectors are the basic data structures in R. The standard arithmetic functions and operators apply to vectors on an element-wise basis, with the usual hierarchy. Below, we state some functions which we need to write R codes for the concepts discussed in this book. The most useful R function for entering small data sets is the

c ('combine') function. This function combines elements together and constructs a vector. For example y=c(10, 20, 30, 40) constructs a vector with four elements $10, 20, 30$ and 40. When a series of observations is stored in R as a vector, the standard arithmetic functions and operators apply to vectors on an element-wise basis. We specify below some such basic functions. One can use any variable name, but care is to be taken as R is case sensitive.

With this non-statistical part of the introduction to R, we now proceed to discuss how it is used for statistical analysis. The discussion is not exhaustive. We restrict to the functions which are repeatedly used in this book.

Code 2.6.1. This code specifies some basic functions to input the vectors, sequence function, division of two vectors etc.

```
x=c(12, 25, 37, 41, 54, 63) # c function to construct a vector
  # with given elements
x # displays x, print(x) also displays the object x
length(x) # specifies a number of elements in x
y=1:6; y # constructs a vector with consecutive elements 1 to
# 6 and prints it, two commands can be given on the same line
# with separator ";"
```

```
u=seq(100,250,50); u # sequence function to create a vector
# with first element 100, last element 250 and increment 50
v=c(rep(1,3),rep(2,2),rep(3,5)); v # rep function to create a
# vector where 1 is repeated thrice, 2 twice and 3 five times
w=c( 10, 20, 30, 40)/c(5, 4, 3, 2 ); w
 # 2 5 10 20
w1=c(10, 20, 30, 40 )/2; w1
 # 5 10 15 20
```

■

The object w stores the results of division of two vectors. The output is 2 5 10 20. It is to be noted that the division is done element wise. Such an operation is useful in the calculation of premiums for n-year term and endowment insurance for various values of n in one function. In $w1$ denominator 2 is repeated 4 times to construct a vector of length 4, each element of which is 2.

To combine various column vectors of the same length in a matrix form or a tabular form we use a `data.frame` function from R. Tables presented in all the chapters are constructed using data.frame function. Suppose we have two vectors x and y of the same length. The function `z = data.frame(x,y)` constructs a table consisting of three columns, the first corresponds to row number, the second corresponds to x and the third corresponds to y. Suppose we want to round off the numbers in the table z up to 2 decimals. The function `round(z,2)` prints the values of z, rounded to the second decimal point. It is to be noted that rounding is mainly for printing purpose, original unrounded values of z are stored in the object z.

In most applications, data sets are large. For example in Chapter 4, we discuss the construction of a life table on the basis of given data on mortality rates. The given data are the values of rates q_x for $x = 0$ to 110. Suppose the Excel file consists of two columns of values of x and q_x for $x = 0$ to 110 and we want to import these data from Excel to R for the construction of a life table. The procedure for this consists of the following steps. First we save the file as tab delimited text file with some name, `qx.txt`, say on the local disk c and then close the file. Suppose the two columns have headings as x and q_x. In R console type the function `z = read.table("c:/qx.txt", header = TRUE)`. This function stores the file `qx.txt` in data object z. The first argument of this function gives the path of the file, which should be enclosed in quotes. The second argument header is logical. It is TRUE, if columns in the data set are named, otherwise it is FALSE. The function `x = z[,1]` extracts all elements of the first column of matrix or table z, which is a column of x values running from 0 to 110 in this illustration and assigns it to object x. Similarly the function `q = z[,2]` extracts all elements of the second column of z, which is a column of q_x values and assigns it to object q.

There are some useful built-in functions. Apart from many built-in functions, one can write a suitable function as required in a specific situation. In subsequent chapters you will find many such functions, written to serve the specific purpose.

Code 2.6.2. In this code we illustrate commonly used functions with a data set stored in variable x.

```
x=rnorm(25,3,2) # random sample of size n=25 from normal
# distribution with mean 3 and standard deviation 2
mean(x); median(x); max(x); min(x); sum(x)
cumsum(x)# cumulative sum
var(x)# divisor is (n-1) and not n
quantile(x,c(.25,.5,.75)) # three quartiles
summary(x) # gives minimum, maximum, three quartiles
          # and mean
# Partial output
 summary(x)
   Min. 1st Qu. Median Mean 3rd Qu. Max.
 -1.387 2.250 3.426 3.356 4.360 7.978
```

∎

Note that your output would be different as your simulated sample would be different. In fact, every time you run the code, the output will be different. There is a function `set.seed` to simulate the same sample in every run. We will not elaborate on it as we do not need it in subsequent chapters.

Graphical features of R software have a remarkable variety. Each graphical function has a large number of options making production of graphics very flexible. Graphical device is a graphical window or a file. There are two kinds of graphical functions—the high-level plotting functions, which create a new graph, and low-level plotting functions, which add elements to an already existing graph. The standard high-level plotting functions are `plot()` function for scatter plot, `hist()` function for histogram and `boxplot()` function for box plot and so on. The lower level plotting functions are `lines()` to impose curves on the existing plot, `abline()` to add a line with given intercept and slope, `points()` to add points at appropriate places. These functions take extra arguments that control the graphic. The graphs are produced with respect to graphical parameters, which are defined by default and can be modified with the function 'par'. If we type `?par` on the R console, we get the description on number of arguments for graphical functions, as documented in R. We explain one among these which is frequently used in the book. It is `par(mfrow=c(2,2))` or `par(mfcol=c(2,2))`. This command divides the graphical window invisibly in 2 rows and 2 columns to accommodate 4 graphs. A function `legend()` is usually added in plots to specify a list of symbols or colors used in the graphs.

Code 2.6.3. This code illustrates `plot` and `lines` functions used to draw Figure 2.1. It also includes a function written to find probability mass function of binomial $B(5, p)$ distribution for various values of p and `par(mfrow=c(3,3))` function.

```
g=function(p)
{
x=0:5
g=dbinom(x,5,p)
return(g)
```

```
}
par(mfrow=c(3,3))
p=seq(.1,.9,.1)
pname=paste("p =",p,sep=" ")
for(i in 1:length(p))
{
x=0:5
plot(x,g(p[i]),type="h",xlab="x",ylab="Probability",main=pname[i],
ylim=c(0,0.7),col="blue",lwd=2)
points(x,g(p[i]),pch=16,col="dark blue")
}
```

■

All these features are used in a variety of graphs drawn in this chapter and the subsequent chapters.

We will use all these and some more functions in the computation of various monetary functions to be studied in the following chapters. Explanation of each function is given in front of the function using # symbol. Any thing that appears after # is ignored by R. It is only for understanding the code. To find out more functions and its usage one can use built-in-help system of R. For example, `?c`, `?read.table`, `?plot` will display information on these functions, its usage with illustration. There are many books on introduction to statistics using R, such as Crawley [2], Dalgaard [3], Purohit et al [14] and Verzani [21]. There is a tremendous amount of information about R on the web at `http://cran.r-project.org/` with a variety of R manuals. The following are some links useful for beginners to learn R software.

1. https://www.datacamp.com/courses/free-introduction-to-r

2. http://www.listendata.com/p/r-programming-tutorials.html

3. http://www.r-tutor.com/r-introduction

4. https://www.r-bloggers.com/list-of-free-online-r-tutorials/

5. https://www.tutorialspoint.com/r/

6. https://www.codeschool.com/courses/try-r

As for any software or programming language, the best way to learn R is to use it for understanding the concepts and solving problems. We hope that this brief introduction will be useful to a reader to be comfortable with R and follow the codes written in subsequent chapters.

Chapter 3

Feasibility of Insurance Business

Key Terms : Concave function, Decreasing marginal utility, Expected value principle, Exponential utility function, Fractional power utility function, Logarithmic utility function, Quadratic utility function, Risk aversion, Utility function.

3.1 Introduction

We have seen in Chapter 1 that an insurance system is a mechanism for providing protection against financial loss due to random events. In non-life insurance, the time of occurrence, the frequency of the occurrence and size of the financial loss are usually random variables; consequently the benefit to be paid is also a random variable. In life insurance, the amount of benefit is usually fixed but the time of payment of the benefit is random. The insurer has to take the decision of determination of the premium to be charged in such uncertain situations.

The implicit trust between the insured and the insurance company is at the center of the interaction. A reasonable mathematical theory of insurance provides a scientific basis for this trust.

The natural and the foremost question that arises in this theory is how to measure the adverse financial impact as a result of random events. In the next section, we discuss a simple principle, known as expected value principle, used to answer this question. After addressing it, we discuss the concept of utility function which is heavily used in economics and insurance. We select a typical utility function to model the mentality of a risk averse individual. A reasonable and natural condition on the utility function of the wealth of both the parties in the insurance contract leads to establish the feasibility of the insurance business.

3.2 Expected Value Principle

One of the solutions to the problem of measuring the financial impact, in the face of uncertainty, is to define the value of an economic project with a random outcome to be its expected value.

By this expected value principle, the distribution of possible outcomes is replaced by a single number, the expected value of the random monetary outcome. In economics, the expected value of random prospects with monetary payments is frequently called the fair or actuarial value of the prospect. The following three examples illustrate how to use this principle to define the value of an economic project with random output.

Illustration 3.2.1: Suppose a decision maker has two random economic prospects available. The outcome of the first prospect, denoted by X, has a normal distribution with mean $\mu = 10$ and variance $\sigma^2 = 2$, denoted by $N(10, 2)$. The second prospect, denoted by Y, has a normal $N(12, 6)$ distribution. With the expected value principle, the fair values of prospects X and Y are 10 and 12 respectively. The second prospect is preferred to the first, if one is interested in the higher expected value of the prospect.

Illustration 3.2.2: The probability that a property will not be damaged by fire in a period of one year is 0.90. The probability density function of positive loss is given by

$$f(x) = 0.0005e^{-0.0005x}, \quad x > 0$$

Here, damage to the property is a random variable which takes a value 0 with a certain probability and positive value according to the specified distribution. In this case, the expected loss is given by,

$$E(X) \quad = \quad (0.90)(0) + (0.10)\int_0^\infty 0.0005xe^{-0.0005x}dx = (0.10)(50000) = 5000 \text{ units}$$

Illustration 3.2.3: Suppose the probability of an accident in a specific time period is 0.02 and X, Y and Z are three random variables associated with the accident which measure the loss incurred to a decision maker. X corresponds to a loss of Rs 1000, Y corresponds to the loss of Rs 20,000 and Z corresponds to the loss of Rs 50,000. If an accident does not occur, then there is no loss. Thus, X, Y and Z each take a value 0 with a probability 0.98. It is easy to compute that $E(X) = 20$, $E(Y) = 400$ and $E(Z) = 1000$. Thus, the expected losses in these three situations are 20, 400 and 1000 respectively.

With the expected value principle, the random loss is summarized by its expected value. We will now see how this principle is to be used in the context of insurance. Suppose an individual is likely to face the potential loss as described by Y in Illustration 3.2.3. By the expected value principle, on the average, an individual has to suffer $E(Y)$ as a loss. So the individual would be indifferent between assuming a random loss Y, say and paying $E(Y)$ in order to be relieved of the possible loss. Thus, an individual will be willing to pay $E(Y) = 400$ Rs as a one time premium to get a cover for potential loss of Rs 20,000. However, the insurance company would not be satisfied with $E(Y) = 400$ as a premium because according to the expected value principle, company also has to suffer loss of $E(Y) = 400$. The company has to have provisions for the expenses of running the business, paying taxes and making a profit. Thus, the company would like to charge more than the expected loss as a premium to cover the random loss. If the individual is not willing to pay more than the expected loss then there seems to be little opportunity for a mutually advantageous insurance contract. Thus, the expected value principle fails to arrive at the agreement between insured and insurer. One would then like to know the theory behind the feasibility of insurance contract. Generally, the individuals are risk averse and are willing to pay more than the expected loss as a premium. This willingness is a consequence of appropriate modelling of the mental behaviour of the risk averse individual. Mathematically, it is proved by imposing some reasonable conditions on utility function. We elaborate on this concept in the next section.

3.3   Notion of Utility

In Illustration 3.2.3, we have seen that expected losses are $20, 400$ and 1000 in three different situations. If an individual has to face these three situations, his reaction towards accepting the insurance coverage will depend upon how much $1000, 20, 000$ and $50, 000$ are important to him. One may not go for insurance cover if the expected loss is Rs 20. In the second and the third case, it depends on the value of $20, 000$ and $50, 000$ for the concerned person. Suppose an individual opts for the insurance and event of accident does not take place. Then, he will lose the money paid in terms of premium. If the event of accident occurs, by paying an insurance premium of say Rs 400, the entire loss of Rs $20, 000$ is covered. Thus, whether to opt for insurance cover or not, depends on the mentality of the individual and also on the importance or utility of the wealth that he wants to insure. Utility theory, which has many applications in economics and other areas, plays a key role in such decision making.

Utility function is a function which measures the value or the significance that an individual attaches to his wealth. Suppose w is the wealth of a decision maker, either a potential insurer or a potential insured. Wealth includes money, property, business, bank deposits, and so on. Suppose the utility that the decision maker attaches to a wealth of amount w, is specified by a function $U(w)$. We now decide the nature of this function. It seems natural to assume that $U(w)$ is an increasing function, more the wealth, better the life! In addition, it has been observed that, each additional equal increment of wealth results in a smaller increment of associated utility. This is the concept of decreasing marginal utility in economics.

To illustrate this concept, suppose Vijay's monthly salary is Rs $80, 000$ while that of Ajay is Rs $30, 000$. An increment of Rs 1000 in the monthly salary will be more important to Ajay than Vijay. Of course, for both, the utility increases as the wealth increases, but for a large value of wealth, the impact of the increment is low.

Another illustration in non-technical language is to consider a simple experiment of drinking water. Suppose Deepak is very thirsty and when he drinks his first glass of water, he gets 10 units of satisfaction. He further drinks a second glass of water. Naturally, the satisfaction he gets now would be less than 10. Suppose he gets 7 units of satisfaction. He then has his third glass of water, the satisfaction he gets now would be still lower, say 4 units. If Deepak's thirst is fully quenched by now, then he would not go in for the fourth glass. But suppose he does, then he might experience trouble, and his satisfaction could be in negative units, say -3. Thus, the marginal utility Deepak derives from drinking water goes on decreasing with each additional glass of water. It is hence said that water is tasty when you are thirsty.

Marginal utility is defined as the satisfaction that a person receives from consuming an additional unit of the same good or service. Whereas, the total utility is defined as the aggregate satisfaction a person receives from the consumption of all the units of the same good or service.

In short, the total utility is the sum of all the marginal utilities from the consumption of all the units of the same good or service. In the illustration of drinking water, the total utility Deepak receives from drinking four glasses of water would be $10 + 7 + 4 - 3 = 18$ units. Thus the total utility increases and reaches its maximum at three glasses and then starts declining.

It is generally observed that people attach their own imaginary utility values to the original monetary values. Economists use the hypothetical term 'util' to denote these utility units (not in wealth terms). People make decisions based on these 'utils' rather than the monetary values. They obviously want to maximize the utility and not the observed monetary gains.

Mathematically, this phenomenon of human mentality can be modelled by a non-decreasing concave function. We first define a concave function.

Concave function: A real valued function $f : D \to R$ is said to be a concave function if $\forall\ x, y \in D$ and $\lambda \in (0, 1)$ with $\lambda x + (1 - \lambda)y \in D$,

$$f(\lambda x + (1 - \lambda)y) \geq \lambda f(x) + (1 - \lambda)f(y)$$

If a function f is differentiable twice, then a necessary and sufficient condition for a function to be concave is that the second derivative $f''(x) < 0$. If a utility function U is differentiable twice, then we would like to have (i) its first derivative $U'(w) > 0$ and (ii) its second derivative $U''(w) < 0$. The condition in (i) states that as wealth increases, utility increases and the second condition implies that for small w, increase in w is more prominent than that for large w. Figure 3.1 exhibits this nature of an increasing concave function.

An individual whose utility function satisfies these two conditions is said to be a risk averse individual. The risk aversion is measured by the coefficient

$$r(w) = -U''(w)/U'(w)$$

The utility function of the insurance company is also of the same nature. Insurance business is a corporate business. Thus, the company works with a huge capital and hence any single claim or a marginal addition via premiums is not going to alter significantly the utility of the wealth of the insurance company. In other words, for a large w, the function is linear in some neighbourhood of w. Consequently, the utility function of the insurance company is also a non-decreasing concave function. The two conditions on the utility function of both the insurer and the insured explain why the insured is willing to pay more than the expected loss and why the insurer wishes to have a one-time premium more than the expected loss, which in turn results in an insurance contract. We prove below these assertions mathematically.

Suppose both the insured and the insurer have their utility functions to be non-decreasing and concave. We apply the expected value principle to the utility of random prospect and establish that the two parties reach an agreement and sign a contract.

Suppose $U(w)$ denotes the utility function of the property owner with wealth w. The owner faces a possible loss due to random events that may damage the property. The distribution of a random loss X is assumed to be known. The insured has two options before him; sign a contract and have the insurance cover or suffer a loss if at all it occurs. If he signs a contract, he has to pay a premium, say a one time premium of amount G, then his wealth will be reduced to $w - G$, where w denotes the current wealth. If he does not opt for insurance coverage and if loss occurs, his current wealth will reduce to $w - X$. Suppose the individual is indifferent

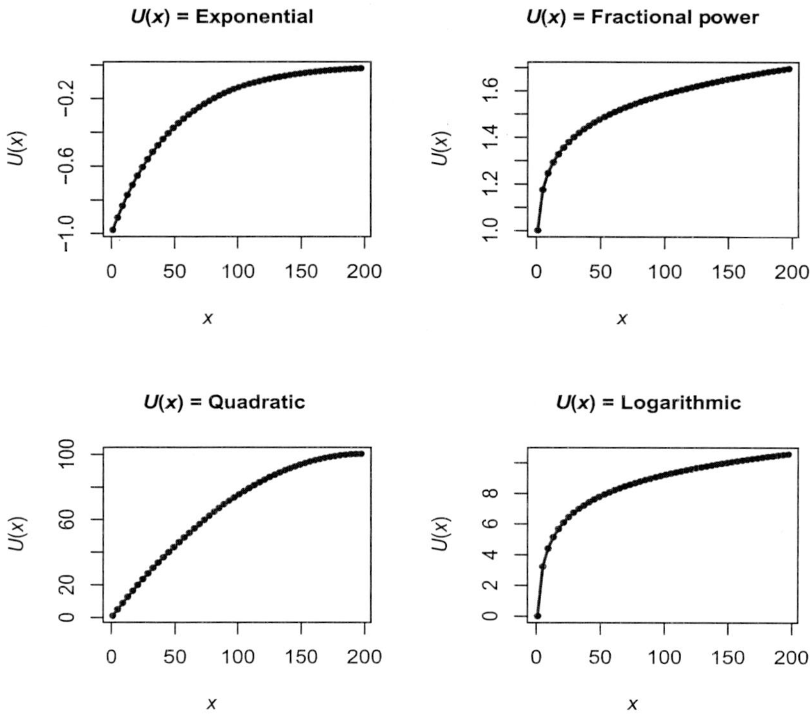

Figure 3.1 Non-decreasing concave utility functions

to these two situations as far as the utility of the wealth is concerned. Consequently, utility of $(w - G)$ is the same as that of expected utility of $w - X$. Mathematically, this can be stated as

$$U(w - G) = E(U(w - X)) \tag{3.3.1}$$

The left hand side of Eq. 3.3.1 represents the utility of paying the premium G for complete financial protection while the right hand side represents the expected utility of not buying insurance. It is assumed that $E(U(w - X))$ is finite. Suppose the utility function $U(w)$ is non-decreasing and concave, then by Jensen's inequality we have,

$$E(U(w - X)) \ \leq \ U(E(w - X)) = U(w - \mu), \quad \text{where} \quad \mu = E(X)$$

Thus, Eq. 3.3.1 leads to the inequality,

$$U(w - G) = E(U(w - X)) \ \leq \ U(w - \mu) \ \iff \ G \geq \mu$$

in view of the non-decreasing nature of U. Thus, if the utility function of a decision maker is non-decreasing and concave, then he is willing to pay an amount greater than the expected loss. If G is at least equal to the premium set by the insurer, then there is an opportunity for a mutually advantageous insurance policy. To test the feasibility of insurance contract, we apply the theory of utility function to the wealth of an insurer. Suppose $U_I(w)$ denotes the utility of wealth of the insurer. The minimum acceptable premium H corresponding to random loss X, is determined by the identity

$$U_I(w) = E(U_I(w + H - X)) \qquad (3.3.2)$$

The left hand side of Eq. 3.3.2 is the utility attached to the insurer's wealth, when a contract is not signed. The right hand side is the expected utility corresponding to the collection of premium H associated with random loss X and paying random loss X. $E(X)$ is assumed to be finite. Equation 3.3.2 states that the insurer is indifferent between not providing insurance and providing insurance for a random loss X by charging the premium of amount H. The concavity of $U_I(w)$ implies that

$$U_I(w) = E(U_I(w + H - X)) \quad \leq \quad U_I(w + H - \mu)$$

Further, the non-decreasing nature of U_I leads to the inequality

$$w \; \leq \; w + H - \mu \qquad \Longleftrightarrow \qquad H \geq \mu$$

Thus, the minimum acceptable premium that the insurer would like to charge is larger than the expected loss, which is as expected. No profit consideration or administrative expenses are taken into account at this stage. Thus, if the utility function of both the parties is non-decreasing and concave, then $G > \mu$ and $H > \mu$. If G is fixed such that $\mu < H < G$, then there is an opportunity for a mutually advantageous insurance policy. The judicious fixation of H in the gap between μ and G is an important task of an actuary, so that the interests of both the parties are safeguarded.

The above discussion conveys that utility as a philosophy and risk aversion as a feature of human psychology lead to the evolution of the insurance business as a means of reducing the financial consequences of unfavorable random events. As an illustration, consider property insurance. Suppose a person wants to insure his home against fire on an annual basis. The risk of destroying the home in a fire is usually very small, however, if at all the event occurs, the financial consequences of losing the home and all its contents could be serious. Thus, a risk averse person will prefer to pay a small amount as a premium, even though it is larger than the expected loss, to cover the potential large loss due to destruction of the home in fire. Decision making using a utility function is thus based on the expected value of the random variable involving utility function. Table 3.1 gives some commonly used non-decreasing concave utility functions and the corresponding coefficient $r(w)$ of risk aversion. For each function in Table 3.1, it is easy to verify that the first derivative is positive and the second derivative is negative and hence each is a non-decreasing concave function. Thus each of these functions serves as a utility function of a risk-averse individual. The third column presents the coefficient $r(w)$ of risk aversion. Note that for the exponential utility function, it is α, free from w. For the fractional power utility function and the logarithmic utility function, it is a decreasing function of w, while for the quadratic utility function, it is an increasing function of w.

Figure 3.1 exhibits the nature of these four types of utility functions. For the exponential utility function, $\alpha = 0.02$, for the fractional power utility function, $\gamma = 0.01$, for the quadratic utility function, $\alpha = 0.00249$ and for the logarithmic utility function, $k = 2$. In all the graphs, for a large x, the graph of the utility function $U(x)$ is almost linear.

Utility function	Form of $U(w)$	$r(w)$
Exponential	$-e^{-\alpha w}, w \in R \ \ \alpha \geq 0$	α
Fractional power	$w^{\gamma}, \ \ w > 0, \ \ 0 < \gamma < 1$	$(1-\gamma)/w$
Quadratic	$w - \alpha w^2, \ \ w < (2\alpha)^{-1}, \ \ \alpha > 0$	$1/((2\alpha)^{-1} - w)$
Logarithmic	$k \ \log w + c, \ \ w > 0, \ \ k > 0 \ \ c \in \mathbb{R}$	$1/w$

Table 3.1 Non-decreasing concave utility functions

We noted earlier that people like to maximize utility. In the case of uncertain payoffs, naturally people would try to maximize the expected utility. In the discussion on the expected value principle, we have seen that the two random economic prospects are compared by their expected values. In Illustration 3.2.1, we have two random prospects X and Y and Y is preferred to X as far as the expected value principle is concerned. Now suppose the utility function of the decision maker is given by $U(w) = -\exp(-2w)$ and the expected value principle is applied to $U(X)$ and $U(Y)$ to compare the random prospects. For X and Y having $N(10,2)$ and $N(12,6)$ distribution respectively,

$$E(U(X)) \ \ = \ \ E\{-\exp(-2X)\} = -M_X(-2) = -e^{-16}$$

where $M_X(t) = \exp\{\mu t + 0.5\sigma^2 t^2\}$ is the moment generating function of X. Similarly, $E(U(Y))$ is given by

$$E(U(Y)) = -M_Y(-2) = -e^{-12}$$

Thus,

$$E(U(X)) = -e^{-16} > E(U(Y)) = -e^{-12}$$

and the random prospect X is preferred to Y. It is in contrast to the result arrived at using expected values of X and Y.

Following examples illustrate how to compute G and H for the utility functions listed in Table 3.1.

Example 3.3.1. The decision maker has wealth of $w = 100$ units and faces a random loss X following the exponential distribution with a mean of 0.4. What is the maximum amount this decision maker will pay for complete insurance against a random loss if a decision maker's utility function is $U(w) = w - 0.003 \ w^2$?

Solution: With $U(w) = w - 0.003w^2$ and Eq. 3.3.1 we get,

$$w - G - 0.003(w - G)^2 = E(w - X) - 0.003E(w - X)^2$$

For the exponential loss random variable X, the mean is 0.4 and $E(X^2)$ is 0.32. Substituting these values and the value of w in this equation, we get the quadratic equation in G as $0.003G^2 +$

$0.4G - 0.16096 = 0$. Solving it, we get two roots. The root with the negative value is discarded and the one with positive value 0.4012 is the value for G. In this case, the decision maker is risk averse as $U''(w) < 0$. Following the discussion after Eq. 3.3.1, we should have $G > E(X)$. In this example, $E(X) = 0.4$ and $G = 0.4012 > E(X) = 0.4$. ∎

Remark 3.3.1. For the quadratic utility function, information about the first two moments is enough to find G, since Eq. 3.3.1 with the quadratic utility function reduces to the quadratic equation

$$\alpha G^2 - (2\alpha w - 1)G + \{(2\alpha w - 1)E(X) - \alpha E(X^2)\} = 0$$

Example 3.3.2. Suppose the probability of damage to the house in a year is 0.1. The probability distribution of positive loss in some units is uniform $U(0, 100)$. The owner of the property has a utility function given by, $U(w) = 2\log_e w$. Calculate the expected loss and the maximum insurance premium G the owner will pay for the complete insurance when wealth of the individual is 500 units and 1000 units. Comment on the values of G for different values of wealth w.

Solution: The expected loss is given by,

$$E(X) = 0.90(0) + 0.10 \times 50 = 5$$

We use Eq. 3.3.1 to determine the maximum premium that the owner will pay for complete insurance. Suppose $w = 500$.

$$U(w - G) = 0.90U(w) + 0.10E(U(w - X))$$

$$\Rightarrow 2\log(w - G) = 0.90 \times 2\log(w) + 0.10/100 \int_0^{100} 2\log(w - x)dx$$

$$= 11.1863 + 0.002 \times 610.7182 = 12.4077$$

$$\Rightarrow G = 500 - \exp(12.4077/2) = 5.35$$

With $w = 1000$, using the same procedure we get $G = 5.16$ which is smaller than G corresponding to $w = 500$. However, in both the cases $G > E(X)$. ∎

Remark 3.3.2. With $U(w) = k\log(w)$, the coefficient of risk aversion is given by $r(w) = 1/w$. Thus, risk aversion is a decreasing function of wealth. As wealth increases, G decreases, as noted in Example 3.3.2.

Example 3.3.3. The insurance company has wealth A units and faces a random loss X which follows the chi-square distribution with a mean of 50 units. What is the minimum acceptable amount this company would like to charge for complete insurance against the random loss if its utility function is $U(w) = -\exp\{-0.05w\}$?

Solution : With $U(w) = -\exp\{-0.05w\}$ and Eq. 3.3.2 we get,

$$-\exp\{-0.05w\} = E[-\exp\{-0.05(w + H - X)\}]$$

It simplifies to

$$\exp\{0.05H\} = M_X(0.05) = (1 - 2 \times 0.05)^{(-50/2)} = (0.90)^{-25} \Rightarrow H = 52.68$$

Note that $H > E(X) = 50$ ∎

Example 3.3.4. An individual has a wealth of 100 units and faces a random loss X following the exponential distribution with a mean of 0.2. (i) What is the maximum amount this individual will pay for the complete insurance against a specified random loss if his utility function is $U(w) = -\exp\{-2w\}$? (ii) Suppose the wealth of the insurance company is 1000000 units. What is the minimum acceptable amount this company would like to charge for complete insurance against a random loss if its utility function is given by $U(w) = -\exp\{-w\}$?

Solution: (i) From Eq. 3.3.1, we have

$$
\begin{aligned}
-\exp\{-2(w-G)\} &= E[-\exp\{-2(w-X)\}] \\
\Rightarrow \exp\{2G\} &= \int_0^\infty \exp\{2x\}5\exp\{-5x\} = 5/3 \int_0^\infty 3\exp\{-3x\} = 5/3 \\
\Rightarrow G &= 0.5\log_e(5/3) = 0.255
\end{aligned}
$$

(ii) With $U(w) = -\exp\{-w\}$ and Eq. 3.3.2 we get,

$$
\begin{aligned}
-\exp\{-w\} &= E[-\exp\{-(w+H-X)\}] \\
\Rightarrow \exp\{H\} &= 5\int_0^\infty \exp\{-4x\} = 1.25 \Rightarrow H = \log_e 1.25 = 0.223
\end{aligned}
$$

In this case both the decision makers are risk averse as the utility functions of both are non-decreasing concave functions. Following the discussion after Eq. 3.3.1, we should have $G > E(X)$. In this example $G = 0.255 > E(X) = 0.2$. Further, from Eq. 3.3.2, we should have $H > E(X)$ and the relation is also satisfied in this case. Further, note that $H < G$ and hence the insurance contract is feasible. ∎

Remark 3.3.3. In Example 3.3.4, note that G does not depend on wealth w. It is true in general also. In fact, G is related to the moment generating function $M_X(\cdot)$ of X. Suppose $U(w) = -e^{-\alpha w}$, then from Eq. 3.3.1, we get

$$
\begin{aligned}
-\exp\{-\alpha(w-G)\} &= E[-\exp\{-\alpha(w-X)\}] \\
\Rightarrow \exp\{\alpha G\} &= E(\exp\{\alpha X\}) = M_X(\alpha) \\
\Rightarrow G &= (\log M_X(\alpha))/\alpha
\end{aligned}
$$

Further, from Example 3.3.2, it is to be noted that H also does not depend on the wealth w. It is also related to the moment generating function of X. Suppose $U(w) = -e^{-\alpha_I w}$, then from Eq. 3.3.2, we get

$$
\begin{aligned}
-\exp\{-\alpha_I w\} &= E[-\exp\{-\alpha_I(w+H-X)\}] \\
\Rightarrow \exp\{\alpha_I H\} &= E(\exp\{\alpha_I X\}) = M_X(\alpha_I) \\
\Rightarrow H &= (\log M_X(\alpha_I))/\alpha_I
\end{aligned}
$$

Example 3.3.5. Find the value of G if the utility function of the individual is $U(w) = -e^{-\alpha w}$ and a random loss X that he may face has normal distribution with a mean μ and variance σ^2. Comment on the nature of G as a function of α. Find the value of H if the utility function of the insurance company is $U(w) = -e^{-\alpha_I w}$. What is the nature of H as a function of α_I? Derive a relation between α and α_I so that the insurance contract is feasible.

Solution: From the formula derived in Remark 3.3.3, $G = (logM_X(\alpha))/\alpha$. The moment generating function of the normal $N(\mu, \sigma^2)$ distribution is $\exp\{\mu\alpha + \alpha^2\sigma^2/2\}$. Hence, $G = \mu + \sigma^2\alpha/2$. Note that G is an increasing function of α, which is the coefficient of risk aversion for the exponential utility function. Thus, the more risk averse the insurer, the higher is the value of G. On similar lines we get $H = \mu + \sigma^2\alpha_I/2$. Here also H is an increasing function of α_I. The insurance contract is feasible if $G > H \iff \alpha > \alpha_I$. $\blacksquare$

Remark 3.3.4. In Example 3.3.4, note that $\alpha = 2$ and $\alpha_I = 1$ and it is shown that the insurance contract is feasible.

Example 3.3.6. (i) Suppose an individual suffers no loss with a probability of 0.1 and suffers a loss of 10 units with a probability of 0.9. Suppose his wealth is 20 units and his utility function is $U(w) = \sqrt{w}$. Find the maximum insurance premium that the individual will pay for complete insurance. (ii) Suppose the wealth of the insurance company is 1000 units. What is the minimum acceptable amount this company would like to charge for complete insurance against a random loss if its utility function is given by $U(w) = -\exp\{-5w\}$?

Solution: (i) For the given distribution of random loss, Eq. 3.3.1 becomes

$$\begin{aligned}
U(w - G) &= 0.1U(w) + 0.9U(w - 10) \\
\Rightarrow \quad \sqrt{(w - G)} &= 0.1\sqrt{w} + 0.9\sqrt{(w - 10)} \\
\Rightarrow \quad G &= 9.15 \text{ with } w = 20
\end{aligned}$$

(ii) From Eq. 3.3.2 we get,

$$\begin{aligned}
U(w) &= E(U(w + H - X)) \\
\Rightarrow \quad -\exp(-5w) &= -0.1\exp(-5(w + H)) - 0.9\exp(-5(w + H - 10)) \\
\Rightarrow \quad H &= \log(0.1 + 0.9\exp(50))/5 = 9.98
\end{aligned}$$

Note that $H > G > E(X) = 9$. However, $H > G$ and hence the insurance contract is not feasible. $\blacksquare$

In the next two examples we use R to compute G.

Example 3.3.7. Each of the following four decision makers faces a random loss of X units, where the possible values of X are $0, 100, 200, 300$ with respective probabilities $0.5, 0.1, 0.3, 0.1$. Find the expected loss. Each decision maker has the same utility function given by $U(w) = w^{1/3}$. The wealth of the decision makers A, B, C, D is $800, 900, 1000, 1100$ units respectively. Using R, find out who will purchase complete insurance at a premium of 105 units.

Solution: The possible values of X are $0, 100, 200, 300$ with respective probabilities $0.5, 0.1, 0.3, 0.1$. Hence we get $E(X) = 100$. As in the above examples, we use Eq. 3.3.1 to determine the maximum premium that the owner will pay for complete insurance. Thus for given w, we need to solve the equation

$$U(w - G) = E(U(w - X)) \Rightarrow (w - G)^{1/3} = \sum_{i=1}^{4}(w - x_i)^{1/3}p_i$$

$$\Rightarrow \quad G = w - \left(\sum_{i=1}^{4}(w - x_i)^{1/3}p_i\right)^3$$

We use Code 3.4.1 to find G for 4 values of w. These are given by $106.03, 105.23,$ $104.62, 104.14$ for $w = 800, 900, 1000, 1100$ units respectively. Note that for $A, B, H = 105 < G$ and hence A and B will purchase complete insurance at a premium of 105 units.   ■

Example 3.3.8. Suppose a decision maker faces a random loss X where the possible values of X are $0, 10, 20, 30$ units, with respective probabilities $0.5, 0.1, 0.3, 0.1$. Find the expected loss. Suppose the utility function of the decision maker is $U(w) = -\exp(-\alpha w)$. Find the coefficient $r(w)$ of risk aversion. Using R, find G corresponding to different values of $r(w)$. Comment on your findings.

Solution: For the given probability distribution, $E(X) = 10$. For the utility function $U(w) = -\exp(-\alpha w)$, the coefficient $r(w)$ of risk aversion is α. It does not depend on w. To find G, we use Eq. 3.3.1. Thus,

$$U(w - G) \;=\; E(U(w - X)) \Rightarrow \; \exp(-\alpha(w - G)) = \sum_{i=1}^{4} \exp(-\alpha(w - x_i))p_i$$

$$\Rightarrow \; G \;=\; \log\left(\sum_{i=1}^{4} \exp(\alpha x_i)p_i\right)/\alpha$$

We use Code 3.4.2 to find G for different values of $r(w) = \alpha$ (Table 3.2). We note that as

α	0.05	0.10	0.15	0.20	0.25	0.30
G	13.14	16.09	18.47	20.30	21.70	22.80

Table 3.2 Exponential utility function: G

the coefficient of risk aversion increases, the value of G also increases, which seems reasonable.   ■

We have noted that in the insurance business, both the insured and insurer are risk averse and their utility functions are non-decreasing concave functions. A majority of people in the society are risk averse most of the time. However, utility theory in general talks about two other types of individuals as well. These are 'risk seekers (lovers)' and 'risk neutral' individuals. Risk lovers are those who play challenging games like hiking, sky-diving and so on. Since risk-seeking behavior shows choices of high risk actions, naturally the utility curve is also opposite to that of a risk averse individual. This is similar to the 'Bull' approach adopted in stock markets. It represents the mind-set of gamblers. For such a person, the utility curve is of increasing type, but it is convex in nature. For a risk-loving person, the graphs of the utility function are shown in Figure 3.2.

Observe that for a risk-loving person, the utility function is a non-decreasing convex function. Hence, by Jensen's inequality, Eq. 3.3.1, given by $U(w - G) = E(U(w - X))$ implies that $G \leq \mu$. It is shown below.

$$U(w - G) = E(U(w - X)) \geq U(w - E(X)) \;\; \Rightarrow \;\; w - G \geq w - E(X) \;\; \Rightarrow \;\; G \leq \mu$$

Thus, a risk-lover is willing to pay the premium which is less than the expected loss. However, the insurer requires the premium to be larger than μ. So in this setup, an insurance contract is not feasible.

A risk neutral individual is indifferent between playing the lottery and not playing it. The utility function for such a person is a linear function of the form $U(w) = aw$, where $a > 0$ is a real number. Such an individual gains a constant marginal utility of wealth irrespective of his belongings from time to time. The utility function of a risk-neutral person is shown in the fourth graph of Figure 3.2. When $U(w) = aw$, Eq. 3.3.1, given by

$$U(w - G) = E(U(w - X)) \quad \Rightarrow \quad a(w - G) = E(a(w - X)) \quad \Rightarrow \quad G = E(X) = \mu$$

Thus, a risk-neutral person is willing to pay the premium which equals the expected loss. However, the insurer requires the premium to be larger than μ. So in this setup also, an insurance contract is not feasible.

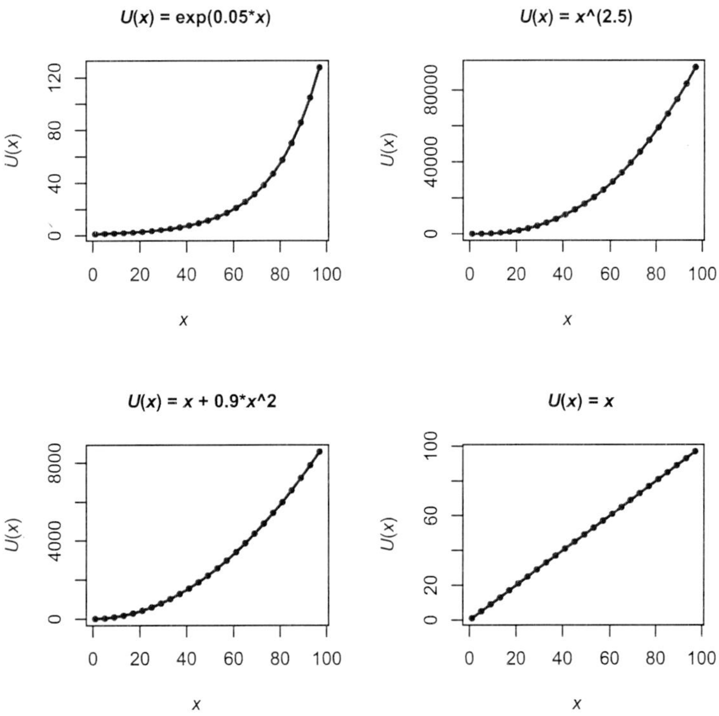

Figure 3.2 Convex and linear utility functions

Figure 3.2 exhibits the nature of three types of utility functions for risk lovers. These are increasing convex curves. The fourth graph is of the utility function $U(x) = x$ corresponding to a risk-neutral person.

We have considered these two types of behaviour for the sake of completeness. In insurance business, only 'risk averse' behaviour is of concern. In fact, since people are risk averse, the insurance business works.

The next section presents two R codes, used to solve two examples in this section.

3.4   R Codes

Code 3.4.1. In Example 3.3.7, we have obtained the values of G for four decision makers facing a random loss X with utility function $U(w) = w^{1/3}$. We use this R code to find out these values.

```
x=c(0,100,200,300); p=c(.5,.1,.3,.1); mu=sum(x*p); mu
w=c(800,900,1000,1100);l=length(w)
u=matrix(nrow=l,ncol=length(x));G=c()
for(i in 1:l)
{
for(j in 1:length(x))
{
u[i,j]=((w[i]-x[j])^(1/3))*p[j]
}
}
y=apply(u,1,sum); y; G=round(w-y^3,2); G
```

■

Code 3.4.2. In Example 3.3.8, we have obtained the values of G corresponding to different values of the coefficient of risk aversion, when the utility function is exponential. We use this R code to find out the values of G.

```
x=c(0,10,20,30); pr=c(.5,.1,.3,.1) # pmf of X
mean=sum(x*pr); mean # mu
alpha=c(.05,.1,.15,.2,.25,.3);l=length(alpha); b=G=c()
for(i in 1:l)
{
b[i]=sum(exp(alpha[i]*x)*pr)
G[i]=log(b[i])/alpha[i]
}
G=round(G,2);d=data.frame(alpha,G); d
```

■

3.5   Conceptual Exercises

3.5.1 A utility function is given by,

$$U(w) = e^{-(w-5)^2}, \quad \text{if } w < 5 \quad \text{and} \quad U(w) = 5 - e^{-(w-5)^2}, \text{if } w \geq 5.$$

Is $U(w)$ a utility function of a risk averse individual? Justify your answer.

3.5.2 An insurance company faces a random loss X with a mean of 2 and variance 5. The utility of its wealth is modelled by a function $U(w) = w - 0.0024w^2$. Find the upper limit on w so that it is a non-decreasing concave function. Suppose $w = 200$. Is it below the upper limit obtained above? If yes, find the minimum acceptable premium to the insurance company for complete insurance.

3.5.3 An individual and the insurance company model the utility of their wealth by a function $U(w) = -\exp(-0.002w)$ and $U_I(w) = -\exp(-0.001w)$ respectively. Both may face a random loss X, which has the exponential distribution with a mean of 50 units. Find the maximum insurance premium that the individual will pay for complete insurance and minimum acceptable premium to the insurance company. Is the insurance contract feasible? Justify your answer.

3.5.4 An individual and the insurance company face a random loss X, which has the gamma distribution with a scale parameter of 1 and shape parameter of 25. Utility of their wealth is given by the functions $U(w) = -\exp(-0.2w)$ and $U(w) = -\exp(-0.02w)$ respectively. Find $E(X)$, the maximum insurance premium that the individual will pay for complete insurance and minimum acceptable premium to the insurance company. Is the insurance contract feasible? Justify your answer.

3.5.5 Suppose in Exercise 3.5.4, X follows gamma distribution with a scale parameter of 1 and shape parameter of 10. Utility of their wealth is given by the functions $U(w) = -\exp(-0.2w)$ and $U_I(w) = -\exp(-0.02w)$ respectively. Examine how the answers change.

3.5.6 A decision maker's utility of wealth is given by $U(w) = -\exp(-w)$. The individual suffers a financial loss which has a normal distribution with a mean of 60 units and standard deviation 6 units. Suppose his wealth is 120 units. (i) Find the maximum insurance premium that the individual will pay for complete insurance. (ii) Suppose the wealth of the insurance company is 1000 units. What is the minimum acceptable amount this company would like to charge for complete insurance against the random loss if its utility function is given by $U_I(w) = \exp(-0.04w)$? Is the insurance feasible?

3.5.7 Each of the following four decision makers faces a random loss with a uniform distribution over the interval $(0, 10)$ and each has the same utility function given by $U(w) = \sqrt{w}$. The wealth of each of the four decision makers A, B, C, D is $10, 15, 20, 25$ units respectively. Who will purchase complete insurance at a premium of 5.2 units?

3.5.8 Each of the following four decision makers faces a random loss of X units, where the possible values of X are $0, 200$ with respective probabilities $0.8, 0.2$. Find the expected loss. Each has the same utility function given by $U(w) = w^{1/3}$. The wealth of the decision makers A, B, C, D is $500, 580, 670, 800$ units respectively. Find out who will purchase complete insurance at a premium of 45 units.

3.5.9 A decision maker has utility function $U(w) = \log_e w$, $w > 0$ and wealth of Rs 10 lakhs. He is willing to pay a premium of Rs $50,000$/- to cover a risk with probabilities $3/4$ of no loss and $1/4$ of loss L. Determine the value of L.

3.6   Computational Exercises

3.6.1 Draw the graph of the following utility function and verify the results in Exercise 3.4.1.

$$U(w) = e^{-(w-5)^2}, \quad \text{if } w < 5 \quad \text{and} \quad U(w) = 5 - e^{-(w-5)^2}, \text{if } w \geq 5$$

3.6.2 Each of the following four decision makers faces a random loss of X units, where the possible values of X are $0, 100, 200, 300, 400$ units with respective probabilities $0.4, 0.1, 0.1, 0.3, 0.1$. Find the expected loss. Each has the same utility function given by $U(w) = w^\gamma$, $0 < \gamma < 1$. The wealth of the decision makers A, B, C, D is $800, 900, 1000, 1100$ units respectively. Using R, find out who will purchase complete insurance at a premium of 165 units, for any three different values of γ.

3.7   Multiple Choice Questions

Note: Unless specified otherwise, you have to identify which of the options is correct. Answers are given in the solutions of conceptual exercises.

3.7.1 Which of the following is not a utility function of a risk averse individual? Suppose $w > 0$.

(a) $U(w) = w - 0.0024w^2$

(b) $U(w) = \exp(-0.02w)$

(c) $U(w) = 2\log w + 3$

(d) $U(w) = w^{1/2}$

3.7.2 For $w > 0$, $U(w)$ is a utility function of a risk averse individual, if
(I) $U(w) = w - 0.0031w^2$, (II) $U(w) = \exp(-0.01w)$, (III)$U(w) = 2\log w + 3$ and (IV) $U(w) = w^4$.

(a) Only (I) is true

(b) Only (I) and (II) are true

(c) Only (I) and (III) are true

(d) Only (I) and (IV) are true

3.7.3 An individual has a wealth $w = 100$ units and faces a random loss X which follows an exponential distribution with a mean of 0.2. What is the maximum amount this individual would pay for complete insurance against the specified random loss if his utility function is $U(w) = -\exp(-2w)$?

(a) 0.223

(b) 0.255

(c) 1.25

(d) 5/3

3.7.4 The following are two statements. (I) For the fractional power utility function, the coefficient of risk aversion is a decreasing function of w. (II) For the logarithmic utility function the coefficient of risk aversion is a decreasing function of w.

(a) Both (I) and (II) are false

(b) Both (I) and (II) are true

(c) (I) is true but (II) is false

(d) (I) is false but (II) is true

3.7.5 The following are two statements. (I) For the exponential utility function, the coefficient of risk aversion is a decreasing function of w. (II) For the quadratic utility function the coefficient of risk aversion is a decreasing function of w.

(a) Both (I) and (II) are false

(b) Both (I) and (II) are true

(c) (I) is true but (II) is false

(d) (I) is false but (II) is true

3.7.6 For the fractional power utility function $U(w) = w^\gamma$, the coefficient of risk aversion is

(a) $(\gamma - 1)/w$

(b) $w/(\gamma - 1)$

(c) $(1 - \gamma)/w$

(d) γ/w

Chapter 4

Future Lifetime Distribution and Life Tables

Key Terms : Analytical laws of mortality, Balducci assumption, Constant force of mortality, Curtate future lifetime, Deferred probabilities, Force of mortality, Life table, Select life table, Select period, Time until death random variable, Ultimate life table, Uniform distribution of deaths in a unit interval.

4.1   Introduction

We have seen in Chapter 3 how a risk averse individual agrees to sign a contract with the insurance company to get a cover for the financial risk due to death. Time of death being uncertain, the policy term is also a random variable, as discussed in Section 1.6. Thus in any insurance product, the time at which the benefit is to be paid, and the period for which the insurance company will receive the premiums are random variables. In annuity products, the period for which the annuitant receives the benefit payments is a random variable. In all these products, the random variable involved is a future lifetime random variable, which is the time from signing the contract to the time of paying the claims. The next section discusses how to derive its distribution.

4.2   Future Life Time Random Variable

Death is a certain event, however, the age at which an individual will die cannot be predicted. Hence, the lifetime of an individual is modelled by a random variable. Suppose a random variable X denotes the new born's age at death, that is, the life length of an individual. Then X is a continuous random variable. Being a life length, X cannot be negative and hence X is a non-negative valued random variable. As a consequence, if F denotes its distribution function, then it is 0 for $x < 0$. Suppose $S(x) = 1 - F(x)$ denotes the survival function. It specifies the probability that the new born will attain age x, that is, the child will survive for at least x years. Further the probability that his life is between ages x and z, that is, the person will not die before age x but will die before age z, is given by,

$$P[x \leq X \leq z] \;=\; F(z) - F(x) \;=\; S(x) - S(z) \;=\; \int_x^z f(u)\, du$$

f being the probability density function of X. We have noted in Section 2.2 that a distribution function F satisfies certain properties. It can also be proved that any function F satisfying those four properties is a distribution function. Similarly any function S satisfying the following four properties is a survival function of a non-negative random variable. (i) $0 \leq S(x) \leq 1$, $\forall\; x \in \mathbb{R}$, (ii) S is non-increasing, (iii) S is right continuous and (iv) $S(0) = 1$ and $S(\infty) = 0$.

In the following example, we verify which of the functions can be labelled as survival functions.

Example 4.2.1. Which of the following functions are survival functions for $x \geq 0$?   (a) $S(x) = \exp(x - 0.7(2^x - 1))$,   (b) $S(x) = (1 + x)^{-2}$ and (c) $S(x) = \exp(-x^2)$.

Solution: (a) Observe that $S(1) = \exp\{1 - 0.7(2 - 1)\} = \exp(0.3) > 1$, hence it cannot serve as a survival function. (b) Observe that $S(0) = 1$, $S(\infty) = 0$, and S is a decreasing function. Further, being a polynomial $(1 + x)^2$ is continuous and hence $S(x) = (1 + x)^{-2}$ is also continuous. Hence S can serve as a survival function. (c) In this case also, $S(0) = 1$, $S(\infty) = 0$, and S is a decreasing function. Further, being an exponential function, it is continuous and hence right continuous. Thus, all the four requirements for a function to be a survival function are satisfied. Hence, it is a survival function.   ∎

It is known that the distribution of a continuous random variable can be specified either in terms of its distribution function or the survival function or the probability density function. There is one more function, often used in actuarial science, demography and reliability theory, which has a one to one relation with all the above three functions. It is known as the force of mortality. We discuss this concept and derive its relation with other functions. The force of mortality at age x, denoted by μ_x, is defined as follows.

$$
\begin{aligned}
\mu_x \;&=\; \lim_{\delta x \to 0} (1/\delta x) P[x \leq X \leq x + \delta x \,|\, X > x] \\
&=\; \lim_{\delta x \to 0} (1/\delta x)(S(x) - S(x + \delta x))/S(x) \\
&=\; (-1/S(x)) \lim_{\delta x \to 0} (S(x + \delta x) - S(x))/\delta x \\
&=\; -S'(x)/S(x) \;=\; -\frac{d}{dx} \log S(x) \\
&=\; f(x)/(1 - F(x)), \quad \text{as} \quad -S'(x) = f(x) = F'(x)
\end{aligned}
$$

From the formula of μ_x, it is clear that it has a conditional probability density interpretation. For each age x, it gives the value of the conditional probability density function of X at the exact age x, given survival to that age. Figure 4.1 shows a typical curve of force of mortality for human life.

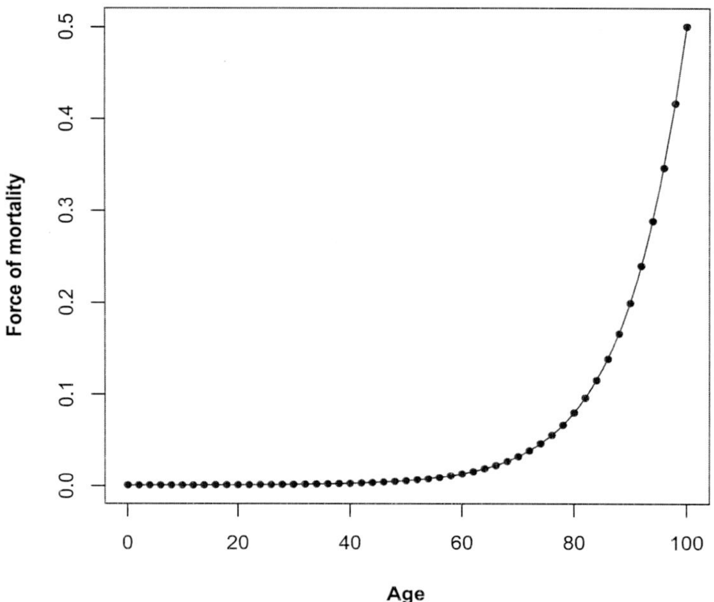

Figure 4.1 Force of mortality μ_x

In Figure 4.1 we note that the force of mortality is more or less constant up to a certain age and then increases rapidly beyond 60. In real life, the force of mortality is high in the first few months after birth, up to year 1. This feature is not revealed in the figure.

In actuarial science and demography, μ_x is called the force of mortality. In reliability theory, μ_x is known as the failure rate or hazard rate or hazard rate function. μ_x essentially specifies the rate at which there is a failure or death at age x given that the individual has survived up to age x. The force of mortality is sometimes referred to as the mortality function or mortality law.

We derive below the relations among the distribution function, survival function, probability density function and the force of mortality. It is known that $F'(x) = f(x) = -S'(x)$. As defined above,

$$\mu_x = f(x)/(1 - F(x)) = f(x)/S(x) = -S'(x)/S(x) = -\frac{d}{dx}\log S(x)$$

$$\Rightarrow \int_0^x \mu_t\,dt = -\log S(x) \quad\text{and}\quad S(x) = \exp\left(-\int_0^x \mu_t\,dt\right) \tag{4.2.1}$$

$$\Rightarrow f(x) = -S'(x) = \mu_x \exp\left(-\int_0^x \mu_t\,dt\right) = \mu_x\,S(x)$$

We know that a function has to satisfy certain conditions for it to qualify as a distribution function or a survival function or a probability density function. Analogously, we have some conditions on μ_x for it to be a proper force of mortality. We derive these conditions below.

We have $S(x) = \exp\{-\int_0^x \mu_s\, ds\}$ and $S(x)$ is a decreasing function, that is $\exp\{-\int_0^x \mu_s\, ds\}$ is a decreasing function, so $\mu_s \geq 0\ \forall\ s \geq 0$. Further,

$$\lim_{x \to \infty} S(x) = 0 \quad \Rightarrow \quad \lim_{x \to \infty} \exp\{-\int_0^x \mu_s\, ds\} = 0 \quad \Rightarrow \quad \int_0^\infty \mu_s ds = \infty$$

Thus, (i) $\mu_x \geq 0\ \forall\ x$ and (ii) $\int_0^\infty \mu_x\, dx = \infty$ are the two conditions on μ_x, for it to be a proper force of mortality.

Example 4.2.2. It is given that $\mu_x = A + e^x$ and $S(0.50) = 0.50$, calculate A.

Solution: Observe that,

$$\begin{aligned}
0.50 &= S(0.50) = \exp\{-\int_0^{0.5} \mu_x dx\} = \exp\{-\int_0^{0.5}(A + e^x)dx\} \\
&= \exp\{-0.5A - e^{0.5} + 1\} \quad \Rightarrow \quad \log(0.5) = -0.5A - \exp(0.5) + 1 \\
\Rightarrow \quad A &= 0.089
\end{aligned}$$

∎

Example 4.2.3. Suppose X follows an exponential distribution with a parameter μ. Find the force of mortality.

Solution: For the exponential distribution with a parameter μ, the probability density function is given by, $f(x) = \mu e^{-\mu x}$, for $x > 0$ and 0 otherwise. The survival function is given by $S(x) = e^{-\mu x}$, for $x > 0$. Hence, the force of mortality is $\mu\ \forall\ x > 0$. Note that it does not depend on x. ∎

Remark 4.2.1. In the class of all continuous distributions, exponential distribution is the only distribution having a constant force of mortality, which is a scale parameter of the distribution. Note that for the exponential distribution, the force of mortality at age 10 is the same as at age 80, which is unrealistic in the case of the human life length random variable. As a consequence, exponential distribution is not a good model for human life length.

Example 4.2.4. For the uniform distribution over $(0, 100)$, find the force of mortality.

Solution: For the uniform distribution over $(0, 100)$, the probability density function is given by, $f(x) = 1/100,\quad x \in (0, 100)$ and the survival function is

$$S(x) = \begin{cases} 1, & \text{if} \quad x < 0 \\ 1 - x/100, & \text{if} \quad 0 \leq x < 100 \\ 0, & \text{if} \quad x \geq 100 \end{cases}$$

Hence, the force of mortality is $1/(100 - x)$ for $0 \leq x < 100$. ∎

The most commonly encountered probability laws to model human life length are (i) De-Moivre' law (1729),(ii) Gompertz's law (1825), (iii) Makeham's law (1860) and Weibull law (1939). For human lives, there have been few observations of age at death beyond 110. Death being a certain event, it is reasonable to assume that there is an upper limit to age. Age w such that $S(x) > 0$ for $x < w$ and $S(x) = 0$ for $x \geq w$, is called the limiting age.

In one of the first attempts to describe algebraically the mortality experience of human lives, Abraham De-Moivre proposed, early in the eighteenth century, that the human life length can be modelled by the uniform distribution over the interval $(0, w)$ with force of mortality $\mu_x = (w - x)^{-1}$ and survival function $S(x) = 1 - x/w$, $0 \leq x \leq w$, w being the limiting age. With this form of survival function,

$$P[x \leq X \leq x + 1] = S(x) - S(x + 1) = 1/w \ \& \ P[x \leq X \leq x + 10] = 10/w$$

Note that the probability of death in a unit interval $(x, x + 1)$ does not depend on x. For the uniform distribution over $(0, 100)$,

$$P[10 \leq X \leq 20] = P[80 \leq X \leq 90] = 1/10$$

Thus, if the human life length is modeled by uniform distribution on $(0, 100)$, then the probability of death in the age group $(10, 20)$ is the same as the chance of death in the age group $(80, 90)$. It is unrealistic, since we observe that more deaths occur in the age group $(80, 90)$ than in the age group $(10, 20)$. Hence, uniform distribution is also not a good model for human life length and hence is rarely used.

Benjamin Gompertz, actuary of the Alliance of London, reasoning on physiological grounds, proposed in 1825, the mathematical formulation of changes in mortality with age. According to him the force of mortality is given by

$$\mu_x = BC^x, \quad B > 0, \ C > 1, \ x \geq 0$$

This formula implies that the force of mortality increases with age. It thus represents the deterioration in the ability of human beings to withstand death. The survival function for Gomperz's law is given by

$$S(x) = \exp[-m(C^x - 1)], \quad \text{where} \quad m = B/\log_e C, \ B > 0, \ C > 1, \ x \geq 0$$

Gompertz' law is often found to be quite accurate, at least as a first approximation, for ages over 25, the value of C usually being between 1.07 and 1.12. With $C = 1$, Gompertz' law reduces to the exponential distribution which is characterized by the fact that the force of mortality is constant.

Some 35 years later, William Makeham, proposed to add to the geometric component of the force of mortality postulated by Gompertz, a constant component intended to reflect the causes of death due to chance and irrespective of age. For example, death due to accident may happen at any age. The force of mortality and the survival function according to Makeham's law are

$$\mu_x = A + BC^x \ \text{ and } \ S(x) = \exp[-Ax - m(C^x - 1)], \quad B > 0, \ A \geq -B, \ C > 1, \ x \geq 0$$

The constant A has been interpreted as the term to capture the accident hazard. Both the Gompertz' and Makeham's formulae produce excessively high mortality rates at ages over 90.

Weibull's law of mortality is given by

$$\mu_x = kx^n \ \text{ and } \ S(x) = \exp(-ux^{n+1}), \ \ u = k/(n+1), \ k > 0, \ n > 0, \ x \geq 0$$

It is usually less successful than those of Gompertz and Makeham to represent human mortality, since its support for any combination of parameters does not cover the human life span. Figure 4.2 and Figure 4.3 depict the force of mortality and the survival function respectively, corresponding to these four laws. The parameters for these four laws are as follows. For De-Moivre's law, $w = 100$, for Gompertz's law $B = 0.5 * 10^{-4}$ and $C = 10^{0.04} \approx 1.096$, for Makeham's law $A = 0.7 * 10^{-3}, B = 0.5 * 10^{-4}$ and $C = 10^{.04} \approx 1.096$ and for Weibull's law $k = 3$ and $n = 2$.

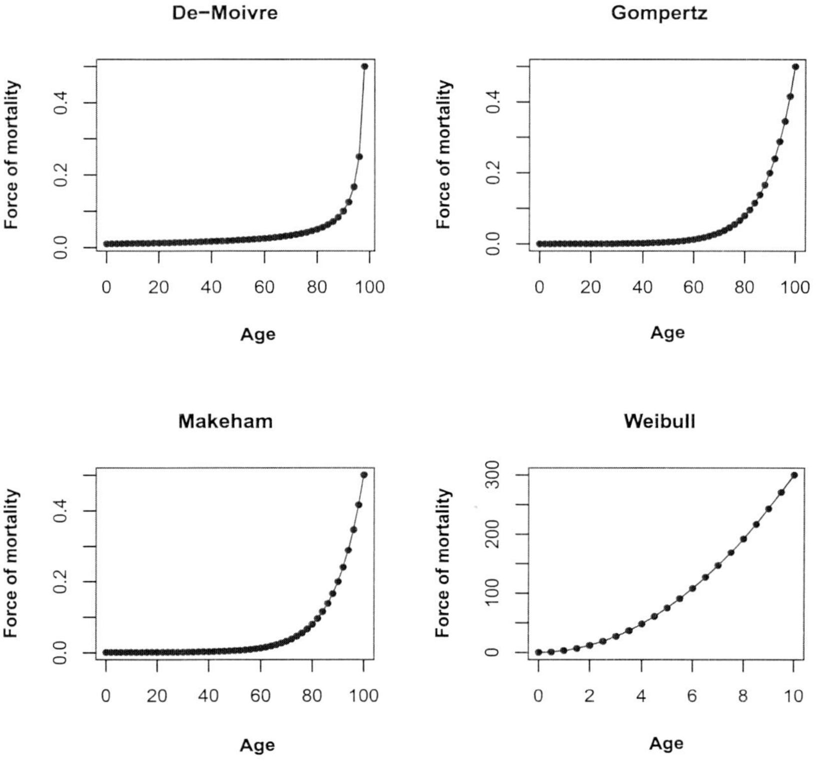

Figure 4.2 Force of mortatlity

Note that force of mortality increases as age increases for all the four laws but the increase is at a high rate for the Weibull law. From the graph of survival function for Weibull law, we see that chance of survival beyond 2 is almost 0 for the specified values of the parameters. Similar results are observed for any other combination of parameters of the Weibull distribution. Consequently the Weibull law is not a good model for modelling human life length. From the graph of survival function for the De-Moivre law, it is clear that this is also not a good model for human life length. Graphs of Gompertz's and Makeham's laws are more or less similar and these two are found to be good models for human life length.

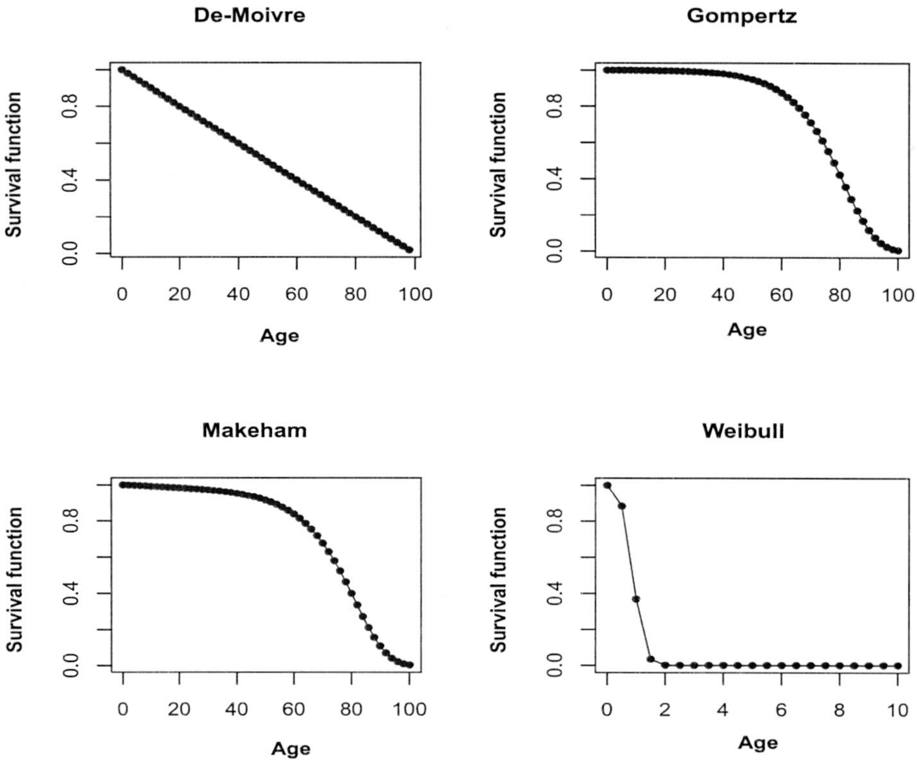

Figure 4.3 Survival function

The probability laws described above are often referred to by the mortality function than the survival function and are known as the analytical laws of mortality.

In the insurance business, the insurer is interested in the time until the death random variable for a person aged x. Symbol (x) is used to denote an individual of age x or in insurance terminology 'a life aged x'. Suppose $T(x)$ denotes the time until death of (x). It is also known as a future lifetime of (x) or residual life of (x) or remaining life of (x). In particular, $T(0) \equiv X$. It is to be noted that the distribution of $T(x)$ is the same as the conditional distribution of $X - x$ given that $X > x$. If $G_x(t)$ denotes the distribution function of $T(x)$, then for $t \geq 0$,

$$\begin{aligned} G_x(t) &= P[T(x) \leq t] = P[X - x \leq t | X > x] \\ &= P[x < X \leq x + t]/P[X > x] \\ &= \{S(x) - S(x+t)\}/S(x) \end{aligned} \qquad (4.2.2)$$

In actuarial science, it is frequently necessary to make probability statements about $T(x)$. Hence, the International Actuarial Congress held in 1898 adopted the international actuarial notation to describe the distribution function and the survival function of $T(x)$. These notations are revised or extended as necessary by the International Actuarial Association's permanent committee on notation. The distribution function and the survival function of $T(x)$ are respectively denoted by $_tq_x$ and $_tp_x$ and are given by,

$$_tq_x = P[T(x) \leq t], \quad t \geq 0 \quad \text{and} \quad _tp_x = 1 - \ _tq_x = P[T(x) > t], \quad t \geq 0$$

$_tq_x$ is the probability that (x) dies in $(x, x+t]$ while $_tp_x$ is the probability that (x) will survive upto age $x + t$. In particular, if $x = 0$, then $T(0) \equiv X$, $_tq_0 = F(t)$, $t \geq 0$ and $_tp_0 = S(t)$, $t \geq 0$. If $t = 1$, as per the international actuarial notation, t is omitted. So that, $_1q_x = q_x$, which denotes the probability that (x) will die within 1 year and $_1p_x = p_x$ is the probability that (x) will attain age $x + 1$. $_tq_x$ and $_tp_x$ are essentially functions of the survival function of X. Using Eq. 4.2.2, we express $_tq_x$ and $_tp_x$ in terms of $S(x)$ as follows.

$$_tq_x \quad = \quad P[T(x) \leq t] = (S(x) - S(x + t))/S(x) = 1 - S(x + t)/S(x)$$
$$\text{and} \ _tp_x \quad = \quad 1 - \ _tq_x = S(x + t)/S(x)$$

Observe that $_tp_x$ can be expressed as

$$_tp_x = (S(x + t)/S(0))(S(0)/S(x)) = \ _{x+t}p_0/_xp_0 \quad \Rightarrow \quad _{x+t}p_0 = \ _xp_0 \ _tp_x$$

The left hand side of the equation $_{x+t}p_0 = \ _xp_0 \ _tp_x$ is the chance that a new born will survive for $x + t$ years, while the right hand side expresses the same probability as a product of two probabilities, the first factor is the probability of survival to age x. The attained age of the individual is then x. The second factor is the probability that the individual of age x survives for t more years. In general for $0 < s < t$,

$$_tp_x = \frac{S(x + t)}{S(x)} = \frac{S(x + s)}{S(x)} \frac{S(x + t)}{S(x + s)} = \ _sp_x \ _{t-s}p_{x+s}$$

Using the formula of $S(x)$ in terms of μ_x, as given in Eq. 4.2.1, $_tp_x$ can be expressed as

$$_tp_x = \frac{S(x + t)}{S(x)} = \frac{\exp(- \int_0^{x+t} \mu_s \ ds)}{\exp(- \int_0^x \mu_s \ ds)} = \exp\left(- \int_x^{x+t} \mu_s \ ds\right) = \exp\left(- \int_0^t \mu_{x+s} \ ds\right)$$

From the survival function of $T(x)$, we obtain the probability density function of $T(x)$ as follows. The probability density function $g_x(t)$ of $T(x)$ is the derivative of its survival function with a minus sign. Thus,

$$g_x(t) = -\frac{d}{dt} \ _tp_x = \exp(- \int_x^{x+t} \mu_s \ ds) \mu_{x+t} = \ _tp_x \ \mu_{x+t}, \quad t \geq 0 \tag{4.2.3}$$

The formula $g_x(t) = \ _tp_x \ \mu_{x+t}$ for the probability density function of $T(x)$ can be interpreted as follows. It is a product of two factors. The first factor is the probability that (x) survives for t units of time. The attained age then will be $x + t$. The second factor is the force of mortality at age $(x + t)$, which is the value of the conditional probability density function of X at the exact age $x + t$, given survival to that age. We heavily use this expression in the subsequent sections and chapters.

By the definition of force of mortality, the force of mortality of $T(x)$ at t is given by, $g_x(t)/_tp_x = \mu_{x+t}$. Thus, if μ_t is the force of mortality of life length random variable X at t, then μ_{x+t} is the force of mortality of $T(x)$ at t. In international actuarial notation $E(T(x))$ is denoted by $\overset{0}{e}_x$.

Example 4.2.5. It is given that, $F(x) = 0$ for $x < 0$ and $F(x) = 1 - 1/(x+1)$ for $x \geq 0$. Which of the following are true? (i) $_x p_0 = 1/(x+1)$, (ii) $\mu_{49} = 0.02$, (iii) $_{10}p_{39} = 0.8$.

Solution: Note that (i) $_x p_0 = S(x)/S(0) = 1/(x+1)$. (ii) By definition

$$\mu_x = f(x)/(1 - F(x)) = (x+1)/(x+1)^2 = 1/(x+1) \quad \Rightarrow \quad \mu_{49} = 1/50 = 0.02.$$

(iii) $_{10}p_{39} = S(49)/S(39) = (1/50)/(1/40) = 4/5 = 0.8$. Thus (i), (ii) and (iii) are true.   ■

Example 4.2.6. Suppose the life length random variable X has survival function $S(x) = 1 - x^2/100$, $0 \leq x \leq 10$. Find (i) $F_X(x)$, (ii) $_5 p_4$, (iii) probability density function of $T(4)$ and (iv) median future lifetime of (5).

Solution: (i) Observe that

$$F(x) = 1 - S(x) = \begin{cases} 0 & \text{if} & x < 0 \\ x^2/100 & \text{if} & 0 \leq x < 10 \\ 1 & \text{if} & x \geq 10 \end{cases}$$

$$
\begin{aligned}
(ii)\ _t p_4 &= S(4+t)/S(4) = (1 - (4+t)^2/100)/0.84 = 1 - (8t + t^2)/84 \\
\Rightarrow\ _5 p_4 &= 1 - 65/84 = 0.2262 \\
(iii)\ g_4(t) &= \mu_{4+t}\ _t p_4 = \left\{ -\frac{d}{dt} \log S(4+t) \right\} \frac{S(4+t)}{S(4)} \\
&= \frac{-\frac{d}{dt}S(4+t)}{S(4+t)} \frac{S(4+t)}{S(4)} = \frac{2(4+t)}{100} \frac{1}{0.84} = \frac{4+t}{42}, \quad 0 < t < 6
\end{aligned}
$$

Note that $\int_0^6 (4+t)/42 \, dt = 1$. The upper limit is 6 in view of the fact that the upper limit of X is 10 and $g_4(t)$ is the probability density function of $T(4)$, the future lifetime of (4). Survival distribution in this example is a suitable model for the life span of some birds and animals. (iv) The median future lifetime of (x) is the solution of the equation

$$_t q_x = \ _t p_x = 0.50 \quad \Longleftrightarrow \quad S(x+t) = (0.5)S(x)$$

With given $S(x)$ and $x = 5$, we solve the equation,

$$1 - (5+t)^2/100 = (1/2)(1 - 5^2/100) \quad \Rightarrow \quad t = 2.9056$$

Thus, the median future lifetime of (5) is 2.9056 years.   ■

Example 4.2.7. Suppose the life length random variable X is modelled by a uniform distribution over $(0, 100)$. Find the probability density function $g(t)$ of $T(25)$.

Solution: To find the probability density function $g(t)$ of $T(x)$, we first find $_t p_x$ and μ_{x+t}. When X has a uniform distribution over $(0, 100)$, then as in Example 4.2.4, its survival function is given by,

$$S(x) = \begin{cases} 1, & \text{if} & x < 0 \\ 1 - x/100, & \text{if} & 0 \leq x < 100 \\ 0, & \text{if} & x \geq 100 \end{cases}$$

Hence,

$$
{}_t p_x = \frac{S(x+t)}{S(x)} = \frac{100 - x - t}{100 - x} \quad \text{and} \quad \mu_x = \frac{f(x)}{S(x)} = \frac{1/100}{(100 - x)/100} = \frac{1}{(100 - x)}
$$

The probability density function $g_x(t)$ of $T(x)$ is given by,

$$
g_x(t) = {}_t p_x \, \mu_{x+t} = 1/(100 - x), \quad 0 \le t \le 100 - x
$$

Hence, $T(x)$ follows uniform distribution over $(0, 100 - x)$, that is, the distribution of $T(25)$ is uniform over $(0, 75)$. ∎

Example 4.2.8. Suppose the life length random variable X is modelled by an exponential distribution with a parameter μ. Find the probability density function $g_x(t)$ of $T(x)$.

Solution: If X is modelled by an exponential distribution with a parameter μ, then its probability density function is $f(x) = \mu \exp(-\mu x)$ for $x > 0$, its survival function is $S(x) = \exp(-\mu x)$ for $x > 0$ and its force of mortality is $\mu_x = \mu$ for $x > 0$. Consequently, the probability density function $g_x(t)$ of $T(x)$ is given by,

$$
g_x(t) = {}_t p_x \, \mu_{x+t} = \mu \exp(-\mu t) \quad \text{for } t > 0
$$

Thus $T(x)$ again has an exponential distribution with a parameter μ. ∎

Remark 4.2.2. From Example 4.2.8, we see that if X has exponential distribution, then $T(x)$ also has exponential distribution with the same parameter and it is the same for all x, again showing that exponential distribution is not a good model for human life length.

Example 4.2.9. Suppose the life length random variable is modeled by a distribution with force of mortality as specified below.

$$
\mu_s \;=\; \begin{cases} 0.01, & \text{if} \quad 0 < s < 15 \\ 0.02, & \text{if} \quad 15 \le s < 25 \\ 0.03, & \text{if} \quad\quad s \ge 25 \end{cases}
$$

Find the corresponding $S(x)$, $f(x)$ and the probability density function $g(t)$ of $T(20)$. Examine whether $f(x)$ and $g(t)$ satisfy the conditions for density functions.

Solution: To find the survival function from given force of mortality, note that for $0 \le x < 15$,

$$
S(x) \;=\; \exp\left\{ -\int_0^x \mu_s \, ds \right\} = \exp\left\{ -\int_0^x 0.01 \, ds \right\} = \exp(-0.01x)
$$

For $15 \le x < 25$,

$$
\begin{aligned}
S(x) &= \exp\left\{ -\int_0^x \mu_s \, ds \right\} = \exp\left\{ -\int_0^{15} 0.01 \, ds - \int_{15}^x 0.02 \, ds \right\} \\
&= \exp(-0.02x + 0.15)
\end{aligned}
$$

For $x \geq 25$,

$$
\begin{aligned}
S(x) &= \exp\left\{-\int_0^x \mu_s \, ds\right\} = \exp\left\{-\int_0^{15} 0.01 \, ds - \int_{15}^{25} 0.02 \, ds - \int_{25}^x 0.03 \, ds\right\} \\
&= \exp(-0.03x + 0.4)
\end{aligned}
$$

Thus, the survival function is given by,

$$
S(x) = \begin{cases}
\exp(-0.01x), & \text{if} \quad 0 \leq x < 15 \\
\exp(-0.02x + 0.15), & \text{if} \quad 15 \leq x < 25 \\
\exp(-0.03x + 0.4), & \text{if} \quad x \geq 25
\end{cases}
$$

To find $f(x)$, we use the formula, $f(x) = -\frac{dS(x)}{dx}$ and we have,

$$
f(x) = \begin{cases}
0.01 \, \exp(-0.01x), & \text{if} \quad 0 \leq x < 15 \\
0.02 \, \exp(-0.02x + 0.15), & \text{if} \quad 15 \leq x < 25 \\
0.03 \, \exp(-0.03x + 0.4), & \text{if} \quad x \geq 25
\end{cases}
$$

It is easy to check that $f(x)$ is always non-negative and $\int_0^\infty f(x) \, dx = 1$. To find the probability density function $g(t)$ of $T(20)$, we use the formula,

$$
g(t) = \mu_{20+t} \; {}_t p_{20} = \mu_{20+t} \; S(20 + t)/S(20)
$$

The form of $S(20 + t)$ changes for $0 < t < 5$ as $(20 + t)$ varies between 20 and 25, and for $t \geq 5$, $(20 + t) > 25$. Thus with appropriate choice of $S(20 + t)$, we get

$$
g(t) = \begin{cases}
0.02 \, \exp(-0.02t), & \text{if} \quad 0 < t < 5 \\
0.03 \, \exp(-0.03t + 0.05), & \text{if} \quad t \geq 5
\end{cases}
$$

It is easy to check that $g(t)$ is non-negative for $t \geq 0$ and $\int_0^\infty g(t) \, dt = 1$. $\blacksquare$

Remark 4.2.3. For the life length random variable modelled by the force of mortality given in Example 4.2.9, it is found that $P[T(20) \leq 100] = 0.9477$. It means that the future lifetime of (20) can be larger than 100. Thus, it may not be a good model for human life length. The following example is similar to Example 4.2.9, but in that setup $P[T(30) \leq 70]$ is almost 1. So it is a feasible model for human life length.

Example 4.2.10. Suppose the life length random variable of (30) is modelled by a distribution with force of mortality as specified below.

$$
\mu_s = \begin{cases}
0.04, & \text{if} \quad 0 \leq s < 15 \\
0.08, & \text{if} \quad 15 \leq s < 25 \\
0.12, & \text{if} \quad 25 \leq s < 35 \\
0.18, & \text{if} \quad s \geq 35
\end{cases}
$$

Find the probability density function $g(t)$ of $T(30)$. Examine whether it satisfies the conditions for a density function.

Solution: Proceeding exactly on similar lines we get the survival function as,

$$S(x) = \begin{cases} \exp(-0.04x)\,, & \text{if} \quad 0 \le x < 15 \\ \exp(-0.08x + 0.6), & \text{if} \quad 15 \le x < 25 \\ \exp(-0.12x + 1.6), & \text{if} \quad 25 \le x < 35 \\ \exp(-0.18x + 3.7), & \text{if} \quad x \ge 35 \end{cases}$$

To find probability density function $g(t)$ of $T(30)$, we use the formula,

$$g(t) = \mu_{30+t}\ {}_tp_{30} = \mu_{30+t}\ S(30+t)/S(30)$$

The form of $S(30+t)$ changes for $0 < t < 5$ as $(30+t)$ varies between 30 and 35, and for $t \ge 5$, $(30 + t) > 35$. Thus with appropriate choice of $S(30 + t)$, we get

$$g(t) = \begin{cases} 0.12\ \exp(-0.12t), & \text{if} \quad 0 < t < 5 \\ 0.18\ \exp(-0.18t + 0.3), & \text{if} \quad t \ge 5 \end{cases}$$

It is easy to check that $g(t)$ is always non-negative and $\int_0^\infty g(t)\ dt = 1$. Further, $P[T(30) \le 55]$ is 0.9999189 and $P[T(30) \le 71]$ is 0.9999954. Thus, the future life span of (30) is at most 71. So it seems to be a feasible model for human life length. $\blacksquare$

Example 4.2.11. Suppose the life length random variable is modelled by (i) Gompertz' force of mortality and (ii) Makeham's force of mortality. In each case find the probability density function $g_x(t)$ of $T(x)$. Examine whether it satisfies the conditions for a density function.

Solution: The force of mortality and survival function for Gompertz' law are

$$\mu_x = BC^x \ \ \& \ \ S(x) = \exp[-m(C^x - 1)], \ B > 0, \ C > 1, \ x \ge 0, \ m = B/log_e C.$$

The probability density function $g_x(t)$ of $T(x)$ is then given by,

$$g_x(t) = \ {}_tp_x\ \mu_{x+t} = \exp[-mC^x(C^t - 1)]\ BC^{x+t}, \quad \text{for} \ \ t > 0$$

It is clearly a non-negative function. To verify that it integrates to 1 over $(0, \infty)$, we substitute, $mC^x(C^t - 1) = y$. Then $mC^xC^t logC dt = dy$, that is $BC^{x+t}dt = dy$. Hence,

$$\int_0^\infty g_x(t)dt \ = \ \int_0^\infty \exp(-y)\ dy = 1$$

The force of mortality and survival function for Makeham's law are

$$\mu_x = A + BC^x \ \ \& \ \ S(x) = \exp[-Ax - m(C^x - 1)], \ B > 0, \ A \ge -B, \ C > 1, \ x \ge 0$$

The probability density function $g_x(t)$ of $T(x)$ for Makeham's law is given by,

$$g_x(t) = \ {}_tp_x\ \mu_{x+t} = \exp[-At - mC^x(C^t - 1)]\ [A + BC^{x+t}], \quad \text{for} \ \ t > 0$$

It is clearly a non-negative function. To verify that it integrates to 1 over $(0, \infty)$, we substitute, $At + mC^x(C^t - 1) = y$. Then $(A + mC^xC^t logC)dt = dy$, that is $(A + BC^{x+t})dt = dy$. Hence,

$$\int_0^\infty g_x(t)dt = \int_0^\infty \exp(-y)\ dy = 1 \qquad \blacksquare$$

We use the probability density functions of $T(x)$, corresponding to Gompertz' law and Makeham's law in subsequent chapters to evaluate premiums, if the underlying mortality pattern is one of these two.

Many insurance and annuity products involve a deferment of benefit. For example, in the 5-year deferred whole life insurance, the benefit is not paid if a claim is made in the first five years. In such cases, probabilities of more general event that (x) will survive t years and die within the following u years are involved. The corresponding probabilities are known as the deferred probabilities. The symbol $t|$ indicates deferment for t years. Suppose $_{t|u}q_x$ denotes the probability that (x) will survive t years and die within the following u years. Then it is derived as follows.

$$
\begin{aligned}
_{t|u}q_x &= \text{probability that } (x) \text{ will die between ages } x+t \text{ and } x+t+u \\
&= P[t < T(x) \le t+u] \;=\; P[t < X - x \le t+u | X > x] \\
&= P[x+t < X \le x+t+u | X > x] \\
&= P[x+t < X \le x+t+u, X > x]/P[X > x] \\
&= P[x+t < X \le x+t+u]/P[X > x] = (S(x+t) - S(x+t+u))/S(x) \\
&= \{(S(x+t) - S(x+t+u))/S(x+t)\}\, \{S(x+t)/S(x)\} = \; _uq_{x+t}\; _tp_x
\end{aligned}
$$

Alternatively,

$$
\begin{aligned}
_{t|u}q_x &= (S(x+t) - S(x+t+u))/S(x) = S(x+t)/S(x) - S(x+t+u)/S(x) \\
&= \;_tp_x - \;_{t+u}p_x \;=\; 1 - \;_tq_x - 1 + \;_{t+u}q_x \;=\; _{t+u}q_x - \;_tq_x
\end{aligned}
$$

If $u = 1$, the prefix u in $_{t|u}q_x$ is deleted and $_{t|u}q_x$ is denoted by $_{t|}q_x$. The relation $_{t|u}q_x = \;_tp_x\;_uq_{x+t}$ can be interpreted as follows. The left hand side of the equation is the probability that (x) will survive for the next t units of time and the death will occur in the next u units of time. The right hand side of the equation is the product of two probabilities. The first factor is the probability that (x) will survive for the next t units of time. The attained age then will be $x + t$. The second factor is the probability that $(x + t)$ will die in the next u units of time.

Example 4.2.12. Suppose the life length random variable X has a survival function $S(x) = 1 - x^2/100, \; 0 \le x \le 10$. Find $_{2|2}q_4$.

Solution: By definition,

$$
\begin{aligned}
_{2|2}q_4 &= \;_2p_4\; _2q_6 = (S(6)/S(4))\,\{1 - S(8)/S(6)\} \\
&= (S(6) - S(8))/S(4) = (0.64 - 0.36)/0.84 = 1/3
\end{aligned}
$$

$\blacksquare$

So far the discussion in this chapter is related to human lives; however, the ideas would be the same for other objects such as equipment, machines and business ventures. The following example illustrates this point.

Example 4.2.13. The life of a television is specified by the following survival function.

$$
S(x) = \begin{cases} 1, & \text{if} \quad x < 0 \\ 1 - x/w, & \text{if} \quad 0 \le x < w \\ 0, & \text{if} \quad x \ge w \end{cases}
$$

and $e_0^0 = 10$. For a proposed new type of television, with the same w, the new survival function is,

$$S^*(x) = \begin{cases} 1, & \text{if } 0 \le x \le 4 \\ (w - x)/(w - 4), & \text{if } 4 < x \le w \end{cases}$$

Calculate the increase in life expectancy at time 0.

Solution : For the old type of television,

$$10 = e_0^0 = \int_0^w S(x)dx = \int_0^w (1 - x/w)dx = w/2$$

Hence $w = 20$. For the proposed new type, the life expectancy is

$$\tilde{e}_0^0 = \int_0^4 dx + \int_4^{20} \left(\frac{20 - x}{16} \right) dx = 4 + \frac{1}{16} \int_4^{20} (20 - x)dx = 4 + 8 = 12$$

Thus, the increase in life expectancy of the proposed television is $12 - 10 = 2$ years.

In the above derivation, expectation is obtained using the survival function. Alternatively, from $S(x)$, one can find the probability density function $f(x)$ as $f(x) = -\frac{d}{dx}S(x)$. Thus, for the old type of television, the probability density function of lifetime is constant over the range $(0, 20)$. Hence its expectation is 10. For the proposed type, the probability density function of lifetime is $f(x) = 1/16$, for $4 < x < 20$. Hence its mean is $(20 + 4)/2 = 12$. ∎

In many problems of life insurance, the exact age at death is not important, but it is enough to know the number of complete one year long periods an individual survives after signing the contract. For example, according to a certain annuity plan, if the insured gets Rs 10,000/- on each birthday between ages 60 and 70, it does not make a difference whether a death occurs at exact age 65.25, that is, 65 years and 3 months, or at 65.75, that is 65 years and 9 months. The only thing that counts is that the death occurs between 65 and 66. In some insurance products the benefit payments are made at the end of the policy year. Thus, policy term is related to the complete years lived after signing the contract. Hence, it is worth studying the number of complete one year long periods an individual aged x survives after signing the contract. Such considerations lead to the concept of curtate future lifetime random variable. It is a discretized version of a random variable $T(x)$ and is discussed in the next section.

4.3   Curtate Future Lifetime

A discrete random variable $K(x)$ associated with future lifetime $T(x)$ is defined as the largest integer strictly smaller than $T(x)$. For example, if $T(x) = 7.5$, then $K(x) = 7$, if $T(x) = 8$, then $K(x) = 7$, if $T(x) = 0.5$ then $K(x) = 0$ and if $T(x) = 1.3$ then $K(x) = 1$. Thus, $K(x)$ is an integer valued random variable with possible values $0, 1, 2, \cdots$, and it specifies the number of complete future years of (x). It is known as the curtate future lifetime at age x. It specifies the remaining life in complete years of a person aged x. It is to be noted that the possible values of $K(x)$ are non-negative integers but x need not be an integer. However, in most of the practical cases and examples in this and subsequent chapters, x is taken to be an integer.

The probability mass function of $K(x)$ is obtained from the distribution of $T(x)$ as shown below.

$$\begin{aligned}
P[K(x) = k] &= P[k < T(x) \le k+1] \\
&= {}_{k+1}q_x - {}_k q_x \;=\; {}_k p_x - {}_{k+1}p_x \;=\; {}_k p_x - {}_k p_x\, {}_1 p_{x+k} \\
&= {}_k p_x\, q_{x+k} = (S(x+k) - S(x+k+1))/S(x) \\
&= {}_{k|}q_x
\end{aligned}$$

where ${}_{k|}q_x$ is the deferred probability as defined in Section 2. Note that the possible values of $K(x)$ depend on the values of $T(x)$. If the limiting age for humans is w, then $T(x) \le w - x$ and hence by the definition of $K(x)$, the maximum possible value of $K(x)$ is the largest integer strictly smaller than $w - x$. Suppose we denote it by w^*. Thus, the support of $K(x)$ is $\{0, 1, \cdots, w^*\}$. For example, if $w = 100$ and $x = 30$, then $w^* = 69$. If $w = 100$ and $x = 40.2$, then $w^* = 59$. If $w = 100$ and $x = 99$, then $w^* = 0$. If $w = 100$ and $x = 99.3$, then also $w^* = 0$. Observe that the possible values of $K(w - 1)$ are from 0 to the largest integer strictly smaller than $w - w + 1$, that is, 0. Thus, the distribution of $K(w - 1)$ is degenerate at 0.

The expectation $E(K(x))$ of $K(x)$ is denoted by e_x and is known as a curtate expectation while e_x^0, defined in Section 2, is referred to as a complete expectation. e_x is often used as a measure of the standard of living and health care in a country.

Example 4.3.1. Using $S(x) = 1 - x^2/100$ for $0 \le x \le 10$, (i) find the distribution of $K(4)$ and its expectation e_4. (ii) Obtain e_9. (iii) Find the distribution of $K(6.3)$ and its expectation $e_{6.3}$.

Solution: (i) We have $P[K(4) = k] = {}_{k|}q_4 = {}_k p_4\, q_{4+k}$, $k = 0, 1, \cdots, w^* = 5$. Hence, for the given survival function,

$$\begin{aligned}
{}_{0|}q_4 &= q_4 = 1 - S(5)/S(4) = 9/84 \\
{}_{1|}q_4 &= {}_1 p_4\, q_5 = (S(5)/S(4))\,(1 - S(6)/S(5)) = (S(5) - S(6))/S(4) = 11/84 \\
{}_{2|}q_4 &= {}_2 p_4\, q_6 = (S(6)/S(4))\,\{1 - S(7)/S(6)\} = 13/84
\end{aligned}$$

Similarly, ${}_{3|}q_4 = 15/84$, ${}_{4|}q_4 = 17/84$, ${}_{5|}q_4 = 19/84$ and ${}_{k|}q_4 = 0$ for $k \ge 6$. From this distribution, we get $e_4 = 2.916$.

(ii) Note that $e_9 = \sum_{k=0}^{0} k\, {}_k p_9\, q_{9+k} = 0$. Thus, curtate expectation of $K(9)$ is 0, as expected, since the distribution of $K(9)$ is degenerate at 0.

(iii) From the given survival function, $P[K(6.3) = k] = {}_k p_x\, q_{x+k}$, for $k = 0, 1, \cdots, w^* = 3$. Thus,

$$\begin{aligned}
P[K(6.3) = 0] &= (S(6.3) - S(7.3))/S(6.3) = 0.2255 \\
P[K(6.3) = 1] &= (S(7.3) - S(8.3))/S(6.3) = 0.2587 \\
P[K(6.3) = 2] &= (S(8.3) - S(9.3))/S(6.3) = 0.2918 \\
P[K(6.3) = 3] &= S(9.3)/S(6.3) = 0.2240
\end{aligned}$$

Note that $\sum_{k=0}^{3} P[K(6.3) = k] = 1$. From the distribution of $K(6.3)$, its expectation $e_{6.3} = 1.5143$.

Using the definition of expectation in terms of survival function, we find below the recurrence relation for e_x. Note that

$$
\begin{aligned}
e_x &= \sum_{k \geq 1} {}_k p_x = p_x + p_x\, p_{x+1} + p_x\, {}_2 p_{x+1} + p_x\, {}_3 p_{x+1} + \cdots \\
&= p_x\{1 + p_{x+1} + {}_2 p_{x+1} + {}_3 p_{x+1} + \cdots\} \\
&= p_x\{1 + \sum_{k \geq 1} {}_k p_{x+1}\} = p_x\{1 + e_{x+1}\}
\end{aligned}
\tag{4.3.1}
$$

This is a recurrence relation between e_x and e_{x+1}.

Example 4.3.2. Suppose $e_{30} = 58.72$ and $\mu_{30+t} = 0.004$ for $0 \leq t \leq 2$. Find e_{32}.

Solution: Using the recurrence relation derived in Eq. 4.3.1, we find e_{31} from e_{30} and e_{32} from e_{31}. Hence, we first find p_{30} and p_{31} using the formula ${}_t p_x = \exp(-\int_0^t \mu_{x+s} ds)$. Thus, $p_{30} = p_{31} = \exp(-0.004) = 0.996008$. Using the recurrence relation, we get $e_{31} = 57.95$ and $e_{32} = 57.19$. ∎

Example 4.3.3. Find the distribution of $K(30)$, when the life length random variable is modelled by a force of mortality as specified below.

$$
\mu_s \;=\; \begin{cases}
0.04, & \text{if} \quad 0 \leq s < 15 \\
0.08, & \text{if} \quad 15 \leq s < 25 \\
0.12, & \text{if} \quad 25 \leq s < 35 \\
0.18, & \text{if} \quad\;\; s \geq 35
\end{cases}
$$

Solution: In Example 4.2.10, we have obtained the survival function corresponding to this mortality law. It is given by,

$$
S(x) \;=\; \begin{cases}
\exp(-0.04x), & \text{if} \quad 0 \leq x < 15 \\
\exp(-0.08x + 0.6), & \text{if} \quad 15 \leq x < 25 \\
\exp(-0.12x + 1.6), & \text{if} \quad 25 \leq x < 35 \\
\exp(-0.18x + 3.7), & \text{if} \quad\;\; x \geq 35
\end{cases}
$$

The probability mass function of $K(30)$ involves $S(30+k)$ and $S(30+k+1)$. Its form changes depending on the value of k. Thus for $k \leq 4$ we have,

$$
\begin{aligned}
P[K(30) = k] &= {}_k p_{30}\ q_{30+k} = (S(30+k) - S(30+k+1))/S(30) \\
&= \frac{\exp[-0.12(30+k) + 1.6] - \exp[-0.12(30+k+1) + 1.6]}{\exp[-0.12(30) + 1.6]} \\
&= \exp[-0.12k] - \exp[-0.12(k+1)]
\end{aligned}
$$

For $k \geq 5$ we have,

$$
\begin{aligned}
P[K(30) = k] &= {}_k p_{30}\ q_{30+k} = (S(30+k) - S(30+k+1))/S(30) \\
&= \frac{\exp[-0.18(30+k) + 3.7] - \exp[-0.18(30+k+1) + 3.7]}{\exp[-0.12(30) + 1.6]} \\
&= \exp[-0.18k + 0.3] - \exp[-0.18k + 0.12]
\end{aligned}
$$

We use the first part of the Code 4.7.1 to obtain the probability distribution of $K(30)$. It is displayed in Table 4.1. Probabilities beyond 60 are 0 up to five decimal places and hence are taken as 0. The sum of the probabilities from 0 to 59 is 0.999977 which is almost 1. Thus, for the specified mortality pattern, the curtate future life of (30) has support $\{0, 1, \cdots, 60\}$ years, approximately.   ∎

k	$P[K(30) = k]$	k	$P[K(30) = k]$	k	$P[K(30) = k]$
0	0.11308	20	0.00608	40	0.00017
1	0.10029	21	0.00507	41	0.00014
2	0.08895	22	0.00424	42	0.00012
3	0.07889	23	0.00354	43	0.00010
4	0.06997	24	0.00296	44	0.00008
5	0.09041	25	0.00247	45	0.00007
6	0.07551	26	0.00206	46	0.00006
7	0.06307	27	0.00172	47	0.00005
8	0.05268	28	0.00144	48	0.00004
9	0.04401	29	0.00120	49	0.00003
10	0.03676	30	0.00100	50	0.00003
11	0.03070	31	0.00084	51	0.00002
12	0.02564	32	0.00070	52	0.00002
13	0.02142	33	0.00059	53	0.00002
14	0.01789	34	0.00049	54	0.00001
15	0.01494	35	0.00041	55	0.00001
16	0.01248	36	0.00034	56	0.00001
17	0.01043	37	0.00028	57	0.00001
18	0.00871	38	0.00024	58	0.00001
19	0.00727	39	0.00020	59	0.00001

Table 4.1 Distribution of $K(30)$

Example 4.3.4. Find the distribution of $K(x)$ when the life length random variable is modelled by Gompertz' law.

Solution: Example 4.2.11 specifies the survival function corresponding to Gompertz' law. Hence, the probability mass function of $K(x)$ for Gompertz' law is given by,

$$
\begin{aligned}
P[K(x) = k] &= {}_kp_x\, q_{x+k} = (S(x+k) - S(x+k+1))/S(x) \\
&= \frac{\exp[-m(C^{x+k} - 1)] - \exp[-m(C^{x+k+1} - 1)]}{\exp[-m(C^x - 1)]} \\
&= \exp[-mC^x(C^k - 1)] - \exp[-mC^x(C^{k+1} - 1)]
\end{aligned}
$$

We use the second part of the Code 4.7.1 to obtain the probability distribution of $K(30)$ under Gompertz' law when $B = 0.0001151$ and $C = 1.096$. It is displayed in Table 4.2. Probabilities beyond 69 are 0 up to five decimal places and hence are taken as 0. Sum of the probabilities from 0 to 69 is 0.9999939 which is almost 1. Thus, for the given Gompertz' mortality pattern, the curtate future life of (30) has support $\{0, 1, \cdots, 69\}$ years approximately.   ∎

k	$_kp_x\ q_{x+k}$	k	$_kp_x\ q_{x+k}$	k	$_kp_x\ q_{x+k}$	k	$_kp_x\ q_{x+k}$
0	0.00188	18	0.00900	36	0.02984	54	0.01489
1	0.00206	19	0.00976	37	0.03100	55	0.01235
2	0.00225	20	0.01057	38	0.03204	56	0.00998
3	0.00246	21	0.01145	39	0.03293	57	0.00783
4	0.00269	22	0.01238	40	0.03363	58	0.00595
5	0.00294	23	0.01337	41	0.03412	59	0.00437
6	0.00322	24	0.01441	42	0.03436	60	0.00308
7	0.00351	25	0.01552	43	0.03432	61	0.00209
8	0.00384	26	0.01668	44	0.03398	62	0.00135
9	0.00419	27	0.01789	45	0.03331	63	0.00083
10	0.00457	28	0.01915	46	0.03231	64	0.00048
11	0.00498	29	0.02046	47	0.03098	65	0.00026
12	0.00543	30	0.02180	48	0.02932	66	0.00014
13	0.00592	31	0.02317	49	0.02736	67	0.00006
14	0.00644	32	0.02454	50	0.02514	68	0.00003
15	0.00701	33	0.02592	51	0.02272	69	0.00001
16	0.00762	34	0.02728	52	0.02015		
17	0.00828	35	0.02859	53	0.01752		

Table 4.2 Gompertz' law: Distribution of $K(30)$

Example 4.3.5. Find the distribution of $K(x)$ when the life length random variable is modelled by Makeham's law.

Solution: Using a similar approach as for Gomperz' law, the probability mass function of $K(x)$ for Makeham's law is given by $P[K(x) = k] = u_k$, where

$$
\begin{aligned}
u_k &= {}_kp_x\ q_{x+k} = (S(x+k) - S(x+k+1))/S(x) \\
&= \frac{\exp[-A(x+k) - m(C^{x+k} - 1)]}{\exp[-Ax - m(C^x - 1)]} \\
&\quad - \frac{\exp[-A(x+k+1) - m(C^{x+k+1} - 1)]}{\exp[-Ax - m(C^x - 1)]} \\
&= \exp[-Ak - mC^x(C^k - 1)] - \exp[-A(k+1) - mC^x(C^{k+1} - 1)]
\end{aligned}
$$

We use the second part of the Code 4.7.1 to obtain the probability distribution of $K(30)$ under Makeham's law when $A = 0.0007, B = 0.0001151$ and $C = 1.096$. It is displayed in Table 4.3. Probabilities beyond 69 are 0 up to five decimal places and hence are taken as 0. Sum of the probabilities from 0 to 69 is 0.9999936 which is almost 1. ∎

Remark 4.3.1. Comparing the values in Table 4.2 and Table 4.3, we note that probabilities under Gompertz' law are slightly higher than those under Makeham's law, but the difference is marginal. $P[K(30) = 42]$ is maximum for Gompertz' law, while for Makeham's law $P[K(30) = 43]$ is maximum. In Table 4.1, we note that $P[K(30) = 0]$ is maximum. Further in Table 4.1, the probabilities decrease from ages 0 to 4, but there is an increase at 5 and then the probabilities

k	u_k	k	u_k	k	u_k	k	u_k
0	0.00118	18	0.00846	36	0.03019	54	0.01543
1	0.00136	19	0.00925	37	0.03143	55	0.01281
2	0.00156	20	0.01009	38	0.03254	56	0.01036
3	0.00177	21	0.01099	39	0.03350	57	0.00814
4	0.00201	22	0.01195	40	0.03427	58	0.00619
5	0.00226	23	0.01298	41	0.03482	59	0.00454
6	0.00254	24	0.01406	42	0.03512	60	0.00321
7	0.00284	25	0.01520	43	0.03513	61	0.00218
8	0.00317	26	0.01641	44	0.03482	62	0.00141
9	0.00353	27	0.01767	45	0.03418	63	0.00087
10	0.00392	28	0.01898	46	0.03320	64	0.00050
11	0.00434	29	0.02034	47	0.03186	65	0.00028
12	0.00480	30	0.02174	48	0.03019	66	0.00014
13	0.00530	31	0.02317	49	0.02821	67	0.00007
14	0.00584	32	0.02461	50	0.02595	68	0.00003
15	0.00642	33	0.02606	51	0.02347	69	0.00001
16	0.00705	34	0.02748	52	0.02084		
17	0.00773	35	0.02887	53	0.01813		

Table 4.3 Makeham's law: Distribution of $K(30)$

again decrease. These changes are in view of the fact that in the mortality pattern specified in Example 4.3.3, force of mortality remains constant after 35. We refer to this mortality law as 'D''. The expected curtate future lifetime $E(K(30))$ is 6.32 for the mortality law 'D'', it is 37.02 for Gompertz' law and 37.57 for Makeham's law.

Figure 4.4 displays the nature of distribution of $K(30)$ under three laws—'D'', Gompertz (G) and Makeham (M). It is drawn using the third part of the Code 4.7.1. From the figure, we note that the distributions of $K(30)$ coincide for Gompertz' and Makeham's law. However, the distribution of $K(30)$ is quite different for law 'D'' than that for 'G'' and 'M''. Such a discrete law 'D'' may not be a good model for curtate future lifetime for humans.

In both the Examples 4.3.4 and 4.3.5, the form of the survival function was specified, hence we could explicitly find out the form of the probability mass function of $K(x)$. In most of the practical applications, the form of the survival function is rarely specified, instead the mortality pattern is usually specified in terms of the values of p_x or q_x for values of x running from 0 to some limiting age w. The next example illustrates the computation in such a setup.

Example 4.3.6. Find the distribution of $K(25)$ corresponding to q_x values as specified in Table 4.8.

Solution: The distribution of $K(x)$ is given by, $P[K(x) = k] = {}_kp_x\, q_{x+k}$. We first find ${}_kp_x$ using the relation ${}_kp_x = p_x\, p_{x+1}\cdots p_{x+k-1}$. We use Code 4.7.2 to find the distribution of $K(25)$ corresponding to given set of q_x values. Table 4.4 presents the distribution. From the distribution of $K(25)$ as displayed in Table 4.4, we find $e_{25} = E(K(25)) = 51.42$. Further, note that the mode of the distribution is at 58, Thus mode is higher than the mean, supporting the

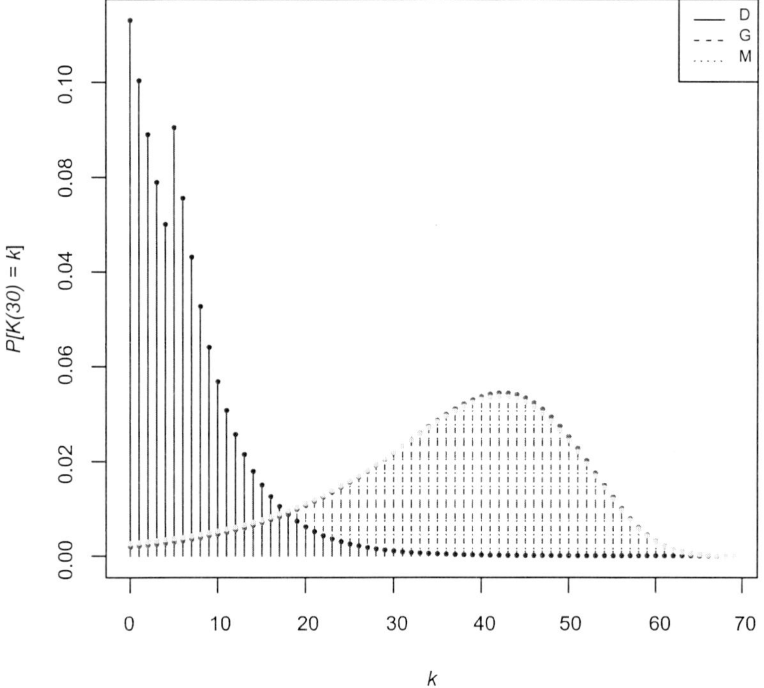

Figure 4.4 Distribution of $K(30)$

negative skewness of the distribution of $K(25)$, as is evident from Figure 4.5, drawn using the last two lines of the Code 4.7.2.　■

In this chapter we demonstrate how a distribution of the time until death random variable can be summarized in a life table. It carries a lot of importance in actuarial science, demography and population studies. In the insurance business, life tables are useful to calculate premiums and reserves. In Section 4.4, we discuss in detail the construction of a life table. One column of this table specifying the values of p_x for integer values of x is sufficient to obtain the distribution of $K(x)$ for any integer x, as is clear from Example 4.3.6. Such tables are useful in many disciplines, where the 'death' is considered as some other phenomenon. For example, failure of a machine, onset of a side effect, discharge from a hospital. Engineers use life tables to study the reliability of complex systems. Biostatisticians use life tables to compare the effectiveness of various treatments for some disease. Demographers use life tables as a tool in population projections. Life tables also help to solve certain problems in hospital, transport and hotel management. Chapters 4 and 5 of the book by Pollard et al [13] discuss many interesting applications of life tables.

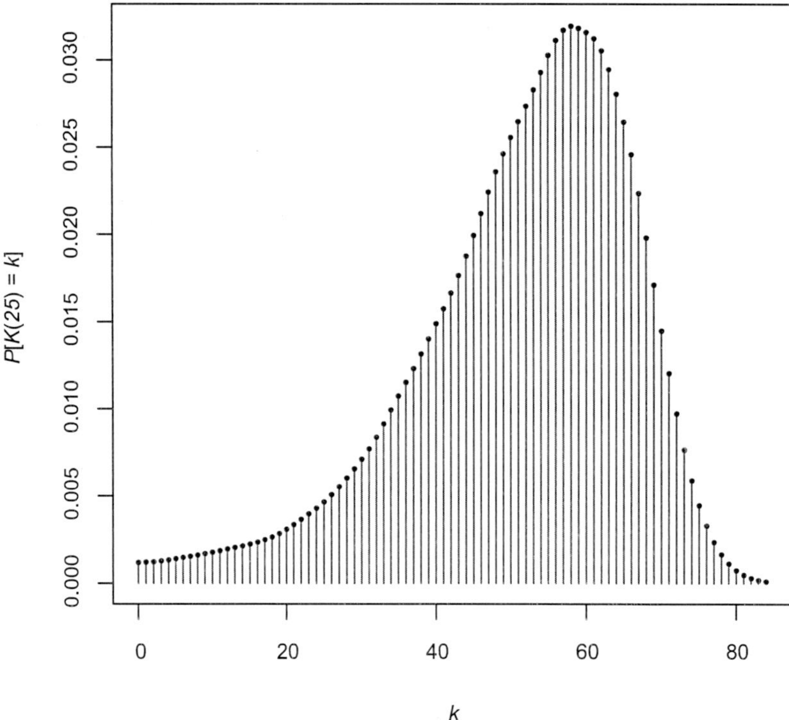

Figure 4.5 Distribution of $K(25)$

4.4   Life Tables

A life table is useful in many actuarial calculations. The importance of life tables in actuarial science is recognized by fixing the date of the beginning of actuarial science as 1693, because in that year, Edmund Halley published a paper entitled 'An Estimate of the Degrees of the Mortality of Mankind, drawn from various Tables of Births and Funerals at the city of Breslau'. The life table contained in Halley's paper is known as the Breslau table. Some modern notation and ideas are surprisingly found in the Breslau table.

A life table describes the mortality experience of a certain group of individuals. It usually contains tabulations by individual ages x, of the basic functions p_x, q_x and some additional derived functions l_x, d_x, L_x, T_x and e_x, which are discussed in this section. Suppose we have a group of l_0 newborns. l_0 is known as the radix and is usually taken as $1,00,000$. It is further assumed that each newborn's age at death random variable X has the same distribution and is specified by the survival function $S(x)$. The group of l_0 newborns, each with survival function $S(x)$, is usually referred to as a random survivorship group.

k	$P[K(25) = k]$	k	$P[K(25) = k]$	k	$P[K(25) = k]$
0	0.00117	28	0.00602	56	0.03111
1	0.00119	29	0.00655	57	0.03170
2	0.00121	30	0.00710	58	0.03193
3	0.00126	31	0.00769	59	0.03181
4	0.00132	32	0.00836	60	0.03157
5	0.00139	33	0.00912	61	0.03121
6	0.00146	34	0.00992	62	0.03052
7	0.00153	35	0.01071	63	0.02943
8	0.00160	36	0.01149	64	0.02802
9	0.00168	37	0.01228	65	0.02642
10	0.00176	38	0.01312	66	0.02457
11	0.00185	39	0.01400	67	0.02235
12	0.00195	40	0.01487	68	0.01979
13	0.00203	41	0.01572	69	0.01710
14	0.00212	42	0.01662	70	0.01447
15	0.00223	43	0.01762	71	0.01202
16	0.00234	44	0.01873	72	0.00971
17	0.00247	45	0.01992	73	0.00764
18	0.00263	46	0.02117	74	0.00588
19	0.00282	47	0.02242	75	0.00444
20	0.00307	48	0.02358	76	0.00327
21	0.00334	49	0.02461	77	0.00234
22	0.00364	50	0.02554	78	0.00163
23	0.00395	51	0.02644	79	0.00111
24	0.00427	52	0.02732	80	0.00072
25	0.00463	53	0.02827	81	0.00046
26	0.00505	54	0.02926	82	0.00028
27	0.00552	55	0.03026	83	0.00016
				84	0.00009

Table 4.4 Probability distribution of $K(25)$

Suppose $\mathbb{L}(x)$ denotes the cohort's number of survivors to age x, the values of which are $0, 1, 2, \cdots$. Suppose l_0 lives are labelled as $j = 1, 2, \cdots, l_0$. The indicator random variable is defined as, $I_j(x) = 1$ if the life j survives to age x and 0 otherwise. Thus, $I_j(x)$ is an indicator of the survival to age x of life j and we have $\mathbb{L}(x) = \sum_{j=0}^{l_0} I_j(x)$. Further $E(I_j(x)) = 1 \times S(x)$. Therefore, $E(\mathbb{L}(x)) = l_0 S(x) = l_x$, say. Under the assumption that indicators $I_j(x)$ are mutually independent, $\mathbb{L}(x)$ follows binomial $B(l_0, S(x))$ distribution for each $x = 0, 1, 2, \cdots$. From it also we have $l_x = l_0 S(x)$. Thus, l_x represents the expected number of survivors to age x from the l_0 newborns. From the relation $l_x = l_0 S(x)$, it is clear that if the values of l_x are specified for all $x \geq 0$, then the distribution of X gets completely specified. In many practical situations, the survival function is specified using l_x instead of $S(x)$. The upper panel of Figure 4.6 shows the graph of (x, l_x). Note that $S(x)$ and hence l_x is a decreasing function of x.

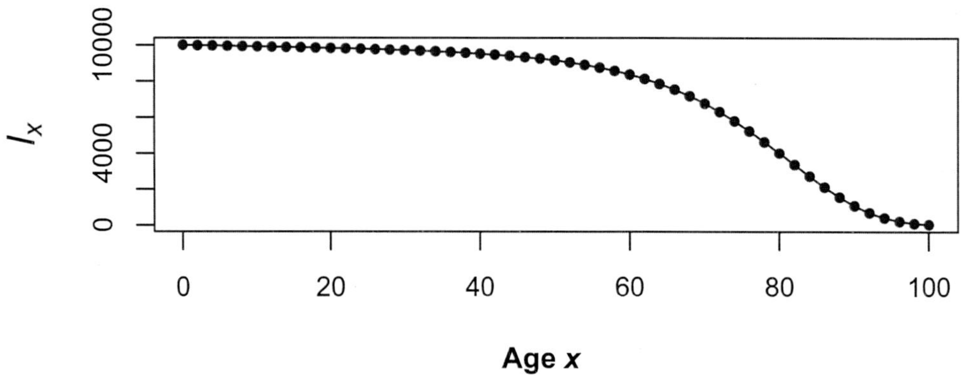

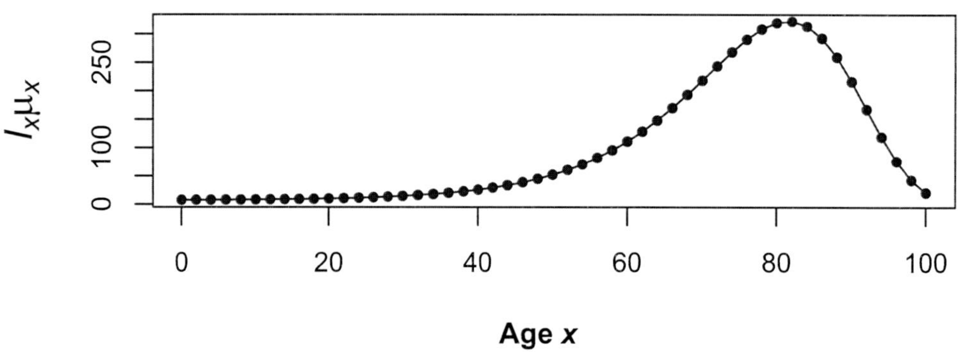

Figure 4.6 Graphs of l_x and $l_x\mu_x$

Example 4.4.1. With $S(x) = 1 - x^2/100$, for $0 \le x \le 10$, and $l_0 = 1,00,000$, find l_1, l_4 and $l_{6.5}$.

Solution: From the formula for l_x, $l_1 = l_0 S(1) = 99,000$, $l_4 = l_0 S(4) = 84,000$ and $l_{6.5} = l_0 S(6.5) = 57,750$. ∎

In analogy with $\mathbb{L}(x)$, suppose $_nD_x$ denotes the number of deaths between ages x and $x + n$ from among the initial l_0 lives. Since a newborn has probability $S(x) - S(x + n)$ of death between ages x and $x + n$ from among the initial l_0 lives, the expected value of $_nD_x$, denoted by $_nd_x$ is given by

$$_nd_x = E(_nD_x) = l_0(S(x) - S(x + n)) = l_x - l_{x+n}$$

When $n = 1$, $_nD_x$ and $_nd_x$ are written as D_x and d_x respectively.

Remark 4.4.1. Suppose the life length is modelled by uniform distribution over the interval $(0, w)$ with force of mortality $\mu_x = (w-x)^{-1}$ and survival function $S(x) = 1 - x/w$, w being the limiting age. With this form of survival function, $l_x = l_0(w-x)/w$. Hence, $d_x = l_x - l_{x+1} = l_0/w$. Thus, the expected number of deaths in a unit interval $(x, x+1)$ does not depend on x, that is, it is expected that equal numbers would die each year. It has of course been found unrealistic and hence uniform distribution is rarely used to model the human life length.

There are interesting inter-relations between μ_x and l_x. We have $l_x = l_0 S(x)$. Hence, $\log l_x = \log l_0 + \log S(x)$. If $S(x)$ is a differentiable function of x, then l_x is also a differentiable function of x. Taking derivative with respect to x, we get $-\frac{d}{dx}\log l_x = -\frac{d}{dx}\log S(x)$, that is, $(-1/l_x)(\frac{d}{dx}l_x) = \mu_x$, which can be expressed as $-dl_x = \mu_x l_x dx$. Thus, decrement in l_x is proportional to the force of mortality and l_x. If force of mortality is high (low), then decrease in l_x is high (low). Similarly if l_x is high(low), then decrease in l_x is high (low). $l_x \mu_x$ is interpreted as the expected density of deaths in the age interval $(x, x+dx)$. The graph of $l_x\mu_x$ against x is called the curve of deaths. One such curve is shown in the lower panel of Figure 4.6. From the figure we note that the curve of deaths increases initially, attains a maximum around 85 and then again decreases. The decreasing nature after certain age is attributable to the fact that after certain age, the expected number of survivors is small.

Example 4.4.2. Suppose a survival model is defined by the values of p_x, as given in Table 4.5.

x	0	1	2	3	4
p_x	0.9	0.8	0.6	0.3	0

Table 4.5 Survival model: Values of p_x

(i) What are the corresponding values of $S(x)$ for $x = 0, 1, 2, 3, 4$ and 5?
(ii) Using a radix $l_0 = 10,000$, find values of l_x and d_x for $x = 0, 1, 2, 3, 4$.
(iii) What is the value of w in this case? (iv) Verify that $\sum_{x=0}^{w-1} d_x = l_0$.
(v) Find $_3d_0$, $_2q_1$, $_3p_1$ and $_3q_2$.

Solution: (i) We know that $p_x = S(x+1)/S(x)$ and $S(0) = 1$. Using the recursive relation $S(x+1) = p_x S(x)$ with $S(0) = 1$, we get $S(1) = 0.9$, $S(2) = 0.72$, $S(3) = 0.432$, and so on.
(ii) We find l_x using the relation $l_x = l_0 S(x)$ and d_x using, $d_x = l_x - l_{x+1}$. The results are presented in Table 4.6.
(iii) $S(x) = 0$ is first reached at $x = 5$. Hence $w = 5$.
(iv) Note that total of the numbers in the last column is $10,000$.
(v) $_3d_0 = l_0 - l_3 = 5680$, $\quad _2q_1 = (l_1 - l_3)/l_1 = 0.52$, $\quad _3p_1 = l_4/l_1 = 0.144$, $_3q_2 = (l_2 - l_5)/l_2 = 1$. ∎

We now discuss in detail the construction of a life table. In a life table, the first column is of age x, the values of which are usually taken as integers $0, 1, 2, \cdots$. The second is of q_x, third

x	p_x	$S(x)$	l_x	d_x
0	0.9	1	10,000	1000
1	0.8	0.9	9,000	1800
2	0.6	0.72	7,200	2880
3	0.3	0.432	4,320	3024
4	0	0.1296	1296	1296
5	-	0	0	0

Table 4.6 Life table functions

gives values of p_x, the fourth column gives values of l_x and the fifth column gives values of d_x. When the survival function $S(x)$ is known, which is assumed to be the same for all individuals, one can compute these columns of the life table using the formulae developed above.

The approach of constructing a life table using the specific survival function is known as a random survivorship approach. The use for such analytic functions and such approach has declined in recent years, since it is difficult to construct the life tables by observing l_0 newborns until the last survivor dies. The other approach to construct a life table is known as the deterministic survivorship approach. This is rooted mathematically in the concept of decrement rates. In the deterministic approach, the functions in the life table are based on the estimates of q_x, which are obtained from the age specific death rates m_x, also known as the central death rate at age x. The age specific death rates are obtained using vital statistics records. Under the assumption of uniformity of deaths in a unit interval, a concept to be discussed in the next section, q_x and m_x are related by the equation

$$q_x = m_x/(1 + m_x/2)$$

Further, it is assumed that all the members of the group, at each age of their lives, are subject to effective annual rates of mortality specified by the values of q_x. With these quantities q_x, the other functions in the life table are determined as follows.

$$\begin{aligned}
l_1 &= l_0(1 - q_0) = l_0 p_0 = l_0 - d_0, \quad \text{since } d_0 = l_0 q_0 \\
l_2 &= l_1(1 - q_1) = l_1 p_1 = l_1 - d_1 = l_0 - (d_0 + d_1) = l_0 p_1 p_0
\end{aligned}$$

In general,

$$\begin{aligned}
l_x &= l_{x-1}(1 - q_{x-1}) = l_{x-1}\, p_{x-1} = l_{x-1} - d_{x-1} = l_0 - \sum_{y=0}^{x-1} d_y \\
&= l_0(1 - {}_x d_0/l_0) = l_0(1 - {}_x q_0) = l_0\, {}_x p_0
\end{aligned}$$

Thus, l_x is the number of lives attaining age x in the survivorship group. Values of x running from 0 to limiting age w, corresponding value of q_x, p_x, l_x and $d_x = l_x q_x$, constitute the first five columns of a life table. There are three more functions derived from l_x, which we discuss below.

The fifth column in the life table is that of L_x. L_x denotes the total expected number of years lived between ages x and $x+1$ by survivors of the initial group of l_0 lives. L_x is computed as the addition of two terms, one corresponding to the number of years lived by l_{x+1} survivors in x to $x+1$ and the second corresponding to the number of years lived by those who die in $(x, x+1)$. Thus, L_x is given by,

$$L_x = 1 \times l_{x+1} + \int_0^1 t\mu_{x+t}l_{x+t}\ dt$$

The first term gives the number of years lived by l_{x+1} survivors in x to $x+1$ and the integral in the second term counts the years lived by those who die between x and $x+1$. Here μ_{x+t} is the instantaneous death rate, l_{x+t} is the number of survivors at $x+t$, who live fraction t of years. Thus, the number of years of these survivors is $t\ l_{x+t}\mu_{x+t}$ where $0 < t < 1$. Integrating t between 0 and 1 gives the number of years lived in $(x, x+1)$ by those who die in $(x, x+1)$. The word 'expected' is in view of the fact that l_x are expected values. Now, we have a result that $\mu_x l_x dx = -dl_x$. Therefore,

$$\frac{d\ l_{x+t}}{dt} = -l_{x+t}\mu_{x+t}$$

Hence, using integration by parts we get,

$$\begin{aligned}
L_x &= l_{x+1} + \int_0^1 tl_{x+t}\mu_{x+t}\ dt = l_{x+1} - \int_0^1 t\frac{d\ l_{x+t}}{dt}\ dt \\
&= l_{x+1} - [tl_{x+t}]_0^1 + \int_0^1 l_{x+t}\ dt = l_{x+1} - l_{x+1} + \int_0^1 l_{x+t}\ dt \\
&= \int_0^1 l_{x+t}dt = \int_x^{x+1} l_t\ dt
\end{aligned}$$

Thus, with the usual interpretation of the definite integral, L_x represents the area below the curve of l_t bounded by the ordinates at x and $x+1$. The upper panel of Figure 4.7 presents L_{40} as the area below the curve of l_x bounded by the ordinates at 40 and 41. In Figure 4.7, the underlying mortality law is taken as Makeham's law with $A = 0.7 \times 10^{-3}$, $B = 0.0001151, C = 10^{.04} = 1.09648$ and $m = B/\log_e C = 0.00125$.

Under the assumption of uniformity of deaths in a unit interval (a concept discussed in the next section), l_t is a straight line in the interval $(x, x+1)$. Hence the region below the curve of l_t bounded by the ordinates at x and $x+1$ becomes a trapezium. Using the formula for the area of trapezium it follows that
$$L_x = (l_x + l_{x+1})/2$$
L_x is usually interpreted as the expected number of individuals in the age group $(x, x+1)$.

The age specific death rate m_x is related to the function l_x as shown below. In terms of l_x, it is defined as,

$$m_x = \left(\int_0^1 l_{x+t}\mu_{x+t}\ dt\right)\left(\int_0^1 l_{x+t}\ dt\right)^{-1}$$

Note that, m_x is the weighted average of μ_{x+t}, weights being l_{x+t}. It can be expressed in terms of d_x and L_x as follows.

$$\int_0^1 l_{x+t}\mu_{x+t}dt = -\int_0^1 \frac{d\ l_{x+t}}{dt}dt = [-l_{x+t}.1]_0^1 + \int_0^1 l_{x+t}\left(\frac{d1}{dt}\right)dt = l_x - l_{x+1}$$

$$\Rightarrow\quad m_x = (l_x - l_{x+1})/L_x = d_x/L_x$$

From the registry of vital statistics m_x is computed using the formula $m_x = d_x/L_x$ and q_x is estimated from m_x as noted above.

The next column is of T_x, which denotes the expected number of years lived beyond age x by the survivorship group with l_0 initial members. It is given by,

$$
\begin{aligned}
T_x &= \int_0^\infty t l_{x+t}\mu_{x+t}\, dt = -\int_0^\infty t\frac{d\, l_{x+t}}{dt}\, dt = \int_0^\infty l_{x+t} dt = \int_x^\infty l_t\, dt \\
&= \int_x^{x+1} l_t\, dt + \int_{x+1}^{x+2} l_t\, dt + \int_{x+2}^{x+3} l_t\, dt + \cdots \\
&= L_x + L_{x+1} + L_{x+2} + \cdots = L_x + T_{x+1}
\end{aligned}
$$

It gives the backward recurrence relation between T_x and T_{x+1}. It is useful in the calculation of T_x values for all x, with $T_w = L_w$. From the formula of T_x, it is clear that it represents the area below the curve of l_t beyond the ordinate at x. The lower panel of Figure 4.7 presents T_{40} as the area below the curve of l_t beyond the ordinate at 40. T_0 indicates the expected total population. Such an interpretation follows from the fact that $T_0 = L_0 + L_1 + \cdots$ and the interpretation of L_x as the expected population in the age group $(x,\ x+1)$.

The last column of the life table is of e_x. It is defined in terms of T_x and l_x as $e_x = T_x/l_x$. Thus, e_x denotes the average number of years of future lifetime of an individual of the l_x survivors of the group at age x. We note below that the formula $e_x = T_x/l_x$ reduces to the formula $e_x = \sum_{k=1}^\infty {}_kp_x$.

$$
\begin{aligned}
e_x &= T_x/l_x = l_x^{-1}\int_x^\infty l_{x+t}\, dt = l_x^{-1}\int_0^\infty l_x\, {}_tp_x\, dt = \int_0^\infty {}_tp_x\, dt \\
&= \sum_{k=1}^\infty {}_kp_x = E(K(x))
\end{aligned}
$$

The expectation e_x^0 analogous to e_x and defined in Section 3 is given by,

$$
e_x^0 = E(T(x)) = \int_0^\infty t\, {}_tp_x\mu_{x+t}\, dt
$$

A life table consists of eight columns specifying $x, q_x, p_x, l_x, d_x, L_x, T_x$ and e_x, where $x = 0, 1, 2, \cdots$. Example 4.4.3 illustrates the construction of a life table corresponding to given set of q_x values and l_0. We summarize below the formulae used in the construction of a life table.

$$
\begin{aligned}
p_x &= 1 - q_x, & l_{x+1} &= l_x p_x \\
d_x &= l_x - l_{x+1}, & L_x &= (l_x + l_{x+1})/2 \\
T_x &= L_x + T_{x+1}, & e_x &= T_x/l_x
\end{aligned}
\tag{4.4.1}
$$

While using the formula for L_x as $L_x = (l_x + l_{x+1})/2$, we make the assumption of uniformity of deaths over a unit interval. The calculation of T_x values is easy if we proceed in the reverse direction, that is, from age w to 0, using the backward recurrence relation derived above. It is to be noted that $T_w = L_w$. Then

$$
T_{w-1} = L_{w-1} + T_w, \quad T_{w-2} = L_{w-2} + T_{w-1}
$$

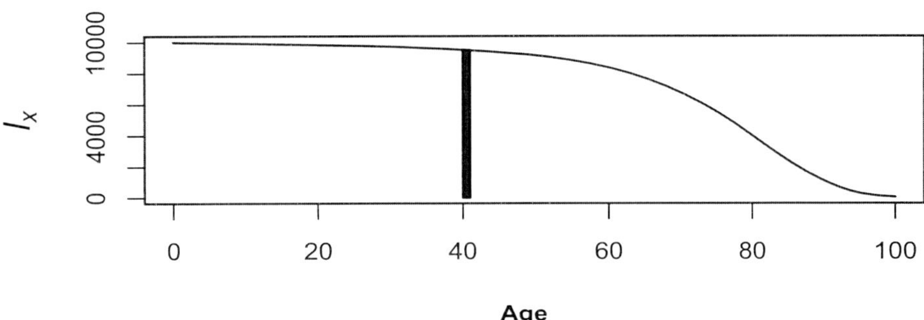

Expected population in (40,41)

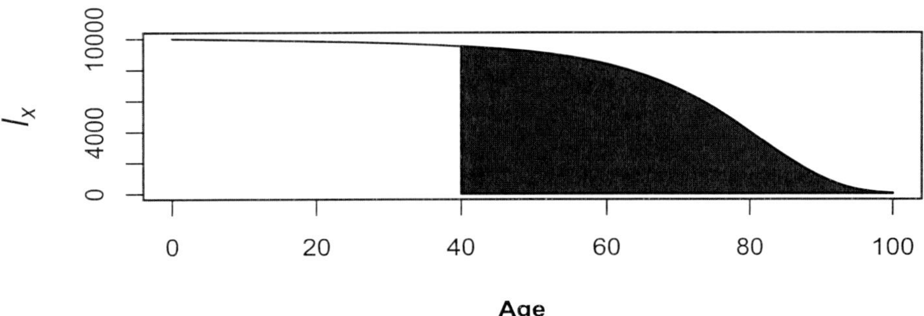

Expected population beyond 40

Figure 4.7 L_{40} and T_{40}

Continuing in this direction we find the T_x column.

Example 4.4.3. For a certain bird population, the probabilities q_x obtained for five weeks are $q_0 = 0.3$, $q_1 = 0.1$, $q_2 = 0.2$, $q_3 = 0.4$, $q_4 = 0.7$ and $q_5 = 1$. Taking $l_0 = 100$, construct the life table with values for l_x, d_x, L_x, T_x and e_x.

Solution: From the given data and the formulae summarized in Eqs 4.4.1, we construct the life table as shown in Table 4.7. We interpret it as follows. If a population of birds, with a life cycle of 6 weeks, follows a survival pattern as depicted by q_x values and if there are 100 new born birds each week, then 30 are expected to die in the first week. At the end of the first week, that is, at the beginning of the second week, there would be 70 birds of one week old; out of these 7 are expected to die in the second week. If initial population consists of 500 birds, the numbers in l_x and d_x columns will get multiplied by 5. Further, $L_0 = 85$ indicates that expected number of birds with age between 0 to 1 week is 85. T_0 always has an interpretation of total population, that is, number of birds of all ages. ∎

Note that given a set of q_x values or given an analytic form of the survival function, one can construct a life table. In the last section, Code 4.7.3 is given to construct a life table

x	q_x	p_x	l_x	d_x	L_x	T_x	e_x
0	0.3	0.7	100	30	85	272.71	2.73
1	0.1	0.9	70	7	66.5	187.71	2.68
2	0.2	0.8	63	12.6	56.7	121.21	1.92
3	0.4	0.6	50.4	20.16	40.32	64.51	1.28
4	0.7	0.3	30.24	21.168	19.656	24.19	0.8
5	1	0	9.072	9.072	4.536	4.54	0.5

Table 4.7 Life table for bird population

corresponding to a given a set of q_x values. Table 4.8 presented below, displays the 8 columns of the life table, corresponding to q_x values as given in the second column of Table 4.8.

x	q_x	p_x	l_x	d_x	L_x	T_x	e_x
0	0.00936	0.99064	100000.00	936.00	99532.00	7537002.16	75.37
1	0.00073	0.99927	99064.00	72.32	99027.84	7437470.16	75.08
2	0.00048	0.99952	98991.68	47.52	98967.93	7338442.32	74.13
3	0.00037	0.99963	98944.17	36.61	98925.86	7239474.39	73.17
4	0.00030	0.99970	98907.56	29.67	98892.73	7141534.43	72.20
5	0.00027	0.99973	98877.89	26.70	98864.54	7041655.81	71.22
6	0.00025	0.99975	98851.19	24.71	98838.83	6942791.27	70.23
7	0.00023	0.99977	98826.48	22.73	98815.11	6843952.44	69.25
8	0.00020	0.99980	98803.75	19.76	98793.87	6745137.33	68.27
9	0.00018	0.99982	98783.98	17.78	98775.09	6646343.46	67.28
10	0.00016	0.99984	98766.20	15.80	98758.30	6547568.37	66.29
11	0.00016	0.99984	98750.40	15.80	98742.50	6448810.07	65.30
12	0.00022	0.99978	98734.60	21.72	98723.74	6350067.57	64.31
13	0.00032	0.99968	98712.88	31.59	98697.09	6251343.83	63.33
14	0.00047	0.99953	98681.29	46.38	98658.10	6152646.74	62.35
15	0.00063	0.99937	98634.91	62.14	98603.84	6053988.64	61.38
16	0.00077	0.99923	98572.77	75.90	98534.82	5955384.80	60.42
17	0.00089	0.99911	98496.87	87.66	98453.04	5856849.98	59.46
18	0.00096	0.99904	98409.21	94.47	98361.97	5758396.94	58.51
19	0.00101	0.99899	98314.74	99.30	98265.09	5660034.97	57.57
20	0.00104	0.99896	98215.44	102.14	98164.37	5561769.88	56.63
21	0.00109	0.99891	98113.29	106.94	98059.82	5463605.51	55.69
22	0.00112	0.99888	98006.35	109.77	97951.47	5365545.69	54.75
23	0.00114	0.99886	97896.58	111.60	97840.78	5267594.23	53.81
24	0.00116	0.99884	97784.98	113.43	97728.27	5169753.44	52.87
25	0.00117	0.99883	97671.55	114.28	97614.41	5072025.18	51.93
26	0.00119	0.99881	97557.27	116.09	97499.23	4974410.77	50.99
27	0.00121	0.99879	97441.18	117.90	97382.23	4876911.54	50.05

Table 4.8 Life table *(continued)*

x	q_x	p_x	l_x	d_x	L_x	T_x	e_x
28	0.00126	0.99874	97323.28	122.63	97261.96	4779529.31	49.11
29	0.00133	0.99867	97200.65	129.28	97136.01	4682267.35	48.17
30	0.00140	0.99860	97071.37	135.90	97003.42	4585131.34	47.23
31	0.00147	0.99853	96935.47	142.50	96864.23	4488127.91	46.30
32	0.00154	0.99846	96792.98	149.06	96718.45	4391263.69	45.37
33	0.00162	0.99838	96643.92	156.56	96565.64	4294545.24	44.44
34	0.00170	0.99830	96487.35	164.03	96405.34	4197979.60	43.51
35	0.00178	0.99822	96323.33	171.46	96237.60	4101574.26	42.58
36	0.00188	0.99812	96151.87	180.77	96061.49	4005336.67	41.66
37	0.00198	0.99802	95971.10	190.02	95876.09	3909275.18	40.73
38	0.00207	0.99793	95781.08	198.27	95681.95	3813399.09	39.81
39	0.00217	0.99783	95582.81	207.41	95479.11	3717717.14	38.90
40	0.00228	0.99772	95375.40	217.46	95266.67	3622238.03	37.98
41	0.00240	0.99760	95157.94	228.38	95043.75	3526971.36	37.06
42	0.00254	0.99746	94929.56	241.12	94809.00	3431927.61	36.15
43	0.00271	0.99729	94688.44	256.61	94560.14	3337118.60	35.24
44	0.00292	0.99708	94431.84	275.74	94293.97	3242558.46	34.34
45	0.00318	0.99682	94156.10	299.42	94006.39	3148264.49	33.44
46	0.00348	0.99652	93856.68	326.62	93693.37	3054258.10	32.54
47	0.00380	0.99620	93530.06	355.41	93352.35	2960564.73	31.65
48	0.00414	0.99586	93174.65	385.74	92981.77	2867212.38	30.77
49	0.00449	0.99551	92788.90	416.62	92580.59	2774230.61	29.90
50	0.00490	0.99510	92372.28	452.62	92145.97	2681650.38	29.03
51	0.00537	0.99463	91919.66	493.61	91672.85	2589504.05	28.17
52	0.00590	0.99410	91426.05	539.41	91156.34	2497831.20	27.32
53	0.00647	0.99353	90886.63	588.04	90592.62	2406674.86	26.48
54	0.00708	0.99292	90298.60	639.31	89978.94	2316082.24	25.65
55	0.00773	0.99227	89659.28	693.07	89312.75	2226103.30	24.83
56	0.00844	0.99156	88966.22	750.87	88590.78	2136790.55	24.02
57	0.00926	0.99074	88215.34	816.87	87806.90	2048199.77	23.22
58	0.01019	0.98981	87398.47	890.59	86953.17	1960392.87	22.43
59	0.01120	0.98880	86507.88	968.89	86023.43	1873439.69	21.66
60	0.01223	0.98777	85538.99	1046.14	85015.92	1787416.26	20.90
61	0.01328	0.98672	84492.85	1122.07	83931.81	1702400.34	20.15
62	0.01439	0.98561	83370.78	1199.71	82770.93	1618468.53	19.41
63	0.01560	0.98440	82171.08	1281.87	81530.14	1535697.60	18.69
64	0.01691	0.98309	80889.21	1367.84	80205.29	1454167.46	17.98
65	0.01827	0.98173	79521.37	1452.86	78794.94	1373962.17	17.28
66	0.01967	0.98033	78068.52	1535.61	77300.71	1295167.22	16.59
67	0.02121	0.97879	76532.91	1623.26	75721.28	1217866.51	15.91
68	0.02297	0.97703	74909.65	1720.67	74049.31	1142145.23	15.25
69	0.02499	0.97501	73188.97	1828.99	72274.47	1068095.93	14.59
70	0.02727	0.97273	71359.98	1945.99	70386.99	995821.45	13.95

Table 4.8 Life table *(continued)*

x	q_x	p_x	l_x	d_x	L_x	T_x	e_x
71	0.02979	0.97021	69413.99	2067.84	68380.07	925434.47	13.33
72	0.03251	0.96749	67346.15	2189.42	66251.44	857054.40	12.73
73	0.03534	0.96466	65156.73	2302.64	64005.41	790802.96	12.14
74	0.03824	0.96176	62854.09	2403.54	61652.32	726797.55	11.56
75	0.04126	0.95874	60450.55	2494.19	59203.45	665145.23	11.00
76	0.04455	0.95545	57956.36	2581.96	56665.38	605941.78	10.46
77	0.04819	0.95181	55374.40	2668.49	54040.16	549276.40	9.92
78	0.05239	0.94761	52705.91	2761.26	51325.28	495236.25	9.40
79	0.05723	0.94277	49944.65	2858.33	48515.48	443910.97	8.89
80	0.06277	0.93723	47086.31	2955.61	45608.51	395395.49	8.40
81	0.06885	0.93115	44130.71	3038.40	42611.51	349786.98	7.93
82	0.07535	0.92465	41092.31	3096.31	39544.15	307175.47	7.48
83	0.08207	0.91793	37996.00	3118.33	36436.84	267631.32	7.04
84	0.08907	0.91093	34877.67	3106.55	33324.39	231194.48	6.63
85	0.09705	0.90295	31771.12	3083.39	30229.42	197870.09	6.23
86	0.10627	0.89373	28687.73	3048.64	27163.41	167640.67	5.84
87	0.11625	0.88375	25639.08	2980.54	24148.81	140477.26	5.48
88	0.12688	0.87312	22658.54	2874.92	21221.08	116328.45	5.13
89	0.13834	0.86166	19783.63	2736.87	18415.19	95107.36	4.81
90	0.15135	0.84865	17046.76	2580.03	15756.74	76692.17	4.50
91	0.16591	0.83409	14466.73	2400.18	13266.64	60935.43	4.21
92	0.18088	0.81912	12066.56	2182.60	10975.26	47668.78	3.95
93	0.19552	0.80448	9883.96	1932.51	8917.70	36693.53	3.71
94	0.21000	0.79000	7951.45	1669.80	7116.54	27775.83	3.49
95	0.22502	0.77498	6281.64	1413.50	5574.89	20659.28	3.29
96	0.24126	0.75874	4868.15	1174.49	4280.90	15084.39	3.10
97	0.25689	0.74311	3693.66	948.86	3219.23	10803.48	2.92
98	0.27175	0.72825	2744.79	745.90	2371.85	7584.26	2.76
99	0.28751	0.71249	1998.90	574.70	1711.55	5212.41	2.61
100	0.30418	0.69582	1424.19	433.21	1207.59	3500.87	2.46
101	0.32182	0.67818	990.98	318.92	831.52	2293.28	2.31
102	0.34049	0.65951	672.06	228.83	557.65	1461.76	2.18
103	0.36024	0.63976	443.23	159.67	363.40	904.11	2.04
104	0.38113	0.61887	283.56	108.07	229.53	540.71	1.91
105	0.40324	0.59676	175.49	70.76	140.11	311.18	1.77
106	0.42663	0.57337	104.72	44.68	82.39	171.08	1.63
107	0.45137	0.54863	60.05	27.10	46.49	88.69	1.48
108	0.47755	0.52245	32.94	15.73	25.08	42.20	1.28
109	0.50525	0.49475	17.21	8.70	12.86	17.12	0.99
110	1.00000	0.00000	8.52	8.52	4.26	4.26	0.50

Table 4.8 Life table

Example 4.4.4. On the basis of Table 4.8, evaluate the probability that
(i) (25) will live to 75 and (ii) (25) die before 80.

Solution: (i) The probability that (25) will live to 75 is
$_{50}p_{25} = S(75)/S(25) = l_{75}/l_{25} = 0.6189$. (ii) The probability that (25) will die before 80 is
$_{55}q_{25} = 1 - S(80)/S(25) = 1 - l_{80}/l_{25} = 0.5179$.   ∎

Life table is a versatile tool which can be applied whenever we face situations involving the
number of survivors out of an original group. Consequently, the life table approach has been
used for the study of animal, bird and fish populations, the staffing of companies, survival
following major operations or following the onset of disease, the wearing out and replacement
of equipments, replacement of stocks and many other similar problems. Thus, there is a wide
range of problems for which the life table technique is useful. The following are some examples
which will be helpful for better understanding of the life table technique. Interested readers
may refer to Chapters 4 and 5 of the book by Pollard et al [13] for more applications.

Example 4.4.5. An aviary of birds which has a constant intake of $1,500$ new-born birds per
year, experiences the mortality rates as specified in Table 4.9. (i) What is the expected total
number of birds in the aviary at any time? (ii) What is the expected number living between
ages 1 and 4? (iii) If the owner wanted the population to be steady at $5,000$, on an average
how many extra new-born birds would he have to add each year?

Age (x)	0	1	2	3	4	5
q_x	0.3	0.1	0.2	0.4	0.7	1.0

Table 4.9 Mortality rates of bird population

Solution: It is given that $l_0 = 1500$. From the given q_x values and the formulae summarized
in Eqs 4.4.1, we construct a life table as presented in Table 4.10.

x	q_x	p_x	l_x	L_x	T_x
0	0.3	0.7	1500	1275	4090.68
1	0.1	0.9	1050	997.5	2815.68
2	0.2	0.8	945	850.5	1818.18
3	0.4	0.6	756	604.8	967.66
4	0.7	0.3	4536	294.84	362.88
5	1	0	136.08	68.04	68.04

Table 4.10 Life table for bird population

(i) The expected total number of birds in the aviary at any time is given by $T_0 = 4090.68$.
(ii) The expected number living between ages 1 and 4 is $T_1 - T_4 = 2452.8$. (iii) For a total
population to be $T_0 = 4090.68$, the intake of birds is 1500, hence to have a population to be
steady at 5000, the intake should be $(5000 \times 1500)/4090.68 = 1833$. So the additional number
of birds is $1833 - 1500 = 333$.   ∎

Example 4.4.6. A shopkeeper takes in a stock of 100 packets of gulabjamun mix. In the first week he sells 20% of them, in the second week 45% of the remaining, in the third week 50% of the remaining and in the fourth week 65% of the remaining. In the fifth week he sells all he has left. (a) What is the chance that one of the original 100 packets (i) is sold in the first week? (ii) is sold in the second week? (iii) is unsold by the fourth week? (b) Suppose that over the years he finds that he sells 500 packets of these each week which are provided every week by a salesman. If the retail price of the packet is Rs 25/- per packet, what is the retail value of his stock at any time?

Solution: From the given information, we construct a life table as shown in Table 4.11. From

x	q_x	p_x	l_x	d_x	L_x	T_x
0	0.20	0.8	100	20	90	203.7
1	0.45	0.55	80	36	62	113.7
2	0.50	0.5	44	22	33	51.7
3	0.65	0.35	22	14.3	14.85	18.7
4	1	0	7.7	7.7	3.85	3.85

Table 4.11 Life table for collection of packets of gulabjamun mix

this table we have, (a) (i) 0.2, (ii) probability that a packet is not sold in the first week and then sold in the second week $= (0.8)(0.45) = 0.36$. The probability in (a) (ii) can also be found as follows: 36 out of 100 packets are sold in the second week, so the required probability is $36/100 = 0.36$.

(iii) By the end of 4th week 7.7 out of 100 packets are left. So the probability is $7.7/100 = 0.077$. Alternatively, the required probability is equal to the probability that it is not sold in the first week, second week, third and fourth week and is given by $(0.8)(0.55)(0.5)(0.35) = 0.077$.

(b) With $l_0 = 100$, $T_0 = 203.7$ which gives the total stock. With intake of 500 packets, his total stock would be 5×203.7. So the retail price of his stock at any time is $5 \times 203.7 \times 25 = 25,462.50$ Rs. ∎

Example 4.4.7. The hypothetical replacement rates per annum of electric light poles are given in Table 4.12. Replacement rate (R) is defined as the probability, at the beginning of the year, that a pole will in the next 12 months be in such a state that it must be replaced.

Year	0	1	2	3	4	5	6	7	8	9
R	0.1	0.2	0.3	0.4	0.5	0.6	0.7	0.8	0.9	1

Table 4.12 Replacement rates of electric light poles

(a) One thousand new poles have just been 'planted'. Prepare a table to show (i) how many of these 'original' poles will be standing in $1, 2, 3, \cdots, 10$ years' time, (ii) the number of these 'original' poles which will need to be replaced during the first, second, third, ..., years.

(b) (i) What is the probability that a pole just 'planted' will be standing in 5 years' time? (ii) What is the probability that a pole 'planted' 3 years ago will last more than 3 years? (iii) What is the probability that a pole 'planted' exactly a year ago will be replaced in the 4th or 5th year?

(iv) How many poles are aged between 2 and 5 years? (v) If it is the policy of the authority to paint poles as soon as they have been standing for 5 years, how much would it cost to them in a year, at Rs 100 per pole, to get a contractor to do the job? (vi) What is the average life of a pole just planted? (vii) What is the average future life of 3 years old poles? Write answers in terms of life table functions.

Solution: From the given information, we have $l_0 = 1000$ and replacement rates as q_x values to construct a life table. Thus answers in terms of life table functions are as follows: (a) (i) l_x column (ii) d_x column. (b) (i) l_5/l_0, (ii) l_6/l_3, (iii) $(d_4 + d_5)/l_1$, (iv) $T_2 - T_5 = L_2 + L_3 + L_4$, (v) $l_5 \times 100$ rupees, (vi) T_0/l_0 and (vii) T_3/l_3 ■

From the illustrative examples discussed so far, it is clear that the life table specifies the probability distribution of $K(x)$ completely. The values of l_x for integer values of x are available from the life table but not for any $x \geq 0$. However, the lack of knowledge of life table functions for non-integer values of x does not cause great difficulty, since in practice, it is usually sufficient to know the curtate future lifetime. In some cases, it still becomes necessary to use the exact age at death and the life table functions at exact ages. Thus, a passage from the distribution of $K(x)$ to that of $T(x)$ is necessary. To specify the distribution of $T(x)$, we must postulate an analytic form for $S(x)$ or adopt a life table and an assumption about the distribution between integer ages. Each of the approximately 100 intervals of age within the typical life table possesses a pattern of mortality within its unit length. Reasonably accurate prediction of this pattern for each such interval is essential for the satisfactory representation of the distribution of $T(x)$. In the following section, we study such patterns for mortality within a unit interval.

4.5   Assumptions for Fractional Ages

Three assumptions for fractional ages are widely used in actuarial science. These can be stated in terms of the survival function $S(x)$ or l_x or $_tq_x$ or $_tp_x$. Assumptions stated in terms of the survival function $S(x)$ are as given below. Suppose x is an integer and $0 \leq t \leq 1$.

$$\text{Uniform distribution of deaths} \quad : \quad S(x+t) = (1-t)S(x) + tS(x+1)$$

$$\text{Balducci assumption} \quad : \quad \frac{1}{S(x+t)} = \frac{(1-t)}{S(x)} + \frac{t}{S(x+1)}$$

$$\text{Constant force of mortality} \quad : \quad S(x+t) = S(x)e^{-\mu t}, \quad \text{where} \quad \mu = -\log p_x$$

The second assumption is named after Balducci, an Italian actuary, who pointed out its role in the traditional actuarial method of constructing life tables. The third assumption is expressible as follows. Observe that,

$$
\begin{aligned}
S(x+t) &= S(x)e^{-\mu t}, \quad \text{where} \quad \mu = -\log p_x \\
\Rightarrow \quad S(x+t) &= S(x)(p_x)^t = S(x)\left(S(x+1)/S(x)\right)^t \\
\Rightarrow \quad \log S(x+t) &= \log S(x) + t\log S(x+1) - t\log S(x) \\
\Rightarrow \quad \log S(x+t) &= (1-t)\log S(x) + t\log S(x+1)
\end{aligned}
$$

With this presentation, these three assumptions state the type of interpolation used in the unit age interval. In the first case, it is a linear interpolation, in the second, it is a harmonic interpolation and in the third, it is a logarithmic interpolation. Balducci assumption is also known as a hyperbolic assumption as in this setup $_tp_x$ is a hyperbolic curve. We now study in detail the consequences of these assumptions on distribution function and survival function of $T(x)$ and also on the force of mortality.

Uniform Distribution of Deaths: Uniform distribution was found to be inappropriate to model human survival patterns. Yet, a spin-off from this hypothesis is in wide use in the context of fractional ages. Under this assumption in a unit age interval $(x, x + 1)$ deaths tend to occur uniformly, where x is an integer. We know that the distribution function of an uniform distribution is a straight line over its support. Analogously, a mathematical statement of the 'uniform distribution of deaths' hypothesis, stated in terms of distribution function, is that the distribution function of $T(x)$ given by $_tq_x$ is linear over the interval $0 \leq t \leq 1$, for each integer x. Under this assumption for $0 \leq t \leq 1$,

$$_tq_x \;=\; a + bt \;\Rightarrow\; _1q_x = q_x = a + b \quad \text{and} \quad _0q_x = 0 = a \;\Rightarrow\; a = 0,\; b = q_x$$

$$\Rightarrow \;\; _tq_x \;=\; t \times q_x$$

Thus, rather than considering the entire survivorship curve as a single straight line, the assumption of uniformity of deaths merely considers it to be a connected series of straight lines, each with its own slope. For the interval $(x,\, x + 1)$, the slope is q_x. This simple assumption states that the probability of death of a person aged x within a quarter of a year, for example, is one-fourth times the rate which would apply over the full one-year interval. The assumption of uniformity of deaths in a unit interval as expressed in terms of the survival function can be obtained from the equation $_tq_x = t\, q_x$. The derivation is as follows.

$$_tq_x = t\, q_x \;\;\Longleftrightarrow\;\; 1 - S(x + t)/S(x) = t\,\{1 - S(x + 1)/S(x)\}$$
$$\Longleftrightarrow\;\; S(x + t)/S(x) = 1 - t \,+\, tS(x + 1)/S(x)$$
$$\Longleftrightarrow\;\; S(x + t) = (1 - t)S(x) \,+\, t\, S(x + 1)$$

This expression for $S(x + t)$ has a nice geometrical interpretation. If $P(x, S(x))$, $Q(x+1, S(x+1))$ and $A(x+t, S(x+t)), 0 < t < 1$, are three points on the survival curve, then the last expression is essentially the result of the section formula which gives the coordinates of the point dividing the line PQ in the ratio $t : (1 - t)$. The assumption of uniformity of deaths in a unit interval guarantees that the survival curve, joining points P and Q is a straight line.

Under the assumption of uniform distribution of deaths in a unit interval, for a fixed integer value of x and for $t \in (0, 1)$, the functions μ_{x+t}, l_{x+t} can be expressed in terms of q_x, l_x, d_x. We proceed with these derivations below. Under this assumption,

$$t\, q_x \;=\; _tq_x \;=\; 1 - S(x + t)/S(x) \;\Rightarrow\; S(x + t) = (1 - tq_x)S(x)$$

$$\Rightarrow\; \frac{d}{dt}S(x + t) \;=\; -S(x)q_x$$

$$\Rightarrow\; \mu_{x+t} \;=\; -\frac{d}{dt}\log S(x + t) = -\frac{1}{S(x + t)}(-S(x)q_x)$$

$$=\; S(x)q_x/((1 - tq_x)S(x)) = q_x/(1 - tq_x)$$

It is easy to verify that,

$$\mu_{x+1} = q_x/(1-q_x) \ > \ q_x = \mu_x$$

Further, the first and also the second derivative of μ_{x+t} with respect to t are positive, which implies that for $t \in (0,1)$ and for any x, μ_{x+t} is an increasing convex function of t. Figure 4.8 displays this nature of μ_{x+t} for a unit interval.

To obtain the expression for l_{x+t}, note that,

$$\begin{aligned}
l_x - l_{x+t} &= l_0(S(x) - S(x+t)) = l_0 S(x)\{1 - S(x+t)/S(x)\} \\
&= l_x \ _tq_x = tl_x q_x = td_x \\
\Rightarrow \ l_{x+t} &= l_x - td_x
\end{aligned}$$

This expression for l_{x+t} can also be obtained from the assumption of uniformity in terms of survival function. Thus,

$$\begin{aligned}
S(x+t) &= (1-t)S(x) + tS(x+1) \quad \Rightarrow \quad l_{x+t} = (1-t)l_x + tl_{x+1} \\
\Rightarrow \ l_{x+t} &= l_x - t(l_x - l_{x+1}) = l_x - td_x \\
\Rightarrow \ _td_x &= l_x - l_{x+t} = td_x
\end{aligned}$$

The last expression shows that the expected number $_td_x$ of deaths in $(x, x+t)$ is proportional to t. Thus, the expected number of deaths in $(x, x+0.25)$ is $0.25d_x$. From the expression for l_{x+t}, it easily follows that,

$$l_{x+1/2} = l_x - d_x/2 = l_x - (l_x - l_{x+1})/2 = (l_x + l_{x+1})/2$$

which is nothing but the trapezoidal rule, justification of which follows from the fact that l_{x+t} for $t \in (0,1)$ is a straight line.

We now prove the already stated result $L_x = (l_x + l_{x+1})/2$, under the assumption of uniformity. By definition of L_x, $L_x = \int_0^1 l_{x+t}dt$. Under the assumption of uniformity, $l_{x+t} = l_x - td_x$. Substituting this expression in the integral we get,

$$L_x = \int_0^1 (l_x - td_x)dt = l_x - 0.5d_x = l_x - 0.5(l_x - l_{x+1}) = (l_x + l_{x+1})/2$$

It is easy to see that under the assumption of uniformity, the probability density function $g_x(t)$ of $T(x)$ for integer x is,

$$g_x(t) = \ _tp_x\mu_{x+t} = (1 - tq_x)(q_x/(1 - tq_x)) = q_x, \qquad 0 \le t \le 1$$

Thus, the probability density function of $T(x)$ for $0 \le t \le 1$ is q_x and it is free from t, but depends on x. Suppose for integer x, a random variable $U = U(x)$ is defined by $T(x) = K(x) + U(x)$, where $T(x)$ and $K(x)$ indicate the future lifetime and the curtate future lifetime of (x) respectively. The random variable U represents the fractional part of a year lived in the year of death. It is a continuous random variable with support $(0,1)$. Under the assumption of uniformity of deaths in a unit interval, it can be proved that $K(x)$ and U are independent random variables and U has uniform distribution on $(0,1)$. We heavily use this result in subsequent chapters.

Following examples illustrate the application of the assumption of uniformity of deaths in a unit interval.

Example 4.5.1. Under the assumption of uniform distribution of deaths in a unit interval, find $_{1.5}p_{30.5}$ and $\mu_{30.5}$ in terms of l_{30}, l_{31} and l_{32}.

Solution: By definition, (i) $_{1.5}p_{30.5} = l_{32}/l_{30.5} = l_{32}/((l_{30} + l_{31})/2)$. (ii) Note that

$$\mu_{30.5} = \frac{q_{30}}{1 - (.5)q_{30}} = \left(1 - \frac{l_{31}}{l_{30}}\right)\left\{1 - \frac{1}{2}\left(1 - \frac{l_{31}}{l_{30}}\right)\right\}^{-1} = \frac{2(l_{30} - l_{31})}{(l_{30} + l_{31})}$$

∎

Example 4.5.2. There are 1000 persons of age 40, of whom 32 are expected to die before age 41. How many will die between $40\frac{3}{8}$ and $40\frac{3}{4}$ under the assumption of uniformity of deaths?

Solution: Under the assumption of uniformity of deaths, the number d of deaths between $40\frac{3}{8}$ and $40\frac{3}{4}$ is given by,

$$d = l_{40\frac{3}{8}} - l_{40\frac{3}{4}} = l_{40} - (3/8)d_{40} - l_{40} + (3/4)d_{40} = (3/8)d_{40} = 12$$

∎

The probability that $(x + t)$ dies in the next $(1 - s)$ time units, when $0 \le t \le s \le 1$ and $0 \le (t + 1 - s) \le 1$, is denoted by $_{1-s}q_{x+t}$. We obtain below its expression in terms of q_x, under the assumption of uniformity of deaths. We use it in Example 4.5.3 after the derivation. By definition,

$$
\begin{aligned}
_{1-s}q_{x+t} &= 1 - S(x + t + 1 - s)/S(x + t) \\
&= [S(x + t) - S(x + t + 1 - s)]/S(x + t) \\
&= [(1 - t)S(x) + tS(x + 1)]/S(x + t) \\
&\quad - [(s - t)S(x) + (t + 1 - s)S(x + 1)]/S(x + t) \\
&= [(1 - s)S(x) - (1 - s)S(x + 1)]/S(x + t) \\
&= (1 - s)(S(x) - S(x + 1))/S(x + t) \\
&= (1 - s)S(x)\,(1 - S(x + 1)/S(x))\,/S(x + t) \\
&= (1 - s)q_x/(S(x + t)/S(x)), \quad 0 \le t \le s \le 1 \\
&= (1 - s)q_x/(1 - {}_tq_x) = (1 - s)q_x/(1 - tq_x) \\
&= (1 - s)\mu_{x+t}, \quad 0 \le t \le s \le 1
\end{aligned}
$$

Example 4.5.3. If $_{1/4}q_x = 0.01$, find the value of $_{1/2}q_{x+3/8}$ when $_tq_x = tq_x$, $0 \le t \le 1$.

Solution: The equation $_tq_x = tq_x$ implies the assumption of uniformity of deaths in a unit interval. Hence, $_{1-s}q_{x+t} = (1 - s)q_x/(1 - tq_x)$. Note that,

$$0.01 = {}_{1/4}q_x = (1/4)q_x \Rightarrow q_x = 0.04$$
$$\Rightarrow {}_{1/2}q_{x+3/8} = (1/2)q_x/(1 - (3/8)q_x) = 4/197$$

∎

Example 4.5.4. It is given that deaths are uniformly distributed over each year of age and values of l_x corresponding $x = 35$ to 39 are given in Table 4.13. Which of the following are true? (i) $_{0.5}p_{36} = 0.091$ and (ii) $_{0.33}q_{35.5} = 0.021$.

x	35	36	37	38	39
l_x	100	99	96	92	87

Table 4.13 Values of l_x

Solution : (i) $_{0.5}p_{36} = 1 - {}_{0.5}q_{36} = 1 - 0.5q_{36} = 1 - 0.5\,(1 - l_{37}/l_{36}) = 0.995$ and (ii) $_{0.33}q_{35.5} = (0.33)q_{35}/(1 - (0.5)q_{35}) = 0.018$. Thus, both (i) and (ii) are false.   ■

Example 4.5.5. Given $q_{60} = 0.3$ and $q_{61} = 0.4$, find the probability that (60.5) will die between (60.5) and (61.5) under the assumption of uniformity of deaths in a unit interval.

Solution: In usual notation, the required probability is,

$$
\begin{aligned}
{}_{1}q_{60.5} &= {}_{0.5}q_{60.5} + {}_{0.5}p_{60.5}\,{}_{0.5}q_{61} \\
&= 0.5q_{60}/(1 - 0.5q_{60}) + (1 - {}_{0.5}q_{60.5})(0.5)q_{61} \\
&= 0.1765 + (0.8235)(0.2) = 0.3412
\end{aligned}
$$

■

As stated in Section 4, under the assumption of uniformity of deaths in a unit interval, q_x and m_x are related by the equation $q_x = m_x/(1 + m_x/2)$. We now derive this relation. As defined in Section 4,

$$
m_x = (l_x - l_{x+1})\Big/ L_x = (1 - l_{x+1}/l_x)\Big/ \int_0^1 l_{x+t}/l_x\,dt = (1 - p_x)\Big/ \int_0^1 {}_tp_x
$$

Under the assumption of uniformity,

$$
\begin{aligned}
\int_0^1 {}_tp_x &= \int_0^1 (1 - tq_x) = 1 - q_x/2 \\
\Rightarrow m_x &= q_x\Big/(1 - q_x/2) \iff q_x = m_x\Big/(1 + m_x/2)
\end{aligned}
$$

We now proceed to discuss the second pattern of mortality, often used as an estimator of the distribution of deaths over a unit age interval.

Balducci Assumption: It was suggested by the Italian actuary Gaetano Balducci in 1920. Although the Balducci hypothesis does not have a simple verbal description, it has proven to be a quite reasonable indicator of the shape of the survivorship curve in many instances. The Balducci hypothesis is already stated in terms of the survival function $S(x)$. It can also be expressed in terms of the distribution function of $T(x)$. Balducci assumed that the function $_{1-t}q_{x+t}$ is linear over the interval $0 \leq t \leq 1$. Thus, under this assumption,

$$
\begin{aligned}
{}_{1-t}q_{x+t} &= a + bt, \ \ 0 \leq t \leq 1 \\
\Rightarrow q_x &= a \ \text{ with } \ t = 0 \ \text{ and } \ {}_0q_{x+1} = 0 = a + b \text{ with } \ t = 1 \\
\Rightarrow a &= q_x \ \text{ and } \ b = -q_x \\
\Rightarrow {}_{1-t}q_{x+t} &= (1 - t)q_x, \ \ 0 \leq t \leq 1
\end{aligned}
$$

This relation can be derived from the relation among $S(x), S(x+1)$ and $S(x+t)$ under the Balducci assumption and vice versa. We establish below this equivalence. Under the Balducci assumption,

$$
\begin{aligned}
{}_{1-t}q_{x+t} &= (1-t)q_x, \quad 0 \le t \le 1 \\
\Longleftrightarrow \quad 1 - S(x+1)/S(x+t) &= (1-t)(1 - S(x+1)/S(x)) \\
&= 1 - S(x+1)/S(x) - t + tS(x+1)/S(x) \\
\Longleftrightarrow \quad S(x+1)/S(x+t) &= S(x+1)/S(x) + t - tS(x+1)/S(x) \\
\Longleftrightarrow \quad 1/S(x+t) &= 1/S(x) + t/S(x+1) - t/S(x) \\
&= (1-t)/S(x) + t/S(x+1)
\end{aligned}
$$

Under this assumption, μ_{x+t}, l_{x+t} and ${}_{1-s}q_{x+t}$ can be expressed in terms of q_x, l_x and d_x as under the assumption of uniformity. We derive these relations below.

$$
\begin{aligned}
{}_{1-t}q_{x+t} = (1-t)q_x \quad \Longleftrightarrow \quad & 1 - S(x+1)/S(x+t) = (1-t)q_x \\
\Longleftrightarrow \quad & S(x+t)/S(x+1) = 1/(1 - (1-t)q_x) \\
\Longleftrightarrow \quad & S(x+t)/S(x) = [S(x+1)/S(x)][1/(1 - (1-t)q_x)] \\
\Longleftrightarrow \quad & {}_tp_x = p_x/(1 - (1-t)q_x) \qquad\qquad (4.5.1) \\
\Rightarrow \quad & \mu_{x+t} = \frac{-1}{{}_tp_x}\frac{d}{dt}\,{}_tp_x = \frac{q_x}{(1 - (1-t)q_x)} = \frac{q_x}{(p_x + tq_x)} \\
\Rightarrow \quad & \mu_x = q_x/p_x \; > \; q_x = \mu_{x+1}
\end{aligned}
$$

It is easy to see that μ_{x+t} is a decreasing function of t for $0 \le t \le 1$. Thus, the pattern of the force of mortality under the Balducci hypothesis is opposite to that encountered under the assumption of a uniform distribution of deaths. Under the Balducci assumption, the curve of μ_{x+t} is a decreasing hyperbola over a unit interval. This is a somewhat confusing result, however it is to be noted that its decreasing nature is evident only over a single-unit age interval. Figure 4.8 displays this nature of μ_{x+t} for a unit interval.

The following are some examples to illustrate the Balducci hypothesis.

Example 4.5.6. (i) Show that $\mu_{x+t}^{A} = \mu_{x+1-t}^{B}$ for $0 \le t \le 1$, where the superscripts A and B indicate the uniformity and the Balducci assumption respectively. (ii) Find the value of t for which force of mortality under A and B is the same.

Solution: (a) Under A, $\mu_{x+t}^{A} = q_x/(1 - tq_x)$ and under B,
$\mu_{x+t}^{B} = q_x/(1 - (1-t)q_x)$. Thus, $\mu_{x+t}^{A} = \mu_{x+(1-t)}^{B}$.
(b) From the expressions of μ_{x+t}^{A} and μ_{x+t}^{B}, it is easy to see that for $t = 1/2$, the force of mortality has the same expression under A and B. ∎

To derive the expression for l_{x+t}, we again begin with the relation,

$$
\begin{aligned}
_{1-t}q_{x+t} &= (1-t)q_x \\
\Rightarrow\ S(x+t) &= S(x+1)/(1-(1-t)q_x) \\
\Rightarrow\ l_{x+t}/l_0 &= \left\{\frac{l_{x+1}}{l_0}\right\}\left\{1-(1-t)(1-\frac{l_{x+1}}{l_x})\right\}^{-1} \\
\Rightarrow\ l_{x+t} &= l_{x+1}\cdot l_x/\{tl_x+(1-t)l_{x+1}\} \\
&= l_{x+1}l_x/(l_{x+1}+td_x), \qquad 0\le t\le 1
\end{aligned}
$$

From this expression, it follows that under the Balducci hypothesis, the survivorship function l_{x+t} is represented graphically by the reciprocal of a straight line, that is, a hyperbola, rather than by the straight line produced under the earlier assumption of uniformity of deaths in a unit interval.

We now obtain the expression for $_{1-s}q_{x+t}$ for $0\le t\le s\le 1$, which is needed in the next example. By definition,

$$
\begin{aligned}
_{1-s}q_{x+t} &= 1 - S(x+t+1-s)/S(x+t) \\
&= S(x+t+1-s)\left\{\frac{1}{S(x+t+1-s)}-\frac{1}{S(x+t)}\right\} \\
&= S(x+t+1-s)\left\{\frac{s-t}{S(x)}+\frac{t+1-s}{S(x+t)}-\frac{1-t}{S(x)}-\frac{t}{S(x+1)}\right\} \\
&= S(x+t+1-s)\left\{\frac{s-1}{S(x)}+\frac{1-s}{S(x+1)}\right\} \\
&= \frac{S(x+t+1-s)}{S(x+1)}\left\{1-\frac{S(x+1)}{S(x)}\right\}(1-s) \\
&= (1-s)q_x\frac{S(x+t+1-s)}{S(x)}\left\{\frac{S(x)}{S(x+1)}\right\} \\
&= (1-s)q_x\ \frac{1-(s-t)p_x}{p_x} \\
&= \frac{(1-s)q_x}{p_x}\ \frac{p_x}{1-(s-t)q_x}\quad \text{by Equation (4.5.1)} \\
&= (1-s)q_x/(1-(s-t)q_x)
\end{aligned}
$$

Example 4.5.7. If $_{5/12}q^A_{x+1/4} = {}_{5/12}q^B_{x+k}$, find k; A and B indicating the uniformity and the Balducci assumption respectively.

Solution: Note that

$$
\begin{aligned}
_{5/12}q^A_{x+1/4} &= {}_{5/12}q^B_{x+k} \\
\Rightarrow\ \frac{5}{12}q_x\left\{1-\frac{1}{4}q_x\right\}^{-1} &= \frac{5}{12}q_x\left\{1-\left(\frac{7}{12}-k\right)q_x\right\}^{-1} \\
\Rightarrow\ (1/4)q_x &= (7/12-k)q_x \\
\Rightarrow\ k &= 1/3 \qquad\qquad\blacksquare
\end{aligned}
$$

Example 4.5.8. Under the Balducci assumption, find q_{51} when it is given that,

$$q_{50} = 0.04 \quad \text{and} \quad {}_{1/2}q_{50+3/4} = (2/3)({}_{3/4}q_{50+1/2}).$$

Solution: Observe that

$$
\begin{aligned}
{}_{1/2}q_{50+3/4} &= P\left[\text{an individual dies in the interval } 50\frac{3}{4} \text{ to } 51\frac{1}{4}\right] \\
&= P\left[\text{an individual dies in } (50\frac{3}{4}, 51) \text{ or} \right. \\
&\qquad \left. \text{survives up to } 51 \text{ and dies in } (51, 51\frac{1}{4})\right] \\
&= {}_{(1/4)}q_{50+(3/4)} + {}_{(1/4)}p_{50+(3/4)} \; {}_{(1/4)}q_{51} \\
&= \frac{(1/4)q_{50}}{1 - ((3/4) - (3/4))q_{50}} + \frac{(1/4)q_{51}}{1 - (1 - (1/4))q_{51}}(1 - {}_{(1/4)}q_{50+(3/4)}) \\
&= 0.01 + (0.99)(0.25)q_{51}/(1 - (3/4)q_{51})
\end{aligned}
$$

Further,

$$
\begin{aligned}
{}_{3/4}q_{50+1/2} &= P\left[\text{an individual dies in the interval } (50\frac{1}{2}, 51\frac{1}{4})\right] \\
&= P\left[\text{an individual dies in } (50\frac{1}{2}, 51) \text{ or} \right. \\
&\qquad \left. \text{survives to 51 and dies in } (51, 51\frac{1}{4})\right] \\
&= {}_{1/2}q_{50+1/2} + {}_{1/2}p_{50+1/2} \; {}_{1/4}q_{51} \\
&= 0.02 + \frac{(1/4)q_{51}}{(1 - (3/4)q_{51})}(0.98)
\end{aligned}
$$

It is given that,

$$
{}_{1/2}q_{50+3/4} = (2/3)\left({}_{3/4}q_{50+1/2}\right)
$$

Hence, substituting from above and solving, we get $q_{51} = 1/26$. ∎

We discuss below one more assumption for the fractional ages, which is in a very real sense intermediate to the two already considered. It is the assumption of a constant force of mortality over a given unit age interval.

Assumption of a Constant Force of Mortality: Under this assumption,

$$\mu_{x+t} = \mu, \quad 0 \leq t \leq 1$$

hence the name constant force of mortality. Figure 4.8 displays the nature of force of mortality in a unit interval under three assumptions discussed in this section.

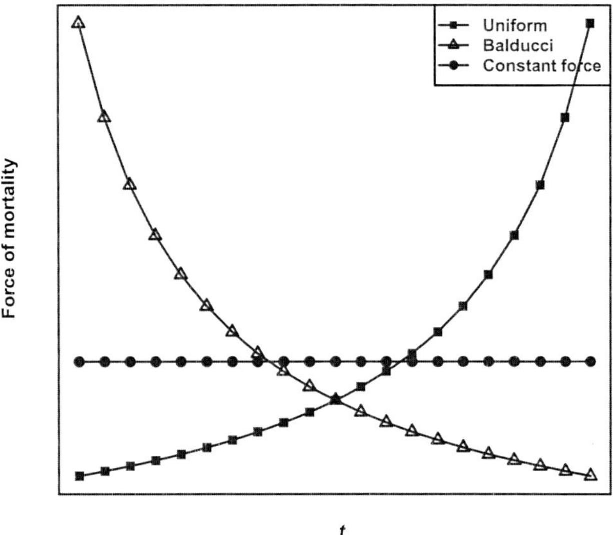

Figure 4.8 Force of mortality in a unit interval

Under the assumption of uniformity, $\mu_{x+t} = q_x/(1-tq_x)$. As stated earlier, μ_{x+t} is an increasing convex function of t and this nature is displayed in Figure 4.8. Under the Balducci assumption, $\mu_{x+t} = q_x/(1 - (1 - t)q_x)$ and the μ_{x+t} curve is a decreasing hyperbola over a unit interval as is evident from Figure 4.8. Under the third assumption, force of mortality is constant over the unit interval, hence curve the corresponding to this assumption is a line parallel to the x-axis. It is to be noted that this constant μ changes from age to age. More precisely, as defined in Section 2,

$$p_x = \exp\left(-\int_0^1 \mu_{x+s}\, ds\right) = \exp(-\mu) \quad \Longleftrightarrow \quad \mu = -\log_e p_x$$

The assumption of constant force of mortality as stated in terms of survival function at the beginning of this section can be derived from the assumption stated in terms of force of mortality and vice versa. Suppose $\mu_{x+t} = \mu,\ 0 \le t \le 1$. Then,

$$S(x+t) = \exp\left(-\int_0^{x+t} \mu_s ds\right) = \exp\left(-\int_0^x \mu_s ds\right) \times \exp\left(-\int_x^{x+t} \mu_s ds\right)$$

$$= S(x)\exp\left(-\int_0^t \mu_{x+s} ds\right) = S(x)\exp(-\mu t)$$

Now assume that $S(x+t) = S(x)e^{-\mu t}$. Then

$$\mu_{x+t} = -\frac{d}{dt}\log S(x+t) = -\frac{d}{dt}\{-\mu t + \log S(x)\} = \mu$$

In the next example, we prove some results which are valid under the assumption of a constant force of mortality.

Example 4.5.9. For $0 \leq t \leq 1$, prove the following relations under the assumption of a constant force of mortality. (i) ${}_tq_x = 1 - \exp(-\mu t)$.
(ii) ${}_{1-t}q_{x+t} = 1 - \exp(-\mu(1-t))$. (iii) $\log({}_tp_x)$ is linear in t. (iv) $p_x = ({}_tp_x)^{1/t}$.
(v) $l_{x+t} = l_x \exp(-\mu t)$. (vi) For any s such that, $s + t < 1$, ${}_tp_{x+s} = {}_tp_x$

Solution: (i) From the expression of $S(x+t)$ we get,

$$ {}_tq_x = 1 - S(x+t)/S(x) = 1 - \exp(-\mu t) \quad \text{and} \quad {}_tp_x = \exp(-\mu t) $$

Thus, for any fixed x and for $0 \leq t \leq 1$, the distribution of $T(x)$ has the same survival function as that of exponential with parameter $\mu = -\log p_x$ for $0 \leq t \leq 1$. However, for $t = 1$, ${}_tp_x = \exp(-\mu) < 1$.

(ii) By definition,

$$ {}_{1-t}q_{x+t} = 1 - S(x+1)/S(x+t) \quad = \quad 1 - e^{-\mu}/e^{-\mu t} = 1 - e^{-\mu(1-t)} $$

(iii) From (i), ${}_tp_x = \exp(-\mu t)$, thus $\log({}_tp_x) = -\mu t$, a linear function of t.
(iv) By taking, t = 1, $p_x = \exp(-\mu)$, hence

$$ {}_tp_x = \exp(-\mu t) = (\exp(-\mu))^t = (p_x)^t \quad \Rightarrow \quad p_x = ({}_tp_x)^{1/t} $$

(v) $l_{x+t} = l_0 S(x+t) = l_x \exp(-\mu t)$, which indicates the exponential decay.
(vi) For any s such that, $s + t < 1$,

$$ {}_tp_{x+s} \quad = \quad S(x+t+s)/S(x+s) \quad = \quad \exp(-\mu t) \quad = \quad {}_tp_x $$

Thus, a characteristic feature of this assumption is that within any unit age interval, the probability of survival is a function of duration only. Age is irrelevant but both the beginning and ending age of the interval defined by the probability consideration should be within the closed unit interval over which the assumption is effective. Thus, ${}_{1/4}q_{x+1/4} = {}_{1/4}q_{x+1/3} = {}_{1/4}q_{x+1/2}$ and so on.   ■

The following example clarifies the consequences of result (iv) and (vi) of the above example.

Example 4.5.10. Suppose $p_x = 0.729$. Under the assumption of constant force of mortality find (i) ${}_{1/3}p_x$, (ii) ${}_{1/3}p_{x+1/3}$ and (iii) ${}_{2/3}p_{x+1/4}$.

Solution: (i) ${}_{1/3}p_x = (p_x)^{1/3} = (0.729)^{1/3} = 0.9$
ii) To find ${}_{1/3}p_{x+1/3}$, we proceed as follows.

$$ {}_{1/3}p_{x+1/3} \quad = \quad S(x+2/3)/S(x+1/3) = e^{-\frac{2}{3}\mu}/e^{-1/3\mu} $$
$$ = \quad e^{-1/3\,\mu} = (e^{-\mu})^{1/3} = (p_x)^{1/3} = 0.9 $$

Alternatively, ${}_{1/3}p_{x+1/3} = {}_{1/3}p_x = 0.9$
(iii) ${}_{2/3}p_{x+1/4} = {}_{2/3}p_x = (p_x)^{2/3} = (0.9)^2 = 0.81$ ■

Figure 4.9 shows the nature of l_{x+t} in a unit interval under three assumptions for fractional ages. The figure clearly displays the intermediate nature of the third assumption of constant force of mortality.

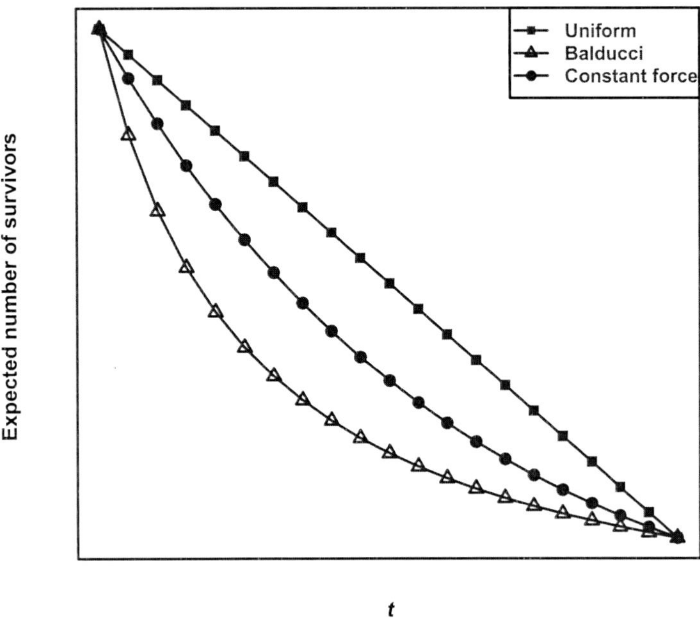

Figure 4.9 Expected number of survivors in a unit interval

Example 4.5.11. If $l_{x+t} = a/(3+t)$, $0 \le t \le 1$, $l_{x+1/2} = 800$, evaluate $l_{x+3/4}$, without finding a.

Solution: Observe that $l_{x+1/2}/l_{x+3/4} = 800/l_{x+3/4}$. Similarly,

$$\frac{l_{x+1/2}}{l_{x+3/4}} = \frac{a}{3+1/2}\frac{3+3/4}{a} = \frac{15}{4}\frac{2}{7} = \frac{15}{14}$$

$$\Rightarrow \quad 800/l_{x+3/4} = 15/14 \quad \Rightarrow \quad l_{x+3/4} = 800 \times (14/15) = 746.67$$

$\blacksquare$

Example 4.5.12. Suppose $q_x = 0.002$. Find the value of $_{1/2}q_x$ under each of the three assumptions. Comment on the results.

Solution: Under uniformity assumption, $_{1/2}q_x = (1/2)q_x = 0.001$. Under the Balducci assumption, $_{1/2}q_x = 0.5q_x/(1 - 0.5q_x) = 0.001001$. Under constant force of mortality assumption, the constant μ is given by

$$\mu = -\log_e(1 - p_x) = -\log_e(0.998) \quad \Rightarrow \quad _{1/2}q_x = 1 - \exp(-0.5\mu) = 0.0010005$$

We note that $_{1/2}q_x$ is the smallest under the uniformity assumption and the largest under the Balducci assumption. Under constant force of mortality, it is in between the two. $\blacksquare$

Under the assumption of constant force of mortality we have two expressions for $S(x+t)$ as given at the beginning of this section. These can be shown to be equivalent as follows. Using the exponential interpolation form we have,

$$S(x+t) = (S(x))^{1-t} \, (S(x+1))^t = S(x) \, (S(x+1)/S(x))^t = S(x)(p_x)^t = S(x) \, e^{-\mu t}$$

Three assumptions about the fractional assumptions discussed above completely specify the distribution of $T(x)$ corresponding to a given life table. In the subsequent chapters, we come across the application of uniformity assumption in a unit age interval in various derivations.

In the next section, we introduce one variation of a life table.

4.6   Select and Ultimate Life Tables

Life tables studied in Section 4, present various functions such as q_x, p_x, l_x, d_x for all non-negative integer values of x where x is age or time since an initial event of birth. The basic quantity in all these functions is the survival function $S(t) = P[X > t]$, where X is the life-length of a new born individual. Thus, the survival function at birth is applied for all future ages while constructing a life table. In such a table, q_x is considered to be a function of age x only and no additional information about the individual of age x is used while defining the probability q_x. However, the chance of death, along with age x, also depends on many other factors such as sex, health condition, occupation and hobbies of the individual. More precisely, the chance q_x of death depends on two factors; one is age and the other is the time (duration) since a certain event, known as selection.

One important example of selection is the acceptance of a proposal for life insurance at normal rates of premium. Before the company accepts the individuals for insurance, it tries to ensure, by medical examination, that they are not in such poor health that they are a 'bad risk'. Consequently, the mortality rates of lives who have been recently accepted for life insurance at normal rates may be expected to be lower than those of a general population.

The second illustration is that individual might have become disabled at age x. This information leads to the requirement that the survival function of individuals in such a group would be different from that of the rest. Thus, using the auxiliary information, a group of individuals can be clustered into homogeneous groups of similar conditions. Such groups are known as select groups. The survival function of individuals in these groups will differ depending on the nature of the auxiliary information forming a select group. In this section, we study in detail how to augment the additional information in a life table to get what is known as a select life table.

Survival models are usually divided into two categories: (i) the aggregate model and (ii) the select model. In the aggregate model, the survival function $S(t)$ specifies the probability of survival for t units of time since the initial event (such as birth) defining initial time $t = 0$. Then using x for attained age in general, we have $x = 0$ when $t = 0$. So, the attained age and elapsed time will run exactly together. In this case, one could use either x or t as a primary variable, with the usual convention being to use x. A life table constructed using such an aggregate survival function is known as an aggregate life table. Thus, all the life tables constructed in Section 4 are aggregate life tables.

To explain the other type of life table, we consider a situation in which a life insurance company has an insurance product for women of age ≥ 50. Two ladies Anita and Geeta of 55 years each apply for the same. Suppose Anita is from a general population whereas Geeta is from a group of women who have recently passed the full physical tests and hence is healthier than the general population. The issue is whether the company can charge the same premium to both Anita and Geeta. It seems reasonable that considering the health status of the women, the group of women who have passed the physical tests, should be charged less premium, since for these women risk of death is lesser as compared to the general population of women of the same age. Such situations are considered in 'select' life tables.

In the select survival model, the survival function is a function of two arguments x and t. The argument x is the age at which the individual is included in a select group decided on the basis of some criteria such as health condition, sex, occupation, nature of hobbies. The argument t denotes the time since selection and is a variable of primary interest. Such a survival function is denoted by $S(t; x)$. A life table developed using a select model of survival function is known as a select life table. The entire construction is parallel to that of the aggregate life table, except the notation. There is a special notation that allows for effect of selection. The essential feature of this notation is the age at selection, x, say, is enclosed in square brackets. Thus, we denote by $[x]$ the age of the individual when accepted (selected) to buy the insurance. In aggregate life tables we have used the symbol (x) to denote a life of age x. Time since selection, t say, is indicated by addition, so that $[x] + t$ means that t years have elapsed since entry to select status at age x. In this context, the age at selection, x, is called a concomitant variable; time since selection, t, being the primary variable of interest. Age at selection is not the only concomitant variable, another important one is sex. This concomitant variable is reflected by having separate survival models for males and females. Still another example would be smokers and non-smokers.

Suppose we begin with a radix $l_{[x]}$ which denotes the number of individuals in a specified select group at age x. Then the subsequent functions are obtained as follows:

$$l_{[x]+t} = l_{[x]} S(t; x)$$

The probability of death in an interval $(x + t, x + t + s)$, when the individual is included in a select group at age x is denoted by $_s q_{[x]+t}$. It is defined as

$$_s q_{[x]+t} \;=\; 1 - S(t + s; x)/S(t; x) = 1 - l_{[x]+(t+s)}/l_{[x]+t}$$

By taking $t = 0, 1, 2, \cdots$, and $s = 1$, we obtain the relationship between $l_{[x]}$ and $q_{[x]}$ functions as follows:

$$\begin{aligned}
l_{[x]+1} &= l_{[x]}(1 - q_{[x]}) \\
l_{[x]+2} &= l_{[x]+1}(1 - q_{[x]+1}) \ \text{ and so on}
\end{aligned}$$

The select life table is usually presented as a two way table, where rows correspond to the age at selection while columns correspond to duration after selection. The basic quantities reported are either $q_{[x]+t}$ or $l_{[x]+t}$ for various x values and for $t = 0, 1, 2, \cdots$. A typical select life table is as shown in Table 4.14.

The first row specifies the probability of death in a unit interval when the individual is a member of a select group at age 25. Thus, $q_{[25]+1}$ is a chance of death in age group $(26, 27)$

Age at selection	Year following selection					
x	0	1	2	3	$\cdots$	$\cdots$
25	$q_{[25]}$	$q_{[25]+1}$	$\cdots$	$\cdots$	$\cdots$	$\cdots$
26	$q_{[26]}$	$q_{[26]+1}$	$\cdots$	$\cdots$	$\cdots$	$\cdots$
27	$\cdots$	$\cdots$	$\cdots$	$\cdots$	$\cdots$	$\cdots$
28	$\cdots$	$\cdots$	$\cdots$	$\cdots$	$\cdots$	$\cdots$
$\vdots$	$\cdots$	$\cdots$	$\cdots$	$\cdots$	$\cdots$	$\cdots$

Table 4.14 Select life table

when the individual is assigned to a select group at age 25. On the other hand, $q_{[26]}$ denotes the probability of death in the same age group $(26, 27)$ but when the individual is assigned to a select group at age 26. Consequently, the two probabilities are different.

From the quantities $q_{[x]+1}$ various other life table functions can be calculated. For example, if the selection is at age x, the conditional probability of death between ages $x + 7$ and $x + 10$ for a person known to be alive at age $x + 5$ is given by

$$2|3 q_{[x]+5} = (l_{[x]+7} - l_{[x]+10})/l_{[x]+5}$$

The impact of selection on the distribution of time-until death goes on decreasing after some years following selection. For example, consider two persons, one selected at age 20 and now of age 30 and the other selected at age 21 and now of age 30. Their respective conditional probabilities of survival for another n years are given by

$$n p_{[20]+10} = l_{[20]+10+n}/l_{[20]+10} \quad \& \quad n p_{[21]+9} = l_{[21]+9+n}/l_{[21]+9}$$

It is generally observed that the selection does not have much effect after 9 or 10 years and both the probabilities may be the same as l_{30+n}/l_{30} and functions of the attained age. If there exists a smallest integer s such that $q_{[x]+t} = q_{x+t}$ for all $t \geq s$ or for which $q_{[x]+s}$ is a function of the attained age, then such an integer s is known as 'select period'. The Society of Actuaries usually takes $s = 15$ years as a period of selection. The $(s + 1)^{\text{th}}$ column of the two dimensional array of the select life table with period s, will be almost the same as the quantities q_{x+s} of the aggregate life table. Eventually, the difference becomes so small that it is not worthwhile, for practical purposes, to distinguish the 'select-group' from the 'non-select group'. Consequently the select life table can be truncated at $(s + 1)^{\text{th}}$ column. The period where the impact of selection is negligible is called the 'ultimate period'. Hence, this last column is referred to as an ultimate life table and the entire table is referred to as select and ultimate life table. In general, the select life table corresponds to the first s columns and the ultimate table corresponds to the $(s + 1)^{\text{th}}$ column. The select and ultimate life table is constructed from ultimate table towards left. The construction is discussed below.

Construction of Select and Ultimate Life Table: To represent a select and ultimate life table, we use a two dimensional array. Each row represents a different age at selection and each column represents the duration since the selection. We illustrate the procedure for a select period of 2 years. The building blocks of the table are the mortality rates given by

$q_y, y = 0, 1, 2, \cdots$, and $q_{[x]+t}$, $x = 0, 1, 2, \cdots$, $t = 0, 1$. Suppose l_0 is fixed, then from given q_y's one can find $l_1, l_2, l_3, \cdots$. Since the select period is two years, $l_{[x]+2} = l_{x+2}$, $\forall \ x \geq 0$. In particular, let $x = 0$. Therefore, $l_2 = l_{[0]+2} = (l_{[0]+1})(1 - q_{[0]+1})$. The value of $q_{[0]+1}$ is given, so that

$$l_{[0]+1} = l_2/(1 - q_{[0]+1})$$

Further, $l_{[0]+1} = l_{[0]}(1 - q_{[0]})$, $\Rightarrow$ $l_{[0]} = l_{[0]+1}/(1 - q_{[0]})$

Suppose $x = 1$ in $l_{[x]+2} = l_{x+2}$. Hence,

$$l_3 = l_{[1]+2} = l_{[1]+1}(1 - q_{[1]+1}) \ \Rightarrow \ l_{[1]+1} = l_3/(1 - q_{[1]+1})$$

Similarly,

$$l_{[1]+1} = l_{[1]}(1 - q_{[1]}) \ \Rightarrow \ l_{[1]} = l_{[1]+1}/(1 - q_{[1]})$$

Proceeding on these lines all the quantities in the select life table can be obtained.

In general, if s is the select period, and $q_{[x]+t}$ are given for $t = 0, 1, 2, \cdots$, $s - 1$, we get

$$l_{[x]+s-1} \ = \ l_{[x]+s}/(1 - q_{[x]+s-1}) \ \& \ l_{[x]+s-2} = l_{[x]+s-1}/(1 - q_{[x]+s-2})$$

The following illustrative examples will help to fix the ideas developed in this section.

Example 4.6.1. A life table has a select period of three years. (i) Find expressions in terms of life table functions $l_{[x]+t}$ and l_y for $q_{[50]}$, $_2p_{[50]}$, $_{2|}q_{[50]}$ and $_{2|3}q_{[50]+1}$.

Solution: From the formulae derived above, we get,

$$q_{[50]} \ = \ 1 - l_{[50]+1}/l_{[50]} \ \& \ _2p_{[50]} = l_{[50]+2}/l_{[50]}$$
$$_{2|}q_{[50]} \ = \ (l_{[50]+2} - l_{[50]+3})/l_{[50]} = (l_{[50]+2} - l_{53})/l_{[50]}$$
$$_{2|3}\,q_{[50]+1} \ = \ (l_{[50]+3} - l_{[50]+6})/l_{[50]+1} = (l_{53} - l_{56})/l_{[50]+1}$$

∎

Example 4.6.2. For a 3-year select life table, calculate $_3p_{53}$ given that $q_{[50]} = 0.01601$, $_2q_{[50]} = 0.096411$, $_{2|}q_{[50]} = 0.02410$ and $_{2|3}q_{[50]+1} = 0.09272$.

Solution: Suppose $l_{[50]} = 1,00,000$. Observe that,

$$l_{[50]} \ = \ 1,00,000 \ \Rightarrow \ l_{[50]+1} = l_{[50]}(1 - q_{[50]}) = 98399$$
$$_2q_{[50]} \ = \ 1 - l_{[50]+2}/l_{[50]} \ \Rightarrow \ l_{[50]+2} = 96411$$
$$_{2|}q_{[50]} \ = \ (l_{[50]+2} - l_{53})/l_{[50]} \ \Rightarrow \ l_{53} = 94007$$
$$_{2|3}q_{[50]+1} \ = \ (l_{53} - l_{56})/l_{[50]+1} \ \Rightarrow \ l_{56} = 84,877$$
$$\Rightarrow \ _3p_{53} \ = \ l_{56}/l_{53} = 0.90294$$

∎

Example 4.6.3. Table 4.15 presents a two year select and ultimate life table. On the basis of the given information, which of the following are true?

(i) $_2p_{[31]} > {}_2p_{[30]+1}$ (ii) $_{1|}q_{[31]} > {}_{1|}q_{[30]+1}$ (iii) $_2q_{[33]} > {}_2q_{[31]+2}$

$[x]$	$l_{[x]}$	$l_{[x]+1}$	$l_{[x]+2}$	$x+2$
30	1000	998	995	32
31	996	994	988	33
32	994	990	982	34
33	987	983	970	35

Table 4.15 Two year select and ultimate life table

Solution:

$$
\begin{aligned}
(i)\,{}_2p_{[31]} &= l_{[31]+2}/l_{[31]} = 988/996 \\
{}_2p_{[30]+1} &= l_{[30]+3}/l_{[30]+1} = l_{33}/l_{[30]+1} = 988/998 < 988/996 \\
\Rightarrow &\quad \text{(i) is true} \\
(ii)\,{}_{1|}q_{[31]} &= (l_{[31]+1} - l_{[31]+2})/l_{[31]} \\
&= (l_{[31]+1} - l_{33})/l_{[31]} = (994 - 988)/996 = 0.006 \\
{}_{1|}q_{[30]+1} &= (l_{[30]+2} - l_{[30]+3})/l_{[30]+1} = (l_{32} - l_{33})/l_{[30]+1} \\
&= (995 - 988)/998 = 0.007 \quad \Rightarrow \quad \text{(ii) is false} \\
(iii)\,{}_2q_{[33]} &= (l_{[33]} - l_{[33]+2})/l_{[33]} = (l_{[33]} - l_{35})/l_{[33]} \\
&= (987 - 970)/987 = 0.0172 \\
{}_2q_{[31]+2} &= {}_2q_{33} = (l_{33} - l_{35})/l_{33} = (988 - 970)/988 = 0.0182 \\
\Rightarrow &\quad \text{(iii) is false}
\end{aligned}
$$

$\blacksquare$

Example 4.6.4. For a ten year select and ultimate life table, calculate $e^0_{[30]}$, given the following information. (i) $l_{[30]+t} = (\sqrt{60}/9)(1 - t/100), \quad 0 \le t < 10$ and
(ii) $l_{30+t} = \sqrt{70 - t}/10, \quad 10 \le t \le 70$.

Solution: Observe that

$$
\begin{aligned}
e^0_{[30]} &= \int_0^{70} {}_tp_{[30]}\,dt = \int_0^{70} \frac{l_{[30]+t}}{l_{[30]}}\,dt = \int_0^{10} \frac{l_{[30]+t}}{l_{[30]}}\,dt + \int_{10}^{70} \frac{l_{30+t}}{l_{[30]}}\,dt \\
&= \int_0^{10} \frac{\frac{\sqrt{60}}{9}\left(1 - \frac{t}{100}\right)}{\frac{\sqrt{60}}{9}}\,dt + \int_{10}^{70} \frac{\frac{\sqrt{70-t}}{10}}{\frac{\sqrt{60}}{9}}\,dt = 9.5 + 36 = 45.5
\end{aligned}
$$

$\blacksquare$

Example 4.6.5. Express the conditional probabilities (i) ${}_{2|4}q_{[20]+1}$ and
(ii) ${}_{2|4}q_{[22]+3}$ in terms of l_x, assuming a 4-year select period.

Solution: (i) By definition,

$$
\begin{aligned}
{}_{2|4}q_{[20]+1} &= (l_{[20]+1+2} - l_{[20]+1+2+4})/l_{[20]+1} = (l_{[20]+3} - l_{[20]+7})/l_{[20]+1} \\
&= (l_{[20]+3} - l_{27})/l_{[20]+1}
\end{aligned}
$$

since the select period is four years and hence $l_{[20]+7} = l_{27}$.

(ii) $_{2|4}q_{[22]+3} = (l_{[22]+3+2} - l_{[22]+3+2+4})/l_{[22]+3} = (l_{27} - l_{31})/(l_{[22]+3})$ ■

Example 4.6.6. A select and ultimate life table with a three year select period begins at selection age 20. Find the radix $l_{[20]}$, given the following information: $l_{26} = 90,000$, $q_{[20]} = 1/6$, $_5p_{[21]} = 4/5$ and $_3p_{[20]+1} = (9/10)_3p_{[21]}$.

Solution: Observe that

$$
\begin{aligned}
4/5 &= \;_5p_{[21]} = l_{26}/l_{[21]} \;\Rightarrow\; l_{[21]} = 1,12,500 \\
_3p_{[20]+1} &= (9/10)_3p_{[21]} \;\Rightarrow\; l_{24}/l_{[20]+1} = (9/10)\,l_{24}/l_{[21]} \\
\Rightarrow\; l_{[20]+1} &= (9/10)\,l_{[21]} = 1,25,000 \\
1/6 &= q_{[20]} \;\Rightarrow\; 5/6 = p_{[20]} = l_{[20]+1}/l_{[20]} \\
\Rightarrow\; l_{[20]} &= (6/5)l_{[20]+1} = 1,50,000
\end{aligned}
$$

■

Example 4.6.7. Construct columns corresponding to $l_{[x]}$ and $l_{[x]+1}$ for years 30 to 34 of a select and ultimate life table with a select period of 2 years. Use the information given in Table 4.16.

x	$1000\ q_{[x]}$	$1000\ q_{[x]+1}$	$1000\ q_{x+2}$	l_{x+2}
30	0.222	0.330	0.422	9901
31	0.234	0.352	0.459	9897
32	0.250	0.377	0.500	9892
33	0.269	0.407	0.545	9887
34	0.291	0.441	0.596	9882

Table 4.16 Select and ultimate life table

Solution: We have $l_{[x]}p_{[x]} = l_{[x]+1}$ and $l_{[x]+1}p_{[x]+1} = l_{[x]+2} = l_{x+2}$. Hence,

$$
l_{[30]+1} = l_{32}\left\{1 - q_{[30]+1}\right\}^{-1} = 9904 \quad \text{and} \quad l_{[30]} = l_{[30]+1}\left\{1 - q_{[30]}\right\}^{-1} = 9906
$$

Continuing in this way, we get Table 4.17.

The next section presents R codes used to solve the examples in this chapter.

4.7   R Codes

Code 4.7.1. We use this code to compute the probability distribution of the curtate future lifetime random variable for the three mortality laws. The first part of the code is illustrated in Example 4.3.3 for a law labeled as 'D'. The second part of the code is illustrated in Example 4.3.4 for Gompertz' law and in Example 4.3.5 for Makeham's law. The third part of the code is used to draw Figure 4.4.

x	$l_{[x]}$	$l_{[x]+1}$
30	9906	9904
31	9902	9900
32	9897	9895
33	9893	9891
34	9889	9886

Table 4.17 Values of $l_{[x]}$

```
# Part I: Discrete law
k=0:4; pr1=exp(-0.12*k)-exp(-0.12*(k+1))
k=5:69; pr2=exp(-0.18*k+0.3)-exp(-0.18*k+0.12)
dpr=c(pr1,pr2); k=0:69; dpr1=round(data.frame(k,dpr),5); dpr1
# Part II: Gompertz' law and Makeham's law
A=0.0007; B=0.0001151; C=1.096; m=B/log(C); k=0:69; x=30
gpr=exp(-m* C^x*(C^k-1))-exp(- m*C^x*(C^(k+1)-1))
mpr=exp(-A*k-m*C^x*(C^k-1))-exp(-A*(k+1)-m*C^x*(C^(k+1)-1))
g=round(data.frame(k,gpr),5); mk=round(data.frame(k,mpr),5)
g; mk
# Part III: Graphs of probability mass functions
plot(k,dpr,"h",xlab="k", ylab="P[K(30)=k]",col="dark blue",lty=1)
points(k,dpr,pch=16,cex=.5,col=" blue")
lines(k,gpr,"h",xlab="k",ylab="P[K(30)=k]",lty=2,col="dark red")
points(k,gpr,pch=16,cex=.5,col="red")
lines(k,mpr,"h",xlab="k",ylab="P[K(30)=k]",lty=3, col="dark green")
points(k,mpr,pch=16,cex=.5,col="green")
legend("topright",legend=c("D","G","M"),lty=c(1,2,3),
col=c("dark blue","dark red","dark green"),cex=.7)
```

∎

Code 4.7.2. We use this code to compute the probability distribution of the curtate future lifetime random variable $K(25)$, when values of q_x are given for $x = 0$ to w^*. The code is illustrated in Example 4.3.6 for q_x values as specified in Table 4.8.

```
z=read.table("F://qx.txt", header=T)
x=z[,1] # column of x values from 0 to 110
q=z[,2] # column of qx values
p=1-q # column of px values
p=p[26:111] # column 'of px values for x=25 to 110
p1=c(p[1],2:85) # dummy vector to store values of kpx for x=25
               # and k=1 to 85, first element being p25
```

```
for (i in 2:85)
{
p1[i]=p1[i-1]*p[i]
}
q=1-p # column of qx values for x=25 to 110
p3=c(q[1],2:84)
{
for(i in 2:85)
p3[i]=p1[i-1]*q[i]
}
p4=sum(p3); p4
# total probability, which is approximately 1(0.9999128)
k=0:84; e25=sum(k*p3);e25;p3=round(p3,5); y1=data.frame(k,p3)
y1 # distribution of K(25) displayed in Table 4.4
plot(k,p3,"h",xlab="k",ylab="P[K(25)=k]",col="blue")
points(k,p3,,pch=16,cex=.5,col="dark blue")
```

■

As discussed in section 4, the mortality pattern is usually available in terms of age specific death rates for a certain period for a certain region, from which we can find q_x values for ages x from 0 to some limiting age. We give below a set of R functions to construct a life table studied in Section 4, corresponding to a given set of q_x values.

Code 4.7.3. In this code we use the set of formulae given in Eq. 4.4.1, to write a set of R functions to construct a life table corresponding to a given set of q_x values. We use the mortality pattern as given by q_x values reported in Siegel and David [18]. Values of q_x for $x = 0$ to 110 are stored in a text file, **qx.txt**, on the local disk F. To import this file in R, we use the function **read.table**. In **qx.txt** file, the first column is labelled as x and the second column is labelled as qx. Hence, the argument **header=TRUE** is used. This code produces the life table as shown in Table 4.8.

```
z=read.table("F://qx.txt", header=TRUE)
x=z[,1]; q=z[,2]; p=1-q; w=length(p); l1=10^5;l=c(l1,2:w)
for (i in 2:w)
{
l[i]=l[i-1]*p[i-1]
}
de=l*q;L=c(1:(w-1),.5*l[w])
for (i in 1:(w-1))
{
L[i]=.5*(l[i]+l[i+1])
}
```

```
T=c(1:(w-1),L[w])
for (i in 1:(w-1))
{
T[w-i]=T[w-i+1]+L[w-i]
}
ex=T/l; y=round(data.frame(l,de,L,T,ex),2)
y1=data.frame(x,q,p,y); y1 # Table 4.8
```

■

In the next chapter we discuss various insurance models wherein the theory developed in this chapter serves as a mathematical foundation.

4.8   Conceptual Exercises

4.8.1 Check whether each of the following functions can serve as a survival function. If yes, find A and the corresponding $\mu(x)$, $f_X(x)$, and $F_X(x)$.

(a) $S(x) = Ax/(1 + Ax)$, $x \geq 0$ (b) $S(x) = 1/(1 + Ax)$, $x \geq 0$

4.8.2 Find a set of values of n for which $\mu(x) = (1+x)^{-n}$, $x \geq 0$ is a force of mortality. Justify your answer. If such a set is non-empty, then for those values of n find corresponding $S(x)$ and $f_X(x)$.

4.8.3 Examine whether $S(x) = x^2 - 2.5x + 1$, $0 \leq x \leq 2$ is a survival function. Justify your answer.

4.8.4 If $S(x) = \exp\{-0.00133((1.07152)^x - 1))\}$, $0 \leq x \leq 100$, find the following probabilities. (a) (40) will die in the next decade. (b) (30) will die in his 60th year of age. (c) (40) will die between his 50th and 60th birthday. (d) (40) will die either between $(45, 50)$ or between $(60, 70)$.

4.8.5 If $\mu(x) = 0.02$, for $40 \leq x \leq 47$, evaluate (a) $_2p_{43}$, (b) $_4q_{41}$ and (c) $_{3|4}q_{40}$.

4.8.6 The changed force of mortality is given by $\hat{\mu}_{x+t} = \mu_{x+t} + c$, $0 \leq t \leq 1$. $\hat{q}_x = 0.02$, where $\hat{q}_x$ is based on the force of mortality $\hat{\mu}_{x+t}$ and $q_x = .01$ where q_x is based on the force of mortality μ_{x+t}. Find c.

4.8.7 It is given that $\mu_{x+t} = 0.03x + 0.001t$, where x is an integer and t is a fraction. Calculate q_{50}.

4.8.8 Calculate the following probabilities on the basis of the life table given in Table 4.8. (a) $_{40}p_{25}$, (b) $_{60|10}q_{25}$, (c) the probability that (30) survives for at least 10 years, (d) the probability that (40) survives to age 65, (e) the probability that (50) dies within 10 years, (f) the probability that (50) fails to reach age 70, (g) the probability that (60) dies between ages 80 and 85, (h) the probability that (60) dies within the first five years after retiring at age 65.

4.8.9 Find the probability that a life aged 30 will (a) survive to age 40, (b) die before reaching age 50, (c) die in his 50th year of age, i.e. between ages 49 and 50, (d) die between his 40th birthday and 50th birthday, (e) die either between exact ages 35 and 45 or between exact ages 70 and 80. Use the life table given in Table 4.8.

4.8.10 For the first 5 years after arrival in a certain country, lives are subject to a constant force of mortality of 0.005. Thereafter, lives are subject to mortality according to Table 4.8. (a) A life aged exactly 30 has just arrived in the country. (i) Find the probability that the life will survive to age 35. (ii) Find the probability that the life will survive to age 60.
(b) What is the probability that a life aged exactly 33 who has been in the country for 3 years will die between ages 50 and 51? (Assume that these lives will remain in the given country.)

4.8.11 For a certain animal population, $l_x = l_0/(1+x)^2$, $(x \geq 0)$. Calculate
(a) the complete expectation of life at birth, (b) the force of mortality at age 1 year and
(c) the chance that a newly-born animal will die between ages 1 and 2 years.

4.8.12 Find l_x and d_x columns of a life table from age 30 to 35 given the following information: The probability at birth of living to age 30 is 0.75,
$q_{30} = 0.01$, $q_{31} = 0.01$, $q_{32} = 0.02$, $q_{33} = 0.03$, $q_{34} = 0.04$, $q_{35} = 0.05$ and $l_0 = 1,00,000$.

4.8.13 Give answers in terms of life table functions for each of the following questions:

(a) What is the proportion of men aged 31, expected to live to age 35?

(b) What proportion of males now aged 32, will die while aged 34 last birthday?

(c) What is the expected number of males who might be expected to die between age 31 and 35 out of 3000 males now aged 30?

(d) What is the chance that a female aged 31 and a male aged 33, will both die within 20 years?

(e) What is the chance that a male child just born to a mother of age 31 and father of age 33, will be alive 20 years later, but orphaned by both his parents?

(f) What is the percentage of population between 0 and 15 years?

(g) What is the percentage of the total population above 65 years?

(h) What is the expected age at death of a group of persons now aged 20?

4.8.14 If $\mu_x = 2/(x+1) + 2/(100-x)$, $0 \leq x < 100$, find the expected number of deaths which occur between ages 1 and 4 in a life table with a radix of $10,000$.

4.8.15 If $l_x = 2500(64 - 0.8x)^{1/3}$, $0 \leq x \leq 80$, find the probability density function of life length random variable X.

4.8.16 If $l_x = 1000\sqrt{100-x}$, $0 \leq x \leq 100$, calculate exact value of $\mu_{36+(1/4)}$ and also find it under the assumption of uniformity of deaths in a unit interval. Compare the two values.

4.8.17 Which of the following relationships are correct under the assumption of uniformity of deaths in a unit interval. (a) $_{.5}q_x < {}_{0.5}q_{x+0.5}$
(b) $_s q_x = {}_{1-s}p_x \, {}_s q_{x+1-s}$, $0 < s < 1$ (c) $\mu_{x+s} > {}_s q_x$

4.8.18 Given the following information, find the median of future lifetime for a person of age 50, under the assumption of uniformity of deaths in a unit interval, $l_{50} = 80,000$, $l_{74} = 42,693$, $l_{75} = 40,280$ and $l_{76} = 37,480$.

4.8.19 If $q_{55} = 0.2$ and $q_{56} = 0.3$, calculate the probability that (55) will die between ages 55.5 and 56.5 under the assumption of (a) uniformity and (b) Balducci within each year of age.

4.8.20 Under the uniform distribution of deaths assumption, calculate m_x when it is given that $\mu_{x+.75} = 0.04$.

4.8.21 It is given that $\mu_{x+t} = 0.06 + 0.001722t$ and $\mu_{y+t} = 0.05 + 0.001244t$. Further, $\mu_{x+t}/\mu_{y+t} = 1.201822$. Calculate p_y under the assumption of uniformity of deaths in a unit interval.

4.8.22 A group of individuals of age 25 is subject to a constant force 0.005 of mortality for first 10 years. For next 20 years the constant force of mortality is 0.008 and for the remaining life it is 0.0085. Find the probability that (25) from this group (a) survives to age 40 and (b) dies in the 70^{the} year of age.

4.8.23 It is given that $\mu_{80.5} = 0.2020$, $\mu_{81.5} = 0.0408$, $\mu_{82.5} = 0.0619$. Under the uniform distribution of deaths assumption, find the probability that (80.5) will die within two years.

4.8.24 If $\mu_{x+t} = k$, $0 \leq t \leq 1$, $_{1/4}q_x = (0.6)\,_{1/2}q_x$, evaluate $_{3/4}q_{x+1/8}$.

4.8.25 If $\log(_{1-t}p_{x+t}) = 0.004(t-1), 0 \leq t \leq 1$, evaluate (a) q_x, (b) $_{1/4}q_{x+3/4}$ (c) and $_{1/8}q_{x+1/8}$.

4.8.26 If μ_{x+t} is linear, $0 \leq t \leq 1$, $q_x = 1 - e^{-0.5}$ and $_{1/3}p_x = (_{2/3}p_x)^{1/6}$, find $\mu_{x+1/3}$.

4.8.27 If $_{3/4}q_x = 0.3$ and $_{3/4-t}q_{x+t}$ is linear for $0 \leq t \leq 3/4$, find $_{1/4}q_x$.

4.8.28 If μ_{x+t} is constant and $_{1/3}p_{x+1/3} = 0.9$, find (a) $_{2/3}p_{x+1/8}$, (b) $_{7/8}q_{x+7/8}$ and (c) $_{1/4}p_x$.

4.8.29 There are 89509 persons now aged 50, of whom 530 are expected to die before age 51. How many will be expected to die between $50 + 3/8$ and $50 + 2/3$ if (a) $\mu_{x+t} = q_x/(1 - tq_x)$ and (b) $\mu_{x+t} = 0.004, 0 < t < 1$?

4.8.30 It is given that $_{0.25}p_x\,_{0.5}q_{x+3/8} = 0.0018$ and mortality follows the uniformity assumption in the unit interval. Calculate $\mu_{x+0.25}$.

4.8.31 Express the following probabilities in terms of select life table functions. The probability that an individual of age 35 and just selected will
(a) survive at least 10 years and (b) die between ages 45 and 50.

4.8.32 Given $l_{[30]+2} = 900$, $q_{[30]+2} = 0.006$, $q_{[30]+3} = 0.007$ and $_2q_{[30]} = 0.005$ evaluate
(a) $_2q_{[30]+2}$ and (b) $_{3|}q_{[30]}$.

4.8.33 For a 2-year select and ultimate mortality table, find $l_{[28]}$, given the following information. (i) $q_{28} = 0.1$, $\quad q_{29} = 0.15$, $\quad q_{30} = 0.2$, $\quad q_{31} = 0.25$, (ii) $q_{[x]} = 0.4\, q_x$ $\;\forall\; x$, (iii) $q_{[x]+1} = 0.3\, q_{x+1}$ $\;\forall\;$ x and (iv) $l_{[30]} = 1000$.

4.8.34 For a 2 year select and ultimate life table, you are given the information in Table 4.18. Which of the following are true?

$[x]$	$l_{[x]}$	$l_{[x]+1}$	l_{x+2}	$x+2$
40	1000	998	996	42
41	997	995	989	43
42	995	990	982	44
43	987	982	971	45

Table 4.18 Two year select and ultimate life table

(i) $_2p_{[41]} >\ _2p_{[40]+1}$, (ii) $_{1|}q_{[41]} >\ _{1|}q_{[40]+1}$ and (iii) $_2q_{[43]} >\ _2q_{[41]+2}$.

4.8.35 Suppose 1000 individuals were selected for life insurance at age 35. If $q_{[35]+k} = (1.1)^k/50$ for $k = 0, 1, 2, \cdots$, calculate the number of deaths between the ages 35 and 37.

4.8.36 For a 2 year select and ultimate life table, you are given the following information. (i) $q_{96} = 0.36$, $q_{97} = 0.465$, $q_{98} = 0.668$. (ii) $q_{[x]} = 0.5q_x$ $\forall\; x$, (iii) $q_{[x]+1} = 0.5q_{x+1}$ $\;\forall\;$ x and (iv) $l_{[96]} = 20000$. Find $l_{[97]}$.

4.8.37 You are given the extract from a 2 year select and ultimate life table as shown in Table 4.19. Complete Table 4.20.

$[x]$	$q_{[x]}$	$q_{[x]+1}$	q_{x+2}	$x+2$
45	0.0092	0.0085	0.0073	47
46	0.0084	0.0067	0.0055	48

Table 4.19 Two year select and ultimate life table

$[x]$	$l_{[x]}$	$l_{[x]+1}$	l_{x+2}	$x+2$
45	10000			47
46				48

Table 4.20 Two year select and ultimate life table

4.9   Computational Exercises

4.9.1 (i) For the following four mortality laws, identify a parameter space which is suitable to model the human life length. (a) De Moivre law, (b) Gompertz' law, (c) Makeham's law and (d) Weibull law. (ii) Draw the graphs of survival function and the force of mortality. Comment on the suitability of these laws to model human life length. (iii) Find the probability mass function of $K(35)$ when the mortality law is Gompertz' law and Makeham's law. Take the parameters that you have fixed in (i). (iv) Draw their graphs. (iii) Find the mean and mode of the distribution of $K(35)$ under both the laws.

4.9.2 Construct a life table with eight columns corresponding to Gompertz' mortality law; with suitable values for the parameters. Draw the survival curve, curve of d_x and curve of force of mortality.

4.9.3 Construct a life table with eight columns corresponding to Makehams' mortality law; with suitable values for the parameters. Draw the survival curve, curve of d_x and curve of force of mortality.

4.9.4 Table 4.21 and Table 4.22 display the values of q_x for $x = 0$ to $x = 100$, based on the mortality experience of USA male population and female population respectively, for the year 2020. Source of both the tables is 'National Center for Health Statistics, National Vital Statistics System, Mortality. USA National Vital Statistics Reports, Vol. 71, No. 1, August 8, 2022'. Construct life tables depicting 8 columns as in Table 4.8. Compare the male and female mortality patterns by drawing the curves of l_x and d_x.

4.9.5 Corresponding to the values of q_x in Table 4.21 and Table 4.22, find the probability mass function of $K(35)$. (ii) Draw its graph. (iii) Find the mean and mode of the distribution.

4.10   Multiple Choice Questions

Note: Unless specified otherwise, you have to identify which of the options is correct. Answers are given in the solutions of conceptual exercises.

4.10.1 The force of mortality μ_x, at age x and the survival function $S(x)$, are related as

(a) $S(x) = \exp(-\int_x^{x+1} \mu_t \, dt)$
(b) $S(x) = \exp(-\int_0^x \mu_t \, dt)$
(c) $S(x) = \exp(-\int_x^\infty \mu_t \, dt)$
(d) $S(x) = \exp(-\mu_x)$

4.10.2 Makeham's law of mortality is defined by the force of mortality $\mu_x = A + BC^x$, $B > 0$, $A \geq -B$, $C > 1$, $x \geq 0$. Suppose $m = B/\log_e C$. Then its survival function is given by,

(a) $S(x) = \exp[-Ax - m(C^x - 1)]$
(b) $S(x) = \exp[-Ax - mC^x]$
(c) $S(x) = \exp[Ax - m(C^x - 1)]$
(d) $S(x) = \exp[Ax - mC^x]$

x	q_x	x	q_x	x	q_x	x	q_x
0	0.005849	26	0.001899	52	0.007004	78	0.053409
1	0.000403	27	0.001966	53	0.007657	79	0.058234
2	0.000259	28	0.002050	54	0.008381	80	0.064014
3	0.000208	29	0.002148	55	0.009115	81	0.070301
4	0.000155	30	0.002251	56	0.009859	82	0.077280
5	0.000144	31	0.002351	57	0.010668	83	0.086551
6	0.000132	32	0.002448	58	0.011568	84	0.095951
7	0.000121	33	0.002539	59	0.012548	85	0.107089
8	0.000109	34	0.002627	60	0.013599	86	0.116675
9	0.000096	35	0.002722	61	0.014668	87	0.130906
10	0.000091	36	0.002827	62	0.015723	88	0.146410
11	0.000107	37	0.002931	63	0.016751	89	0.163192
12	0.000159	38	0.003033	64	0.017793	90	0.181227
13	0.000255	39	0.003140	65	0.018910	91	0.200462
14	0.000385	40	0.003264	66	0.020241	92	0.220810
15	0.000529	41	0.003411	67	0.021617	93	0.242150
16	0.000676	42	0.003580	68	0.023122	94	0.264330
17	0.000831	43	0.003769	69	0.024700	95	0.287167
18	0.000991	44	0.003983	70	0.026327	96	0.310455
19	0.001152	45	0.004231	71	0.028145	97	0.333969
20	0.001320	46	0.004515	72	0.030318	98	0.357477
21	0.001483	47	0.004831	73	0.032487	99	0.380747
22	0.001620	48	0.005181	74	0.036455	≥ 100	1.000000
23	0.001717	49	0.005570	75	0.039507		
24	0.001785	50	0.005985	76	0.043893		
25	0.001840	51	0.006450	77	0.048013		

Table 4.21 Life table of USA male population for year 2020

4.10.3 The distribution function $_tq_x$ of $T(x)$ can be expressed as

(a) $_tq_x \;=\; 1 - \exp(-\int_t^{x+t} \mu_s \, ds)$

(b) $_tq_x \;=\; 1 - \exp(-\int_x^t \mu_s \, ds)$

(c) $_tq_x \;=\; 1 - \exp(-\int_x^{x+t} \mu_s \, ds)$

(d) $_tq_x \;=\; 1 - \exp(-\int_0^{x+t} \mu_s \, ds)$

4.10.4 Suppose the distribution of life length random variable X follows Gompertz' law with the force of mortality $\mu_x = BC^x$, $B > 0$, $C > 1$. Suppose $m = B/log_e C$. Then probability density function of $T(x)$ for $t > 0$ is

(a) $\exp[-mC^x(C^t - 1)] \, BC^t$

(b) $\exp[-mC^x(C^t - 1)] \, BC^{x+t}$

(c) $\exp[-m(C^{x+t} - 1)] \, BC^{x+t}$

(d) $\exp[-mC^t(C^x - 1)] \, BC^{x+t}$

x	q_x	x	q_x	x	q_x	x	q_x
0	0.004918	26	0.000750	52	0.004097	78	0.037086
1	0.000310	27	0.000798	53	0.004475	79	0.041213
2	0.000199	28	0.000853	54	0.004885	80	0.045945
3	0.000161	29	0.000914	55	0.005300	81	0.051104
4	0.000123	30	0.000978	56	0.005725	82	0.057111
5	0.000115	31	0.001044	57	0.006192	83	0.064163
6	0.000103	32	0.001111	58	0.006717	84	0.072353
7	0.000094	33	0.001176	59	0.007296	85	0.081451
8	0.000087	34	0.001243	60	0.007928	86	0.090029
9	0.000089	35	0.001315	61	0.008578	87	0.101896
10	0.000089	36	0.001393	62	0.009217	88	0.115015
11	0.000098	37	0.001471	63	0.009834	89	0.129437
12	0.000117	38	0.001548	64	0.010462	90	0.145190
13	0.000147	39	0.001629	65	0.011129	91	0.162282
14	0.000185	40	0.001720	66	0.011932	92	0.180688
15	0.000230	41	0.001826	67	0.012871	93	0.200353
16	0.000276	42	0.001946	68	0.014000	94	0.221184
17	0.000324	43	0.002079	69	0.015265	95	0.243052
18	0.000371	44	0.002226	70	0.016693	96	0.265792
19	0.000418	45	0.002393	71	0.018272	97	0.289207
20	0.000468	46	0.002579	72	0.020046	98	0.313074
21	0.000519	47	0.002780	73	0.021730	99	0.337151
22	0.000570	48	0.002996	74	0.024519	≥ 100	1.000000
23	0.000617	49	0.003231	75	0.026862		
24	0.000662	50	0.003484	76	0.029942		
25	0.000705	51	0.003768	77	0.033037		

Table 4.22 Life table of USA female population for year 2020

4.10.5 Suppose the survival function of the life length random variable X is
$S(x) = 1 - x^2/100, \ \ 0 \leq x \leq 10$. Then $_{2|2}q_4$ is

(a) 2/3
(b) 3/7
(c) 16/21
(d) 1/3

4.10.6 The following are two statements. (I) $\mu_x = (3.5)^x$ for $x > 0$ is a force of mortality.
(I) $\mu_x = 3(0.5)^x$ for $x > 0$ is a force of mortality.

(a) Both (I) and (II) are false
(b) Both (I) and (II) are true
(c) (I) is true but (II) is false
(d) (I) is false but (II) is true

4.10.7 Suppose $S(x) = 1 - x^2/225$ for $0 \le x \le 15$. Then $P[K(5) = 5]$ is

    (a) $1/9$
    (b) $8/9$
    (c) $21/200$
    (d) $21/225$

4.10.8 Suppose $S(x) = 1 - x^2/225$ for $0 \le x \le 15$. Then e_7^0 is

    (a) $\int_0^7 S(x)\, dx$
    (b) $\int_0^{15} S(x)\, dx$
    (c) $\int_0^8 {}_x p_7\, dx$
    (d) $\int_0^{15} {}_7 p_x\, dx$

4.10.9 Suppose $S(x) = 1 - x^2/225$ for $0 \le x \le 15$. Then e_5^0 is

    (a) $35/6$
    (b) 10
    (c) 5.4
    (d) $230/27$

4.10.10 Suppose $S(x) = 1 - x^2/225$ for $0 \le x \le 15$. Then e_{14} is

    (a) 1
    (b) 0
    (c) 14
    (d) 4

4.10.11 Suppose the limiting age $w = 97.6$. Then the maximum possible value of $K(78.9)$ is

    (a) 17
    (b) 18.7
    (c) 19
    (d) 18

4.10.12 Suppose the distribution function of a life length random variable X is given by $F(x) = x^3/216$, for $0 \le x \le 6$ and $l_0 = 1000$. Then $l_2 - l_5$ is

    (a) 333.33
    (b) 541.67
    (c) 300
    (d) 700

4.10.13 Suppose $e_{50} = 40$ and $\mu_{50+t} = 0.006$ for $0 \le t \le 2$. Then e_{51} is

    (a) $40e^{0.006} + 1$
    (b) $40e^{-0.006} - 1$
    (c) $40e^{0.006} - 1$
    (d) $40e^{-0.006} + 1$

4.10.14 If $\mu(x) = 0.02$, for $40 \leq x \leq 47$, then $_2p_{43}$

    (a) is 0.96
    (b) is 0.04
    (c) is 0.40
    (d) cannot be computed in view of insufficient information

4.10.15 The expected number of males who might be expected to die between age 50 and 55 out of 10000 males now aged 45, in terms of life table functions is given y,

    (a) $10000(l_{50} - l_{55})$
    (b) $10000(l_{55} - l_{50})/l_{45}$
    (c) $10000(l_{50} - l_{55})/l_{45}$
    (d) $l_{50} - l_{55}$

4.10.16 Under the assumption of uniform distribution of deaths in a unit interval, for a fixed integer value of x and for $t \in (0, 1)$

    (a) $\mu_{x+t} = q_x/(1 - tq_x)$
    (b) $\mu_{x+t} = q_x/(1 + tq_x)$
    (c) $\mu_{x+t} = p_x/(1 - tq_x)$
    (d) $\mu_{x+t} = q_x/(1 - tp_x)$

4.10.17 Under the assumption of uniform distribution of deaths in a unit interval, for a fixed integer value of x and for $t \in (0, 1)$ (I) $l_{x+t} = l_x - td_x$ (II) $l_{x+t} = l_x - tl_x q_x$.

    (a) (I) is true but (II) is false
    (b) (I) is false but (II) is true
    (c) Both (I) and (II) are true
    (d) Both (I) and (II) are false

4.10.18 Under the assumption of uniformity,

    (a) $q_x = m_x \big/ (1 - m_x/2)$

    (b) $q_x = m_x \big/ (1 + m_x/2)$

    (c) $q_x = 2m_x \big/ (1 + m_x)$

    (d) $q_x = 2m_x \big/ (1 - m_x)$

4.10.19 Under the Balducci assumption for fractional ages, for $0 \leq t \leq 1$

    (a) $_{1-t}p_{x+t} = (1 - t)p_x$
    (b) $_{1-t}q_{x+t} = (1 - t)q_x$
    (c) $_{t}q_{x+t} = (1 - t)q_x$
    (d) $_{1-t}q_{x+t} = t\,q_x$

4.10.20 There are 1000 persons of age 50, of whom 40 are expected to die before age 51. Expected number of deaths between 50 $\frac{3}{8}$ and 50$\frac{3}{4}$ under the assumption of uniformity of deaths is

   (a) 12
   (b) 30
   (c) 15
   (d) 20

4.10.21 Under the assumption of a constant force of mortality, for $0 \leq t \leq 1$, which of the following is NOT true?

   (a) $\log({}_t p_x)$ is a linear function of t
   (b) $\log({}_t q_x) = 1 - \mu t$
   (c) $S(x+t) = S(x)e^{-\mu t}$
   (d) $\log\ S(x+t) = (1-t)\ \log S(x) + t\ \log S(x+1)$

4.10.22 Under the assumption of a constant force of mortality, for $0 \leq t \leq 1$, $\mu_{x+t} = \mu$, where

   (a) μ is a constant which does not depend x
   (b) $\mu = -\log_e p_x$
   (c) $\mu = \log_e p_x$
   (d) $\mu = -\log_e q_x$

4.10.23 The following are three statements. Under the assumption of a constant force of mortality, if $p_x = 0.0625$, then (I) ${}_{1/4}p_x = 0.5$, (II) ${}_{1/4}p_{x+1/2} = 0.5$ and (III) ${}_{1/4}p_{x+1/3} = 0.5$

   (a) Only (I) is true
   (b) Only (II) is true
   (c) Only (III) is true
   (d) All three are true

4.10.24 The following are three statements. If $p_{65} = 0.81$, then ${}_{1/2}p_{65}$ under the assumption of (I) a constant force of mortality is 0.9, (II) Balducci is 0.895 and (III) uniformity is 0.905

   (a) Only (I) is true
   (b) Only (II) is true
   (c) Only (III) is true
   (d) All three are true

4.10.25 A mortality table has a select period of three years. The expression for ${}_{2|}q_{[50]}$ in terms of life table functions is given by,

   (a) $(l_{[50]+1} - l_{[50]+2})/l_{[50]}$
   (b) $(l_{[50]+2} - l_{52})/l_{[50]}$
   (c) $(l_{[50]+2} - l_{53})/l_{[50]}$
   (d) $(l_{[50]+2} - l_{53})/l_{[52]}$

Chapter 5

Actuarial Present Value of Benefit

Key Terms: Actuarial present value, Benefit function, Compound interest, Deferred insurance, Discount factor, Discount function, Endowment insurance, Force of interest, Net single premium, Term life insurance, Varying benefit insurance, Whole life insurance.

5.1 Introduction

In Section 1.6, we have noted that the fundamental principle in the determination of premiums is the equivalence principle. It is given by,

$$\text{Actuarial present value of outflow} = \text{Actuarial present value of inflow}$$

where outflow and inflow correspond to an insurance company. To decide the actuarial present values, that is, the expected present values, we utilize the distribution theory of the future lifetime random variable $T(x)$ and $K(x)$ of (x), studied in Chapter 4. In the present chapter we use it to decide the left hand side of the equivalence principle for some standard life insurance models, designed to reduce the financial impact of the random event of untimely death.

In life insurance products discussed in this chapter, the benefit amount is fixed in most of the products and the time of benefit payment depends on the time of death of the insured. Hence, the products are developed as functions of the insured's future lifetime random variable, either $T(x)$ or $K(x)$. As a consequence, all the life insurance models are specified by a benefit function b_t and a discount function v_t. In such models, v_t is the discount factor from the time of payment back to the time of policy issue, t is the length of the interval from policy issue to the payment of benefit and that is the only random quantity.

To understand the notion of a discount factor, we study the notion of time value of the money. It is a fundamental concept throughout the business and financial world. It is widely used in actuarial science to obtain the present value of the benefit payment and premiums to be received in future. The aim is to find the money to be invested, to accumulate after a specified time to a given sum, at a specified rate of interest. The process of solving this problem is known as finding the present value of the given sum. In the calculation of present values, money flows are discounted, that is, valued in the current time frame by taking into account the time value of the money.

Actuaries use the concept of time value of the money together with the random mortality laws, in the calculation of the actuarial present values of benefit to be paid. Theory of compound interest is heavily used to find the time value of money. We briefly discuss this concept in the next section. We also study a notion of force of interest, which is analogous to the notion of the force of mortality studied in Chapter 4.

5.2   Compound Interest and Discount Factor

There are many situations in everyday life when we come across the concept of interest. For example, a deposit in a bank account earns interest. When one borrows money from a bank, money has to be returned after some period with interest. Thus, the payment of interest as a reward for the use of capital is an established part of the economic life of any country. Interest rates reflect the economic conditions of a country. The most common schemes of interest are (i) simple and (ii) compound interest. Simple interest is paid on the principal while compound interest is computed on both the principal and the accumulated interest. We discuss below why the compound interest scheme is preferred to the simple interest scheme. Suppose $a(n)$ denotes the accumulated value of 1 rupee at the end of n years. Then the effective rate i_n of interest in the nth year is defined as the actual rate of increase per unit invested during that time. It is given by,

$$i_n = \frac{a(n) - a(n-1)}{a(n-1)} = \frac{I(n)}{a(n-1)}$$

where $I(n)$ denotes the interest in the nth year. With simple interest,
$a(n) = 1 + in$, where i denotes the rate of interest per rupee per annum. Thus, with simple interest, effective rate i_n of interest in the nth year is given by,

$$i_n = \frac{1 + in - 1 - i(n-1)}{1 + i(n-1)} = \frac{i}{1 + i(n-1)}$$

It is clear that i_n is a decreasing function of n, that is, as the number of years of investment increases, the effective rate of interest decreases. This is certainly not desirable. Consequently, the simple interest scheme is rarely used in practice. With compound interest,

$$a(n) = (1+i)^n \Rightarrow i_n = \frac{(1+i)^n - (1+i)^{n-1}}{(1+i)^{n-1}}$$

$$\Rightarrow i_n = \frac{(1+i)^{n-1}(1+i-1)}{(1+i)^{n-1}} = i, \ \forall \ n \geq 1$$

Thus, with compound interest, the effective rate of interest remains the same in every year. This is definitely a noteworthy feature of the compound interest scheme and that is why the compound interest scheme is always preferred to the simple interest scheme.

With compound interest, the sequence $\{a(n) = (1+i)^n, n \geq 1\}$ is in geometric progression with a common ratio $(1+i)$. Thus, the principal at the end of the jth year is $(1+i)^j$, when it is multiplied by $(1+i)$, we get the principal at the end of the $(j+1)^{th}$ year. In this sense, $(1+i)$ is known as an accumulation factor. Thus, with the compound interest rate i, the accumulated value A, known as amount, of principal P after n years is $A = P(1+i)^n$.

The system of compound interest can be viewed from the angle of discount. To get the amount $1+i$ at the end of one year, one has to invest 1 rupee at the beginning of the year. Thus the discount on $1+i$ is i or discounted value of $1+i$ is 1. So, the discounted value of 1 is $(1+i)^{-1}$. This is known as the present value of 1, due at the end of the year and is denoted by v. Thus $v = (1+i)^{-1}$ and is known as the discount factor. Sometimes it is convenient to work with a rate of discount. Analogous to the effective rate of interest in the nth year, one can define effective rate of discount in the nth year. To get the amount $a(n)$ at the end of the nth year, one has to invest $a(n-1)$ at the beginning of the nth year. Thus the discount is $a(n) - a(n-1)$ on the amount $a(n)$. So the effective rate d_n of discount in the nth year is defined as follows.

$$d_n = \frac{a(n) - a(n-1)}{a(n)} = \frac{(1+i)^n - (1+i)^{n-1}}{(1+i)^n}$$

$$= (1+i)^{-n}(1+i)^{n-1}(1+i-1) = i(1+i)^{-1}$$

Thus $d_n = i(1+i)^{-1}$, for all $n \geq 1$. As it is free from n, we denote it by d. Thus $d = i(1+i)^{-1}$. Since the interest and discount are merely different ways of looking at the same problem, it follows that for every given rate of interest there is a corresponding rate of discount and vice versa. The following is a list of relations among v, d and i.

(i) $v = (1+i)^{-1}$

(ii) $d = i/(1+i) = iv$

(iii) $1 - d = v = (1+i)^{-1}$

(iv) $i = d/(1-d)$

(v) $i = (1-v)/v$

From (ii) we note that $d < v$ and $d < i$. To show that $i < v$, observe that for $i = 0.6$, $i^2 + i = 0.96 < 1$. Thus, for i from 0 to 0.6,

$$i + i^2 < 1 \Rightarrow iv + i^2 v < v$$
$$\Rightarrow d + di < v \Rightarrow (i/(1+i))(1+i) < v \Rightarrow i < v$$

for these values of i. In practice, the rate of interest i ranges from 0.03 to maximum 0.15, so practically $i < v$ always.

Many a times, the interest is paid several times a year. Monthly, quarterly or six monthly interest schemes are common schemes of interest payment. In such cases, it is necessary to find the rate of interest for the fraction of the year. The common financial practice is to express the rate of interest as rate per annum, even though the interest is paid more frequently than once per year. Suppose $i^{(m)}$ denotes the rate of interest when it is paid m times per annum, and i is the effective rate of interest in a year. The rate $i^{(m)}$ is referred to as a nominal rate of interest.

Suppose the interest is paid at the end of each mth part of the year. Then, $i^{(m)}/m$ denotes the rate of interest paid in the mth part. It is to be noted that $i^{(m)}/m$ cannot be i/m as the interest in the previous m^{th} part is added in the principal for the next m^{th} part. Thus, $i^{(m)}$ is slightly less than i. To find its relation with i we proceed as follows. With the interest rate $i^{(m)}/m$, the accumulated value of 1 at the end of m parts, that is one year, is $\left(1 + i^{(m)}/m\right)^{m}$. With an annual rate of interest i, the accumulated value of 1 at the end of one year is $1 + i$. To find relation between i and $i^{(m)}$, observe that

$$\left(1 + i^{(m)}/m\right)^{m} \;=\; 1 + i \;\;\Rightarrow\;\; i^{(m)} = m\left\{(1+i)^{1/m} - 1\right\}$$

Table 5.1 specifies $i^{(2)}$, $i^{(4)}$ and $i^{(12)}$ for various values of i. These are obtained using Part I of Code 5.6.1.

i	$i^{(2)}$	$i^{(4)}$	$i^{(12)}$
0.05	0.0494	0.0491	0.0489
0.06	0.0591	0.0587	0.0584
0.07	0.0688	0.0682	0.0678
0.08	0.0785	0.0777	0.0772
0.09	0.0881	0.0871	0.0865
0.10	0.0976	0.0965	0.0957

Table 5.1 Effective rate i and nominal rate of interest $i^{(m)}$

It is to be noted that $i^{(12)} < i^{(4)} < i^{(2)} < i$, as expected. In Chapter 6, we prove algebraically that $i^{(m)} < i \;\forall\; m$ and $i^{(m)}$ is a decreasing function of m.

Suppose we extend such a division of one year in m parts, for the payment of interest. In the limit, the nominal rate becomes the annual rate of continuous growth and is known as the force of interest or instantaneous rate of interest. We proceed to define the concept of instantaneous rate of interest at time t, in some sense parallel to the concept of force of mortality. Suppose the instantaneous rate of interest at time t is denoted by δ_t. It is defined below. With $a(t) = (1+i)^t$,

$$\delta_t \;=\; \lim_{h\downarrow 0} \frac{a(t+h) - a(t)}{h\, a(t)} = \frac{1}{a(t)} \frac{d}{dt} a(t)$$

$$=\; \frac{1}{a(t)} \frac{d}{dt}(1+i)^t = \frac{1}{a(t)}(1+i)^t \log_e(1+i) = \log_e(1+i),\; \forall\, t$$

Since δ_t is free from t, we denote it by δ. Thus,

$$\delta_t = \delta = \log(1+i) \;\;\Rightarrow\;\; e^{\delta} \;=\; (1+i) \;\text{ or }\; e^{-\delta} = (1+i)^{-1} = v$$

that is, the discount factor and the instantaneous rate of interest are related by the equation $v = e^{-\delta}$. In the next section, we use this relation heavily while finding the moments of the distribution of present value random variable corresponding to benefit payment. Note that,

$$\lim_{m\to\infty} i^{(m)} \;=\; \lim_{m\to\infty} \left\{ \frac{(1+i)^{1/m} - 1}{1/m} \right\} \;=\; \log_e(1+i) \;=\; \delta$$

as is expected. To illustrate the limiting behaviour, in Table 5.2 values of $i^{(50)}$, $i^{(100)}$, $i^{(200)}$ and $i^{(400)}$ are presented corresponding to the values of i used in Table 5.1. These are obtained using Part II of Code 5.6.1. The last column of the table displays values of δ obtained using the formula $\delta = \log_e(1+i)$.

i	$i^{(50)}$	$i^{(100)}$	$i^{(200)}$	$i^{(400)}$	δ
0.05	0.04881	0.04880	0.04880	0.04879	0.04879
0.06	0.05830	0.05829	0.05828	0.05827	0.05827
0.07	0.06770	0.06768	0.06767	0.06766	0.06766
0.08	0.07702	0.07699	0.07698	0.07697	0.07696
0.09	0.08625	0.08621	0.08620	0.08619	0.08618
0.10	0.09540	0.09536	0.09533	0.09532	0.09531

Table 5.2 Force of interest

Note that $i^{(50)} = i^{(100)} = i^{(200)} = i^{(400)}$, up to three decimals. Thus, rounding values of $i^{(400)}$ to 4 decimals, we get $\lim_{m \to \infty} i^{(m)}$ approximately. These values coincide up to four decimal places, with the values of δ displayed in the last column, obtained using the formula $\delta = \log_e(1+i)$. Further, note that $i > \delta$ for various values of i. This inequality can be proved as shown below, using the Taylor series expansion of e^x.

$$\delta = \log(1+i) \quad \Rightarrow \quad i = e^{\delta} - 1 = 1 + \delta + \delta^2/2! + \cdots - 1$$
$$\Rightarrow \quad i = \delta + \delta^2/2 + \cdots + \quad \Rightarrow \quad i > \delta$$

To sum up in financial words, the continuous rate at which a sum of money grows under the operation of interest, is the force of interest, and the increase per unit due to the effect of the operation of this force of interest in any given period, is the effective rate of interest.

Following examples illustrate all the concepts related to interest rate. Unless stated otherwise all the amounts/values are in rupees.

Example 5.2.1. Find the amount of Rs 10,000 after 10 years if the rate of interest is (a) 5% as the force of interest, (b) 5% as the effective rate of interest and (c) 5% per annum payable quarterly.

Solution: (a) $\delta = 0.05 \Rightarrow i = e^{0.05} - 1 = 0.051271$ (slightly higher than δ). Hence the amount of Rs. $10,000$ after 10 years will be

$$10,000(1+i)^{10} = 10,000(e^{0.05})^{10} = 10,000 \, e^{0.5} = 16,487$$

(b) It is given that $i = 0.05$, hence the amount after 10 years will be

$$10^4(1+0.05)^{10} = 16,289$$

(c) We are given that $i^{(4)} = 0.05$. So the amount at the end of 10 years will be

$$10^4 \times (1+0.05/4)^{40} = 16,436$$

Note that accumulation of 10,000 after 10 years is highest with 5% as the force of interest and least with 5% as the effective rate of interest. ∎

Example 5.2.2. Find the amount to which Rs 1,000 will accumulate at
(a) 4% per annum payable quarterly for 10 years, (b) 6% per annum payable half yearly for 5 years, (c) the rate of interest corresponding to an effective rate of discount at 3% per annum for 8 years and (d) 5% effective for 10 years, 4% effective for 5 years and 2.5% effective for 3 years.

Solution: (a) The effective rate of interest is $0.04/4$ per quarter and the money is invested for 40 quarters, hence the answer is

$$1000(1.01)^{40} \; = \; 1488.90$$

(b) As in (a) above, the answer is, $1000(1.03)^{10} \; = \; 1343.90$
(c) We are given that $d \; = \; 0.03$. Hence the effective rate of interest i is obtained using the relation $v = 1 - d$. Hence,

$$1/(1+i) = 1 - 0.03 \;\; \Rightarrow \;\; (1+i) = 1/0.97 \;\; \Rightarrow \;\; 1000\,(1/0.97)^{8} \; = \; 1275.90$$

is the corresponding amount.
(d) After 10 years the amount is $1000(1.05)^{10}$. This accumulates for 5 years at 4% and therefore becomes $1000(1.05)^{10}(1.04)^{5}$. This further accumulates for 3 years at 2.5% and therefore amounts to,

$$1000(1.05)^{10}(1.04)^{5}(1.025)^{3} \; = \; 2134.20$$

∎

Example 5.2.3. In how many years will a sum of money double itself at compound interest with an effective rate $i = 0.05$?

Solution: If the effective rate is $i \; = \; 0.05$, the value of the required number n of years, is given by the equation $(1+i)^{n} = 2$ and hence $n = \log 2 / \log(1.05) \; = \; 14.21$ years. ∎

With the compound interest we have seen that 1 rupee accumulates to $1 + i$ if the effective rate of interest is i per rupee per annum. Thus, the present value of $1 + i$ to be received after 1 year is 1. Further, 1 rupee accumulates to $(1 + i)^{2}$ after two years. Thus, the present value of $(1 + i)^{2}$ to be received after 2 years is 1. Continuing in this manner, the present value of $(1+i)^{n}$ to be received after n years is 1. Using the simple rule of threes, if the present value of $(1+i)^{n}$ to be received after n years is 1, the present value of 1 to be received after n years is $(1+i)^{-n} = v^{n}$. In other words, if v^{n} is deposited today at the effective rate of interest i, then after n years it will accumulate to 1. Thus v^{n} is the discounted value of 1 to be received after n years. v^{n} is known as the present value of 1 to be received after n years and v, as stated earlier, is known as the discount factor. In general, the present value can be obtained from the formula $A = P(1+i)^{n}$, i being the annual interest rate and A is the accumulated value of principal P after n years. From this formula,

$$P = A(1+i)^{-n} \; = \; A\,v^{n} \tag{5.2.1}$$

is the present value of amount A to be received after n years with an effective rate of interest i. In all subsequent sections and chapters, we heavily use the formula

$$\text{Present value of 1 unit to be received after } n \text{ years} \; = \; v^{n} = (1+i)^{-n} = e^{-n\delta}.$$

Table 5.3 displays the present value of 1000 for values of n from 1 to 10 and values of $i = 0.05$ to 0.10, with an increment of 0.01. These are obtained using Part III of Code 5.6.1 which uses the formula $P = A(1+i)^{-n}$, with A = 1000.

n \ i	0.05	0.06	0.07	0.08	0.09	0.10
1	952.38	943.40	934.58	925.93	917.43	909.09
2	907.03	890.00	873.44	857.34	841.68	826.45
3	863.84	839.62	816.30	793.83	772.18	751.31
4	822.70	792.09	762.90	735.03	708.43	683.01
5	783.53	747.26	712.99	680.58	649.93	620.92
6	746.22	704.96	666.34	630.17	596.27	564.47
7	710.68	665.06	622.75	583.49	547.03	513.16
8	676.84	627.41	582.01	540.27	501.87	466.51
9	644.61	591.90	543.93	500.25	460.43	424.10
10	613.91	558.39	508.35	463.19	422.41	385.54

Table 5.3 Present value of 1000 rupees

From Table 5.3, we note that the present value of 1000, to be received after 1 year is 952.38, if the rate of interest is 5% and 909.09, if it is 10%. It is clear that as the rate of interest increases, the present value decreases. It also decreases when n increases. Observe that if 385.54 is deposited today and if the rate of interest is 10%, then in 10 years it will accumulate to 1000. We also note that if the rate of interest is 8%, then in 9 years, the amount deposited will double itself.

Example 5.2.4. How much money should be deposited today so that after 5 years the investor will get 10,000 rupees? Suppose the effective rate of interest is 0.06.

Solution: From the formula $P = A(1+i)^{-n} = A v^n$ we get
$P = 10000(1+0.06)^{-5} = 7472.58$. Thus, 7472.58 invested today will accumulate to 10000 after 5 years. ∎

Example 5.2.5. Suppose a dealer wants to propose a scheme of sale of mobiles costing 60,000 in 10 equal annual instalments. The schedule of instalments is such that the first instalment will be paid at the time of purchase, the second will be paid after 1 year and similarly subsequent 8 instalments will be paid. What will be the instalment if the current rate of interest is 0.06?

Solution: It is obvious that the instalment cannot be 6000, as it is not at all affordable to the dealer. Instalment I has to be such that the total present value of ten instalments should equal 60000. The present value of the first instalment to be paid at the time of delivery of the mobile is I itself. The present value of the second instalment to be paid after 1 year is Iv, the present value of the third instalment to be paid after 2 years is Iv^2. Continuing in this manner, the total present value of ten instalments is $I + I v + I v^2 + \cdots + I v^9$. Equating it to 60000, we get $I = 7690.64$. The actual instalment will be 7690.64 plus some amount to account for administrative expenses. ∎

Remark 5.2.1. A procedure similar to that outlined in Example 5.2.5 can be used to find the EMI (equal monthly instalments) for housing loans. We discuss this in detail in Chapter 6.

Example 5.2.6. Find the present value on 1 April 2022 of the following cash flow: 5000 to be received on 1 April 2023, 7000 to be paid on 1 April 2024, 4000 to be received on 1 April 2025, 3000 to be paid on 1 April 2026 and 6000 to be received on 1 April 2027, when the effective rate of interest is 0.04.

Solution: The present value B on 1 April 2022 of the given cash flow is

$$\begin{aligned} B &= 5000/(1.04) - 7000/(1.04)^2 + 4000/(1.04)^3 - 3000/(1.04)^4 \\ &+ 6000/(1.04)^5 = 4258.93 \end{aligned}$$

■

In financial mathematics, one studies cash flows of payments whose timings and sizes are fixed in advance. Hence, it is simple to compute the present value of the money as noted in the above examples. These methods are not directly applicable if the period of the cash flow is a random variable. For example, when payments depend on death or survival of the concerned individual. Such cases arise in many situations in practice. For example, a retired employee receives regular pension payment until death and the employer has to decide quite in advance how much to pay in the pension fund in order to provide the employee with life pension. In such situations, the exact number of payments cannot be foreseen.

We have noted in Section 4.1 that in all the insurance products, the period for which the policy is in force is a random variable. Periodical premiums highly depend on the benefit amount, the period for which the policy is in force, along with the interest rate. The main consideration in fixing the premium instalments is to decide how much amount has to be collected from the insured which will grow to a sufficient fund to pay the benefit promised in the contract. Further these premium instalments are to be fixed at the time of issue of the policy, that is, at the time of signing the contract. As a consequence, premiums are mainly based on the present value of the benefit. The important factor involved in deciding the present value is the rate of interest. However, complexity increases when the period for interest calculations is not fixed but is a random variable governed by the future life time of the insured. Hence, it is absolutely essential to take into account random mortality pattern along with the interest rate in the calculations of premiums.

Suppose an insurance product is such that the benefit is paid to the beneficiary at the moment of death of the insured. In this setup, in the computation of the present value of the benefit, the continuous random variable $T(x) \equiv T$ studied in Section 4.2 comes handy. In some products, the benefit is paid at the end of the year of death. In such insurance products, the underlying mathematics is based on a curtate future life time random variable $K(x) \equiv K$ studied in Section 4.3.

The next section elaborates on the concept of the actuarial present value of benefit payable at the moment of death, for three standard insurance products.

5.3   Benefit Payable at the Moment of Death

Suppose b_t is the benefit to be paid after t units of time. To begin with, we assume that t is fixed. From Eq. 5.2.1, its present value is given by $b_t v^t$. Suppose Z_t denotes the present value at policy issue, of the benefit payment to be made after t units of time. We define the present value function Z_t by

$$Z_t = b_t v^t \tag{5.3.1}$$

where b_t is the benefit function and v^t is the discount function. Equation 5.3.1 is the basic equation in the calculation of the expected present values of benefit for a variety of insurance products. In any insurance product, the time for which the policy is in force is a random variable. If the benefit is to be paid at the moment of death, then the time from policy issue to the death of the insured is the insured's future lifetime random variable T, defined in Section 4.2. We use the function Z_t to define the present value random variable Z_T. Thus, the present value, at policy issue, of the benefit payment is a random variable Z_T defined as

$$Z_T = b_T v^T = b_T e^{-\delta T} = b_T (1+i)^{-T}$$

Whenever clear from the context, the suffix T in Z_T will be omitted. From the definition of Z_T, it is clear that to study the random variable Z_T, we need information on three aspects— (i) benefit function, depending on the insurance product, (ii) interest rate reflecting the financial status of the country and (iii) distribution of the future life time random variable reflecting the mortality pattern of the population under study. In practice, the distribution of Z_T is always summarized by two main characteristics—mean and variance of the distribution. The expectation $E(Z_T)$ of the present value random variable Z_T is called the actuarial present value. It is also termed as the net single premium, as it is expected that this amount will accumulate over the period T to the required benefit amount at the specific rate of interest. The adjective 'net' in net single premium conveys that it does not take into account the expenses and profit. It is a single premium in contrast to the annual, semi-annual, quarterly, monthly or other types of premiums that are acceptable in the life insurance industry. With this view, the purchase price of the insurance is at least the net single premium. The actuarial present value $E(Z_T)$ of the benefit can be interpreted as the average present value of the cash outflow of the insurance company corresponding to the number of policies issued to many individuals of the same age.

For any portfolio, it is essential to know how much the present value deviates from the expected present value. $\text{Var}(Z_T)$ is used to gain the knowledge about such spread. Hence, for each insurance product we find $E(Z_T)$ and $\text{Var}(Z_T)$. Notation for the net single premium differs from product to product. We use the international actuarial notation to denote the net single premiums.

All life insurance policies pay a benefit if death of the insured occurs while the policy is in force. However, the features of life insurance policies vary, depending on the type of policy. In this chapter, we discuss the following three major types of life insurance policies.

(i) Term life insurance

(ii) Endowment insurance

(iii) Whole life insurance

We also discuss variations of these models when there is a deferment period and also when the benefit amount is the same for the entire period of policy (known as level benefit in the insurance literature) or when it varies with time. The first step is to define the benefit function b_t. It is free from t if it is a level benefit. In the entire book, it is assumed that the present value is based on compound interest at some constant annual effective rate of interest, say i, so that the discount function is taken as $v^t = (1+i)^{-t} = e^{-\delta t}$.

Term Life Insurance: As noted in Section 1.5, the term life insurance provides coverage for a specified period of time, called the policy term. The policy benefit is payable only if the insured makes a claim, that is, dies during the specified term. The length of the term varies considerably from policy to policy. The term may be as short as the time required to complete an airplane trip or as long as 40 years or more. The most common plan of term insurance is level term life insurance, which provides a death benefit that remains the same amount over the term of the policy. For example, under a five-year level term policy that provides Rs 1,00,000 coverage, the insurer agrees to pay Rs 1,00,000 if the insured makes the claim at any time during the five year period. If a claim is not made in 5 years, the insurer has no liability to pay the benefit. The amount of each renewal premium payable for a level term life insurance policy usually remains the same throughout the stated term of coverage. Term life insurance is also known as the temporary insurance and many times is simply referred to as term insurance.

Term life insurance costs less than other types of life insurance for the same amount of coverage. For this reason, many insurance experts recommend term insurance for people on a limited budget or who require coverage for only a short time. Many businesses provide life insurance to their employees through a group life insurance plan. Most such plans offer term insurance. We define below the present value random variable corresponding to level benefit, n-year term life insurance.

n-Year Term Life Insurance: An n-year term life insurance provides benefit only if the insured dies within the n-year term of an insurance commencing at issue. If a unit is payable at the moment of death of (x), then the benefit function is given by,

$$b_t = \begin{cases} 1, & \text{if } t \le n \\ 0, & \text{if } t > n \end{cases}$$

The present value random variable is then defined as,

$$Z_T = \begin{cases} v^T, & \text{if } T \le n \\ 0, & \text{if } T > n \end{cases}$$

$b_t = 1$ means 1 unit, which may be 10,000 or 1 lakh or its multiple. In the theoretical derivations, we take $b_t = 1$ and adopt appropriate modifications in the examples as will be seen in Examples 5.3.1 and 5.3.7. In n-year term life insurance, a claim is made if the insured dies in the specified term. In practice, death is not the only reason to make the claim. An occurrence of a random event leading to financial loss may be taken as sufficient reason to make the claim. However, such events need to be clearly defined and stated in the policy contract. Random variable T in such cases will be the time to occurrence of such events from the issue of the policy. It is necessary to model its distribution. For example, in a foreign tour, loss of passport or baggage, accident leading to serious injury, illness requiring hospitalization, and so on, are some random, unpredictable events resulting in financial loss. It is to be noted that these situations are covered in general or non-life insurance. In the present book we discuss only the life insurance policies and hence the event of interest is taken as the death of the insured.

The net single premium for the n-year term insurance with a unit benefit payable at the moment of death of (x) is $E(Z)$. In international actuarial notation it is denoted by $\overline{A}\,{}^{1}_{x:\overline{n}|}$ and is given by,

$$\overline{A}\,{}^{1}_{x:\overline{n}|} = E(Z) = E(Z_T) = \int_0^{\infty} z_t g(t) dt = \int_0^{n} v^t \, {}_tp_x \mu_{x+t} \, dt \qquad (5.3.2)$$

Here the letter 'A' stands for 'assurance' and the bar above 'A' indicates that the amount is payable at the moment on death of (x). We use the probability density function $g(t)$ of T given in Eq. 4.2.3. If the benefit amount is b, then the net single premium will be $b \times \overline{A}\,{}^{1}_{x:\overline{n}|}$. The second moment of the distribution of Z is given by,

$$E(Z^2) = \int_0^{n} (v^t)^2 \, {}_tp_x \mu_{x+t} \, dt = \int_0^{n} e^{-2\delta t} \, {}_tp_x \mu_{x+t} \, dt$$

The second integral shows that the second moment of Z is equal to the net single premium for an n-year term insurance for a unit benefit payable at the moment of death of (x), calculated at a force of interest equal to two times the given force of interest, that is, 2δ. Thus the variance of Z_T is given by

$$Var(Z) = {}^{2}\overline{A}\,{}^{1}_{x:\overline{n}|} - (\overline{A}\,{}^{1}_{x:\overline{n}|})^2 \qquad (5.3.3)$$

where ${}^{2}\overline{A}\,{}^{1}_{x:\overline{n}|}$ denotes the net single premium for an n-year term insurance for a unit amount, calculated at the force of interest 2δ. In the following example we find $\overline{A}\,{}^{1}_{30:\overline{5}|}$ when the mortality pattern is as specified in Example 4.2.10.

Example 5.3.1. Suppose the life length random variable is modelled by a distribution with force of mortality as specified below.

$$\mu_s = \begin{cases} 0.04, & \text{if} \quad 0 \le s < 15 \\ 0.08, & \text{if} \quad 15 \le s < 25 \\ 0.12, & \text{if} \quad 25 \le s < 35 \\ 0.18, & \text{if} \quad\quad s \ge 35 \end{cases}$$

(i) Find $\overline{A}\,{}^{1}_{30:\overline{5}|}$. Take $\delta = 0.06$. (ii) Find the net single premium for n-year term insurance for benefit of 1000 to be paid at the moment of death, for $n = 1, 2, \cdots, 10$ and for $\delta = 0.05, 0.06, 0.07, 0.08, 0.09, 0.10, 0.11, 0.12$. Comment on the changes in values as n changes and as δ changes.

Solution: As derived in Example 4.2.10, the probability density function $g(t)$ of $T(30)$ is given by,

$$g(t) = \begin{cases} 0.12 \, \exp(-0.12t), & \text{if} \quad 0 < t < 5 \\ 0.18 \, \exp(-0.18t + 0.3), & \text{if} \quad\quad t \ge 5 \end{cases}$$

(i) By definition of $\overline{A}\,^{1}_{x:\overline{n}|}$,

$$\overline{A}\,^{1}_{30:\overline{5}|} = \int_{0}^{.5} v^{t} g(t)\,dt = 0.12 \int_{0}^{.5} \exp(-0.06t - 0.12t)\,dt$$

$$= 0.12(1 - \exp(-0.9))/0.18 = 0.3956$$

Thus, the actuarial present value of 1 unit benefit is 0.3956 in the 5-year term insurance when the force of interest is 0.06. It is interpreted as, if (30) signs a 5-year term insurance contract for a benefit of 1 unit and if he is willing to pay a one time premium at the time of policy issue, then the net single premium is 0.3956. If 1 unit is 10,000 rupees, then the net single premium is Rs 3956.

(ii) To find the net single premium in n-year term insurance for benefit of 1000, for various values of n and δ, we first find the formula for $\overline{A}\,^{1}_{30:\overline{n}|}$ as a function of n and δ. Note that the form of $g(t)$ changes depending on whether $t \leq 5$ or $t > 5$. So formula for $\overline{A}\,^{1}_{30:\overline{n}|}$ also changes accordingly. For $n \leq 5$,

$$\overline{A}\,^{1}_{30:\overline{n}|} = 0.12 \int_{0}^{n} \exp(-\delta t - 0.12t)\,dt = \frac{0.12(1 - \exp(-(\delta + 0.12)n))}{\delta + 0.12}$$

$$\overline{A}\,^{1}_{30:\overline{n}|} = 0.12 \int_{0}^{5} \exp(-\delta t - 0.12t)\,dt$$

$$+ \ 0.18 \int_{5}^{n} \exp(-\delta t + 0.3 - 0.18t)\,dt, \qquad \text{for } n > 5$$

$$= \frac{0.12(1 - \exp(-5\delta - 0.6))}{\delta + 0.12}$$

$$+ \ \frac{0.18(\exp(-5\delta - 0.6) - \exp(-n\delta - 0.18n + 0.3))}{\delta + 0.18}$$

Values of $1000\,\overline{A}\,^{1}_{30:\overline{n}|}$ for various values of n and δ are obtained using Part I of Code 5.6.2. The output is displayed in Table 5.4.

n	0.05	0.06	0.07	0.08	0.09	0.10	0.11	0.12
1	110.35	109.82	109.29	108.76	108.24	107.72	107.20	106.69
2	203.46	201.55	199.67	197.81	195.97	194.16	192.37	190.61
3	282.00	278.17	274.40	270.71	267.09	263.54	260.05	256.62
4	348.27	342.17	336.21	330.40	324.74	319.21	313.82	308.55
5	404.18	395.62	387.32	379.27	371.46	363.89	356.54	349.40
6	472.91	460.68	448.92	437.58	426.67	416.15	406.01	396.24
7	527.51	511.86	496.88	482.54	468.81	455.64	443.03	430.94
8	570.90	552.12	534.24	517.21	500.97	485.50	470.73	456.65
9	605.37	583.79	563.34	543.94	525.53	508.06	491.46	475.69
10	632.76	608.71	586.00	564.55	544.28	525.11	506.97	489.80

Table 5.4 $1000\,\overline{A}\,^{1}_{30:\overline{n}|}$

With $\delta = 0.05$ for a 1-year term insurance, the net single premium or the purchase price is 110.35 for a benefit of 1000 rupees. It is to be noted that as the force of interest increases, the value of the net single premium decreases, as expected. More specifically, if P_1 and P_2 are two values of principal which accumulate to the same amount A in the same period, with interest rate i_1 and i_2 respectively and if $i_2 > i_1$ then $P_2 < P_1$. The decrease in the net single premium with increase in force of interest is however not significant. For example, if the force of interest is doubled, the values of net single premium are not halved. Further as the term of insurance contract increases, the net single premium increases. It is in view of the fact that as the term increases, the chance of claim for benefit increases and to compensate it, the premium amount increases.   ∎

Endowment Insurance: This product provides a specified benefit amount when either the insured lives to the end of the term or dies during that term. Each endowment policy specifies a maturity date, which is the date on which the insurer will pay the benefit amount to the policy owner, if the insured survives to that date. If the insured dies before that date, then the benefit amount is paid to the beneficiary at the moment of death. Thus, an n-year endowment insurance provides the specified benefit to the insured at the end of the n-year term if the insured is alive, or the benefit is paid to the beneficiary if death occurs before the term of n years ends.

In the endowment insurance, premiums are usually level throughout the term of the policy. A policy holder can purchase an endowment policy with a single premium or with a series of premiums over a limited period of time. Endowment insurance is mainly a means of saving money. Policy holders often use endowment policies to finance the education of their children.

If the insurance is for a unit amount and the death benefit is payable at the moment of death, then the present value random variable is given by,

$$Z_T = \begin{cases} v^T, & \text{if} \quad T \leq n \\ v^n, & \text{if} \quad T > n \end{cases}$$

It can be rewritten as,

$$Z_T = \begin{cases} v^T + 0, & \text{if} \quad T \leq n \\ 0 + v^n, & \text{if} \quad T > n \end{cases}$$

Note that distribution of Z_T is a mixture of continuous and discrete distributions, since with probability $_np_x$, it takes a value v^n and with probability $1 - {_np_x}$, it is distributed like v^T. To obtain its expected value and variance we define a random variable Z_T^P as follows.

$$Z_T^P = \begin{cases} 0, & \text{if} \quad T \leq n \\ v^n, & \text{if} \quad T > n \end{cases} \tag{5.3.4}$$

This random variable can be interpreted as the present value random variable corresponding to benefit function b_t defined as follows.

$$b_t = \begin{cases} 0, & \text{if} \quad t \leq n \\ 1, & \text{if} \quad t > n \end{cases}$$

More specifically, a benefit of 1 unit is paid at the end of the n years if and only if the insured survives at least n years from the time of policy issue. No benefit is paid if the insured's death occurs during the term of n years. An insurance product of this type is known as an n-year pure endowment insurance. Due to its feature of paying no benefit if death occurs during the specified term, it is not a popular insurance product. We study some of its properties mainly for theoretical developments in the endowment insurance product.

Note that the n-year pure endowment insurance product is complementary to the n-year term insurance, in the sense that no benefit is paid before the specified term of n years is over, while in the n-year term insurance, the benefit is not paid after n years.

The random variable Z_T^P defined in Eq. 5.3.4 is the present value random variable corresponding to the n-year pure endowment insurance. Its net single premium is denoted by $A_{x:\overline{n}|}^{\ 1}$ or by $_nE_x$ and is given by,

$$A_{x:\overline{n}|}^{\ 1} \quad = \quad _nE_x \quad = \quad E(Z_T^P) \quad = \quad v^n \, _np_x \tag{5.3.5}$$

Variance of Z_T^P is given by,

$$
\begin{aligned}
Var(Z_T^P) \quad &= \quad v^{2n} \, _np_x - (v^n \, _np_x)^2 \quad = \quad v^{2n} \, _np_x \, _nq_x \\
&= \quad ^2A_{x:\overline{n}|}^{\ 1} - (A_{x:\overline{n}|}^{\ 1})^2
\end{aligned}
\tag{5.3.6}
$$

We thus note that the n-year endowment insurance is a combination of the n-year pure endowment insurance and the n-year term insurance. Using the definitions of the present value random variable for the n-year term insurance and n-year pure endowment insurance, it is easy to note that,

$$Z_T(\text{Endowment}) \quad = \quad Z_T(\text{Term}) + Z_T^P(\text{ Pure endowment}) \tag{5.3.7}$$

We use this relation to find the formula for the net single premium for the n-year endowment insurance. It is denoted by $\overline{A}_{x:\overline{n}|}$ and is given by the formula

$$\overline{A}_{x:\overline{n}|} \quad = \quad \overline{A}_{x:\overline{n}|}^{\ 1} + A_{x:\overline{n}|}^{\ 1} \tag{5.3.8}$$

We study the n-year pure endowment essentially to find this formula. $Var(Z_T)$ is also obtained from this relation, which simplifies to,

$$Var(Z_T) \quad = \quad ^2\overline{A}_{x:\overline{n}|} - (\overline{A}_{x:\overline{n}|})^2 \tag{5.3.9}$$

In the following example, we find the net single premium for the n-year endowment insurance for various values of n, and for the mortality and interest pattern as used in Example 5.3.1.

Example 5.3.2. Suppose the life length random variable is modelled by a distribution with force of mortality as specified in Example 5.3.1. Find the net single premium for an n-year endowment insurance purchased by (30), for a benefit of 1000, to be paid at the moment of death, for $n = 1, 2, \cdots, 10$ and for $\delta = 0.05, 0.06, 0.07, 0.08, 0.09, 0.10, 0.11$ and 0.12. Comment on the changes in values as n changes and as δ changes.

Solution: As derived in Eq. 5.3.8 the net single premium $\overline{A}_{30:\overline{n}|}$ is

$$\overline{A}_{30:\overline{n}|} \quad = \quad \overline{A}_{30:\overline{n}|}^{\ 1} + A_{30:\overline{n}|}^{\ 1}$$

In Example 5.3.1 we have obtained $\overline{A}\,{}^{1}_{30:\overline{n}|}$ for the same values of n and δ. We now obtain $A^{1}_{30:\overline{n}|}$ for various values of n and δ. As derived in Example 4.2.10, the probability density function $g(t)$ of $T(30)$ is given by,

$$g(t) \;=\; \begin{cases} 0.12\ \exp(-0.12t)\,, & \text{if}\quad 0 < t < 5 \\ 0.18\ \exp(-0.18t + 0.3)\,, & \text{if}\quad\ \ t \geq 5 \end{cases}$$

To find the net single premium for an n-year pure endowment for benefit of 1000, for various values of n and δ, we first find the formula for $A^{1}_{30:\overline{n}|} = v^{n}\ {}_{n}p_{30}$ as a function of n and δ. Note that the form of $g(t)$ changes depending on whether $t < 5$ or $t \geq 5$. So formulae for ${}_{n}p_{30}$ and $A^{1}_{30:\overline{n}|}$ also change accordingly.

$$\begin{aligned} {}_{n}p_{30} \;&=\; 1 - {}_{n}q_{30} = 1 - \int_{0}^{n} g(t)dt = \exp(-0.12n), \quad \text{for}\quad n < 5 \\[2mm] &=\; \int_{n}^{\infty} g(t)dt = \exp(-0.18n + 0.3), \qquad \text{for}\quad n \geq 5 \end{aligned}$$

Using these formulae for ${}_{n}p_{30}$, we get,

$$\begin{aligned} A^{1}_{30:\overline{n}|} \;&=\; \exp(-\delta n - 0.12n), \qquad \text{for } n < 5 \\[2mm] &=\; \exp(-\delta n - 0.18n + 0.3), \quad \text{for } n \geq 5 \end{aligned}$$

Values of $1000\,A^{1}_{30:\overline{n}|}$ for various values of n and δ are obtained using Part I of Code 5.6.2. The output is displayed in Table 5.5.

n	0.05	0.06	0.07	0.08	0.09	0.10	0.11	0.12
1	843.66	835.27	826.96	818.73	810.58	802.52	794.53	786.63
2	711.77	697.68	683.86	670.32	657.05	644.04	631.28	618.78
3	600.50	582.75	565.53	548.81	532.59	516.85	501.58	486.75
4	506.62	486.75	467.67	449.33	431.71	414.78	398.52	382.89
5	427.41	406.57	386.74	367.88	349.94	332.87	316.64	301.19
6	339.60	319.82	301.19	283.65	267.14	251.58	236.93	223.13
7	269.82	251.58	234.57	218.71	203.93	190.14	177.28	165.30
8	214.38	197.90	182.68	168.64	155.67	143.70	132.66	122.46
9	170.33	155.67	142.27	130.03	118.84	108.61	99.26	90.72
10	135.34	122.46	110.80	100.26	90.72	82.08	74.27	67.21

Table 5.5 $1000\,A^{1}_{30:\overline{n}|}$

Note that as the force of interest increases, values of the net single premium decrease, as expected. However, the decrease in the net single premium with the increase in force of interest is not significant. For example, if the force of interest is doubled, the values of net single premium are not halved. Further as the term of insurance contract increases, the net single premium decreases in contrast to that for the n-year term insurance. It is in view of the fact that as the term increases, the chance of survival in the term decreases and the liability of the company also decreases as the benefit is not to be paid before the term ends.

We now add the values of $\overline{A}\,{}^{1}_{30:\overline{n}|}$ and $A_{30:\frac{1}{n}|}$ to find $\overline{A}_{30:\overline{n}|}$ for various values of n and δ. These are obtained using Part I of Code 5.6.2. The output is displayed in Table 5.6.

n	0.05	0.06	0.07	0.08	0.09	0.10	0.11	0.12
1	954.02	945.09	936.25	927.49	918.82	910.24	901.73	893.32
2	915.23	899.23	883.53	868.13	853.02	838.20	823.66	809.39
3	882.50	860.92	839.93	819.52	799.68	780.39	761.62	743.37
4	854.89	828.92	803.88	779.73	756.45	733.99	712.34	691.44
5	831.59	802.19	774.06	747.15	721.40	696.76	673.17	650.59
6	812.50	780.50	750.11	721.24	693.80	667.73	642.94	619.37
7	797.33	763.44	731.45	701.25	672.73	645.78	620.32	596.24
8	785.28	750.02	716.93	685.85	656.65	629.20	603.39	579.11
9	775.71	739.47	705.61	673.97	644.37	616.67	590.72	566.41
10	768.10	731.16	696.80	664.81	635.00	607.19	581.24	557.01

Table 5.6 $1000\ \overline{A}_{30:\overline{n}|}$

Net single premiums for an n-year endowment insurance are high as compared to those for the n-year term insurance and for the n-year pure endowment because, in the n-year endowment insurance, the insurer has to pay the benefit amount, may be before the term ends or if not, definitely at the end of the term. Net single premiums are high for a low force of interest and for a small term, as expected. ∎

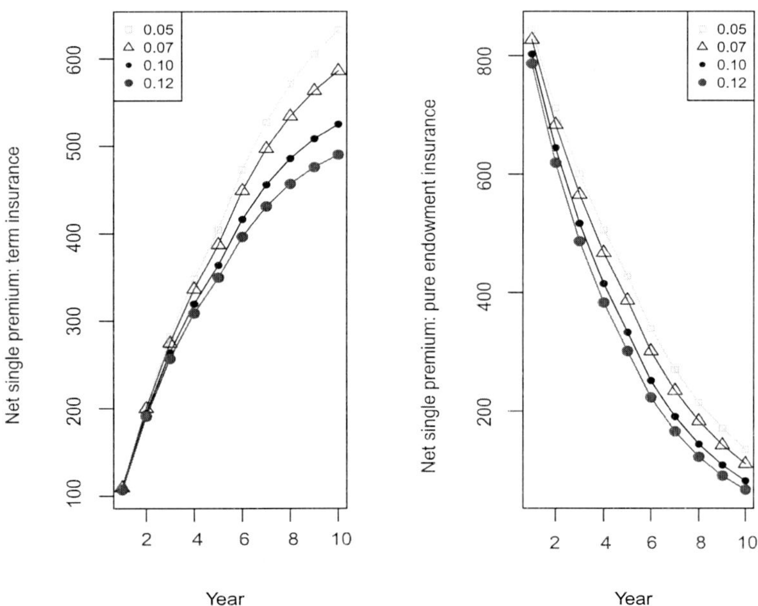

Figure 5.1 $1000\ \overline{A}\,{}^{1}_{30:\overline{n}|}$ and $1000A_{30:\frac{1}{n}|}$

Figure 5.1 is drawn using Part I and II of Code 5.6.2. The left panel of Figure 5.1 displays four curves corresponding to the net single premium in the n-year term insurance with $\delta = 0.05, 0.07, 0.10$ and 0.12, for values of n from 1 to 10. Observe that all the four curves have an increasing trend as n increases, that is, as the term n of insurance contract increases, the net single premium increases. The curve corresponding to $\delta = 0.12$ is below all the others and the curve corresponding to $\delta = 0.05$ is above all the other curves. Further it is to be noted that all the four curves more or less coincide for $n \leq 2$. Thus for small terms, the effect of force of interest is negligible.

The right panel of Figure 5.1 presents four curves corresponding to net single premium for the n-year pure endowment insurance with $\delta = 0.05, 0.07, 0.10$ and 0.12 for values of n from 1 to 10. Observe that all the four curves have a decreasing trend as n increases, that is, as the term n of the insurance contract increases, the net single premium decreases. The curve corresponding to $\delta = 0.12$ is below all the others and the curve corresponding to $\delta = 0.05$ is above all the other curves. Further, it is to be noted that the nature of all the four curves is opposite to that corresponding to the n-year term insurance. It is in view of the fact that these two insurance products are complementary to each other.

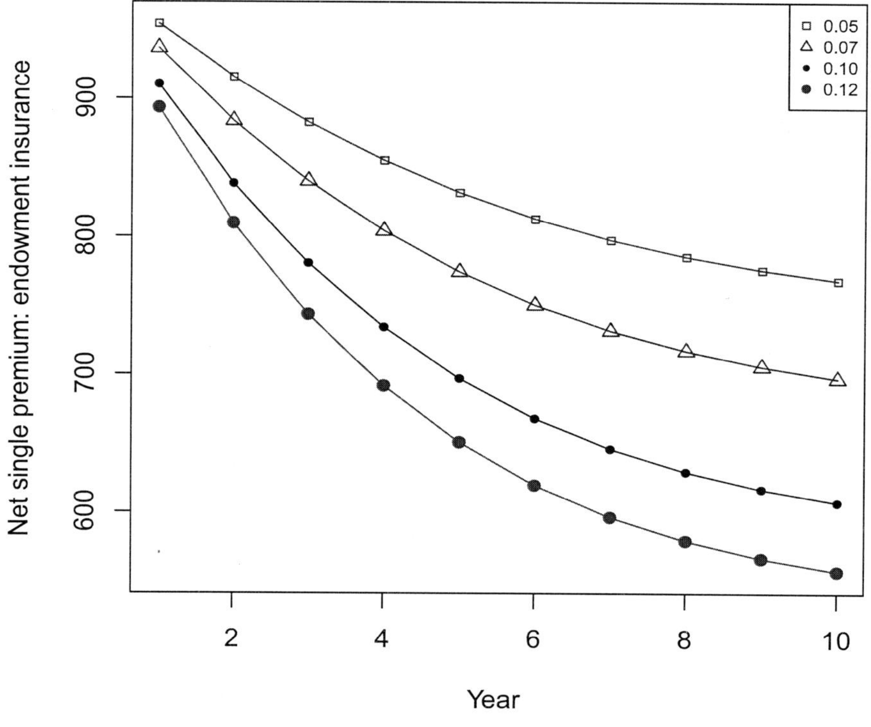

Figure 5.2 $1000\ \overline{A}_{30:\overline{n}|}$

Figure 5.2 is drawn using Part I and III of Code 5.6.2. It shows four curves corresponding to the net single premium of the n-year endowment insurance with $\delta = 0.05, 0.06, 0.10$ and 0.12 and for values of n from 1 to 10. Observe that all the four curves have a decreasing trend as n increases. The curve corresponding to $\delta = 0.12$ is below all the others and the curve corresponding to $\delta = 0.05$ is above all the other curves, as expected.

If we compare a column of Table 5.4 with that of Table 5.6 for the same value of δ, we note that for any year n, the actuarial present value of benefit 1000 is small for the term insurance as compared to that for the endowment insurance. It is expected since the actuarial present value of the term insurance is one of the two terms in the endowment insurance. Such a difference is reasonable in the light of the nature of assurance in these two products. For the term insurance, the insurance company has no liability after the term of the insurance is over while in the endowment insurance, the insurance company has to pay the benefit amount of 1000 at some time during the term or definitely at the end of the term. Thus for an insurance company, it is essential to have larger funds for the portfolio of the endowment insurance products than that for the term insurance products.

We now discuss the most common model of life insurance which is the whole life insurance.

Whole Life Insurance: In this product, the benefit is paid to the beneficiary at the moment of death of the insured. Whole life insurance product offers lifetime coverage, that is, provides payment following the death of the insured at any time in the future after signing the contract. This is in contrast with the term life insurance which provides protection for a certain period of time and provides no benefits after that period ends.

Whole life policies can be classified on the basis of the length of the policy's premium payment period such as (i) continuous premium policies or (ii) limited payment policies. Under a continuous premium whole life policy, premiums are payable until the death of the insured and as such the amount of each premium payment is lower than the premium amount required under any other premium payment schedule. A limited payment whole life policy is a policy for which premiums are payable only until some stated period or until the insured's death, whichever occurs first. If the insured dies before the end of the specified premium payment period, the insurer will pay the death benefit to the named beneficiary and no further premiums are payable. Limited payment policies are designed to meet a policy owner's need for whole life insurance protection that is funded over a limited time period. The policy owner, for example, may expect that his income will drop considerably when he retires and he will still need life insurance coverage after retirement.

The present value random variable for the whole life insurance with benefit of 1 unit, payable at the moment of death of (x), is given by

$$Z_T = v^T, \quad \text{for} \quad T \geq 0$$

The net single premium is denoted by $\overline{A}_x$ and is given by,

$$\overline{A}_x = E(Z) = E(v^{T(x)}) = E(e^{-\delta T(x)}) = \int_0^\infty v^t \, {}_tp_x \, \mu_{x+t} \, dt \tag{5.3.10}$$

Whole life insurance is a limiting case of the n-year term insurance, as $n \to \infty$. It is to be noted that

$$\overline{A}^{\,1}_{x:\overline{n}|} \;\to\; \overline{A}_x \quad \text{as} \quad n \to \infty$$

$$\text{Further,} \quad \overline{A}_{x:\overline{n}|} \;=\; \overline{A}^{\,1}_{x:\overline{n}|} + v^n \, {}_np_x \;\to\; \overline{A}_x \quad \text{as} \quad n \to \infty$$

since v^n converges to 0. From the expression of $\overline{A}_x$, it is to be noted that $\overline{A}_x$ is the moment generating function of $T(x)$ at $(-\delta)$.

Example 5.3.3. Suppose the life length random variable X is modelled by a uniform distribution over $(0, 100)$. Find $50000\overline{A}_x$ for ages $x = 25, 30, 35, 40, 45, 50$ and $\delta = 0.05$.

Solution: As shown in Example 4.2.7, if the life length random variable X has a uniform distribution over $(0, 100)$, then $T(x)$ also has uniform distribution over $(0, 100 - x)$. For the uniform distribution over $(0, a)$, the moment generating function $M_X(t)$ is given by,

$$M_X(t) = E(e^{tX}) = (1/a) \int_0^a e^{tx}\, dx = (e^{at} - 1)/at$$

Hence, $\overline{A}_x = M_{T(x)}(-\delta) = (1 - e^{-\delta(100-x)})/\delta(100 - x)$

Table 5.7 displays the values of $50000\overline{A}_x$ for various values of x.

Age x	25	30	35	40	45	50
$50000\overline{A}_x$	13019.76	13854.32	14788.09	15836.88	17019.49	18358.30

Table 5.7 $50000\ \overline{A}_x$ when $T(x) \sim U(0, 100 - x)$

Note that as age increases, the net single premium for whole life increases, as expected. ∎

Example 5.3.4. Suppose an individual of age 30 signs a whole life insurance contract. He is insured for a death benefit of 1 lakh, payable at the moment of death. Assuming that he is subject to a constant force of mortality $\mu = 0.04$, find the net single premium. Also find the variance of the corresponding present value random variable. The benefit payments are to be withdrawn from an investment fund earning interest at the rate $\delta = 0.06$.

Solution: From the given information we have,

$$Z_T = 10^5 v^T, \quad \text{for} \quad T \geq 0$$

For a constant force of interest δ and constant force of mortality μ, the net single premium for the unit benefit whole life insurance is

$$\overline{A}_x = \int_0^\infty e^{-\delta t} e^{-\mu t} \mu\, dt = \mu/(\mu + \delta) \quad \text{and} \quad {}^2\overline{A}_x = \mu/(\mu + 2\delta)$$

Thus, with $\mu = 0.04$ and $\delta = 0.06$

$$
\begin{aligned}
E(Z_T) &= 10^5 \overline{A}_x = 10^5 (0.04/0.1) = 40000 = 4 \times 10^4 \\
E(Z_T^2) &= 10^{10}\ {}^2\overline{A}_x = 10^{10}\ 0.04/(0.04 + 2(0.06)) = 25 \times 10^8 \quad \text{and} \\
Var(Z_T) &= 9 \times 10^8
\end{aligned}
$$

Thus, the individual has to pay Rs 40000 as a net single premium. Note that the variability of the present value random variable is quite high. ∎

Remark 5.3.1. From Example 5.3.4, we note that when the mortality is modelled by constant force of mortality, the net single premium $\overline{A}_x$ is $\mu/(\mu+\delta)$. It is free from x. Thus, the premium for an individual of age 30 is the same as that of age 80, which is not reasonable. It again shows that constant force of mortality is not a good model for modelling human life length.

Example 5.3.5. Suppose the life length random variable is modelled by a distribution with force of mortality as specified in Example 5.3.1. Find $1000\overline{A}_{30}$ for $\delta = 0.05, 0.06, 0.07, 0.08, 0.09, 0.10, 0.11$.

Solution: For the given force of mortality, we have obtained the probability density function $g(t)$ of $T(30)$ in Example 4.2.10. Using it we find $\overline{A}_{30}$ as shown below.

$$
\begin{aligned}
\overline{A}_{30} &= \int_0^{.5} e^{-\delta t} g(t)dt + \int_5^{\infty} e^{-\delta t} g(t)dt \\[2mm]
&= 0.12 \int_0^{.5} e^{-(\delta+0.12)t}dt + 0.18e^{0.3} \int_5^{\infty} e^{-(\delta+0.18)t}dt \\[2mm]
&= \frac{0.12(1 - e^{-5\delta-0.6})}{0.12+\delta} + \frac{0.18e^{-5\delta-0.6}}{\delta+0.18}
\end{aligned}
$$

Note that this expression is the limit as $n \to \infty$ of $\overline{A}\,^{1}_{30:\overline{n}|}$ obtained in Example 5.3.1. Table 5.8 displays the values of $1000\overline{A}_{30}$ for various values of δ. From Table 5.8, we note that the

δ	0.05	0.06	0.07	0.08	0.09	0.10	0.11
$1000\overline{A}_{30}$	738.68	700.55	665.78	633.96	604.76	577.88	553.07

Table 5.8 Effect of change of force of interest on $1000\overline{A}_{30}$

net single premium decreases as the force of interest increases. However, these are rather high, due to given mortality pattern, in which the values of μ are high. ∎

Example 5.3.6. Suppose the life length random variable is modelled by a distribution with force of mortality as specified in Example 5.3.1. Find $1000\overline{A}_x$ for $x = 25, 30, 35$ and 40. Take $\delta = 0.05, 0.06, 0.07, 0.08, 0.09, 0.10, 0.11, 0.12$. Comment on the effect of age on the values of a net single premium.

Solution: In Example 4.2.10 for the given force of mortality, we have derived the survival function $S(x)$ and the probability density function $g(t)$ of $T(30)$. Proceeding on similar lines, we obtain distribution of $T(25), T(35)$ and $T(40)$. These are as follows:

$$
g_{T(25)}(t) = \begin{cases} 0.12 \, \exp(-0.12t)\,, & \text{if} \quad 0 < t < 10 \\ 0.18 \, \exp(-0.18t + 0.6), & \text{if} \quad t \geq 10 \end{cases}
$$

$$
g_{T(30)}(t) = \begin{cases} 0.12 \, \exp(-0.12t), & \text{if} \quad 0 < t < 5 \\ 0.18 \, \exp(-0.18t + 0.3), & \text{if} \quad t \geq 5 \end{cases}
$$

$$g_{T(35)}(t) \;=\; g_{T(40)}(t) = \; 0.18 \, \exp(-0.18t), \quad \text{if} \;\; t > 0$$

Distribution of $T(35)$ and $T(40)$ is the same as the force of mortality remains the same after 35 in this example. In fact, distribution of $T(x)$ for $x > 35$ is the same and it is the exponential distribution with parameter 0.18. Using these distributions we obtain $\overline{A}_x$ for $x = 25, 30, 35$ and 40 adopting similar procedure as in Example 5.3.5. These are given below.

$$\overline{A}_{25} \;=\; \int_0^{10} e^{-\delta t} g(t)dt + \int_{10}^{\infty} e^{-\delta t} g(t)dt$$

$$= \; 0.12 \int_0^{10} \exp(-(\delta + 0.12)t)dt + 0.18 \exp(0.3) \int_{10}^{\infty} \exp(-(\delta + 0.18)t)dt$$

$$= \; \frac{0.12(1 - \exp(-10\delta - 1.2))}{0.12 + \delta} + \frac{0.18 \exp(-10\delta - 1.2)}{0.18 + \delta}$$

$$\overline{A}_{30} \;=\; \frac{0.12(1 - \exp(-5\delta - 0.6))}{0.12 + \delta} + \frac{0.18 \exp(-5\delta - 0.6)}{0.18 + \delta}$$

$$\overline{A}_{35} \;=\; \overline{A}_{40} = \int_0^{\infty} e^{-\delta t} g(t)dt = 0.18 \int_0^{\infty} \exp(-(\delta + 0.18)t)dt$$

$$= \; 0.18/(0.18 + \delta)$$

Values of $\overline{A}_x \;\; \forall \;\; x > 35$ remain the same. It is in fact $\mu/(\mu + \delta)$, as proved in Example 5.3.4. Table 5.9 and Figure 5.3 display the values of $1000\overline{A}_x$ for $x = 25, 30$ and 35 and for various values of δ. From Table 5.9 we note that as the force of interest increases, the net

δ \ Age	25	30	35
0.05	719.90	738.68	782.61
0.06	680.44	700.55	750.00
0.07	644.80	665.78	720.00
0.08	612.49	633.96	692.31
0.09	583.09	604.76	666.67
0.10	556.25	577.88	642.86
0.11	531.66	553.07	620.69
0.12	509.07	530.12	600.00

Table 5.9 Effect of age on $1000 \, \overline{A}_x$

single premium decreases for all the ages. Further as age increases, the net single premium also increases, as the time span for which the policy will be in force decreases with increase in age. These features are clearly revealed in Figure 5.3. It shows three curves corresponding to the net single premium of whole life insurance for three ages $25, 30, 35$. Observe that all the three curves have a decreasing trend as δ increases. The curve corresponding to age 25 is below all the others and the curve corresponding to age 35 is above all the other curves, as expected. It is drawn using Part II of Code 5.6.3. ∎

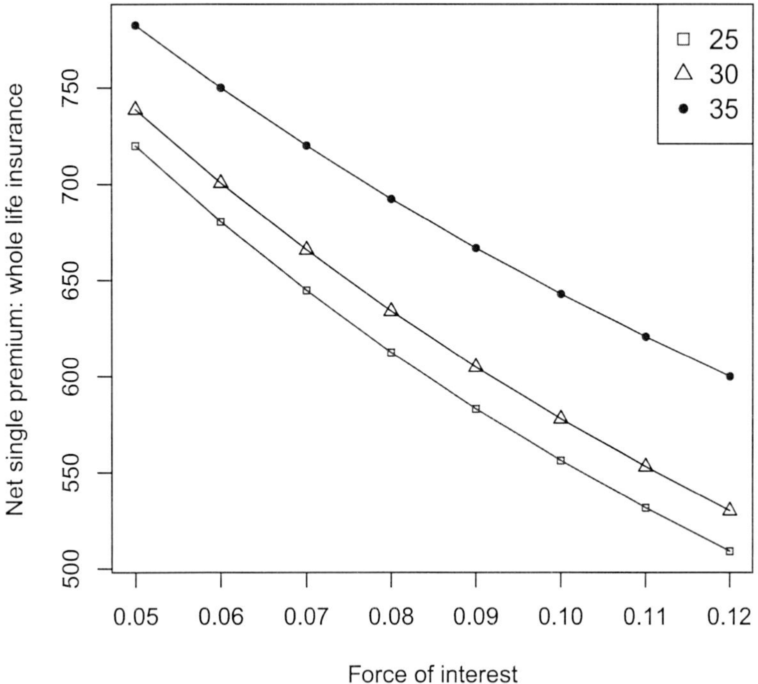

Figure 5.3 Effect of age on $1000\ \overline{A}_x$

Example 5.3.7. Suppose the force of mortality follows Gompertz' law given by $\mu_x = BC^x$. (i) Find the formulae for actuarial present values of unit benefit to be paid at the moment of death for the n-year term insurance, n-year endowment insurance and whole life insurance. (ii) Find the values of the net single premium for benefit of 1000 for n-year term insurance, n-year endowment insurance for $n = 1, 2, \cdots, 10$ and whole life insurance for ages $25, 30, 35, 40$. Take $\delta = 0.05$, $B = 0.0001151$ and $C = 1.096$.

Solution: (i) In Example 4.2.11, we have obtained the distribution of $T(x)$ when the force of mortality follows Gompertz' law. The probability density function $g(t)$ of $T(x)$ for $t > 0$ is given by,

$$g(t) = \ _tp_x\ \mu_{x+t} = e^{-mC^x(C^t-1)}\ BC^{x+t} = e^{-mC^x(C^t-1)}\ m(\log C)C^x C^t$$

where $m = B/\log C$. Suppose mC^x is denoted by α_x. Then $g(t)$ is expressible as,

$$g(t) = e^{-\alpha_x C^t}e^{\alpha_x}\ \log C\ \alpha_x C^t = \alpha_x\ e^{\alpha_x}\log C\ e^{-\alpha_x C^t}C^t, \quad \text{for}\ \ t > 0$$

$$\text{Hence,}\quad \overline{A}\ _{x:\overline{n}|}^{\,1} = \int_0^n v^t g(t)dt = \alpha_x\ e^{\alpha_x}\log C \int_0^n e^{-\delta t - \alpha_x C^t}C^t dt$$

To obtain its expression, we substitute $C^t = y$, which implies $e^{-\delta t} = y^{-\delta/\log C}$, $C^t\ \log C\ dt = dy$ and range of integration changes to 1 to C^n. Suppose $(-\delta/\log C) + 1 = \lambda$.

Then $\overline{A}\,{}^{1}_{x:\overline{n}|}$ simplifies as follows.

$$
\begin{aligned}
\overline{A}\,{}^{1}_{x:\overline{n}|} &= \alpha_x\, e^{\alpha_x} \int_1^{C^n} e^{-\alpha_x y} y^{\lambda-1} dy = \alpha_x\, e^{\alpha_x} \Gamma(\lambda)\, \alpha_x^{-\lambda} \int_1^{C^n} \frac{\alpha_x^{\lambda}}{\Gamma(\lambda)} e^{-\alpha_x y} y^{\lambda-1} dy \\
&= \alpha_x\, e^{\alpha_x} \Gamma(\lambda)\, \alpha_x^{-\lambda}\, P[1 \le W \le C^n]
\end{aligned}
$$

where W follows gamma distribution with shape parameter λ and scale parameter α_x. The probability $P[1 \le W \le C^n]$ is obtained using the **pgamma** function from R, which gives the value of the distribution function of the gamma distribution at specified $x \in \mathbb{R}$, corresponding to the specified values of the parameters. Thus,

$$
\overline{A}\,{}^{1}_{x:\overline{n}|} = e^{\alpha_x} \Gamma(\lambda)\, \alpha_x^{1-\lambda} \Big(\texttt{pgamma}(C^n, \lambda, \alpha_x) - \texttt{pgamma}(1, \lambda, \alpha_x) \Big)
$$

For n-year endowment insurance,

$$
\overline{A}_{x:\overline{n}|} = \overline{A}\,{}^{1}_{x:\overline{n}|} + A_{x:\overline{n}|}^{1}
$$

Using the survival function of Gompertz' law we have,

$$
A_{x:\overline{n}|}^{1} = v^n\, {}_n p_x = e^{-\delta n - \alpha_x (C^n - 1)}
$$

For whole life insurance, the formula for the actuarial present value is obtained by allowing n to tend to ∞ in that of n-year term insurance. Note that since $C > 1$, $\lim_{n \to \infty} C^n = \infty$. Thus, $P[1 \le W \le C^n] \to P[W \ge 1]$. Hence we have,

$$
\overline{A}_x = e^{\alpha_x} \Gamma(\lambda)\, \alpha_x^{1-\lambda}\, P[W \ge 1] = e^{\alpha_x} \Gamma(\lambda)\, \alpha_x^{1-\lambda} \Big(1 - \texttt{pgamma}(1, \lambda, \alpha_x) \Big)
$$

(ii) We find the values of the net single premium for benefit of 1000 for all these products for ages $25, 30, 35, 40$ using the formulae derived above and R software. δ is taken as 0.05. These are computed using Code 5.6.4. Part I of the code corresponds to the term and endowment insurance while Part II is concerned with whole life insurance. Table 5.10 displays the net single premium for benefit of 1000 in the n-year term insurance for $n = 1, 2, \cdots, 10$ and for ages $25, 30, 35, 40$. The net single premium is 1.16 for 1-year term insurance for the benefit of 1000 for (25). It is much less as the insurer has no liability after 1 year and the chance of death in one year $(25 - 26)$ for (25) is 0.0111 under Gompertz' law. From Table 5.10 we note that the net single premium increases as the age increases for all n and it also increases as n increases for all ages, as expected. Table 5.11 displays the net single premium for benefit of 1000 in n-year endowment insurance for $n = 1, 2, \cdots, 10$ and for ages $25, 30, 35, 40$. The net single premium is 951.26 for 1-year endowment insurance for the benefit of 1000 for (25). It is much higher as compared to that for the 1-year term insurance. In the 1-year endowment insurance, the insurer has liability to pay 1000 at the moment of death if it occurs in that year (with probability 0.0111) and if the individual survives 1 year (with probability 0.9889), then insurer has to pay 1000 at the end of one year. Thus, the insurer will decide the single premium such that it will approximately accumulate to 1000 in one year with the given rate of interest. In this example $\delta = 0.05$, hence the corresponding effective rate of interest i is given by $i = \exp(\delta) - 1 = 0.0513$. With this rate, the principal 951.26 in 1 year will accumulate to

n \ Age	25	30	35	40
1	1.16	1.84	2.90	4.59
2	2.37	3.75	5.92	9.35
3	3.63	5.74	9.06	14.28
4	4.94	7.80	12.31	19.40
5	6.31	9.95	15.69	24.69
6	7.73	12.19	19.20	30.18
7	9.21	14.51	22.84	35.85
8	10.74	16.93	26.62	41.71
9	12.34	19.43	30.53	47.77
10	14.00	22.04	34.58	54.01

Table 5.10 Gompertz' law: $1000\, \overline{A}\, {}^{1}_{x:\overline{n}|}$

n \ Age	25	30	35	40
1	951.26	951.27	951.30	951.34
2	904.95	905.02	905.12	905.28
3	860.96	861.11	861.34	861.71
4	819.18	819.44	819.86	820.50
5	779.50	779.91	780.55	781.56
6	741.83	742.41	743.34	744.78
7	706.06	706.86	708.11	710.07
8	672.11	673.15	674.78	677.33
9	639.90	641.21	643.26	646.48
10	609.33	610.95	613.48	617.43

Table 5.11 Gompertz' law: $1000\, \overline{A}\, {}_{x:\overline{n}|}$

$951.26(1 + i) = 1000.06$ which is close to the benefit amount. It is easy to check that all the premium values reported in this table approximately accumulate to 1000 in the specified time period. For example, the net single premium 609.33 for (25) for the 10-year term accumulates to 1004.89 in 10 years. In these calculations, we have ignored the chances of death in any one year interval in (25–35), but it will not have a significant effect as the chance of death in (25–35) is very small (0.0183). The net single premium 617.43 for (40) for the 10-year term accumulates to 1018.25 in 10 years. Thus, if (40) survives the term of 10 years, he will get the benefit of 1000, the accumulated value is slightly higher in this case. It is to be noted that a net single premium is an expected value of the present value random variable of the benefit payment. From Table 5.11 we note that the net single premium decreases as n increases for all ages, as the expected period for which the insurer will earn interest on this fund increases. For the given mortality pattern, changes in net single premiums across ages is marginal for small n. Table 5.12 displays the net single premium for benefit of 1000 for whole life insurance for ages $25, 30, 35, 40$.

Age	25	30	35	40
$1000\overline{A}_x$	152.54	189.12	232.70	283.72

Table 5.12 Gompertz' law: $1000\,\overline{A}_x$

From Table 5.12, we note that the net single premium increases as age increases, as the expected period for which the insurer will earn interest on this fund will decrease. ∎

Remark 5.3.2. Net single premiums for the term insurance, the endowment insurance and the whole life insurance obtained in Example 5.3.7 under Gompertz' law are quite different than the corresponding net single premiums obtained under mortality law specified in Example 5.3.1. These two examples clearly exhibit how the actuarial present values change as the underlying mortality patterns change.

Example 5.3.8. Suppose the force of mortality follows Makeham's law given by $\mu_x = A + BC^x$. Find the formulae for actuarial present values of unit benefit to be paid at the moment of death for n-year term insurance, n-year endowment insurance and whole life insurance.

Solution: In Example 4.2.11, we have obtained the distribution of $T(x)$ when the force of mortality follows Makeham's law. The probability density function $g(t)$ of $T(x)$ in this case is given by,

$$g(t) \;=\; {}_tp_x\,\mu_{x+t} \;=\; \{\exp[-At - mC^x(C^t - 1)]\}\,[A + BC^{x+t}], \qquad t > 0$$

Suppose mC^x is denoted by α_x. Then $g(t)$ is expressible as,

$$g(t) = e^{\alpha_x}\exp[-At - \alpha_x C^t][A + \alpha_x\,\log C\; C^t], \qquad t > 0$$

Hence with $\delta + A = \delta_1$, we have

$$\overline{A}\,{}^{1}_{x:\overline{n}|} \;=\; \int_0^n v^t g(t)dt = \int_0^n \exp(-\delta t)\,g(t)dt$$

$$=\; Ae^{\alpha_x}\int_0^n e^{-\delta_1 t - \alpha_x C^t}\,dt + \alpha_x\,e^{\alpha_x}\log C \int_0^n e^{-\delta_1 t - \alpha_x C^t}\,C^t dt$$

As in the case of Gompertz's law, with $C^t = y$, we get $e^{-\delta_1 t} = y^{-\delta_1/\log C}$, $C^t\,\log C\,dt = dy$ and range of integration changes to 1 to C^n. Hence,

$$\overline{A}\,{}^{1}_{x:\overline{n}|} \;=\; Ae^{\alpha_x}\int_1^{C^n} y^{-\delta_1/\log C}e^{-\alpha_x y}\,(\log C\; y)^{-1}dy$$

$$+\; \alpha_x\,e^{\alpha_x}\int_1^{C^n} y^{-\delta_1/\log C}e^{-\alpha_x y}\,dy$$

Suppose $(-\delta_1/\log C) + 1 = \lambda_1$. Then $\overline{A}\,{}^{1}_{x:\overline{n}|}$ simplifies as follows.

$$\overline{A}\,{}^{1}_{x:\overline{n}|} \;=\; \frac{Ae^{\alpha_x}}{\log C}\int_1^{C^n} e^{-\alpha_x y}y^{\lambda_1 - 2}\,dy \;+\; \alpha_x\,e^{\alpha_x}\int_1^{C^n} e^{-\alpha_x y}y^{\lambda_1 - 1}\,dy$$

$$=\; \frac{Ae^{\alpha_x}\Gamma(\lambda_1 - 1)}{\log C\; \alpha_x^{\lambda_1 - 1}}\,P[1 \le W_1 \le C^n] \;+\; \frac{\alpha_x e^{\alpha_x}\Gamma(\lambda_1)}{\alpha_x^{\lambda_1}}\,P[1 \le W_2 \le C^n]$$

where W_1 follows gamma distribution with shape parameter $\lambda_1 - 1$ and scale parameter α_x, while W_2 follows gamma distribution with shape parameter λ_1 and scale parameter α_x. The probability $P[1 \leq W_1 \leq C^n]$ is obtained using the `pgamma` function from R as

$$P[1 \leq W_1 \leq C^n] = \texttt{pgamma}(C^n, \lambda_1 - 1, \alpha_x) - \texttt{pgamma}(1, \lambda_1 - 1, \alpha_x)$$

The probability $P[1 \leq W_2 \leq C^n]$ is obtained using the `pgamma` function from R as

$$P[1 \leq W_2 \leq C^n] = \texttt{pgamma}(C^n, \lambda_1, \alpha_x) - \texttt{pgamma}(1, \lambda_1, \alpha_x)$$

To find the actuarial present value for n-year endowment insurance, we first find $_nE_x$. Using the survival function of Makeham's law we have,

$$_nE_x \;=\; v^n \; _np_x = e^{-\delta n - An - \alpha_x(C^n - 1)} = e^{-\delta_1 n - \alpha_x(C^n - 1)}$$

We can then compute

$$\overline{A}_{x:\overline{n}|} \;=\; \overline{A}_{\,x:\overline{n}|}^{\,1} + A_{x:\overline{n}|}^{\,1}$$

For whole life insurance, the formula for the actuarial present value is obtained by allowing n to tend to ∞ in that of n-year term insurance. As noted in Example 5.3.7, since $C > 1$, $\lim_{n \to \infty} C^n = \infty$. Thus, $P[1 \leq W_i \leq C^n] \to P[W_i \geq 1]$, $i = 1, 2$. Hence we have,

$$
\begin{aligned}
\overline{A}_x \;&=\; \frac{Ae^{\alpha_x}\Gamma(\lambda_1 - 1)}{\log C \; \alpha_x^{\lambda_1 - 1}} \, P[W_1 \geq 1] \;+\; \frac{\alpha_x e^{\alpha_x}\Gamma(\lambda_1)}{\alpha_x^{\lambda_1}} \, P[W_2 \geq 1] \\[2mm]
&=\; \frac{Ae^{\alpha_x}\Gamma(\lambda_1 - 1)}{\log C \; \alpha_x^{\lambda_1 - 1}} \left(1 - \texttt{pgamma}(1, \lambda_1 - 1, \alpha_x)\right) \\[2mm]
&\;+\; \frac{\alpha_x e^{\alpha_x}\Gamma(\lambda_1)}{\alpha_x^{\lambda_1}} \left(1 - \texttt{pgamma}(1, \lambda_1, \alpha_x)\right)
\end{aligned}
$$

$\blacksquare$

Remark 5.3.3. Once we have the formulae for $\overline{A}_{\,x:\overline{n}|}^{\,1}$, $\overline{A}_{x:\overline{n}|}$, and $\overline{A}_x$, when the underlying mortality law is Makeham's law, we can find their values for specified values of n and δ, as in Example 5.3.7. (See computational exercises 5.8.2 to 5.8.5.)

Example 5.3.9. A senior citizens' association has 70 members, each of age 60. It is decided to contribute a certain amount P to establish a fund to pay a death benefit of Rs 25,000 to the beneficiary of each member. Benefits are to be payable at the moment of death. Suppose for the specific pattern of mortality governing this group and for the specific interest rate $\overline{A}_{60} = 0.05$ and $^2\overline{A}_{60} = 0.02$. Find P so that with probability 0.95 the fund will be sufficient to pay the death benefit. State the assumptions that you may make.

Solution: Suppose Z_j denotes the present value random variable for the benefit of 1 rupee for individual j, $j = 1, 2, \cdots, 70$. Then $S = 25000 \sum_{j=1}^{70} Z_j$ is the present value random variable corresponding to the total death benefit. So the contribution from each member should constitute the fund which will match S in some sense. Thus, we find P so that,

$$Pr[S \leq 70P] \;=\; 0.95$$

S is a sum of 70 independent (assuming that 70 lives are independent) random variables having the same mean and variance. Hence, using the Central Limit Theorem, the distribution of S can be approximated by the normal distribution with

$$\text{mean } \mu \;=\; E(S) \;=\; 70 \times 25,000 \times \overline{A}_{60} \;=\; 87500 \quad \text{and}$$

$$\text{standard deviation } sd(S) \;=\; (70 \times 0.0175 \times 25000^2)^{1/2} \;=\; 27669.93$$

since $Var(Z_j) = {}^{2}\overline{A}_{60} - \overline{A}_{60}^{2} = 0.0175$. The collected fund is $70P$. To find P we standardize S. 95th percentile of the standard normal distribution is 1.645. Hence,

$$Pr\left[\frac{S - E(S)}{sd(S)} \le \frac{70P - E(S)}{sd(S)}\right] = 0.95 \quad \Rightarrow \quad (70P - 87500)/27669.93 \;=\; 1.645$$

$$\Rightarrow \quad P = 1900.24$$

Thus, $P = 1900.24$ rupees is the contribution per individual. ∎

We now discuss all the above models of life insurance when the benefit is postponed or deferred for a certain period. These are known as deferred insurance products.

Deferred Insurance: A deferred insurance provides a benefit at the moment of the death of (x) if it occurs after a certain fixed period called the deferment period. For example, an m-year deferred insurance provides a benefit following the death of the insured only if the insured dies at least m years after the issue of policy. The conditions for the payment of benefit, that is, insurance products, may be any of those discussed above. For example, an m-year deferred whole life insurance with a unit amount payable at the moment of death has the following benefit function and the present value random variable of the benefit.

$$b_t \;=\; \begin{cases} 1 & \text{if } \; t > m \\ 0 & \text{if } \; t \le m \end{cases}$$

$$Z \;=\; \begin{cases} v^T & \text{if } \; T > m \\ 0 & \text{if } \; T \le m \end{cases}$$

The actuarial present value is denoted by ${}_{m|}\overline{A}_x$ and is given by,

$$_{m|}\overline{A}_x \;=\; \int_{m}^{\infty} v^t \; {}_{t}p_x\mu_{x+t} \, dt \;=\; v^m \; {}_{m}p_x \int_{0}^{\infty} v^u \; {}_{u}p_{x+m}\mu_{x+m+u} \, du$$

$$=\; v^m \; {}_{m}p_x \; \overline{A}_{x+m} \;=\; A_{x:\overline{m}|}^{\;\;1} \overline{A}_{x+m} \;=\; {}_{m}E_x \; \overline{A}_{x+m}$$

The second integral in the above equation is obtained by the substitution $t - m = u$. Recall that the formula for Z_T for m-year term insurance is given by,

$$Z_T \;=\; \begin{cases} v^T, & \text{if } \; T \le m \\ 0, & \text{if } \; T > m \end{cases}$$

Suppose we denote it by $Z_T(\text{Term})$. Similarly suppose we denote the present value random variable for m-year deferred whole life insurance by $Z_T(\text{Deferred whole life})$. Then observe that

$$Z_T(\text{Term}) + Z_T(\text{Deferred whole life}) \;=\; Z_T(\text{Whole life})$$

$$\text{Consequently,} \quad _{m|}\overline{A}_x \;=\; \overline{A}_x - \overline{A}\,^{1}_{x:\overline{m}|}$$

This formula enables us to find the net single premium for deferred whole life insurance from the knowledge of those for whole life insurance and term insurance. It also shows that $_{m|}\overline{A}_x < \overline{A}_x$. Such a relation is in view of the fact that in deferred whole life insurance, liability of insurer is less than that in whole life insurance.

Term insurance, endowment insurance discussed above can be modified analogously to take into account the deferment of the benefit. For example, present value random variable in m-year deferred n-year term is given by,

$$Z_T \;=\; \begin{cases} v^T, & \text{if} \quad m \le T \le m+n \\ 0, & \text{if} \quad T \le m \text{ or } T > m+n \end{cases}$$

The actuarial present value is denoted by $_{m|}\overline{A}\,^{1}_{x:\overline{n}|}$ or by $_{m|n}\overline{A}_x$ and is obtained as follows.

$$_{m|}\overline{A}\,^{1}_{x:\overline{n}|} \;=\; \int_m^{m+n} v^t \; {}_tp_x\mu_{x+t} \; dt$$

$$=\; \int_0^{m+n} v^t \; {}_tp_x\mu_{x+t} \; dt - \int_0^{m} v^t \; {}_tp_x\mu_{x+t} \; dt = \overline{A}\,^{1}_{x:\overline{m+n}|} - \overline{A}\,^{1}_{x:\overline{m}|}$$

It shows that $_{m|}\overline{A}\,^{1}_{x:\overline{n}|} < \overline{A}\,^{1}_{x:\overline{m+n}|}$.

m-year deferred n-year endowment insurance is defined by the following present value random variable.

$$Z_T \;=\; \begin{cases} 0, & \text{if} \quad T \le m \\ v^T, & \text{if} \quad m \le T \le m+n \\ v^n, & \text{if} \quad T > m+n \end{cases}$$

Note that the benefit is not paid at the end of the m year period if death occurs in the first m years. The actuarial present value is denoted by $_{m|}\overline{A}_{x:\overline{n}|}$ and is given by,

$$_{m|}\overline{A}_{x:\overline{n}|} \;=\; \int_m^{m+n} v^t \; {}_tp_x\mu_{x+t} \; dt \; + v^{m+n} \; {}_{m+n}p_x$$

$$=\; \overline{A}\,^{1}_{x:\overline{m+n}|} - \overline{A}\,^{1}_{x:\overline{m}|} + A_{x:\overline{m+n}|}^{\;1} = \overline{A}_{x:\overline{m+n}|} - \overline{A}\,^{1}_{x:\overline{m}|}$$

It also shows that $_{m|}\overline{A}_{x:\overline{n}|} < \overline{A}_{x:\overline{m+n}|}$, as expected. Using the formula for $_{m|}\overline{A}\,^{1}_{x:\overline{n}|}$ as derived above $_{m|}\overline{A}_{x:\overline{n}|}$ can also be expressed as

$$_{m|}\overline{A}_{x:\overline{n}|} \;=\; _{m|}\overline{A}\,^{1}_{x:\overline{n}|} + A_{x:\overline{m+n}|}^{\;1}$$

Example 5.3.10. Suppose the force of mortality follows Gompertz' law given by $\mu_x = BC^x$. Find the values of net single premium for benefit of 1000 to be paid at the moment of death in 5-year deferred n-year term insurance, 5-year deferred n-year endowment insurance for $n = 1, 2, \cdots, 5$ and 5-year deferred whole life insurance for ages $25, 30, 35, 40$. Take $\delta = 0.05$, $B = 0.0001151$ and $C = 1.096$.

Solution: We use the relation,

$$_{5|}\overline{A}\,^1_{x:\overline{n}|} = \overline{A}\,^1_{x:\overline{5+n}|} - \overline{A}\,^1_{x:\overline{5}|}$$

and the formulae for the term insurance for Gompertz' law as derived in Example 5.3.7 to find the net single premium for a 5-year deferred n-year term insurance. We have computed $\overline{A}\,^1_{x:\overline{n}|}$ in Table 5.10. Thus, to compute $_{5|}\overline{A}\,^1_{x:\overline{n}|}$, from the last 5 rows of the Table 5.10, we have to subtract the fifth row of the same table. Part III of the Code 5.6.4 uses this approach to compute the required values. Table 5.13 displays these values for ages $25, 30, 35, 40$.

n \ Age	25	30	35	40
1	1.42	2.24	3.51	5.48
2	2.90	4.56	7.15	11.16
3	4.44	6.97	10.92	17.02
4	6.03	9.48	14.83	23.07
5	7.70	12.08	18.89	29.32

Table 5.13 Gompertz' law: $1000\ _{5|}\overline{A}\,^1_{x:\overline{n}|}$

From the table we note that the net single premium increases as the age increases for all n and it also increases as n increases for all ages, as expected. However, all the values are less than the corresponding values for term insurance, with term $5 + n$, as reported in Table 5.10. The decrease is in view of the fact that benefit would not be paid for the first five years after signing the contract.

To find the net single premium for 5-year deferred n-year endowment insurance for $n = 1$ to 5, we use the relation,

$$_{5|}\overline{A}_{x:\overline{n}|} = \overline{A}_{x:\overline{5+n}|} - \overline{A}\,^1_{x:\overline{5}|}$$

and the formulae for the term insurance and the endowment insurance for Gompertz' law as derived in Example 5.3.7. We have computed $\overline{A}_{x:\overline{n}|}$ in Table 5.11. Thus, to compute $_{5|}\overline{A}_{x:\overline{n}|}$, from the last 5 rows of the Table 5.11 we have to subtract the fifth row of the Table 5.10. Part IV of the Code 5.6.4 uses this method to compute the required values. Table 5.14 displays these values for ages $25, 30, 35, 40$.

n \ Age	25	30	35	40
1	735.52	732.46	727.64	720.09
2	699.76	696.90	692.42	685.38
3	665.81	663.20	659.09	652.64
4	633.59	631.25	627.57	621.79
5	603.03	600.99	597.79	592.74

Table 5.14 Gompertz' law: $1000\ _{5|}\overline{A}_{x:\overline{n}|}$

From Table 5.14 we note that the net single premium decreases as n increases for all ages, as the period of accumulation increases. Here also all the values are less than corresponding values for endowment insurance, with term $5 + n$, as reported in Table 5.11, however the decrease is marginal.

To find the net single premium for benefit of 1000 for 5-year deferred whole life insurance, we use the relation

$$_{5|}\overline{A}_x = \overline{A}_x - \overline{A}\,^1_{x:\overline{5|}}$$

We have computed $\overline{A}_x$ in Table 5.12. Thus, to compute $_{5|}\overline{A}_x$, from the row of Table 5.12, we have to subtract the fifth row of the Table 5.10. Part V of the Code 5.6.4 uses this method to compute the required values. Table 5.15 displays these values for ages $25, 30, 35, 40$.

Age	25	30	35	40
$1000\ _{5|}\overline{A}_x$	146.23	179.17	217.01	259.03

Table 5.15 Gompertz' law: $1000\ _{5|}\overline{A}_x$

From Table 5.15 we note that the net single premium increases as age increases, however, when we compare this table with Table 5.12, we note that for each age, the net single premium for benefit of 1000 for 5-year deferred whole life insurance is less than that for the whole life insurance. The decrease is in view of the fact that benefit would not be paid in the first five years after signing the contract. ∎

In the models studied so far, the benefit amount is fixed, which we label as level benefit. However, in practice the amount of benefit may vary. We discuss below the modifications needed in all the standard models, to take into account the varying benefit amounts.

Varying Benefit Insurance: In this setup, the basic equation

$$Z_t = b_t v^t$$

as given in Eq. 5.3.1 remains the same. The benefit function will change as the nature of benefit changes. For example, the level of the benefit may either increase or decrease over all or a part of the term of the insurance. The most common model amongst these is an annually increasing whole life insurance. It provides the benefit of amount 1 at the moment of death during the first year, 2 at the moment of death in the second year, and so on. Thus, the benefit function and the corresponding present value random variable is specified as follows.

$$\begin{aligned} b_t &= \lfloor t + 1 \rfloor & \text{if } \ t \geq 0 \\ Z &= \lfloor T + 1 \rfloor v^T & \text{if } \ T \geq 0 \end{aligned}$$

where the symbol $\lfloor \ \rfloor$ denotes the greatest integer function. The actuarial present value for such an insurance is denoted by $(I\overline{A})_x$ and is defined as follows.

$$(I\overline{A})_x = E(Z) = \int_0^\infty \lfloor t + 1 \rfloor v^t \ _tp_x \ \mu_{x+t} dt$$

We will study the evaluation of this function in Section 4. The following example illustrates the application of general model as given in Eq. 5.3.1.

Example 5.3.11. Suppose (30) signs a 10-year term insurance policy with a death benefit function $b_t = 1000(1+i)^t$, payable at the moment of death. Find the actuarial present value of the benefit with a force of interest $\delta = 0.06$ and force of mortality as given in Example 5.3.1.

Solution: Suppose Z_T is the present value random variable for a 10-year term insurance payable at the moment of death of (30) with a benefit function $b_t = 1000(1+i)^t$. Then

$$Z_T \;=\; b_T v^T = 1000(1+i)^T (1+i)^{-T} \;=\; 1000 \;\text{ for }\; T \le 10$$
$$\text{and }\; Z_T \;=\; 0 \;\text{ for }\; T > 10$$

Thus $Z_T = 1000$ with a probability $_{10}q_{30}$ and zero with a probability $_{10}p_{30}$. Hence, $E(Z) = 1000 \; _{10}q_{30}$. Using the force of mortality as specified in Example 5.3.1,

$$_{10}q_{30} \;=\; 1 - P[T(30) > 10] \;=\; 1 - e^{-1.5} \;=\; 0.7768698$$

Hence, the actuarial present value of the benefit is $1000 \times 0.7768698 = 776.87$. From Example 5.3.1, the actuarial present value of the level benefit of 1000 for the 10-year term insurance payable at the moment of death, with the same force of mortality and with a force of interest 0.06 is 608.71. When the benefit increases over time, to compensate for this increase, the actuarial present value also increases. ∎

The increase in the benefit of the insurance can occur more frequently than once per year. For an m-thly increasing whole life insurance, the benefit would be $1/m$ at the moment of death during the first m^{th} part of a year of the term of the insurance, $2/m$ at the moment of death during the second m^{th} part of a year, during the term of the insurance, and so on, increasing by $1/m$ at m-thly intervals throughout the term of the insurance. For such a whole life insurance, the functions are,

$$b_t \;=\; \frac{\lfloor tm+1 \rfloor}{m} \quad \text{if }\; t > 0$$
$$Z \;=\; \frac{v^T \lfloor Tm+1 \rfloor}{m} \quad \text{if }\; T \ge 0$$

Note that if death occurs in the first m^{th} part then $t < 1/m$, that is $mt < 1$. Hence, $\lfloor tm + 1 \rfloor$ is 1, so that $b_t = 1/m$. Similarly, if death occurs in the second m^{th} part then $1/m < t < 2/m$, that is $1 < mt < 2$. Hence, $\lfloor tm + 1 \rfloor$ is 2, so that $b_t = 2/m$. The actuarial present value is denoted by,

$$E(Z_T) \;=\; (I^{(m)}\overline{A})_x$$

The limiting case, as $m \to \infty$ in the m-thly increasing whole life insurance, is an insurance paying t at the time of death. Its functions are,

$$b_t \;=\; t, \quad t \ge 0 \quad \text{and} \quad Z_T = T v^T \quad T \ge 0$$

The actuarial present value symbol for such an increasing whole life insurance is $(\overline{IA})_x$ and it is given by,

$$(\overline{IA})_x \;=\; \int_0^\infty t v^t \; _t p_x \; \mu_{x+t} dt$$

Example 5.3.12. Find $(\overline{IA})_{30}$ with $\delta = 0.06$ and force of mortality as given in Example 5.3.1.

Solution: By definition

$$(\overline{IA})_{30} = \int_0^\infty t v^t \, g(t) dt$$

where $g(t)$ is as derived in Example 5.3.1. Thus,

$$(\overline{IA})_{30} = 0.12 \int_0^5 t e^{-0.18t} dt + 0.18 e^{0.3} \int_5^\infty t e^{-0.24t} dt = 3.6378 \quad \blacksquare$$

Remark 5.3.4. The actuarial present value $(\overline{IA})_{30}$ obtained in Example 5.3.12 can be interpreted as follows. Suppose (30) lives for 60 years beyond 30. Then at the moment of death the benefit amount to be given is 60 units. Its present value for the given force of mortality and given force of interest is 3.6378. If a unit is taken as 1000, then the benefit amount will be $60 \times 1000 = 60000$ and its actuarial present value is $3.6378 \times 1000 = 3637.8$.

If for m-thly increasing life insurances, the benefit is payable only if death occurs within a term of n years, the insurance is an m-thly increasing n-year term life insurance.

Complementary to the annually increasing n-year term life insurance, is the annually decreasing n-year term life insurance providing n at the moment of death during the first year, $n - 1$ at the moment of death during the second year and so on, with coverage terminating at the end of the nth year. Such an insurance has the following functions.

$$b_t = \begin{cases} n - \lfloor t \rfloor & \text{if} \quad t \le n \\ 0 & \text{if} \quad t > n \end{cases}$$

$$Z = \begin{cases} v^T (n - \lfloor T \rfloor) & \text{if} \quad T \le n \\ 0 & \text{if} \quad T > n \end{cases}$$

The actuarial present value for this insurance is

$$(D\overline{A})^1_{x:\overline{n}|} = \int_0^n v^t (n - \lfloor t \rfloor) \, {}_tp_x \, \mu_{x+t} dt$$

This insurance is complementary to the annually increasing n-year term insurance in the sense that the sum of their benefit functions is the constant $n + 1$ for the n-year term. One may wonder what the use of the decreasing benefit function is. The following example illustrates how the model of decreasing benefit insurance is applicable in practice. Further, it also illustrates how the theory developed for insurances on human lives is also applicable for other objects such as machines, loans, warranties and so on.

Example 5.3.13. Washing machines are sold by a company under the following terms. (i) The price of the washing machine is Rs 10,000. (ii) The company provides a two year warranty. (iii) In the event of failure of the washing machine at time t, for $t \le 2$, the company will refund $(1 - 0.5t)$ times the price. (iv) The refund is payable at the moment of failure. (v) Washing machines are subject to a constant force 0.02 of failure. Calculate the actuarial present value of the warranty when $\delta = 0.08$.

Solution: In the price 10000 of a machine, some part is accounted for the warranty. It is of interest to decide it at the time of fixing the price of a machine. We wish to find the actuarial present value of the warranty. From the given information, the company will refund $(1 - 0.5t)$ times the price if the washing machine fails at time t, for $t \leq 2$. Thus, we define the benefit function as $b_t = 10000(1 - 0.5t)$, $t \leq 2$. Note that it is a decreasing function. If the machine fails at the instant of the purchase, that is if $t = 0$, then the company has to refund the entire amount 10000. If the machine fails after a year, then $t = 1$ and the company has to refund 5000. If $t = 2$, the refund amount is 0, as the company has provided only a two year warranty. Using the given information about the force of interest and the force of failure, we find the actuarial present value of the warranty. It is given by,

$$\int_0^2 10000(1 - 0.5t)e^{-0.08t}e^{-0.02t}(.02)dt \;=\; 200 \int_0^2 e^{-0.1t}dt$$

$$- \; 100 \int_0^2 te^{-0.1t}dt = 538.2$$

Thus, 538.20 out of 10000 is the expected cost of warranty. $\blacksquare$

In the next section, we discuss how to find the actuarial present value of the benefit in a variety of models for life insurance when death benefits are payable at the end of year of death.

5.4   Benefit Payable at the End of the Year of Death

In the previous section, we noted that to find the actuarial present value of the benefit when it is payable at the moment of death, the basic random quantity is $T(x)$, the future life time random variable of the insured of age x. In practice, in some insurance products, the benefits are payable at the end of the year of death, that is, at the end of the policy year after death. As explained in Chapter 1, suppose the policy is written on 20 August 1990 by an individual of age 30. The first policy year is taken as 20 August 1990 to 19 August 1991. If the insured died on 15 May 2022, the policy year ended on 19 August 2022. In this case the benefit was paid on 19 August 2022, that is, at the end of the policy year. Thus, the time period for which the policy is in force is 20 August 1990 to 19 August 2022, that is 32 years. This period is the same as the curtate future life time $K(30)$ of the insured +1 year. Thus, in insurance products where the benefit is payable at the end of year of death, the underlying mathematics is based on the curtate random variable $K(x)$ studied in Section 4.3.

In this section, we discuss all the insurance products studied in Section 3, with the appropriate modification to account for the change in the time of payment of the benefit. The main modification is that the size and time of payment of the death benefits depend only on the number of complete years lived by the insured from policy issue to the time of death. These insurances are referred to as payable at the end of the year of death. The insurance model in such a setup is in terms of functions of the curtate future lifetime of the insured. Suppose b_{k+1} is the benefit function, representing the benefit amount payable at $k + 1$. We again assume that the theory of compound interest is used to find the present value of the money. The basic equation defining the present value random variable Z_{K+1}, at policy issue, of the benefit payment is given by,

$$Z_{K+1} = b_{K+1}v^{K+1} = b_{K+1}e^{-\delta(K+1)} \tag{5.4.1}$$

Note that this equation is similar to Eq. 5.3.1, with the only change being that T is replaced by $K+1$. As in the previous section, we obtain the actuarial present value of the benefit corresponding to the n-year term insurance, n-year endowment insurance, whole life insurance, with level benefit or varying benefit, with or without deferment. Terms and conditions of all these insurance products remain the same except that the benefit is paid at the end of the year of death. For example, for an n-year term insurance with the benefit of a unit amount payable at the end of the year of death, we have

$$b_{k+1} = \begin{cases} 1, & \text{if } k = 0, 1, \cdots, n-1 \\ 0, & \text{elsewhere} \end{cases}$$

$$Z_{K+1} = \begin{cases} v^{K+1}, & \text{if } K = 0, 1, \cdots, n-1 \\ 0, & \text{elsewhere} \end{cases}$$

To find the expectation and variance of this present value random variable, we have to use the distribution of curtate future life time random variable, studied in Section 4.3. It is a discrete random variable with probability mass function given by,

$$P[K(x) = k] = {}_k p_x\, q_{x+k}, \quad \text{for } k = 0, 1, \cdots, w^*$$

where w^* is as defined in Section 4.4. The net single premium for this insurance is denoted by, $A^{\,1}_{x:\overline{n}|}$ and is given by the following formula.

$$A^{\,1}_{x:\overline{n}|} = E(Z_{K+1}) = \sum_{k=0}^{n-1} v^{k+1} P[K(x) = k] = \sum_{k=0}^{n-1} v^{k+1}\, {}_k p_x\, q_{x+k} \tag{5.4.2}$$

Further, $Var(Z) = {}^2A^{\,1}_{x:\overline{n}|} - (A^{\,1}_{x:\overline{n}|})^2$, where, ${}^2A^{\,1}_{x:\overline{n}|} = \sum_{k=0}^{n-1} e^{-2\delta(k+1)}\, {}_k p_x q_{x+k}$

In the formula for actuarial present values and variance of present value random variables, the integrals are replaced by summation. Further note that there is no bar above A, to distinguish it from the net single premium when benefit is payable at the moment of death.

Formulae for net single premiums for other insurance products discussed in Section 3, when benefit is payable at the end of year of death, can be derived on similar lines. Hence, we summarize the benefit function and corresponding international notation for actuarial present value, along with its formula in Table 5.16 for all the insurance products studied in Section 3.

From the formula for $A_{x:\overline{n}|}$, it is to be noted that the net single premium for n-year endowment insurance is again the addition of that for n-year term insurance and n-year pure endowment insurance. Thus,

$$A_{x:\overline{n}|} = A^{\,1}_{x:\overline{n}|} + A^{\,\,1}_{x:\overline{n}|}$$

Observe that $A_{x:\overline{1}|} = vq_x + vp_x = v\ \forall\ x$.

From the formula for net single premium for a whole life insurance issued to (x), it is clear that it may be obtained by letting $n \to \infty$ in the formula for n-year term insurance model.

From the formula for A_x, we find a recursion relation for A_x as follows.

$$
\begin{aligned}
A_x &= \sum_{k=0}^{\infty} v^{k+1}\,{}_kp_x q_{x+k} = vq_x + \sum_{k=1}^{\infty} v^{k+1}\,{}_kp_x q_{x+k} \\
&= vq_x + vp_x \sum_{k=1}^{\infty} v^k\,{}_{k-1}p_{x+1} q_{x+k} = vq_x + vp_x \sum_{j=0}^{\infty} v^{j+1}\,{}_jp_{x+1} q_{x+1+j} \\
&= vq_x + vp_x A_{x+1}
\end{aligned}
\tag{5.4.3}
$$

Using the backward recurrence relation $A_x = vq_x + vp_x A_{x+1}$ and with $A_w = 1$, we can find A_x values for all x.

The following examples illustrate how to compute net single premiums for the three insurance products studied in Section 3, when the benefit is payable at the end of year of death.

Example 5.4.1. Prove that (i) ${}_{m|}A\,^{1}_{x:\overline{n}|} = {}_mE_x\,A\,^{1}_{x+m:\overline{n}|}$
(ii) ${}_{m|}A_{x:\overline{n}|} = {}_mE_x\,A_{x+m:\overline{n}|}$ (iii) ${}_{m|}A_x = {}_mE_x\,A_{x+m}$

Solution: (i) From Table 5.16 we have

$$
\begin{aligned}
{}_{m|}A\,^{1}_{x:\overline{n}|} &= \sum_{k=m}^{m+n-1} v^{k+1}\,{}_kp_x q_{x+k} = \sum_{r=0}^{n-1} v^{m+r+1}\,{}_{m+r}p_x q_{x+m+r} \\
&= v^m\,{}_mp_x \sum_{r=0}^{n-1} v^{r+1}\,{}_rp_{x+m} q_{x+m+r} = {}_mE_x\,A\,^{1}_{x+m:\overline{n}|}
\end{aligned}
$$

(ii) From Table 5.16, we write,

$$
\begin{aligned}
{}_{m|}A_{x:\overline{n}|} &= {}_{m|}A\,^{1}_{x:\overline{n}|} + v^{m+n}\,{}_{m+n}p_x = {}_mE_x\,A\,^{1}_{x+m:\overline{n}|} + v^m\,{}_mp_x\,v^n\,{}_np_{x+m} \\
&= {}_mE_x\,A\,^{1}_{x+m:\overline{n}|} + {}_mE_x\,{}_nE_{x+m} = {}_mE_x\,(A\,^{1}_{x+m:\overline{n}|} + {}_nE_{x+m}) \\
&= {}_mE_x\,A_{x+m:\overline{n}|}
\end{aligned}
$$

(iii) can be obtained from (i) by allowing n to tend to ∞ or by adopting similar steps as in (i) and by changing the upper limit to ∞. ∎

Example 5.4.2. Suppose the force of mortality follows Makeham's law given by $\mu_x = A + BC^x$, where $A = 0.0007, B = 0.0001151$ and $C = 1.096$. (i) Find the values of net single premium for benefit of 1000 to be paid at the end of year of death, in whole life insurance for ages $30, 31, \cdots, 50$ using the backward recurrence relation, with $A_{50} = 0.40356$ and $\delta = 0.05$. (ii) Find the value of the net single premium for benefit of 1000 to be paid at the end of year of death, in the 5-year deferred whole life insurance issued to (40).

Solution: (i) The survival function corresponding to Makeham's law is derived in Example 4.2.11. For the specified values of the parameters, we find the survival function $S(x)$ for values of $x = 30, 31, \cdots, 50$ and then the values of p_x for $x = 30, 31, \cdots, 49$. It is given that $A_{50} = 0.40356$. We then use the backward recurrence relation $A_x = vq_x + vp_x A_{x+1}$ derived in

Product	Benefit function b_{k+1}	Actuarial present value
n-Year term	$1,\ k = 0, 1, \cdots, n-1$ $0,\ \ k = n, n+1, \cdots$	$A^{\,1}_{x:\overline{n}\rvert} = \sum\limits_{k=0}^{n-1} v^{k+1}\ {}_k p_x q_{x+k}$
n-Year endowment	1	$A_{x:\,\overline{n}\rvert} = \sum\limits_{k=0}^{n-1} v^{k+1}\ {}_k p_x q_{x+k} + v^n\ {}_n p_x$ $= A^{\,1}_{x:\overline{n}\rvert} + A_{x:\overline{n}\rvert}^{\ \ 1}$
whole life	1	$A_x = \sum\limits_{k=0}^{\infty} v^{k+1}\ {}_k p_x q_{x+k}$
m-Year deferred n-Year term	$1,\ k = m, m+1, \cdots,$ $m+n-1$ $0,\ \ k = 0, \cdots, m-1,$ $\quad k = m+n, \cdots$	${}_{m\rvert}A^{\,1}_{x:\overline{n}\rvert} = \sum\limits_{k=m}^{m+n-1} v^{k+1}\ {}_k p_x q_{x+k}$
m-Year deferred n-Year endowment	$1,\ k = m, m+1, \cdots,$ $m+n-1$ $0,\ \ k = 0, \cdots, m-1.$	${}_{m\rvert}A_{x:\,\overline{n}\rvert} = \sum\limits_{k=m}^{m+n-1} v^{k+1}\ {}_k p_x q_{x+k}$ $+ v^{m+n}\ {}_{m+n} p_x$
m-Year deferred Whole life	$1,\ \ k = m, m+1, \cdots,$ $0,\ \ k = 0, \cdots, m-1.$	${}_{m\rvert}A_x = \sum\limits_{k=m}^{\infty} v^{k+1}\ {}_k p_x q_{x+k}$
n-Year term Increasing annually	$k+1,\ \ k = 0, 1, \cdots,$ $n-1$ $0, \qquad k = n, n+1, \cdots$	$(IA)^1_{x:\overline{n}\rvert} = \sum\limits_{k=0}^{n-1} (k+1) v^{k+1}\ {}_k p_x\ q_{x+k}$
n-Year term Decreasing annually	$n-k,\ \ k = 0, 1, \cdots,$ $n-1$ $0,\ \ k = n, n+1, \cdots$	$(DA)^1_{x:\overline{n}\rvert} = \sum\limits_{k=0}^{n-1} (n-k) v^{k+1}\ {}_k p_x q_{x+k}$
Whole life Increasing annually	$k+1,\ \ k = 0, 1, \cdots$	$(IA)_x = \sum\limits_{k=0}^{\infty} v^{k+1}(k+1) {}_k p_x q_{x+k}$

Table 5.16 Benefit payable at the end of the year of death

Age x	p_x	$1000A_x$	Age x	p_x	$1000A_x$
30	0.9988	180.30	40	0.9960	276.62
31	0.9986	188.58	41	0.9955	287.95
32	0.9984	197.15	42	0.9950	299.59
33	0.9982	206.02	43	0.9945	311.54
34	0.9980	215.19	44	0.9939	323.80
35	0.9977	224.65	45	0.9933	336.36
36	0.9974	234.43	46	0.9926	349.23
37	0.9971	244.51	47	0.9918	362.38
38	0.9968	254.90	48	0.9909	375.83
39	0.9964	265.61	49	0.9900	389.56

Table 5.17 p_x and $1000A_x$ for Makeham's law

Eq. 5.4.3 to find the net single premium for the benefit of 1000 to be paid at the end of year of death, in whole life insurance for ages 30 to 49. We use Code 5.6.5 to compute these values. These are reported in Table 5.17.

(ii) We have to find $1000\ _{5|}A_{40}$. From Example 5.4.1, we have $_{5|}A_{40} = \ _5E_{40}\ A_{45}$. From (i) we have $A_{45} = 0.33636$. Further, $_5E_{40} = v^5\ _5p_{40} = 0.759525$. Hence, $1000\ _{5|}A_{40} = 255.48$. ∎

Example 5.4.3. It is given that $q_{28} = 0.135$, $q_{29} = 0.146$, $q_{30} = 0.159$, $q_{31} = 0.173$, $q_{32} = 0.188$ and $i = 0.05$. (i) Find the actuarial present value of benefit of 1000 in a 5-year term insurance, 5-year endowment insurance, issued to (28). (ii) Find the actuarial present value of benefit of 1000 in a 2-year deferred 2-year term insurance, 2-year deferred 2-year endowment insurance, issued to (28).

Solution: (i) By definition

$$
\begin{aligned}
A^{\,1}_{28:\overline{5}|} &= \sum_{k=0}^{4} v^{k+1}\ _k p_x q_{x+k} \\
&= v\, q_{28} + v^2 p_{28}q_{29} + v^3 p_{28}p_{29}q_{30} + v^4 p_{28}p_{29}p_{30}q_{31} \\
&\quad + v^5 p_{28}p_{29}p_{30}p_{31}q_{32} = 0.5086848 \\
A_{28:\overline{5}|} &= A^{\,1}_{28:\overline{5}|} + v^5\ _5p_{28} \\
&= A^{\,1}_{28:\overline{5}|} + v^5\ p_{28}p_{29}p_{30}p_{31}p_{32} = 0.8355623
\end{aligned}
$$

Hence, $1000\ A^{\,1}_{28:\overline{5}|} = 508.68\ \ \&\ \ 1000\ A_{28:\overline{5}|} = 835.56$

(ii)We use results of Example 5.4.1 to find the actuarial present value of benefit in deferred insurance. Thus,

$$_{2|}A\,^{1}_{28:\overline{2}|} \;=\; _{2}E_{28}\,A\,^{1}_{30:\overline{2}|} \quad \& \quad _{2|}A_{28:\overline{2}|} \;=\; _{2}E_{28}\,A_{30:\overline{2}|}$$

$$\text{Now } _{2}E_{28} \;=\; v^{2}\,_{2}p_{28} = v^{2}\,p_{28}\,p_{29} = 0.6700317$$

$$\text{Further } A\,^{1}_{30:\overline{2}|} \;=\; v\,q_{30} + v^{2}p_{30}q_{31} = 0.283395$$

$$\& \quad A_{30:\overline{2}|} \;=\; A\,^{1}_{30:\overline{2}|} + v^{2}\,_{2}p_{30} = 0.9142404$$

$$\text{Hence, } 1000\,_{2|}A\,^{1}_{28:\overline{2}|} \;=\; 189.88 \quad \& \quad 1000\,_{2|}A_{28:\overline{2}|} = 612.57$$

■

Example 5.4.4. Suppose the force of mortality follows Gompertz' law given by $\mu_x = BC^x$. Find the values of net single premium for the benefit of 1000 to be paid at the end of year of death, in n-year term insurance, n-year endowment insurance for $n = 1, 2, \cdots, 10$ and whole life insurance for ages $25, 30, 35, 40$. Take $\delta = 0.05, B = 0.0001151$ and $C = 1.096$.

Solution: From Table 5.16, we note that to find the values of the net single premium for the benefit to be paid at the end of year of death, for various insurance products, for various ages, we need to know the probability distribution of $K(x)$. When the force of mortality follows Gompertz' law, we have derived it in Example 4.3.4. Using this distribution and the formulae for actuarial present values as given in Table 5.16, we get the values of net single premium for benefit of 1000 for various insurance products. These are obtained using Code 5.6.6 and reported in Table 5.18, Table 5.19 and Table 5.20.

Age	25	30	35	40
$1000A_x$	148.78	184.47	226.97	276.73

Table 5.18 Gompertz' law: $1000A_x$

n \ Age	25	30	35	40
1	1.13	1.79	2.83	4.48
2	2.31	3.66	5.78	9.12
3	3.54	5.60	8.83	13.93
4	4.82	7.61	12.01	18.92
5	6.15	9.71	15.31	24.09
6	7.54	11.89	18.73	29.44
7	8.98	14.16	22.28	34.97
8	10.48	16.51	25.96	40.69
9	12.04	18.96	29.78	46.60
10	13.66	21.50	33.73	52.69

Table 5.19 Gompertz' law: $1000A\,^{1}_{x:\overline{n}|}$

n \ Age	25	30	35	40
1	951.23	951.23	951.23	951.23
2	904.89	904.92	904.98	905.06
3	860.87	860.97	861.12	861.36
4	819.06	819.25	819.55	820.03
5	779.35	779.67	780.17	780.96
6	741.64	742.12	742.87	744.05
7	705.84	706.50	707.55	709.20
8	671.85	672.74	674.13	676.31
9	639.60	640.73	642.52	645.31
10	608.99	610.41	612.64	616.11

Table 5.20 Gompertz' law: $1000A_{x:\overline{n}|}$

Note that in Table 5.20 $A_{x:\overline{1}|} = vq_x + vp_x = v = 0.95123$ for $x = 25, 30, 35, 40$. Comparing these net single premiums with those reported in Example 5.3.7, when the benefit is paid at the moment of death, we note that all the features noted in Example 5.3.7 are observed here also. The values of net single premium for the benefit of 1000 to be paid at the end of year of death are slightly lower than those when the benefit is paid at the moment of death. It seems reasonable as the period for which policy is in force, is slightly more when the benefit is paid at the end of year of death than that when the benefit is paid at the moment of death.   ∎

Example 5.4.5. For a 2-year term insurance of 10 units issued to (60), it is given that benefits are payable at the end of the year of death and $i = 0.05$. Calculate $Var(Z)$, from the select and ultimate life table with select period 2, given in Table 5.21.

Solution: From the given information we find $E(Z)$ and $E(Z^2)$.

x	$l_{[x]}$	$l_{[x]+1}$	l_{x+2}	$x+2$
60	78,900	77,200	75,100	62
61	76,400	74,700	72,500	63
62	73,800	72,000	69,800	64

Table 5.21 Select and ultimate life table with select period 2

$$
\begin{aligned}
E(Z) &= 10A^1_{[60]:\overline{2}|} = 10(vq_{[60]} + v^2\,{}_{1|}q_{[60]}) \\
&= 10\left[v\frac{l_{[60]} - l_{[60]+1}}{l_{[60]}} + v^2\frac{l_{[61]+1} - l_{[60]+2}}{l_{[60]}}\right] = 0.4466
\end{aligned}
$$

$$
\begin{aligned}
E(Z^2) &= 100\,({}^2A^1_{[60]:2}) = 100[v^2q_{[60]} + v^4\,{}_{1|}q_{[60]}] \\
&= 100\left[v^2\frac{l_{[60]} - l_{[60]+1}}{l_{[60]}} + v^4\frac{l_{[60]+1} - l_{[60]+2}}{l_{[60]}}\right] \\
&= 4.1440 \quad \Rightarrow \quad Var(Z) = 3.94
\end{aligned}
$$

∎

Example 5.4.6. Z is a present value random variable for an insurance to (x) defined by

$$Z = \begin{cases} (6-k)v^{k+1} & \text{if } k = 0,1,2,\cdots,5, \\ 0 & \text{if } k \geq 6 \end{cases}$$

where v^{k+1} is calculated at force of interest δ. Which of the following are true? (i) Z is a present value random variable for a 5-year decreasing term insurance payable at the end of the year of death of (x). (ii) $E(Z) = \sum_{k=0}^{\infty}(6-k)_{k|}A\,{}^{1}_{x:\overline{1}|}$. (iii) $E(Z^2)$ calculated at force of interest δ equals $E(Z)$ calculated at force of interest 2δ.

Solution: (i) is false as Z is a present value random variable for a 6-year decreasing term insurance.

(ii) The actuarial present value symbol for this insurance is $(DA)^{1}_{x:\overline{6}|}$. We derive different expressions for this actuarial present value. By definition of $(DA)^{1}_{x:\overline{n}|}$,

$$(DA)^{1}_{x:\overline{n}|} = \sum_{k=0}^{n-1}(n-k)v^{k+1}\,{}_{k}p_x\,q_{x+k} = \sum_{k=0}^{n-1}(n-k)\,{}_{k|}A\,{}^{1}_{x:\overline{1}|}$$

since by definition of actuarial present value of a unit benefit in m-year deferred n-year term insurance $_{m|}A\,{}^{1}_{x:\overline{n}|} = \sum_{k=m}^{m+n-1}v^{k+1}\,{}_{k}p_x\,q_{x+k}$. With $m = k$ and $n = 1$ it reduces to $v^{k+1}\,{}_{k}p_x q_{x+k}$. Hence (ii) is true. One more expression for $(DA)^{1}_{x:\overline{n}|}$ can be derived as follows. (Although is not needed in this solution.) In the above expression, we substitute $n - k = \sum_{j=0}^{n-k-1}(1)$, to get,

$$(DA)^{1}_{x:\overline{n}|} = \sum_{k=0}^{n-1}\sum_{j=0}^{n-k-1}(1)v^{k+1}\,{}_{k}p_x q_{x+k} = \sum_{j=0}^{n-1}\sum_{k=0}^{n-j-1}(1)v^{k+1}\,{}_{k}p_x\,q_{x+k}$$

by interchanging the order of summation. By comparing the inner summation to the formula for the actuarial present value in the term insurance, we can write $(DA)^{1}_{x:\overline{n}|} = \sum_{j=0}^{n-1}A\,{}^{1}_{x:\overline{n-j}|}$.

(iii) is false as, $b_t^2 \neq b_t$ for all t. ∎

Example 5.4.7. A whole life insurance is issued to (50), with the benefit payable at the end of year of death. It is given that $v = 0.925$, the net single premium for this insurance is 4 if $q_{50} = 0.1$. If P denotes the net single premium for this insurance if $q_{50} = 0.2$, find P.

Solution: If B denotes the actuarial present value at age 51 of the death benefit, then the actuarial present value at age 50 is $vq_{50} + vp_{50}\,B$. With $v = 0.925$ and $q_{50} = 0.1$, we have

$$4 = (0.925)(0.1) + (0.925)(0.9)B \quad \Rightarrow \quad B = 4.094$$

The above expression with $q_{50} = 0.2$ can be written as,

$$P = (0.925)(0.2) + (0.925)(0.8)B \quad \Rightarrow \quad P = 3.66$$

∎

Example 5.4.8. It is given that $A_{35:\overline{1|}} = 0.9434$, $A_{35} = 0.13$, $p_{35} = 0.9964$. Calculate A_{36}.

Solution: Observe that $A_{35:\overline{1|}} = vq_{35} + vp_{35} = v \Rightarrow v = 0.9434$. Hence, $A_{35} = 0.13 = (0.9434)(0.0036) + (0.9434)(0.9964)A_{36} \Rightarrow A_{36} = 0.1347$, using the recurrence relation, $A_x = vq_x + vp_x A_{x+1}$. ∎

Table 5.22 summarizes the international actuarial notation for the net single premiums for the models discussed in Sections 3 and 4.

Insurance product	Benefit at the end of year of death	Benefit at the moment of death
Whole life	A_x	$\overline{A}_x$
n-Year term	$A^{\,1}_{x:\overline{n|}}$	$\overline{A}^{\,1}_{x:\overline{n|}}$
n-Year pure endowment	$A_{x:\overline{n|}}^{\;\;1}$	$A_{x:\overline{n|}}^{\;\;1}$
n-Year endowment	$A_{x:\overline{n|}}$	$\overline{A}_{x:\overline{n|}}$
m-Year deferred n-Year term	$_{m|n}A_x$ or $_{m|}A^{\,1}_{x:\overline{n|}}$	$_{m|n}\overline{A}_x$ or $_{m|}\overline{A}^{\,1}_{x:\overline{n|}}$
m-Year deferred n-Year endowment	$_{m|}A_{x:\overline{n|}}$	$_{m|}\overline{A}_{x:\overline{n|}}$
n-Year term increasing annually	$(IA)^1_{x:\overline{n|}}$	$(I\overline{A})^1_{x:\overline{n|}}$
n-Year term decreasing annually	$(DA)^1_{x:\overline{n|}}$	$(D\overline{A})^1_{x:\overline{n|}}$
Whole life increasing m-thly	$(I^{(m)}A)_x$	$(I^{(m)}\overline{A})_x$

Table 5.22 International actuarial notation for net single premium

5.5   Relation Between A and $\overline{A}$

In all the examples in the previous section, it is to be noted that to calculate the actuarial present value of the benefit payable at the end of the year of death, it is sufficient to have the information on p_x or q_x for various values of x. As stated in Section 4.4, these values are usually estimated from the age specific death rates, which are decided on the basis of vital statistics registry of a state or the region under study. As a consequence, the data on p_x or q_x for integer values of x is easily available from life tables, to model the random mortality law of the individuals and can be revised periodically. Thus, calculation of the actuarial present value of the benefit payable at the end of year of death is possible for a variety of insurance products.

On the other hand, calculation of the actuarial present value of the benefit payable at the moment of death requires the knowledge of probability density function of $T(x)$ and in turn a suitable model for life length X. In practice, given the data on life lengths of various individuals from a specific region, one can estimate $S(x)$ using either the empirical survival function or Kaplan–Meier estimator or Anderson–Nelson estimator or by estimating parameters in some suitable parametric model. Such an estimator then can be used to model the mortality pattern of the individuals in that region and in the calculation of the actuarial present values. However, estimating $S(x)$ is comparatively difficult to estimating p_x or q_x.

Hence, attempts are made to find a relationship between net single premiums corresponding to the insurance payable at the moment of death and at the end of the year of death. Of course, such a relation is possible under certain assumptions and it is the assumption of pattern of deaths at fractional ages. In the present section, we derive the relation between the two under the assumption of a uniform distribution of deaths over a unit interval. Under the assumption of uniformity, for an integer value of $(x + k)$ we have the following result as derived in Section 4.5.

$$_s p_{x+k}\,\mu_{x+k+s} \;=\; q_{x+k}, \qquad 0 \le s \le 1 \tag{5.5.1}$$

We use it to derive the relation between net single premiums corresponding to the insurance payable at the moment of death and at the end of the year of death for the insurance products discussed in Sections 3 and 4. We begin with the n-year term insurance. The actuarial present value of a unit benefit payable at the moment of death is given by,

$$
\begin{aligned}
\overline{A}^{\,1}_{x:\overline{n}|} &= \int_0^n v^t \,_t p_x \,\mu_{x+t}\; dt = \sum_{k=0}^{n-1} \int_k^{k+1} v^t \,_t p_x \,\mu_{x+t}\; dt \\[2mm]
&= \sum_{k=0}^{n-1} \int_0^1 v^{k+s} \,_{k+s} p_x \,\mu_{x+k+s}\; ds \\[2mm]
&= \sum_{k=0}^{n-1} v^{k+1} \,_k p_x \int_0^1 v^{s-1} \,_s p_{x+k} \,\mu_{x+k+s}\; ds \\[2mm]
&= \sum_{k=0}^{n-1} v^{k+1} \,_k p_x q_{x+k} \int_0^1 (1+i)^{1-s} ds \qquad\qquad \text{by (5.5.1)} \\[2mm]
&= \sum_{k=0}^{n-1} v^{k+1} \,_k p_x q_{x+k} \left(-\frac{[(1+i)^{1-s}]_0^1}{\log(1+i)} \right) \\[2mm]
&= A^{\,1}_{x:\overline{n}|}\,(i/\log(1+i)) = (i/\delta)\, A^{\,1}_{x:\overline{n}|} \tag{5.5.2}
\end{aligned}
$$

Allowing n to tend to ∞ in Eq. 5.5.2, we get a similar relation for the whole life insurance. Thus,

$$\overline{A}_x = (i/\delta)\, A_x \tag{5.5.3}$$

For the n-year endowment insurance we have a relation,

$$\overline{A}_{x:\overline{n}|} = \overline{A}\,{}^{1}_{x:\overline{n}|} + A_{x:\overline{n}|}^{\;\;1} = (i/\delta)\, A\,{}^{1}_{x:\overline{n}|} + A_{x:\overline{n}|}^{\;\;1} \tag{5.5.4}$$

Example 5.5.1. Calculate the actuarial present value and the variance of the corresponding random variable for a 10,000 benefit, a 30-year endowment insurance providing the death benefit at the moment of death of a male of age 35 at the issue of the policy. Use the assumption of uniform distribution of deaths over each year of age and $i = 0.06$. Further it is given that $A\,{}^{1}_{35:\overline{30}|} = 0.06748$, $A_{35:\overline{30}|}^{\;\;1} = 0.1392408$ and ${}^{2}\overline{A}_{35:\overline{30}|} = 0.05516$.

Solution: Using Eq. 5.5.4, we calculate the actuarial present value as follows:

$$\begin{aligned}
\overline{A}_{35:\overline{30}|} &= (i/\delta)A\,{}^{1}_{35:\overline{30}|} + A_{35:\overline{30}|}^{\;\;1}\\[4pt]
&= (1.0297087)[0.06748179] + 0.1392408 = 0.208727
\end{aligned}$$

From Eq. 5.3.8, we get the variance as

$$Var(Z) = {}^{2}\overline{A}_{35:\overline{30}|} - (\overline{A}_{35:\overline{30}|})^2 = 0.05516 - (0.208727)^2 = 0.011606$$

For the 10,000 sum insured,
$10{,}000\,\overline{A}_{35:\overline{30}|} = 2{,}087.27$ and $(10{,}000)^2 Var(Z) = 1{,}160{,}600$ ∎

Example 5.5.2. It is given that $S(40) = 0.5$, $S(41) = 0.475$, $i = 0.06$ and $\overline{A}_{41} = 0.54$. Calculate A_{40} under the assumption of uniformity of deaths over each year of age.

Solution: We use the following two relationships: $A_x = vp_x + vp_x A_{x+1}$, which is always valid and $\overline{A}_x = (i/\delta)A_x$, which is valid under the assumption of uniformity in a unit age interval. Thus,

$$\begin{aligned}
0.54 &= \overline{A}_{41} = (i/\delta)A_{41} \;\Rightarrow\; A_{41} = 0.5244\\
p_{40} &= S(41)/S(40) = 0.95 \;\Rightarrow\; A_{40} = vq_{40} + vp_{40}A_{41} = 0.517
\end{aligned}$$

∎

Example 5.5.3. It is given that $A_{x+20} = 0.40$, $A_x = 0.25$, $A_{x:\overline{20}|} = 0.55$ and $i = 0.03$. Calculate $1000\overline{A}_{x:\overline{20}|}$ under the assumption of uniformity in each unit age interval.

Solution: Observe that the two relations

$$A_x = A\,{}^{1}_{x:\overline{20}|} + v^{20}\,{}_{20}p_x\, A_{x+20} \quad \text{and} \quad A_{x:\overline{20}|} = A\,{}^{1}_{x:\overline{20}|} + v^{20}\,{}_{20}p_x$$

$$\Rightarrow 0.25 = A\,{}^{1}_{x:\overline{20}|} + v^{20}\,{}_{20}p_x(0.4) \quad \text{and} \quad 0.55 = A\,{}^{1}_{x:\overline{20}|} + v^{20}\,{}_{20}p_x$$

$$\Rightarrow 0.30 = (0.6)v^{20}\,{}_{20}p_x \;\Rightarrow\; v^{20}\,{}_{20}p_x = 0.5$$

subtracting the first from the second. Further,

$$0.25 \;=\; A^{\,1}_{x:\overline{20|}} + (0.5)(0.4) \;\Rightarrow\; A^{\,1}_{x:\overline{20|}} \;=\; 0.05$$

$$\Rightarrow\; \overline{A}_{x:\overline{20|}} \;=\; (i/\delta)A^{\,1}_{x:\overline{20|}} + v^{20}\,{}_{20}p_x = 0.551$$

under the assumption of uniformity. ∎

Example 5.5.4. Show that under the assumption of a constant force of mortality between integral ages,

$$\overline{A}_x = \sum_{k=0}^{\infty} v^{k+1}\, {}_kp_x\, \mu_{x+k}\, \frac{i + q_{x+k}}{\delta + \mu_{x+k}}, \quad \text{where} \quad \mu_{x+k} = -\log p_{x+k}$$

Solution: Under the assumption of a constant force of mortality between integral ages,

$$\mu_{x+k+s} \;=\; \mu_{x+k} = -\log p_{x+k}$$
$$_sp_{x+k} \;=\; S(x+k+s)/S(x+k) = e^{-\mu s} = (e^{-\mu_{x+k}})^s = (p_{x+k})^s$$

Allowing $n \to \infty$ in the first step of Eq. 5.5.2, we have

$$\overline{A}_x \;=\; \sum_{k=0}^{\infty} v^{k+1}\, {}_kp_x \int_0^1 v^{s-1}\, {}_sp_{x+k}\, \mu_{x+k+s}\, ds$$

$$=\; \sum_{k=0}^{\infty} v^{k+1}\, {}_kp_x\mu_{x+k} \int_0^1 v^{s-1}\, (p_{x+k})^s\, ds$$

$$=\; \sum_{k=0}^{\infty} v^{k+1}\, {}_kp_x\mu_{x+k}\, v^{-1} \int_0^1 (vp_{x+k})^s ds$$

$$=\; \sum_{k=0}^{\infty} v^{k+1}\, {}_kp_x\mu_{x+k}\, \frac{v^{-1}[vp_{x+k} - 1]}{\log(v\, p_{x+k})}$$

$$=\; \sum_{k=0}^{\infty} v^{k+1}\, {}_kp_x\mu_{x+k}\, \frac{[p_{x+k} - (1+i)]}{-\delta + \log p_{x+k}}$$

$$=\; \sum_{k=0}^{\infty} v^{k+1}\, {}_kp_x\mu_{x+k}\, \frac{i + q_{x+k}}{\delta + \mu_{x+k}}$$

since $\log v = -\delta$, $v = (1+i)^{-1}$ and $\mu_{x+k} = -\log p_{x+k}$. ∎

As stated in Section 4.7, the mortality pattern is usually specified by a set of q_x values for integer values of x starting from 0 to some limiting age w. Code 4.7.2 computes the probability distribution of the curtate future life time random variable $K(25)$ when the values of q_x are as specified in Table 4.8. In the following example we find the actuarial present values of the benefit involved in term insurance, endowment insurance, and whole life insurance issued to (25), corresponding to a given set of q_x values. Given a set of q_x values, we cannot find the exact values of actuarial present values of the benefit to be paid at the moment of death. We find these values from the discrete setup, under the assumption of uniformity in unit age interval.

Example 5.5.5. Suppose the mortality law is specified by the values of q_x as specified in Table 4.8. Compute the actuarial present values of the benefit involved in the n-year term insurance, n-year endowment insurance for $n = 1, 2, \cdots, 10$ and whole life insurance issued to (25), when the benefit of 1000 is payable (i) at the end of the year of death and (ii) at the moment of death, under the assumption of uniformity in unit age interval. Suppose the effective rate of interest is $i = 0.06$.

Solution: To compute the actuarial present values of the benefit, we need the probability distribution of $K(25)$. Code 4.7.2 computes the probability distribution of $K(25)$ when the values of q_x are specified. We append this code with the functions to compute the actuarial present values in Code 5.6.7. To find the actuarial present values of the benefit when it is payable at the moment of death, we use the relations derived in Eq. 5.5.2 and Eq. 5.5.3. The output for the term and the endowment insurance is organized in Table 5.23.

n	Term insurance discrete	Term insurance continuous	Pure endowment	Endowment insurance discrete	Endowment insurance continuous
1	1.10	1.14	942.29	943.40	943.43
2	2.16	2.23	887.90	890.06	890.12
3	3.18	3.27	836.63	839.80	839.89
4	4.17	4.29	788.27	792.44	792.57
5	5.16	5.31	742.67	747.83	747.98
6	6.14	6.32	699.65	705.79	705.97
7	7.11	7.32	659.07	666.18	666.40
8	8.07	8.31	620.81	628.88	629.12
9	9.02	9.28	584.72	593.74	594.01
10	9.95	10.25	550.69	560.64	560.94

Table 5.23 Actuarial present values for benefit of 1000

Values of $1000A_{25}$ and $1000\overline{A}_{25}$, are as follows.

$$1000A_{25} = 73.88 \quad 1000\overline{A}_{25} = 76.08$$

■

Example 5.5.6. Suppose the force of mortality follows Gompertz' law given by $\mu_x = BC^x$. Find the values of the net single premium for benefit of 1000 for (i) the whole life insurance for ages $25, 30, 35, 40$ and (ii) for the n-year term insurance for $n = 1, 2, \cdots, 10$ to (25), when the benefit is payable at the moment of death and when the benefit is payable at the end of the year of death. Comment on the values for both the insurance products, under both the modes of payment. Take $\delta = 0.05$, $B = 0.0001151$ and $C = 1.096$.

Solution: Using the formulae used in Examples 5.3.7 and 5.4.1, we find $1000A_x$, $1000(i/\delta)A_x$ and $1000\overline{A}_x$. The values are reported in Table 5.24.

From the table we note that the values in the third and fourth columns are almost the same. Thus, the relation $\overline{A}_x = (i/\delta)A_x$ is approximately satisfied when the mortality pattern is

Age	$1000A_x$	$1000(i/\delta)A_x$	$1000\overline{A}_x$
25	148.78	152.56	152.54
30	184.47	189.15	189.12
35	226.97	232.74	232.70
40	276.73	283.76	283.72

Table 5.24 Gompertz' law: Relation between $1000A_x$ and $1000\overline{A}_x$

governed by Gompertz' law with the specified values of the parameters. It is easy to verify that $_tq_x \neq t\,q_x$ under Gompertz' law, but still we have, $\overline{A}_x = (i/\delta)A_x$ approximately. Table 5.25 reports the values for the term insurance. Here also we note that the relation $\overline{A}\,^1_{x:\overline{n}|} = (i/\delta)\,A\,^1_{x:\overline{n}|}$ holds good approximately. ∎

n	$1000A\,^1_{25:\overline{n}|}$	$1000(i/\delta)A\,^1_{25:\overline{n}|}$	$1000\overline{A}\,^1_{25:\overline{n}|}$
1	1.13	1.16	1.16
2	2.31	2.37	2.37
3	3.54	3.63	3.63
4	4.82	4.94	4.94
5	6.15	6.31	6.31
6	7.54	7.73	7.73
7	8.98	9.21	9.21
8	10.48	10.75	10.74
9	12.04	12.35	12.34
10	13.66	14.01	14.00

Table 5.25 Gompertz' law: Relation between $1000A\,^1_{x:\overline{n}|}$ and $1000\overline{A}\,^1_{x:\overline{n}|}$

We now find the actuarial present value of the annually increasing n-year term insurance payable at the moment of death defined in Section 3. For this insurance, the present-value random variable is

$$Z = \begin{cases} \lfloor T+1\rfloor v^T, & \text{if } T < n \\ 0, & \text{if } T \geq n \end{cases}$$

Note that $\lfloor T+1\rfloor = K+1$. Further, we use the relation stated in Section 4.5, given by $T = K + U$ to rewrite Z as,

$$Z = \begin{cases} (K+1)v^{K+1}v^{U-1}, & \text{for } K = 0,1,\cdots,n-1 \\ 0, & \text{for } K = n,n+1,\cdots \end{cases}$$

It is known that K and U are independent and under the assumption of uniformity for unit age interval $U \sim U(0,1)$. Suppose W denotes the present value random variable for the annually increasing n-year term insurance payable at the end of the year of death, then

$$W = \begin{cases} (K+1)v^{K+1} & K = 0,1,\cdots,n-1 \\ 0 & K = n,n+1,\cdots \end{cases}$$

Thus, $Z = W(1+i)^{1-U}$ and $E(Z) = E[W(1+i)^{1-U}]$. Since W is a function of $K+1$ alone and $K+1$ and $1-U$ are independent, W and $1-U$ are independent. Further, $U \sim U(0,1) \Rightarrow 1-U \sim U(0,1)$. Hence,

$$E[(1+i)^{1-U}] = \int_0^1 (1+i)^x \, dx = i/\log(1+i) = i/\delta \tag{5.5.5}$$

$$\Rightarrow E(Z) = E(W)E[(1+i)^{1-U}] = (i/\delta) \, (IA)^1_{x:\overline{n}|}$$

Thus, we have,

$$(I\overline{A})^1_{x:\overline{n}|} = (i/\delta) \, (IA)^1_{x:\overline{n}|}$$

We use the relation $T = K + U$ in the general model to find a relation between the actuarial present values of the benefit paid at the moment of death and paid at the end of year of death. Under the assumption of a uniform distribution of deaths over each year of age, we have the independence of K and U and $1-U \sim U(0,1)$. Suppose $b_T = b^*_{K+1}$, we can write the present value random variable Z for these insurances as

$$Z = b_T v^T = b^*_{K+1} v^{K+1} (1+i)^{1-U}$$
$$\&\ E(Z) = E[b^*_{K+1} v^{K+1} (1+i)^{1-U}]$$
$$= E(b^*_{K+1} v^{K+1})(i/\delta) \ \text{by Eq. 5.5.5}$$

Example 5.5.7. For an individual of age 50, calculate the actuarial present value for an annually decreasing 5-year term insurance, paying 5,000 at the moment of death, in the first year, 4,000 in the second year, and so on. It is given that $q_{50} = 0.0123$, $q_{51} = 0.0138$, $q_{52} = 0.0151$, $q_{53} = 0.0165$, $q_{54} = 0.018$. Use the assumption of uniform distribution of deaths over each year of age and $i = 0.06$.

Solution: For an annually decreasing 5-year term insurance, we have with 1 unit equal to 1000,

$$b_t = \begin{cases} 5 - \lfloor t \rfloor & \text{if } t \leq 5 \\ 0 & \text{if } t > 5 \end{cases}$$

It is a function of only k, the integral part of t, and hence we may write it as

$$b_k = \begin{cases} 5 - k & k = 0, 1, 2, 3, 4, 5 \\ 0 & k = 6, \cdots \end{cases}$$

The discount function is v^t, so we have

$$(D\overline{A})^1_{50:\overline{5}|} = \frac{i}{\delta}(DA)^1_{50:\overline{5}|}$$

$$= (1.0297087) \sum_{k=0}^4 (5-k)v^{k+1} \, {}_kp_{50} \, q_{50+k} = 0.1867452$$

$$\Rightarrow 1,000(D\overline{A})^1_{50:\overline{5}|} = 186.75$$

∎

The next section presents the R codes used to compute the variety of monetary functions discussed in this chapter.

5.6   R Codes

Code 5.6.1. This code consists of three parts. We use the first part of this code to compute $i^{(m)}$ for $m = 2, 4, 12$ for various values of effective rate of interest i, which are reported in Table 5.1. The second part of this code is used to compute $i^{(m)}$ for $m = 50, 100, 200, 400$ and δ for the same set of values of effective rate of interest i. These are reported in Table 5.2. Table 5.3 displays the present value of 1000 for values of n from 1 to 10 and values of $i = 0.05$ to 0.10, with an increment of 0.01. These are obtained using the third part of this code, in which we use the formula $P = A(1+i)^{-n}$, with A = 1000.

```
int=c(.05,.06,.07,.08,.09,.1)
# Part I: Table 5.1
m=c(2,4,12); E=matrix(nrow=length(int),ncol=length(m))
for (i in 1:length(int))
{
for (j in 1:length(m))
{
E[i,j]=round(m[j]*((1+int[i])^(1/m[j])-1),4)
}
}
E
# Part II: Table 5.2
t=c(50,100,200,400)
F=matrix(nrow=length(int),ncol=length(t))
for (i in 1:length(int))
{
for (j in 1:length(t))
{
F[i,j]=round(t[j]*((1+int[i])^(1/t[j])-1),5)
}
}
del=round(log((1+int)),5);del
F1=cbind(F,del);F1
# Part III: Table 5.3
n=1:10; P=matrix(nrow=length(n),ncol=length(int))
for (i in 1:length(n))
{
for (j in 1:length(int))
{
P[i,j]=round(1000*(1+int[j])^(-n[i]),2)
}
}
```

```
}
P
```

■

Code 5.6.2. In Example 5.3.1, the life length random variable is modelled by a distribution with the specific force of mortality. Corresponding to this mortality law, we find the net single premium for n-year term insurance, pure endowment insurance and for endowment insurance. In each case, the benefit amount is 1000, to be paid at the moment of death, according to the conditions of these three insurance products. We compute these net single premiums for various values of n and various values of δ. We use the following code to find these values. The matrices T_1 and T_2 store the values of net single premium for n-year term insurance for $n = 1$ to 5 and $n = 6$ to 10 respectively. The matrix T stores the values of the net single premium for all n. These are displayed in Table 5.4. The matrices P_1 and P_2 store the values of the net single premium for n-year pure endowment insurance for $n = 1$ to 5 and $n = 6$ to 10 respectively. The matrix P stores the values of net single premium for all n. These are displayed in Table 5.5. Further, it is proved that the net single premium for the endowment insurance is the addition of those for the n-year term insurance and n-year pure endowment insurance. Hence, $E = T + P$ stores the values of the net single premium for n-year endowment insurance for all n. These are displayed in Table 5.6. Part II and Part III of the code are used to draw Figure 5.1 and Figure 5.2 respectively.

```
n=c(1,2,3,4,5); del=c(.05,.06,.07,.08,.09,.10,.11,.12)
T1=T2=U=V=P1=P2=matrix(nrow=length(n),ncol=length(del))
# Part I
for(i in 1:length(n))
{
for(j in 1:length(del))
{
T1[i,j]=(.12/(del[j]+.12))*(1-exp(-(del[j]+.12)*n[i]))
U[i,j]=exp(-5*del[j]-.6)-exp(-(5+n[i])*del[j]-.18*(5+n[i])+.3)
V[i,j]=(.18/(del[j]+.18))*U[i,j]
T2[i,j]=(.12/(del[j]+.12))*(1-exp(-(5*del[j]+.6)))+V[i,j]
P1[i,j]=exp(-(del[j]+.12)*n[i])
P2[i,j]=exp(-(5+n[i])*del[j]-.18*(5+n[i])+.3)
}
}
T3=rbind(T1,T2);T=round(1000*T3,2);T # Term Insurance
P3=rbind(P1,P2);P=round(1000*P3,2);P # Pure Endowment Insurance
E=T+P; E # Endowment Insurance
# Part II
m=1:10; par(mfrow=c(1,2))
plot(m,T[,1], "o",pch = 22,cex=.7,xlab="Year",
```

```
ylab="Net Single Premium:Term Insurance",
ylim= range(min(T[,8]),max(T[,1])),col="green")
lines(m,T[,3],"o",pch=2,col="blue")
lines(m,T[,6],"o",pch=20,col="dark blue")
lines(m,T[,8],"o",pch=19,col="purple")
legend("topleft",legend=c("0.05","0.07","0.10","0.12"),
pch=c(22, 2, 20,19),
col=c("green","blue","dark blue","purple"),cex=.7)
plot(m,P[,1], "o",pch = 22,cex=.7,xlab="Year",
ylab="Net Single Premium:Pure Endowment Insurance",
ylim= range(min(P[,8]),max(P[,1])),col="green")
lines(m,P[,3],"o",pch=2,col="blue")
lines(m,P[,6],"o",pch=20,col="dark blue")
lines(m,P[,8],"o",pch=19,col="purple")
legend("topright",legend=c("0.05","0.07","0.10","0.12"),
pch=c(22, 2, 20,19),
col=c("green","blue","dark blue","purple"),cex=.7)
# Part III
par(mfrow=c(1,1))
plot(m,E[,1], "o",pch = 22,cex=.7,xlab="Year",
ylab="Net Single Premium:Endowment Insurance",
ylim= range(min(E[,8]),max(E[,1])),col="dark green")
lines(m,E[,3],"o",pch=2,col="blue")
lines(m,E[,6],"o",pch=20,col="dark blue")
lines(m,E[,8],"o",pch=19,col="purple")
legend("topright",legend=c("0.05","0.07","0.10","0.12"),
pch=c(22, 2, 20,19),
col=c("dark green","blue","dark blue","purple"),cex=.7)
```

■

Code 5.6.3. In Example 5.3.6, the life length random variable is modelled by a distribution with a specific force of mortality. Corresponding to this mortality law, we find the net single premium for the whole life insurance issued to (25), (30) and (35). For each age, the benefit amount is 1000, to be paid at the moment of death. We compute these for various values of δ. We use Part I of the following code to find these values. These are displayed in Table 5.9. Part II of the code is used to draw Figure 5.3.

```
# Part I
de=c(.05,.06,.07,.08,.09,.1,.11,.12) # values of delta
y1=(.12/(.12+de))*(1-exp(-10*de-1.2))+(.18/(.18+de))*exp(-10*de-1.2)
y2=(.12/(.12+de))*(1-exp(-5*de-.6))+(.18/(.18+de))*exp(-5*de-.6)
```

```
y3=.18/(.18+de)
w=round(1000*data.frame(y1,y2,y3),2)
d=data.frame(de,w); d
# Part II
par(mfrow=c(1,1))
plot(de,w[,1], "o",pch = 22,cex=.7,xlab="Force of Interest",
ylab="Net Single Premium: Whole Life Insurance",
ylim= range(min(w[,1]),max(w[,3])),col="dark green")
lines(de,w[,2],"o",pch=2,col="dark blue")
lines(de,w[,3],"o",pch=20,col="dark red")
legend("topright",legend=c("25","30","35"),pch=c(22, 2, 20),
col=c("dark green","dark blue","dark red"),cex=1.2)
```

■

Code 5.6.4. In Example 5.3.7, the life length random variable is modelled by Gompertz' law given by $\mu_x = BC^x$, $x \geq 0$. We use this code to compute the values of the net single premium for the benefit of 1000 for the n-year term insurance, n-year endowment insurance for $n = 1, 2, \cdots, 10$ and whole life insurance for ages $25, 30, 35, 40$. We take $\delta = 0.05, B = 0.0001151$ and $C = 1.096$. To compute $P[a \leq W \leq b]$, where $W \sim G(\alpha_x, \lambda)$ distribution, we use the pgamma function. Part I of the code is concerned with the term insurance and endowment insurance. The results are displayed in Table 5.10 and Table 5.11 respectively. Part II is for the whole life insurance and the results are reported in Table 5.12. In Example 5.3.10, we compute the net single premium for the 5-year deferred n-year term insurance, n-year endowment insurance for $n = 1, 2, \cdots, 5$ and for the whole life insurance for ages $25, 30, 35, 40$, when the mortality law is Gompertz' law. Part III, IV and V of the code compute these net single premiums. These are displayed in Table 5.13, Table 5.14 and Table 5.15 respectively.

```
# Part I: Term and Endowment insurance
B=0.0001151; C=1.096; m=B/log(C); del=0.05; n=1:10
la=(-del/log(C))+1; x=c(25,30,35,40); alx=m*C^x
T=P=S=matrix(nrow=length(n),ncol=length(x))
for(i in 1:length(n))
{
for(j in 1:length(x))
{
S[i,j]=pgamma(C^n[i],la,alx[j])-pgamma(1,la,alx[j])
T[i,j]=exp(alx[j])*gamma(la)*alx[j]^(1-la)*S[i,j]
P[i,j]=exp(-del*n[i]-alx[j]*(C^n[i]-1))
}
}
T1=round(1000*T,2); T1 # Term insurance
E=T+P; E1=round(1000*E,2); E1 # Endowment insurance
```

```
# Part II: Whole life insurance
w=c()
for(j in 1:length(x))
{
w[j]=exp(alx[j])*gamma(la)*alx[j]^(1-la)*(1-pgamma(1,la,alx[j]))
}
w1=round(1000*w,2);w1 # Whole life insurance
# Part III: 5-year Deferred term insurance
e=matrix(c(1,1,1,1,1),nrow=5,ncol=1)
DT=T1[6:10,]-e%*%T1[5,]; DT
# Part IV: 5-year Deferred endowment insurance
DE=E1[6:10,]-e%*%T1[5,]; DE
# Part V: 5-year Deferred whole life insurance
DW=w1-T1[5,]; DW
```

■

Code 5.6.5. In Example 5.4.2, the life length random variable is modelled by Makeham's law $\mu_x = A + BC^x, x \geq 0$, with $A = 0.0007, B = 0.0001151$ and $C = 1.096$. We use this code to compute the values of the net single premium for the benefit of 1000 for the whole life insurance for ages $30, 31, \cdots, 49$ given the value of A_{50}. We use the backward recurrence relation derived in Eq. 5.4.3 with $\delta = 0.05$. The results are displayed in Table 5.17.

```
A=0.0007; B=0.0001151; C=1.096; m=B/log(C); del=0.05
v=exp(-del); x=30:50; y=exp(A*x -m*(C^x-1)); l=length(x);p=1:(l-1)
for (i in 1:(l-1))
{
p[i]=y[i+1]/y[i]
}
p1=1-p; A1=c(1:(l-1),.403561)
for(i in 1:20)
{
A1[l-i]=v*p1[l-i] + v*p[l-i]*A1[l-i+1]
}
A2=round(1000*A1[-l],2); p2=round(p,4)
x1<-30:49; d1=data.frame(x1,p2,A2); d1
```

■

Code 5.6.6. In Example 5.4.4, the force of mortality follows Gompertz' law given by $\mu_x = BC^x$, with $B = 0.0001151$ and $C = 1.096$. Using this code, we find the values of the net single premium for the benefit of 1000 to be paid at the end of the year of death, in n-year term insurance, n-year endowment insurance for $n = 1, 2, \cdots, 10$ and whole life insurance for ages $25, 30, 35, 40$. We take $\delta = 0.05$. The results are displayed in Table 5.18, Table 5.19 and Table 5.20 respectively. In the following code, we use two functions, `sum(x)` and `cumsum(x)`. If x is a vector of observations, then `sum(x)` gives the sum of the elements of vector x and `cumsum(x)` creates a vector of cumulative sums of the elements of x. For example, suppose `x =c(2,3,5,7)`. Then the functions

```
x=c(2,3,5,7); y=sum(x); z=cumsum(x);
```

result in `y=sum(x)=17` and `z=cumsum(x)` $= (2, 5, 10, 17)$. The `cumsum(x)` function is useful to find $A^{1}_{x:\overline{n}|}$ for various values of n.

```
B=0.0001151; C=1.096; m=B/log(C); del=0.05; v=exp(-del)
x=c(25,30,35,40);k=0:(100-min(x))
y=matrix(nrow=length(k),ncol=length(x))# pmf of K(x)
for(i in 1:length(k))
{
for(j in 1:length(x))
{
y[i,j]=exp(-m*C^x[j]*(C^k[i]-1))-exp(-m*C^x[j]*(C^(k[i]+1)-1))
}
}
b=v^(k+1);w=c(); t=matrix(nrow=length(k),ncol=length(x))
for(j in 1:length(x))
{
w[j]=1000*sum(y[,j]*b)
t[,j]=1000*cumsum(y[,j]*b)
}
wh=round(w,2);wh # Whole life insurance
T=round(t[1:10,],2);T # Term insurance
n=1:10; P=matrix(nrow=length(n),ncol=length(x))
for(i in 1:length(n))
{
for(j in 1:length(x))
{
P[i,j]=1000*v^n[i]*exp(-m*C^x[j]*(C^n[i]-1))
}
}
Pu=round(P,2); E=T+Pu; E # Endowment insurance
```

∎

Code 5.6.7. In Example 5.5.5, we compute the actuarial present values of the benefit involved in term insurance, endowment insurance and whole life insurance issued to (25), when the mortality law is specified by values of q_x for integer values of x. We use this code to compute the values, when the effective rate of interest is $i = 0.06$. The results are displayed in Table 5.23.

```
z=read.table("F://qx.txt", header=T);
x=z[,1] # column of x values running from 0 to 110
q=z[,2] # column of qx values
pr=1-q # column of px values
p=pr[26:111] # column of px values for x = 25 to 110
p1=c(p[1],2:85) # kpx for x=25 and k=1 to 85,
# first element being p25
for (i in 2:85)
{
p1[i]=p1[i-1]*p[i]
}
q=1-p; # column of qx values for x=25 to 110
p3=c(q[1],2:84)
for(i in 2:85)
{
p3[i]=p1[i-1]*q[i]
}
int=0.06; v=(1+int)^(-1); del=log(1+int)
k=0:84; n=1:10; b=v^(k+1); v1=v^n
w25=cumsum(p3*b); t25=w25[1:10] # Term insurance
tbar25=(int/del)*t25; pu25=p1[1:10]*v1
e25=t25+pu25; ebar25=tbar25+pu25
d=data.frame(t25,tbar25, pu25,e25,ebar25)
d1=round(data.frame(n,1000*d),2);d1
wh25=1000*sum(p3*b);wbar25=(int/del)*wh25
wh25=round(wh25,2); wbar25=round(wbar25,2);wh25; wbar25
```

∎

5.7 Conceptual Exercises

5.7.1 (a) Suppose the effective rate of interest per annum is 0.04. Find $d, v, i^{(2)}, i^{(4)}$, $i^{(12)}, i^{(365)}$ and δ. (b) Suppose the force of interest per annum is 0.06. Find $i, d, v, i^{(2)}, i^{(4)}, i^{(12)}$ and $i^{(365)}$. Comment on the results.

5.7.2 Find the accumulated amount of 50000 if it is invested for (a) one day, with nominal interest rate 4% per annum convertible daily, (b) one month with nominal interest rate 4.5% per annum convertible monthly and (c) one year with interest rate 5% per annum.

5.7.3 How much money has to be deposited in an account which has 5% annual effective rate of interest on 1 June 2024, if we want to make a withdrawal of 5000 on 1 June 2025, of 10000 on 1 June 2026 and 15000 on 1 June 2027?

5.7.4 Find the present value on 1 January 2024 of the following cash flow: 6000 to be received on 1 January 2027, 4000 to be paid on 1 January 2028, 5000 to be received on 1 January 2029, 6000 to be paid on 1 January 2030 and 7000 to be received on 1 January 2031, when the effective rate of discount is 5%.

5.7.5 The force of interest is 0.06. Find the nominal rate of interest per annum on deposits of term of seven days, one month and six months.

5.7.6 Suppose the force of interest per annum is 0.06. Suppose Rs 40000 are invested on 1 January 2023. Find its accumulated value on (a) 1 February 2023, (b) 21 July 2023 and (c) 1 January 2027.

5.7.7 Assume that future life time distribution of (30) is a gamma distribution with a shape parameter α and scale parameter β. Suppose $\delta = 0.05$. Write expressions for the following, in terms of the distribution function of a gamma distribution with appropriate parameters. (a) $\bar{A}\,{}^{1}_{30:\overline{25}|}$. (b) Actuarial present value for a 25-year term insurance with benefit $b_t = e^{0.05t}$ payable at death, for a person of age 30 at policy issue. (c) $\bar{A}_{30}$ and $(\overline{IA})_{30}$.

5.7.8 Z is the present value random variable for a whole life insurance of 1 payable at the moment of death of (x). It is given that $\mu_{x+t} = 0.05$ for $t \geq 0$ and $\delta = 0.10$. Which of the following are true?

   (a) $\frac{d}{dx}\bar{A}_x = 0$, (b) $E(Z) = 1/3$, (c) $Var(Z) = 1/5$.

5.7.9 If $A_x = 0.25$, $A_{x+20} = 0.40$ and $A_{x:\overline{20}|} = 0.55$, calculate (a) $A\,{}^{1}_{x:\overline{20}|}$ and (b) $A\,{}^{1}_{x:\overline{20}|}$.

5.7.10 Z is a present value random variable for a discrete one-year term insurance of 1000 issued to (x). It is given that $E(Z) = 19$ and $Var(Z) = 17,689$. Calculate i.

5.7.11 On the basis of the information $A_{60} = 0.58896$, $A_{60:\overline{1}|}^{\ 1} = 0.9506$ and $A_{61} = 0.60122$, calculate q_{60}.

5.7.12 It is given that (i) deaths are uniformly distributed over each year of age, (ii) $i = 0.05$, (iii) $q_{35} = 0.01$ and $\bar{A}_{36} = 0.185$. Calculate A_{35}.

5.7.13 Given that $A_{60} = 0.34487$, $A_{61} = 0.35846$, $p_{60} = 0.98624$, calculate i.

5.7.14 Given that $p_{60} = 0.985$, $p_{61} = 0.98$, $i = 0.05$ and $A_{62} = 0.6$, evaluate A_{60} and A_{61}.

5.7.15 A 3-year temporary insurance to (40) is purchased for Rs 494.47. Find the sum insured payable at the end of year of death if $i = 0.06$, $q_{40} = 0.0028$, $q_{41} = 0.003$ and $q_{42} = 0.0032$.

5.7.16 It is given that, $q_{25} = 0.145, q_{26} = 0.141, q_{27} = 0.156, q_{28} = 0.174, q_{29} = 0.185, q_{30} = 0.194$ and $i = 0.06$. (a) Find the actuarial present value of benefit of 1000 in a 6-year term insurance, 6-year pure endowment insurance and 6-year endowment insurance, issued to (25), when the benefit is payable either at the moment of death or at the end of year of death. (b) Find the actuarial present value of benefit of 1000, payable at the end of year of death, in a 2-year deferred 2-year term insurance, 2-year deferred 2-year endowment insurance, issued to (25). State the assumptions, if any.

5.7.17 Suppose the force of mortality follows Makeham's law given by $\mu_x = A + BC^x$, with $A = 0.0007, B = 0.0001151$ and $C = 1.096$. Take $\delta = 0.05$. (a) Find the values of the net single premium for a benefit of 1000 to be paid at the end of the year of death, for whole life insurance for ages $25, 30, 35$ and 40. (b) Using these values and the assumption of uniformity of deaths in unit age interval, find the values of net single premium for the benefit of 1000 to be paid at the moment of death for whole life insurance for ages $25, 30, 35$ and 40. (c) Find the values of the net single premium for a benefit of 1000 to be paid at the end of year of death, in n-year term insurance, n-year pure endowment insurance and n-year endowment insurance for $n = 1, 2, \cdots, 10$.

5.7.18 Express in actuarial notation the net single premium corresponding to the following insurances: (a) A death benefit of 30000 in a 20-year endowment is payable at the end of year of death of (30). (b) An insurance to (40) pays 1 lakh at the moment of death if death occurs after the age of 50. (c) A death benefit of the insurance to (40) is 30000 payable at the moment of death if death occurs between the ages of 55 and 65.

5.7.19 A whole life insurance is issued to (x), with the benefit payable at the moment of death. Suppose $\mu_{x+t} = 0.05, t > 0, \delta = 0.06$. The death benefit at a time t is $b_t = \exp(0.04t)$. Calculate $Var(Z)$, where Z is the present value random variable for this insurance at issue.

5.7.20 For a whole life insurance issued to (41), with a benefit of 1 unit payable at the end of year of death, calculate $Var(Z)$ given the following information. $i = 0.06, p_{40} = 0.9892$, $A_{41} - A_{40} = 0.0079$ and $^2A_{41} - {}^2A_{40} = 0.0045$.

5.7.21 If $\mu(x) = \mu$, a positive constant, for all $x > 0$, show that
$\overline{A}_{x:\overline{n}|} = (\mu + \delta e^{-(\delta+\mu)n})/(\mu + \delta)$.

5.7.22 Suppose $\mu(x) = 1/x$, for all $x > 0$. Show that
$\overline{A}\,{}^1_{x:\overline{n}|} = 1 - xe^{-\delta n}/(x+n) - \delta x \int_0^n e^{-\delta t}/(x+t)\ dt$.

5.7.23 Show that $\overline{A}_x \geq v^{\overset{0}{e}_x}$, for all x. Verify the relation for $\mu_x = \mu = 0.05$ for all $x \geq 0$ and $\delta = 0.05$.

5.7.24 Suppose the mortality pattern is described by Gompertz' law and δ denotes the force of interest. Suppose the benefit function is given by $b_t = e^{\alpha t}$. Show that the actuarial present value for the whole life insurance and 5-year endowment insurance issued to (30), when benefit is paid at the moment of death, can be expressed in terms of distribution function of gamma distribution, with the shape parameter and scale parameter to be stated. State the conditions that you need to impose on the values of δ, α and the parameters of Gompertz' law. Calculate the actuarial present value for suitable values of parameters of mortality law and for δ and α.

5.8   Computational Exercises

5.8.1 Suppose the values of effective rate of interest i are $0.04, 0.05, 0.06, 0.07, 0.08$, 0.09 and 0.10. Prepare a table to display the corresponding values of $i^{(2)}$, $i^{(4)}$, $i^{(12)}$ and δ. Comment on the ordering among these monetary functions.

Note: In all the following exercises, use the mortality pattern of the individual as specified by Makeham's law, with parameters as decided in the computational exercise 4.9.1. Take $\delta = 0.05$.

5.8.2 Find $\overline{A}^{1}_{x:\overline{n}|}$, $_nE_x$ and $\overline{A}_{x:\overline{n}|}$ for $n = 1, 2, \cdots, 10$ and $x = 25, 30, 35, 40$. Also find $\overline{A}_x$. Display the results in a tabular form. Draw figures to view the changes in these monetary functions as n changes for fixed x, and as x changes for fixed n. Comment on the results. Compare these values with the corresponding values when the mortality pattern is specified by Gompertz' law, computed in Example 5.3.7.

5.8.3 Find $A^{1}_{x:\overline{n}|}$ and $A_{x:\overline{n}|}$ for all values of $n = 1, 2, \cdots, 10$, and $x = 25, 30, 35, 40$. Also find A_x. Display the results in a tabular form. Draw figures to show changes in these monetary functions as n changes for fixed x, and as x changes for fixed n. Comment on the results. Compare these values with the corresponding values when the mortality pattern is specified by Gompertz' law, computed in Example 5.4.4.

5.8.4 For $x = 25, 30, 35, 40$, find $(i/\delta)A_x$ and compare with $\overline{A}_x$. Comment on your findings.

5.8.5 For $x = 25, 30, 35, 40$ and $n = 1, 2, \cdots, 10$, find $(i/\delta)A^{1}_{x:\overline{n}|}$ and compare with $\overline{A}^{1}_{x:\overline{n}|}$. Comment on the results.

5.9   Multiple Choice Questions

Note: Unless specified otherwise, you have to identify which of the options is correct. Answers are given in the solutions of conceptual exercises.

5.9.1 Following are four statements about the relationship between the annual effective rate of interest i and the discount factor v. (I) $i = (1 - v)/v$. (II) $i/(1 + i) = v$.
(III) $i/(1 - i) = v$. (IV) $v = (1 + i)^{-1}$.

   (a) only (I) and (II) are true
   (b) only (II) and (III) are true
   (c) only (I), (II) and (III) are true
   (d) only (I) and (IV) are true

5.9.2 Following are four statements about the relationship between the annual effective rate of interest i and the effective rate of discount d.
   (I) $i = d/(1 - d)$. (II) $i/(1 + i) = d$. (III) $i/(1 - i) = d$. (IV) $i = (1 - d)^{-1}$.

   (a) only (I) and (II) are true
   (b) only (II) and (III) are true

(c) only (I), (II) and (III) are true
(d) only (I) and (IV) are true

5.9.3 Following are four statements about the relationship among the discount factor v, the annual effective rate of interest i and the effective rate of discount d. (I) $d = i/(1+i) = iv$. (II) $1 - d = v = (1 + i)^{-1}$. (III) $i = d/(1 - d)$. (IV) $i = (1 - v)/v$.

(a) only (I) and (II) are true
(b) only (II) and (III) are true
(c) only (I), (II) and (III) are true
(d) all four are true

5.9.4 Following are three inequalities among d, v and i. (I) $d < v$, (II) $d < i$ and (III) $i < v$.

(a) only (I) is true
(b) only (III) is true
(c) both (I) and (II) are true
(d) all three are true

5.9.5 The nominal rate of interest $i^{(m)}$ is defined as

(a) $i^{(m)} = m\left\{(1 - i)^{1/m} + 1\right\}$
(b) $i^{(m)} = m\left\{(1 - i)^{1/m} - 1\right\}$
(c) $i^{(m)} = m\left\{(1 + i)^{1/m} - 1\right\}$
(d) $i^{(m)} = m\left\{(1 + i)^{1/m} + 1\right\}$

5.9.6 Following are three statements. (I) $i = 0.06$, $i^{(2)} = 0.0591$, (II) $i = 0.06$, $i^{(4)} = 0.0587$, $i^{(12)} = 0.0584$ and (III) $i = 0.06$, $i^{(4)} = 0.0584$, $i^{(12)} = 0.0587$.

(a) only (I) is true
(b) only (III) is true
(c) both (I) and (II) are true
(d) all three are true

5.9.7 Following are three statements about relations among the instantaneous rate of interest δ, effective rate of interest i, nominal rate of interest $i^{(m)}$ and the discount factor v. (I) $\delta = \log(1 + i)$, (II) $e^{-\delta} = v$ and (III) $\delta = \lim_{m \to \infty} i^{(m)}$.

(a) only (I) is true
(b) only (III) is true
(c) both (I) and (II) are true
(d) all three are true

5.9.8 Following are four inequalities among d, v, δ and i. (I) $d < v$, (II) $d < i$, (III) $i < v$ and (IV) $i > \delta$.

(a) only (I) is true
(b) only (II) and (III) is true
(c) both (I) and (II) are true
(d) all four are true

5.9.9 The amount to which Rs $1,000$/- will accumulate at an effective rate of discount 3% per annum for 8 years is

(a) Rs 1280.24
(b) Rs 1301.20
(c) Rs 1275.93
(d) Rs 1211.44

5.9.10 How much money should be deposited today so that after 5 years investor will get 10,000 rupees? Suppose the effective rate of interest is 0.06.

(a) Rs 7663.42
(b) Rs 7227.50
(c) Rs 7430.70
(d) Rs 7472.58

5.9.11 If the rate of interest is 5% per annum payable quarterly, Rs $10,000$/- after 10 years will accumulate to

(a) $10^4(1+0.05)^{10}$
(b) $10^4(1+0.05/4)^{40}$
(c) $10^4(1+0.05)^{40}$
(d) $10^4(1+0.05/4)^{10}$

5.9.12 The present value random variable for an n-year endowment life insurance, when a unit benefit is payable at the moment of death is defined as,

(a)
$$Z_T = \begin{cases} v^T, & \text{if} \quad T \le n \\ v^n, & \text{if} \quad T > n \end{cases}$$

(b)
$$Z_T = \begin{cases} v^T, & \text{if} \quad T \le n \\ 0, & \text{if} \quad T > n \end{cases}$$

(c)
$$Z_T = \begin{cases} v^n, & \text{if} \quad T \le n \\ 0, & \text{if} \quad T > n \end{cases}$$

(d) $Z = v^T \quad \text{if} \quad T \le n$

5.9.13 The net single premium $_{m|}\overline{A}_x$ for an m-year deferred whole life insurance when a unit benefit is payable at the moment of death is given by

(I) $A_{x:\overline{m}|}^{\;\;1}\overline{A}_{x+m}$. (II) $\overline{A}_x - \overline{A}_{x:\overline{m}|}^{\;\;1}$. (III) $\overline{A}_x + \overline{A}_{x:\overline{m}|}^{\;\;1}$. (IV) $\overline{A}_{x:\overline{m}|}^{\;\;1}\overline{A}_{x+m}$.

(a) only (I) is true
(b) only (II) and (III) is true
(c) only (I) and (II) are true
(d) all four are true

5.9.14 The net single premium for an n-year term insurance when a unit benefit is payable at the end of year of death is denoted by

(a) $A_{x:\overline{n}|}$

(b) $A^1_{x:\overline{n}|}$

(c) $A_{x:\overset{1}{\overline{n}|}}$

(d) $\overline{A}_{x:\overline{n}|}$

5.9.15 Following are three statements about the net single premium $\overline{A}_x$ for the whole life insurance of 1 unit benefit. (I) $\overline{A}\,^1_{x:\overline{n}|} \to \overline{A}_x$ as $n \to \infty$. (II) $\overline{A}_{x:\overline{n}|} \to \overline{A}_x$ as $n \to \infty$. (III) $\overline{A}_x$ is the moment generating function of $T(x)$ at $(-\delta)$.

(a) only (I) is true

(b) only (II) is true

(c) both (I) and (II) are true

(d) all three are true

5.9.16 Following are three statements. (I) $\overline{A}_x$ increases as age increases. (II) $\overline{A}_{x:\overline{n}|}$ decreases as n increases. (III) $\overline{A}\,^1_{x:\overline{n}|}$ decreases as n increases.

(a) only (I) is true

(b) only (II) is true

(c) both (I) and (II) are true

(d) all three are true

5.9.17 Following are three statements. (I) A_x increases as age increases. (II) $A_{x:\overline{n}|}$ decreases as n increases. (III) $A\,^1_{x:\overline{n}|}$ decreases as n increases.

(a) only (I) is true

(b) only (II) is true

(c) both (I) and (II) are true

(d) all three are true

5.9.18 A whole life insurance is issued to (80), with benefit payable at the end of year of death. It is given that $v = 0.925$, the net single premium for this insurance is 4 if $q_{80} = 0.1$. If P denotes the net single premium for this insurance if $q_{80} = 0.2$, the value of P

(a) is 34.24

(b) is 7.07

(c) is 3.66

(d) cannot be calculated in view of insufficient information

5.9.19 It is given that $p_{40} = 0.95$, $i = 0.06$, $\overline{A}_{41} = 0.54$. Under the assumption of uniformity of deaths over each year of age A_{40} is

(a) 0.57

(b) 0.63

(c) 0.67

(d) 0.52

5.9.20 Which of the following identities is correct?

(a) $e_x = q_x(1 + e_{x+1})$
(b) $e_{x+1} = p_x(1 + e_x)$
(c) $e_x = p_x(1 - e_{x+1})$
(d) $e_x = p_x(1 + e_{x+1})$

5.9.21 Mr Gore's life insurance policy provides a death benefit of Rs 1 lakh if his death occurs during the 3-year period in which the policy is in force. His annual premium payment remains the same throughout this period. At the end of the period his coverage will expire.
Among the following, choose the term that correctly matches the description of the above insurance product.

(a) limited payment whole life insurance
(b) continuous premium whole life insurance
(c) level term life insurance
(d) level endowment life insurance

5.9.22 It is given that $A_{35:\overline{1}|} = 0.9434$, $A_{35} = 0.13$, $p_{35} = 0.9964$. Then A_{36} is

(a) 0.940
(b) 1.069
(c) 0.135
(d) 0.133

5.9.23 For a whole life insurance of 1 unit benefit, payable at the moment of death, with $\delta = 0.06$, $\overline{A}_x = 0.6$. Suppose μ_{x+t} is increased by 0.03 $\forall$ t and δ is decreased by 0.03. Then the revised value of $\overline{A}_x$ is

(a) 0.57
(b) 0.60
(c) 0.63
(d) 0.30

5.9.24 Following are three statements. Under the assumption of uniformity in a unit interval for integer value of x, (I) $\overline{A}_x = (i/\delta)A_x$, (II) $\overline{A}_{x:\overline{n}|} = (i/\delta)A_{x:\overline{n}|}$ and (III) $\overline{A}\,^1_{x:\overline{n}|} = (i/\delta)A\,^1_{x:\overline{n}|}$.

(a) only (I) is true
(b) only (II) is true
(c) both (I) and (III) are true
(d) all three are true

5.9.25 It is given that $i = 0.06$ and $\overline{A}_{51} = 0.54$. Under the assumption of uniformity of deaths over each year of age, A_{51} is

(a) 0.5722
(b) 0.5244
(c) 0.5163
(d) 0.5145

5.9.26 It is given that $i = 0.05$. Following are three statements. (I) $\delta = 0.0488$, (II) $d = 0.0476$ and (III) $v = 0.0462$. Then

(a) only (I) is true
(b) only (II) is true
(c) both (I) and (II) are true
(d) all three are true

Chapter 6

Annuities

Key Terms: Annuity certain, Annuity due, Annuity immediate, Certain and whole life annuity, Continuous life annuity, Deferred annuity, Discrete life annuity, Temporary life annuity.

6.1 Introduction

Determination of premiums for a variety of insurance products which are acceptable to both the insured and the insurer is one of the major tasks of an actuary. As stated in Chapter 1, the main guiding principle in the determination of premiums for the insurance company is,

$$\text{Actuarial present value of outflow} = \text{Actuarial present value of inflow}$$

In Chapter 5, we have discussed how to find the left hand side of this equation for a variety of insurance products, when the benefit is payable either at the moment of death or at the end of the year of death. In this chapter we will discuss how to find the right hand side of this equation for various modes of premium payments. We have noted that the actuarial present value of the benefit can be interpreted as a net single premium or the lump-sum purchase price of the insurance product. In practice however, many times, premiums are paid in instalments, either annually or quarterly or monthly. The number of instalments may be fixed or it may be a random variable if the premiums are paid till the individual is alive. Thus, the payment of premiums is an annuity, as defined in Chapter 1. For ready reference we repeat the definition below.

Definition 6.1.1. *Annuity: A series of periodic payments or a special type of cash flow, where payments occur at regular intervals, is known as an annuity.*

Annuities are common in our day to day transactions. Most of us make and receive such periodic payments; for example, monthly rent, monthly telephone bills, salaries paid on a regular basis, monthly scholarships, equal monthly instalments for repayment of housing loan or car loan. The payment of premiums by the insured to the insurer is an annuity.

As noted in Section 1.5, in the financial services industry, the term annuity is also used as a financial contract. An annuity contract is as defined below.

Definition 6.1.2. *Annuity contract: It is a saving plan sold by the insurance companies to provide regular income to the insured.*

For example, an individual after retirement invests in the annuity contract, to get regular, monthly or annual income from the insurer. Thus, in such a contract, the insurer promises to make a series of periodic payments to the insured, in exchange of a single premium or a series of premiums from the insured. In the annuity contract, benefit payment is again the outflow of the insurance company, but it is a periodic outflow. Techniques discussed in this chapter enable us to find its actuarial present value. An insurer uses a combination of factors to calculate the amount of the periodic annuity benefit payments that it will be liable to pay under an annuity policy. Every annuity calculation, however, is based on the following basic mathematical concept. *A certain amount of money, known as the principal, that is invested for a certain period of time at a certain rate of interest, can be paid out in a series of periodic payments over a stated period of time.*

In this chapter, the term annuity is referred to as a periodic inflow to the insurance company via premiums and also a periodic outflow of the insurance company via payment of benefits in the annuity contract.

There are various forms of annuities, when the term annuity is designated for a series of periodical payments from one party to the other. Depending on the nature of the period of annuity, it is dichotomized as annuity certain and life annuity.

1. **Annuity Certain:** If the period of annuity payments is fixed, then such an annuity is known as an *annuity certain.* For example, monthly scholarship for 24 months of post graduate study forms the annuity certain. In some insurance products the number of premiums are fixed. In such situations, premium payments form the annuity certain.

2. **Life Annuity:** An annuity in which number of payments depend on the survival of the individual is known as a *life annuity.* Whole life insurance is usually purchased by paying the premiums till the insured is alive. Premium payments then form the life annuity. Monthly pension that a government employee receives after retirement till he is alive, forms the life annuity.

   Thus, in life annuity, the period of annuity payments is a random variable while in annuity certain it is deterministic. Each of these two types of annuities are further classified as continuous or discrete annuities depending on the mode of payments.

3. **Continuous Annuity:** An annuity in which payments are made continuously at a rate of 1 unit per annum is known as a *continuous annuity.*

4. **Discrete Annuity:** An annuity in which payments are made at discrete time points at a rate of 1 unit per annum is known as a *discrete annuity.*

   The discrete time points may be the beginning of a year or the end of a year. Such a distinction gives rise to the following two types of discrete annuities.

5. **Annuity Due:** An *annuity due*, also known as annuity payable annually in advance, is an annuity in which payments are made at the beginning of the year.

6. **Annuity Immediate:** An *annuity immediate,* also known as annuity payable annually in arrear, is an annuity in which payments are made at the end of the year.

   The terms, annuity due and annuity immediate do not seem to be logical but these are conventionally used in the insurance field as defined above. The first payment of an annuity immediate is not made immediately at the beginning of first payment period, rather it is due at the end of it. The other corresponding terms, annuity payable annually in advance and annuity payable annually in arrear, for annuity due and immediate respectively, are more logical.

   An annuity period may be a month or a quarter or six months instead of a year. Such division of a year leads to the following type of annuity.

7. m-**thly Annuity:** An annuity which is payable in m parts in a year with a rate of 1 unit per year, that is $1/m$ in each mth part, is known as m-thly annuity. With $m = 2$, it is six monthly, with $m = 4$, it is quarterly and with $m = 12$, it is a monthly annuity. With $m = 1$, it is of course yearly discrete annuity. If the payments are made at the beginning of the mth part, it is m-thly annuity due while if payments are made at the end of the mth part, it is m-thly annuity immediate. Equal monthly instalments in the repayment of a housing loan form a monthly annuity due. Monthly salary of a working person forms monthly annuity immediate. Quarterly interest on the savings in a bank or post office forms a quarterly annuity immediate. It may be a certain or a life annuity.

8. **Deferred Annuity:** An annuity in which periodic payments begin after certain period is known as a *deferred annuity.* People often purchase deferred annuities during their working years in anticipation of the need for income after retirement.

9. **Level Annuity:** An annuity in which periodic payments are of the same amount is known as a *level annuity.* In the level annuity, it is assumed that the payment is of 1 unit per annum. Any other level annuity can be obtained from this by a simple multiplication. Annuities which are not level, are known as *varying annuities.*

Life annuities are further classified as n-year temporary life annuity, whole life annuity and n-year certain and whole life annuity depending on the terms and conditions of payments. We study these in detail in Sections $3, 4$ and 5.

Periodical premiums in life insurance form an annuity where payments are from the insured to the insurer. In an annuity contract, the insurer promises to make a series of periodic payments to the annuitant. Annuities may be purchased as single premium annuities by paying a single lump-sum premium, or by paying periodic premiums over a period of years. Periodic premiums can be paid on either a level premium basis or a flexible premium basis. Under a periodic level premium annuity, the policy holder pays equal premium for the annuity at regularly scheduled intervals. Under a flexible premium annuity, the amount of each premium payment can vary between a fixed minimum amount and a fixed maximum amount. In such a contract, the periodical benefit payments from the insurer to the insured form an annuity and periodical premium payments from the insured to the insurer also form an annuity. In any setup, to decide the purchase value of the annuity contract, it is absolutely essential for the insurer to find out the expected present value of the annuity.

With the main aim of calculation of premiums, to be fixed at the issue of the policy, our entire discussion in this chapter is focused on finding the present value of the various types of annuities in annuity certain and expected present value of the annuities in life annuity.

In the next section, we first discuss level annuity certain and then in the following sections proceed to level life annuities. Life annuity theory is analogous to the theory of annuity certain but brings in survival as a condition for payment. The theory of varying annuities is not discussed in this book.

6.2   Annuities Certain

In annuity certain, an annuity is payable for a fixed period of time. The fixed period of time is known as 'period certain'. At the end of the period certain, annuity payments cease. The theory of compound interest is used to find the present value of the annuity. We derive below the formulae for the present value of discrete annuity immediate and due, with annual payments and m-thly payments and then for continuous annuity.

Annuity Certain Immediate: Suppose the period certain is n years and the payment of 1 unit is to be made at the end of each year for n years. The present value of these payments is denoted by a symbol $a_{\overline{n}|}$, the general symbol for the present value of any type of annuity being a. We now derive the formula for $a_{\overline{n}|}$, under the assumption of a constant effective annual rate of interest i or equivalently the constant force of interest δ. Note that the present value of a unit paid at the end of first year is v, that of paid at the end of second year is v^2 and so on. Thus,

$$\begin{aligned} a_{\overline{n}|} &= v + v^2 + v^3 + \cdots + v^n \\ &= v(1 - v^n)/(1 - v) = (1 - v^n)/i \end{aligned} \tag{6.2.1}$$

is a present value of the n level payments, each of one unit for n years paid at the end of the year, that is, $a_{\overline{n}|}$ is the present value of annuity certain immediate. If annuity certain for n years is purchased by a single lump-sum premium, the insurer has to charge at least $a_{\overline{n}|}$ amount of money to the annuitant. $a_{\overline{n}|}$, thus, represents the purchase price of n-year annuity immediate with 1 unit payment per annum. This amount, along with the interest on this, with effective annual rate of interest i, will accumulate to a fund from which the insurer is able to pay one unit at the end of each year for n years. Table 6.1 shows the value of $1000a_{\overline{n}|}$, the present value of annuity certain immediate with payment of 1000 per annum, for values of n from 1 to 10 and $i = 0.05,\ 0.06, \cdots, 0.10$. We use Code 6.6.1 to compute these values.

From Table 6.1, we note that if the rate of interest is 5%, then the present value of 1-year annuity immediate of 1000 is 952.38, while for 10 years the value is 7721.73. Thus, the amount 952.38 will accumulate to 1000 in 1 year, while the amount 7721.73 will accumulate in 10 years to a fund from which it is possible to pay 1000 at the end of every year for 10 years. Thus, the purchase price of 10-year annuity immediate of 1000 per year is 7721.73. Note that as the rate of interest increases, the present value of the annuity decreases. With high interest rate a lesser amount will accumulate to a fund that will be sufficient to pay periodical payments. Further as n increases, the present value also increases as the period of payments increases.

n \ i	0.05	0.06	0.07	0.08	0.09	0.10
1	952.38	943.40	934.58	925.93	917.43	909.09
2	1859.41	1833.39	1808.02	1783.26	1759.11	1735.54
3	2723.25	2673.01	2624.32	2577.10	2531.29	2486.85
4	3545.95	3465.11	3387.21	3312.13	3239.72	3169.87
5	4329.48	4212.36	4100.20	3992.71	3889.65	3790.79
6	5075.69	4917.32	4766.54	4622.88	4485.92	4355.26
7	5786.37	5582.38	5389.29	5206.37	5032.95	4868.42
8	6463.21	6209.79	5971.30	5746.64	5534.82	5334.93
9	7107.82	6801.69	6515.23	6246.89	5995.25	5759.02
10	7721.73	7360.09	7023.58	6710.08	6417.66	6144.57

Table 6.1 $1000\, a_{\overline{n}|}$ Present value of annuity certain immediate of 1000 per annum

The present value of a series of payments in a certain time period considers the value of cash flow in that period at the beginning of the period. Sometimes (in the calculation of reserves discussed in Chapter 8) we need to find out the value of the cash flow at the end of the period, that is, we want to find the amount to which all the payments would accumulate at a given rate of interest, at the end of the period. The general symbol for accumulated amount is S. There are two approaches to find the accumulated amount. We first find the present value of the cash flow in the given period, its accumulation over the specific period will give the accumulated value. Thus, $a_{\overline{n}|}$ is the present value of the n level payments, each of one unit for n years paid at the end of the year. The accumulated value of $a_{\overline{n}|}$ at a rate of interest i, at the end of n years is $(1+i)^n a_{\overline{n}|}$. It is denoted by $S_{\overline{n}|}$. Thus

$$S_{\overline{n}|} = (1+i)^n a_{\overline{n}|}$$

This expression can be derived alternatively as follows. The accumulated value of 1 unit paid at the end of the first year is $(1+i)^{n-1}$, that paid at the end of the second year is $(1+i)^{n-2}$ and so on, the accumulated value of 1 unit paid at the end of the nth year is 1. Thus,

$$
\begin{aligned}
S_{\overline{n}|} &= (1+i)^{n-1} + (1+i)^{n-2} + \cdots + (1+i) + 1 \\
&= ((1+i)^n - 1)/i = (1+i)^n(1-(1+i)^{-n})/i \\
&= (1+i)^n(1-v^n)/i = (1+i)^n\, a_{\overline{n}|}
\end{aligned}
\tag{6.2.2}
$$

$S_{\overline{n}|}$ thus denotes the value of the series of 1 unit payments at the time of the nth payment. It is clear that $S_{\overline{1}|} = 1$.

Note that $a_{\overline{n}|}$ represents the value of the cash flow in the period of n years at the beginning of the period, while $S_{\overline{n}|}$ represents the value of the cash flow in the period of n years at the end of the period.

Example 6.2.1. Find the present value and the accumulated value of a 10-year annuity immediate of Rs 1000 per annum, if the effective rate of interest is 5%.

Solution: With the usual notation, $i = 0.05$. The present value $1000a_{\overline{10}|}$ and the accumulated value $1000S_{\overline{10}|}$ are given by,

$$1000a_{\overline{10}|} = 1000(1 - v^{10})/i = 1000(1 - (1 + 0.05)^{-10})/0.05 = 7721.73$$

$$1000S_{\overline{10}|} = 1000\left(\frac{(1+i)^{10} - 1}{i}\right) = \frac{1000}{0.05}\{(1 + 0.05)^{10} - 1\} = 12577.89$$

Note that, $1000S_{\overline{10}|} = 12577.89 = (1 + i)^{10}\, 7721.73 = (1 + i)^{10}\, 1000\, a_{\overline{10}|}$.

Thus, the value of the cash flow in the period of 10 years, at the beginning of the period is 7721.73 and at the end of the period is 12577.89. $\blacksquare$

The effective rate of interest may not remain the same over a period of 10 years. The above procedure of finding the present value and the accumulated value can be modified appropriately to compute these values. It is illustrated in the next example.

Example 6.2.2. Find the present value and the accumulated value of a 10 year annuity immediate of 1000 per annum if the effective rate of interest is 5% for the first 6 years and 6% for the next four years.

Solution: Note that the rate of interest changes after 6 years, hence the present value of 1 paid at the end of the 7th year is $v_1 * v^6$, where v corresponds to $i = 0.05$ and v_1 corresponds to $i = 0.06$. We have to make similar changes for the rest of the years. Thus, the present value is given by

$$1000a_{\overline{10}|} = 1000\left\{(1 - v^6)/i + v^6\{v_1 + v_1^2 + v_1^3 + v_1^4\}\right\} = 7661.41$$

Accumulated value for the first six years will be $S_{\overline{6}|}$ with a rate of 5% and this will get accumulated for the next four years with a rate of 6%. The accumulated value for the following four years will be $S_{\overline{4}|}$ with a rate of 6%. Thus, the accumulated value for a period of 10 years is given by,

$$1000\{S_{\overline{6}|}(1 + 0.06)^4 + S_{\overline{4}|}\} = 1000\left[[(1 + 0.05)^6 - 1]/0.05\right][1 + 0.06]^4$$
$$+ 1000\left[[(1 + 0.06)^4 - 1]/0.06\right] = 12961.82$$

$\blacksquare$

Example 6.2.3. An annuity of Rs 12000 per year is payable for 15 years. What is the price of this annuity purchased on 1 January 2023 and if the first payment is to be made on 31 December 2023? Suppose the effective annual rate of interest is 5%.

Solution: The purchase price is nothing but the present value of the annuity. Since the first payment is made at the end of the year, it is annuity immediate for 15 years. Hence, the purchase price is gven by,

$$12000\, a_{\overline{15}|} = 12000\left[1 - (1 + 0.05)^{-15}\right]/0.05 = 124555.90$$

$\blacksquare$

Annuity Certain Due: For an n-year annuity certain due, the payments are made at the beginning of each year for n years. The present value of this annuity is denoted by $\ddot{a}_{\overline{n}|}$. To find its expression, note that the present value of the first payment of 1 unit at the beginning of the year is 1 itself, that of second payment is v and so on. The present value of the last payment of 1 unit at the beginning of the nth year is v^{n-1}. Hence $\ddot{a}_{\overline{n}|}$ is given by,

$$\begin{aligned} \ddot{a}_{\overline{n}|} &= 1 + v + v^2 + \cdots + v^{n-1} \\ &= (1 - v^n)/(1 - v) = (1 - v^n)/d \end{aligned} \qquad (6.2.3)$$

It can be easily verified that,

$$\ddot{a}_{\overline{1}|} = 1, \quad a_{\overline{n}|} = v\ddot{a}_{\overline{n}|} \quad \& \quad \ddot{a}_{\overline{n}|} = 1 + a_{\overline{n-1}|}, \quad \text{with} \quad a_{\overline{0}|} = 0$$

From the second relation, it is clear that

$$a_{\overline{n}|} = v\ddot{a}_{\overline{n}|} \quad < \quad \ddot{a}_{\overline{n}|}, \quad \forall \ n \geq 1 \qquad (6.2.4)$$

The inequality in Eq. 6.2.4 appeals logically also, since in annuity immediate, payments are made at the end of the year, while in annuity due, these are made at the beginning of the year. Table 6.2 shows the values of $1000\ddot{a}_{\overline{n}|}$ for various values of n and the annual rate of interest $i = 0.05, 0.06, \cdots, 0.10$. We use Code 6.6.1 to compute these values.

n \ i	0.05	0.06	0.07	0.08	0.09	0.10
1	1000.00	1000.00	1000.00	1000.00	1000.00	1000.00
2	1952.38	1943.40	1934.58	1925.93	1917.43	1909.09
3	2859.41	2833.39	2808.02	2783.26	2759.11	2735.54
4	3723.25	3673.01	3624.32	3577.10	3531.29	3486.85
5	4545.95	4465.11	4387.21	4312.13	4239.72	4169.87
6	5329.48	5212.36	5100.20	4992.71	4889.65	4790.79
7	6075.69	5917.32	5766.54	5622.88	5485.92	5355.26
8	6786.37	6582.38	6389.29	6206.37	6032.95	5868.42
9	7463.21	7209.79	6971.30	6746.64	6534.82	6334.93
10	8107.82	7801.69	7515.23	7246.89	6995.25	6759.02

Table 6.2 $1000\ \ddot{a}_{\overline{n}|}$ Present value of annuity certain due of 1000 per annum

From Table 6.2, we observe that $1000\ddot{a}_{\overline{1}|} = 1000$ for all the values of i, as it is annuity due and the first payment is made at the beginning of the one year interval. The interpretation of all the values is similar to that for annuity immediate. Thus, if the annual rate of interest is 5%, then the present value of 10-year annuity due of 1000 is 8107.82. This amount will accumulate in 10 years to a fund, from which it is possible to pay 1000 at the beginning of every year for 10 years. It is to be noted that the present value of n-year annuity due of 1000 is larger than that of n-year annuity immediate of 1000 for all n, verifying the relation given in Eq. 6.2.4. Note that here also as the rate of interest increases, the present value of annuity decreases, as expected.

The equation $\ddot{a}_{\overline{n}|} = (1 - v^n)/d$ can be rewritten as $1 = d\ddot{a}_{\overline{n}|} + v^n$. This identity has the following nice interpretation. Suppose unit 1 is invested for n years at the beginning of the first year. The annual interest of d units is paid at the beginning of each year for n years. Thus, interest payments form an annuity due and its present value is $d\ddot{a}_{\overline{n}|}$. At the end of year n, the outstanding capital is still unit 1 and its present value is v^n. Thus, originally invested 1 unit is equivalent to the addition of present values of annuity payment of interest and the capital of 1 unit.

The following example illustrates the practical application of the present value of annuity due to decide the equal instalments to repay a loan.

Example 6.2.4. A loan of Rs 50,000 is taken on 1 January 2020. It has to be repaid in 5 equal instalments payable yearly at the beginning of the year. Based on the 6% annual rate of interest, determine the amount of the instalment.

Solution: Suppose the annual instalment is I. The value of I should be such that the total present value of all the 5 instalments equals 50000. Hence, we get the equation as,

$$50000 = I\ \ddot{a}_{\overline{5}|} = (1 - (1.06)^{-5})\ I/d = 4.465106\ I \quad \Rightarrow \quad I = 11197.94$$

■

The accumulated value corresponding to the annuity due is denoted by $\ddot{S}_{\overline{n}|}$ and is given by,

$$
\begin{aligned}
\ddot{S}_{\overline{n}|} &= (1+i)^n + (1+i)^{n-1} + \cdots + (1+i) = (1+i)\{(1+i)^n - 1\}/i \\
&= ((1+i)^n - 1)/d = (1+i)^n\{1 - (1+i)^{-n}\}/d = (1+i)^n \ddot{a}_{\overline{n}|}. \qquad (6.2.5) \\
&= (1+i)^n + (1+i)^{n-1} + \cdots + (1+i) + 1 - 1 = S_{\overline{n+1}|} - 1
\end{aligned}
$$

$\ddot{S}_{\overline{n}|}$ thus denotes the value of the series of 1 unit payments, one unit time after the last payment. Further, $\ddot{S}_{\overline{n}|} = S_{\overline{n+1}|} - 1$.

Example 6.2.5. Rs 3000 is to be deposited in a bank on 1 January of each year from 2021 to 2029. What is the accumulated value of this fund on 31 December 2029 at 3% annual rate of interest?

Solution: The term of this annuity-due is 9 years. Hence, the accumulated value on 31 December 2029 is

$$3000\ \ddot{S}_{\overline{9}|} = 3000\ (1.03)((1.03)^9 - 1)/0.03 = 31391.64$$

■

Example 6.2.6. Mr Bapat deposited Rs 5000 in a saving bank account on 21 July of each year from 2003 to 2018, both years inclusive. On 21 July 2022 he withdrew his savings. Assuming that the rate of interest for all these years is 6%, find the amount withdrawn by Mr Bapat.

Solution: Mr Bapat deposited Rs 5000 for 16 times from 2003 to 2018. Hence, the accumulated value of series of 16 payments on 21 July 2019 is $5000\ddot{S}_{\overline{16}|} = 136064.40$. From

2019 to 2022, this amount earned interest at 6% rate, hence the accumulated amount is $5000\ddot{S}_{\overline{16}|}\,(1.06)^3 = 162054.90$. Thus, the amount withdrawn by Mr Bapat is 162054.90.   ∎

We now proceed to derive the present value of m-thly annuity immediate and due.

m-thly Annuity Certain: Complexity in the derivation of the present value of annuities increases, when the annuity payment is made m times a year, with a rate of 1 unit per annum, for n years. If the amount of annuity is 1 unit, then $1/m$ unit amount is paid in every m^{th} part. In n years, there are mn, m-parts. The present value of m-thly annuity immediate is denoted by $a_{\overline{n}|}^{(m)}$ and is given by,

$$
\begin{aligned}
a_{\overline{n}|}^{(m)} &= \frac{1}{m}\sum_{k=1}^{mn}(v^{1/m})^k = \frac{1}{m}\left[v^{1/m} + v^{2/m} + \cdots + (v^{1/m})^{mn}\right] \\
&= \frac{1}{m}v^{1/m}\left\{1 + v^{1/m} + \cdots + (v^{1/m})^{mn-1}\right\} \\
&= \frac{1}{m}v^{1/m}\left\{1 - (v^{1/m})^{mn}\right\}/(1 - v^{1/m}) \\
&= \frac{1}{m}(1 - v^n)/\{(v^{-1})^{1/m} - 1\} \\
&= \frac{1 - v^n}{m\{(1+i)^{1/m} - 1\}} = \frac{1 - v^n}{i^{(m)}} = \frac{i}{i^{(m)}}\,a_{\overline{n}|}
\end{aligned}
\tag{6.2.6}
$$

where $i^{(m)} = m\{(1+i)^{1/m} - 1\}$ is the nominal rate of interest when interest is paid m-thly, as defined in Section 5.2. Note that $a_{\overline{n}|}^{(m)}$ denotes the present value of periodical payments where one unit is paid per annum in m parts at the end of mth part and for n years.

The present value of an annuity due, where $1/m$ unit amount is paid at the beginning of every m^{th} part for n years, is denoted by $\ddot{a}_{\overline{n}|}^{(m)}$ and is obtained as follows.

$$
\begin{aligned}
\ddot{a}_{\overline{n}|}^{(m)} &= \frac{1}{m}\sum_{k=0}^{mn-1}(v^{1/m})^k = \frac{1}{m}\{1 + v^{1/m} + \cdots (v^{1/m})^{mn-1}\} \\
&= \frac{1}{m}\left\{\frac{1 - (v^{1/m})^{mn}}{1 - v^{1/m}}\right\} = \frac{1 - v^n}{m(1 - v^{1/m})} \\
&= (1 - v^n)/d^{(m)} = (1+i)^{1/m}\,a_{\overline{n}|}^{(m)} = (i/d^{(m)})\,a_{\overline{n}|} \\
&= (d/d^{(m)})\,\ddot{a}_{\overline{n}|}
\end{aligned}
\tag{6.2.7}
$$

Here $d^{(m)} = m(1 - v^{1/m})$ is the nominal rate of discount paid m-times a year. In this setup also, the amount paid per year is 1 unit, it is paid in m parts at the beginning of the mth part. Note that,

$$
\ddot{a}_{\overline{1}|}^{(m)} = \frac{d}{d^{(m)}} \quad \text{and} \quad \ddot{a}_{\overline{\infty}|}^{(m)} = \frac{d}{d^{(m)}}\,\ddot{a}_{\overline{\infty}|} = \frac{d}{d^{(m)}}\frac{1}{d} = \frac{1}{d^{(m)}}
$$

Example 6.2.7. Prove that $a_{\overline{n}|} < a_{\overline{n}|}^{(m)} < \ddot{a}_{\overline{n}|}^{(m)} < \ddot{a}_{\overline{n}|}$ and hence, $i > i^{(m)} > d^{(m)} > d$.

Solution: From Eq. 6.2.7 we have,

$$\ddot{a}_{\overline{n}|}^{(m)} = (1+i)^{1/m} \, a_{\overline{n}|}^{(m)} > a_{\overline{n}|}^{(m)} \tag{6.2.8}$$

Now to prove that $a_{\overline{n}|} < a_{\overline{n}|}^{(m)}$, suppose $u_m = \sum_{k=1}^{m} v^{k/m}$. Note that $0 < v < 1$ and $0 < k/m < 1$ for $k = 1, 2, \cdots, m$. Hence, $v^{k/m} > v \ \forall \ k = 1, 2, \cdots, m$. As a consequence, $u_m > mv$. Grouping the terms for each of n years, the expression for $a_{\overline{n}|}^{(m)}$ given by $\sum_{k=1}^{mn} v^{k/m}/m$ can be rewritten as follows.

$$a_{\overline{n}|}^{(m)} = \frac{1}{m} \sum_{k=0}^{n-1} v^k \, u_m > \frac{1}{m} \sum_{k=0}^{n-1} v^k \, v \, m = \sum_{k=1}^{n} v^k = a_{\overline{n}|} \tag{6.2.9}$$

To prove that $\ddot{a}_{\overline{n}|}^{(m)} < \ddot{a}_{\overline{n}|}$, suppose $w_m = \sum_{k=0}^{m-1} v^{k/m}$. Note that $0 < v < 1$. Hence, each term in w_m except the first, is < 1. As a consequence, $w_m < m$. Grouping the terms for each of n years, expression for $\ddot{a}_{\overline{n}|}^{(m)}$ given by $\sum_{k=0}^{mn-1} v^{k/m}/m$ can be rewritten as

$$\ddot{a}_{\overline{n}|}^{(m)} = \frac{1}{m} \sum_{k=0}^{n-1} v^k \, w_m < \frac{1}{m} \sum_{k=0}^{n-1} v^k \, m = \sum_{k=0}^{n-1} v^k = \ddot{a}_{\overline{n}|} \tag{6.2.10}$$

Combining the results in Eqs 6.2.8, 6.2.9 and 6.2.10, we get the following inequality among the present values of annuities discussed so far.

$$a_{\overline{n}|} < a_{\overline{n}|}^{(m)} < \ddot{a}_{\overline{n}|}^{(m)} < \ddot{a}_{\overline{n}|} \tag{6.2.11}$$

From the formulae of these present values we note that all of them have the same numerator $(1 - v^n)$, denominator is i, $i^{(m)}$, $d^{(m)}$ and d respectively, according to the order in Eq. 6.2.11. Hence we have,

$$i > i^{(m)} > d^{(m)} > d \tag{6.2.12}$$

∎

To investigate the nature of $a_{\overline{n}|}^{(m)}$ and $\ddot{a}_{\overline{n}|}^{(m)}$ as a function of m, in the next example we find out the nature of $i^{(m)}$ and $d^{(m)}$ as a function of m.

Example 6.2.8. (i) Prove that $i^{(2)} > i^{(4)} > i^{(12)}$ and hence, $a_{\overline{n}|}^{(2)} < a_{\overline{n}|}^{(4)} < a_{\overline{n}|}^{(12)}$.
(ii) Prove that $d^{(2)} < d^{(4)} < d^{(12)}$ and hence, $\ddot{a}_{\overline{n}|}^{(2)} > \ddot{a}_{\overline{n}|}^{(4)} > \ddot{a}_{\overline{n}|}^{(12)}$.

Solution: (i) From Section 5.2 we have $i^{(m)} = m\{(1+i)^{1/m} - 1\}$. Hence,

$$\begin{aligned} i^{(2)} - i^{(4)} &= 2\{(1+i)^{1/2} - 1\} - 4\{(1+i)^{1/4} - 1\} \\ &= 2\{(1+i)^{1/2} - 2(1+i)^{1/4} + 1\} \\ &= 2\{(1+i)^{1/4} - 1\}^2 > 0 \end{aligned}$$

$$\begin{aligned}
i^{(4)} - i^{(12)} &= 4\{(1+i)^{1/4} - 1\} - 12\{(1+i)^{1/12} - 1\} \\
&= 4\{(1+i)^{1/4} - 3(1+i)^{1/12} + 2\} \\
&= 4\{x^3 - 3x + 2\} \quad \text{with} \ (1+i)^{1/12} = x \\
&= 4(x-1)^2(x+2) \ > \ 0
\end{aligned}$$

Thus, $i^{(2)} > i^{(4)} > i^{(12)}$. The ordering $a_{\overline{n}|}^{(2)} < a_{\overline{n}|}^{(4)} < a_{\overline{n}|}^{(12)}$ among the present values of annuities certain follows immediately, since in all these present values, the numerator $(1 - v^n)$ is the same and the denominator is $i^{(2)}$, $i^{(4)}$, $i^{(12)}$ respectively.

(ii) Further we have, $d^{(m)} = m\{1 - v^{1/m}\}$. Hence,

$$\begin{aligned}
d^{(4)} - d^{(2)} &= 2\{v^{1/2} - 2v^{1/4} + 1\} = 2(v^{1/4} - 1)^2 > 0 \\
d^{(12)} - d^{(4)} &= 4\{v^{1/4} - 3v^{1/12} + 2\} = 4(v^{1/12} - 1)^2(v^{1/12} + 2) > 0
\end{aligned}$$

The ordering $\ddot{a}_{\overline{n}|}^{(2)} > \ddot{a}_{\overline{n}|}^{(4)} > \ddot{a}_{\overline{n}|}^{(12)}$ follows immediately. ■

Values in Table 6.3 exhibit these relations and also the relation that $a_{\overline{n}|}^{(m)} < \ddot{a}_{\overline{n}|}^{(m)}$. We use Code 6.6.2 to compute these values.

Annuity n	$a_{\overline{n}|}^{(2)}$	$a_{\overline{n}|}^{(4)}$	$a_{\overline{n}|}^{(12)}$	$\ddot{a}_{\overline{n}|}^{(2)}$	$\ddot{a}_{\overline{n}|}^{(4)}$	$\ddot{a}_{\overline{n}|}^{(12)}$
1	964.14	970.06	974.01	987.95	981.96	977.98
2	1882.37	1893.92	1901.65	1928.85	1917.16	1909.39
3	2756.87	2773.79	2785.11	2824.95	2807.83	2796.45
4	3589.73	3611.76	3626.50	3678.38	3656.09	3641.27
5	4382.94	4409.83	4427.82	4491.17	4463.95	4445.86

Table 6.3 $\ddot{a}_{\overline{n}|}^{(m)}$ and $a_{\overline{n}|}^{(m)}$: Present values of m-thly annuity certain due and immediate of 1000 per annum

Example 6.2.9. A loan of Rs 10 lakhs is taken on 1 January 2020. It has to be repaid by equal monthly instalments payable at the beginning of the month for 10 years. Based on the 6% annual rate of interest, determine the amount of the instalment.

Solution: Denoting the monthly instalment by I, $12 * I$ is the annual payment made in 12 parts. Its present value for 10 years is $12I \, \ddot{a}_{\overline{10}|}^{(12)}$. Hence, to find I, we solve the equation,

$$10,00000 = 12I \, \ddot{a}_{\overline{10}|}^{(12)} = 12I(7.5972) \quad \Rightarrow \quad I = 10969.01$$

■

Example 6.2.10. A loan of Rs 10 lakhs is to be repaid over 10 years by a level annuity payable monthly in arrear. The nominal rate of interest $i^{(12)}$ is 0.06. Find the monthly instalment.

Solution: The nominal rate of interest $i^{(12)}$ is 0.06, hence interest rate per month is $i^{(12)}/12 = 0.005$. The total number of monthly instalments payable in arrear in 10 years is $10 \times 12 = 120$.

Hence we get the following equation to find the monthly instalment I.

$$10,00000 \; = \; I a_{\overline{120}|} = I\frac{1 - v^{120}}{i^{(12)}/12} = I\frac{(1 - (1.005)^{-120})}{0.005} \; = \; I(90.07345)$$

Solving the above equation, $I = 11102.05$ is the monthly instalment. ∎

Remark 6.2.1. In Example 6.2.10, we are given nominal rate of interest. Hence we have calculated the rate of interest per month and the total number of monthly payments. Thus, the unit of time is taken as 1 month. The discount factor v in the above formula is taken as $v = (1+.005)^{-1}$. Note that the instalments in Examples 6.2.9 and 6.2.10 are close, the marginal difference is attributable to change in the nature of interest rate and nature of annuity, due and immediate.

The following are the formulae for the accumulated values corresponding to the m-thly annuities.

$$S_{\overline{n}|}^{(m)} \; = \; (1+i)^n \, a_{\overline{n}|}^{(m)} \; = \; \frac{i}{i^{(m)}} \, S_{\overline{n}|}$$

$$\ddot{S}_{\overline{n}|}^{(m)} \; = \; (1+i)^n \, \ddot{a}_{\overline{n}|}^{(m)} \; = \; \frac{d}{d^{(m)}} \ddot{S}_{\overline{n}|} \; = \; (1+i)^{(1/m)} \, S_{\overline{n}|}^{(m)}$$

From Eq. 6.2.11, multiplying each term in the inequality by $(1+i)^n$, we get the following relation among the accumulated values, similar to that in Eq. 6.2.11.

$$S_{\overline{n}|} \; < \; S_{\overline{n}|}^{(m)} \; < \; \ddot{S}_{\overline{n}|}^{(m)} \; < \; \ddot{S}_{\overline{n}|} \tag{6.2.13}$$

Example 6.2.11. An annuity of Rs 25000 per annum payable monthly in advance is purchased for 5 years. Find its price if the annual rate of interest is 4%. Find its accumulated value at the end of 5 years.

Solution: Purchase price of the annuity contract is the present value of the m-thly annuity due, with $m = 12$. It is given by

$$25000 \, \ddot{a}_{\overline{5}|}^{(12)} \; = \; 25000\frac{d}{d^{(12)}} \, \ddot{a}_{\overline{5}|} \; = \; 25000 \, \frac{0.03846}{0.03956} \, (4.63008) \; = \; 112533.40$$

The corresponding accumulated value is given by,

$$25000 \, \ddot{S}_{\overline{5}|}^{(12)} = 25000 \, (1+i)^5 \, \ddot{a}_{\overline{5}|}^{(12)} = (1.04)^5 \, 112533.40 = 136914.10$$

∎

We now proceed to discuss how to find the present value of continuous annuity certain.

Continuous Annuity Certain: Suppose the partitioning of a period of 1 year in m parts is done finer and finer by increasing the value of m and payments are done in each mth part with a rate of 1 unit per annum. With the finer partition, the beginning and end of the mth part does not make any difference. Hence to find the present value of continuous annuity certain, we allow m to tend to ∞, in the expression for $a_{\overline{n}|}^{(m)}$ or $\ddot{a}_{\overline{n}|}^{(m)}$. The present value of annuity payment made continuously at the rate of 1 unit per year for n years is denoted by $\overline{a}_{\overline{n}|}$ and is given by,

$$
\begin{aligned}
\overline{a}_{\overline{n}|} &= \int_0^n v^t dt = \left[v^t / \log v\right]_0^n = (1 - v^n)/(-\log v) \\
&= (1 - v^n)/\delta = (i/\delta)\, a_{\overline{n}|} = (d/\delta)\, \ddot{a}_{\overline{n}|}
\end{aligned}
\tag{6.2.14}
$$

Note that as $m \to \infty$,

$$
i^{(m)} \to \delta \;\Rightarrow\; a_{\overline{n}|}^{(m)} = (1 - v^n)/i^{(m)} \;\to\; (1 - v^n)/\delta = \overline{a}_{\overline{n}|}
$$

$$
\&\; d^{(m)} \to \delta \;\Rightarrow\; \ddot{a}_{\overline{n}|}^{(m)} = (1 - v^n)/d^{(m)} \;\to\; (1 - v^n)/\delta = \overline{a}_{\overline{n}|}
$$

Table 6.4 shows the values of $1000\overline{a}_{\overline{n}|}$ for $i = 0.05,\ 0.06, \cdots, 0.10$ and $n = 1$ to 10. We use Code 6.6.1 to compute these values.

n \ i	0.05	0.06	0.07	0.08	0.09	0.10
1	976.00	971.42	966.92	962.49	958.12	953.82
2	1905.52	1887.86	1870.59	1853.68	1837.13	1820.94
3	2790.78	2752.42	2715.13	2678.86	2643.57	2609.22
4	3633.88	3568.05	3504.43	3442.91	3383.41	3325.84
5	4436.83	4337.51	4242.09	4150.37	4062.17	3977.32
6	5201.55	5063.41	4931.49	4805.42	4684.89	4569.57
7	5929.86	5748.23	5575.79	5411.95	5256.18	5107.97
8	6623.48	6394.28	6177.94	5973.56	5780.31	5597.44
9	7284.07	7003.76	6740.69	6493.56	6261.16	6042.40
10	7913.21	7578.75	7266.64	6975.04	6702.30	6446.92

Table 6.4 $1000\overline{a}_{\overline{n}|}$ Present value of continuous annuity certain of 1000 per annum

In annuities payable continuously, the distinction between annuity-due and annuity-immediate has no meaning. The concept of annuity payable continuously, at the rate of one per year, is of course an abstraction but makes use of familiar mathematical tools and closely approximates annuities payable on a monthly basis. We need to go for such an abstraction because in life annuity, the future life time random variable is a continuous random variable.

To find a relation between $\overline{a}_{\overline{n}|}$ and $a_{\overline{n}|}$, we note that as shown in Chapter 5,

$$
\begin{aligned}
i/\delta &= (e^\delta - 1)/\delta = (1 + \delta + \delta^2/2 + \cdots - 1)/\delta \\
&= 1 + \delta/2 + \delta^2/6 + \cdots > 1 \;\Rightarrow\; \overline{a}_{\overline{n}|} = (i/\delta)\, a_{\overline{n}|} > a_{\overline{n}|}
\end{aligned}
\tag{6.2.15}
$$

The inequality is reasonable as $\overline{a}_{\overline{n}|}$ is the present value of annuity payment made continuously while $a_{\overline{n}|}$ is the present value of annuity payment made at the end of the year. To find a relation between $\overline{a}_{\overline{n}|}$ and $\ddot{a}_{\overline{n}|}$, we have to check whether $d/\delta > 1$ or ≤ 1. To decide it, we define a function f as follows and study its nature.

$$f(\delta) \;=\; d - \delta \;=\; 1 - e^{-\delta} - \delta \;\Rightarrow\; f(0) \;=\; 0$$
$$f'(\delta) \;=\; e^{-\delta} - 1 < 0 \;\Rightarrow\; f \text{ is a decreasing function}$$
$$\text{Hence,} \quad \delta \;>\; 0 \;\Rightarrow\; f(\delta) = d - \delta < 0 \;\Rightarrow\; d/\delta < 1$$
$$\Rightarrow\; \overline{a}_{\overline{n}|} \;=\; (d/\delta)\,\ddot{a}_{\overline{n}|} \;<\; \ddot{a}_{\overline{n}|} \tag{6.2.16}$$

This inequality is reasonable as $\overline{a}_{\overline{n}|}$ is the present value of annuity payment made continuously while $\ddot{a}_{\overline{n}|}$ is the present value of annuity payment made at the beginning of a year. The following is the formula for the accumulated value corresponding to annuities payable continuously.

$$\overline{S}_{\overline{n}|} \;=\; \int_0^n (1+i)^t \, dt \;=\; \frac{(1+i)^n - 1}{\log(1+i)} \;=\; \frac{(1+i)^n - 1}{\delta}$$
$$=\; (1+i)^n\,\overline{a}_{\overline{n}|} \;=\; (i/\delta)S_{\overline{n}|} \;=\; (d/\delta)\ddot{S}_{\overline{n}|} \tag{6.2.17}$$

Combining Eqs 6.2.15, 6.2016 and 6.2.17 we have

$$a_{\overline{n}|} \;<\; \overline{a}_{\overline{n}|} \;<\; \ddot{a}_{\overline{n}|} \quad\text{and}\quad S_{\overline{n}|} \;<\; \overline{S}_{\overline{n}|} \;<\; \ddot{S}_{\overline{n}|} \tag{6.2.18}$$

Table 6.5 shows the values of $1000a_{\overline{n}|}$, $1000a_{\overline{n}|}^{(m)}$, $1000\overline{a}_{\overline{n}|}$, $1000\ddot{a}_{\overline{n}|}^{(m)}$ and $1000\ddot{a}_{\overline{n}|}$ for $m = 12$, for various values of n and $i = 0.05$. We use Code 6.6.3 for these computations.

n	$10^3 a_{\overline{n}|}$	$10^3 a_{\overline{n}|}^{(12)}$	$10^3 \overline{a}_{\overline{n}|}$	$10^3 \ddot{a}_{\overline{n}|}^{(12)}$	$10^3 \ddot{a}_{\overline{n}|}$
1	952.38	974.01	976.00	977.98	1000.00
2	1859.41	1901.65	1905.52	1909.39	1952.38
3	2723.25	2785.11	2790.78	2796.45	2859.41
4	3545.95	3626.50	3633.88	3641.27	3723.25
5	4329.48	4427.82	4436.83	4445.86	4545.95
6	5075.69	5190.99	5201.55	5212.13	5329.48
7	5786.37	5917.81	5929.86	5941.92	6075.69
8	6463.21	6610.02	6623.48	6636.95	6786.37
9	7107.82	7269.27	7284.07	7298.89	7463.21
10	7721.73	7897.13	7913.21	7929.31	8107.82

Table 6.5 Comparison of present values of annuities

From Table 6.5, note that $1000a_{\overline{n}|} < 1000a_{\overline{n}|}^{(12)}$, as proved above, in view of the nature of periodical payments. In annuity immediate, payment is made once at the end of the interval while in m-thly annuity immediate, payments are made at the end of the mth part of the interval. On the other hand $10^3\ddot{a}_{\overline{n}|} > 10^3\ddot{a}_{\overline{n}|}^{(12)}$, because in annuity due, the payment is made once at the beginning of the interval while in m-thly annuity due, the payments are made at the beginning of the mth part of the interval. Thus the m payments are spread over the interval. Hence its present value is less than that of payment made once at the beginning of the interval. Column 3 and column 5 denote the present values of periodical payment of 1000, when 1000 rupees are paid in 12 instalments. Thus, equal monthly instalments will be 1000/12. Further note that $1000\ddot{a}_{\overline{n}|}^{(12)}$ and $1000\overline{a}_{\overline{n}|}$ are close to each other for all n. This observation is consistent with the remark made above that annuities payable continuously, at the rate of one per year closely approximates annuities payable on a monthly basis.

From Table 6.5, we note that $a_{\overline{n}|}^{(12)} < \overline{a}_{\overline{n}|} < \ddot{a}_{\overline{n}|}^{(12)}$. We prove it algebraically for all m, in the following example.

Example 6.2.12. Prove that $a_{\overline{n}|}^{(m)} < \overline{a}_{\overline{n}|} < \ddot{a}_{\overline{n}|}^{(m)}$.

Solution: From Eq. 6.2.14 we have,

$$\overline{a}_{\overline{n}|} = \int_0^n v^t dt = \sum_{k=0}^{mn-1} \int_{k/m}^{(k+1)/m} v^t dt = \sum_{k=0}^{mn-1} v^{k/m} \int_0^{1/m} v^y dy$$

where the last step follows by the substitution $t - k/m = y$. Note that $0 < v < 1$ and $0 < y < 1/m$. Hence, $v^y > v^{1/m}$. Further, $\int_0^{1/m} dy = 1/m$. Hence we get,

$$\overline{a}_{\overline{n}|} > \frac{1}{m} \sum_{k=0}^{mn-1} v^{k/m} v^{1/m} = \frac{1}{m} \sum_{k=0}^{mn-1} v^{(k+1)/m} = \frac{1}{m} \sum_{k=1}^{mn} v^{k/m} = a_{\overline{n}|}^{(m)}$$

Further note that

$$0 < y < 1/m \text{ and } v^y < 1 \Rightarrow \overline{a}_{\overline{n}|} < \frac{1}{m} \sum_{k=0}^{mn-1} v^{k/m} = \ddot{a}_{\overline{n}|}^{(m)}$$

Thus we have proved that $a_{\overline{n}|}^{(m)} < \overline{a}_{\overline{n}|} < \ddot{a}_{\overline{n}|}^{(m)}$.   ■

Combining results in Example 6.2.7 and Example 6.2.12, we get the following inequality among the present values of annuities discussed so far.

$$a_{\overline{n}|} < a_{\overline{n}|}^{(m)} < \overline{a}_{\overline{n}|} < \ddot{a}_{\overline{n}|}^{(m)} < \ddot{a}_{\overline{n}|} \tag{6.2.19}$$

From the formulae of these present values we note that all of them have the same numerator $(1-v^n)$, denominator is i, $i^{(m)}$, δ, $d^{(m)}$ and d respectively according to the order in Eq. 6.2.19. Hence we have

$$i > i^{(m)} > \delta > d^{(m)} > d \tag{6.2.20}$$

We have following analogous inequality among the accumulated values, obtained by multiplying all the terms of Eq. 6.2.19 by $(1+i)^n$.

$$S_{\overline{n}|} < S_{\overline{n}|}^{(m)} < \overline{S}_{\overline{n}|} < \ddot{S}_{\overline{n}|}^{(m)} < \ddot{S}_{\overline{n}|} \tag{6.2.21}$$

Example 6.2.13. A loan of Rs 10 lakhs is to be repaid over 10 years by a continuous annuity. The force of interest is 0.06. Find the annual instalment.

Solution: We are given $\delta = 0.06$. Hence corresponding value of v is $v = \exp(-\delta) = 0.9417645$. Suppose I is the annual instalment to be paid continuously at a rate of I per annum. Then the equation

$$10^6 = I\,\overline{a}_{\overline{10}|} = I\,(1 - v^{10})/\delta = I\,(7.519806) \quad \Rightarrow \quad I = 132982.20$$

■

Remark 6.2.2. In Example 6.2.13 the instalment is $I = 132982.20$ for one year. The instalments per year for the same setup, obtained in Examples 6.2.9 and 6.2.10 are 131628.10 and 133224.60 respectively. Note that these instalments are close to each other.

We derive below the formulae for the present value for all the annuities studied so far when the payments are deferred for certain period.

Deferred Annuity: Present values for the deferred annuities are obtained on similar lines. Suppose payment of 1 unit is made for n years at the end of the interval, after a deferment period of m years. The present value of such deferred annuity immediate is denoted by $_{m|}a_{\overline{n}|}$ and given by,

$$_{m|}a_{\overline{n}|} = v^{m+1} + \cdots + v^{m+n} = v^{m+1}\left(\frac{1 - v^n}{1 - v}\right) = v^m a_{\overline{n}|} = a_{\overline{m+n}|} - a_{\overline{m}|}$$

The present value of m-year deferred annuity due, denoted by, $_{m|}\ddot{a}_{\overline{n}|}$, is derived on similar lines. Thus,

$$_{m|}\ddot{a}_{\overline{n}|} = v^m + \cdots + v^{m+n-1} = v^m\left(\frac{1 - v^n}{1 - v}\right) = v^m \ddot{a}_{\overline{n}|} = \ddot{a}_{\overline{m+n}|} - \ddot{a}_{\overline{m}|}$$

In m-year deferred continuous annuity certain, annuity payments are deferred for m years and then made continuously at the rate of 1 unit per year for n years. Its present value is denoted by $_{m|}\overline{a}_{\overline{n}|}$ and is given by,

$$_{m|}\overline{a}_{\overline{n}|} = \int_m^{m+n} v^t dt = \frac{v^m - v^{m+n}}{\delta} = v^m\,\overline{a}_{\overline{n}|} = \frac{i}{\delta}\,_{m|}a_{\overline{n}|} = \frac{d}{\delta}\,_{m|}\ddot{a}_{\overline{n}|}$$

From the above formulae, it is clear that

$$_{m|}a_{\overline{n}|} = v^m a_{\overline{n}|} < a_{\overline{n}|}\ , \quad _{m|}\ddot{a}_{\overline{n}|} = v^m \ddot{a}_{\overline{n}|} < \ddot{a}_{\overline{n}|} \ \text{ and } \ _{m|}\overline{a}_{\overline{n}|} = v^m \overline{a}_{\overline{n}|} < \overline{a}_{\overline{n}|}$$

Such relations seem reasonable as in the m-year deferred annuity insurer has no liability of payment in first m years. From Eq. 6.2.18, it follows that

$$_{m|}a_{\overline{n}|} \quad < \quad _{m|}\overline{a}_{\overline{n}|} \quad < \quad _{m|}\ddot{a}_{\overline{n}|}$$

Accumulated values corresponding to the above three deferred annuities are obtained by multiplying the present values by $(1 + i)^{m+n}$. These are thus given by,

$$_{m|}S_{\overline{n}|} \quad = \quad S_{\overline{n}|}, \quad _{m|}\overline{S}_{\overline{n}|} \quad = \quad \overline{S}_{\overline{n}|}, \quad _{m|}\ddot{S}_{\overline{n}|} \quad = \quad \ddot{S}_{\overline{n}|}$$

There is no effect of deferment on the accumulated values as there are no payments in the first m years. The present value for the deferred annuities with m-thly payments can be obtained on similar lines.

Example 6.2.14. An amount of Rs 10000 is payable on 31 December for 10 years. The first payment is due in 2023. Find the purchase price of this annuity on 1 January 2013 with 6% annual effective rate of interest.

Solution: The first payment is deferred by 10 years. Thus it is a 10-year deferred, 10-year annuity certain immediate. Hence the purchase price is given by,

$$10000 \; {}_{10|}a_{\overline{10|}} \; = \; 10000 \; v^{10}a_{\overline{10|}} \; = \; 10000 \; (1.06)^{-10}\left(\frac{1-(1.06)^{-10}}{0.06}\right) \; = \; 41098.34$$

Thus, 41098.34 invested on 1 January 2013 will accumulate to a fund from which payment of 10000 can be made for 10 years from 2023. ∎

So far we have discussed calculation of present values of a variety of level annuities. The same arguments are useful to find present values of varying annuities. The following example illustrates the procedure.

Example 6.2.15. Calculate the present value, at 4% force of interest, of a continuous annuity for 10 years, under which the payment at time t is at the rate of t^2 per annum.

Solution: The present value is given by,

$$\int_0^{10} t^2 e^{-t\delta}\, dt \; = \; \frac{-100v^{10}}{\delta} - \frac{20v^{10}}{\delta^2} - \frac{2v^{10}}{\delta^3} + \frac{2}{\delta^3} = 247.70$$

∎

Table 6.6 summarizes the international actuarial notation for present value of annuity certain and corresponding accumulated value for various annuities studied in this section, for payment of 1 unit per annum.

In the next section, we discuss the theory of life annuities, in which the phenomenon 'period certain' is replaced by a random quantity, as the period of annuity payments depends on the survival of the individual.

Life annuities are also classified into two categories—continuous life annuities and discrete life annuities, depending on the mode of periodical payments. If the life annuities are paid continuously at a rate of 1 unit per annum, then these are termed as continuous life annuities. In this case, the period of payments is governed by the future life time random variable $T(x)$. A life annuity where the payments are made either at the beginning of the payment intervals or at the end of the payment intervals till the individual survives, is known as discrete life annuity. For continuous annuities, there is no distinction between payments at the beginning of payment intervals or at the end of the intervals, that is, between annuity due and annuity immediate. For discrete annuities, the distinction is meaningful. In discrete life annuities, the period of payments is governed by the curtate future life time random variable $K(x)$. Discrete life annuity due has a more prominent role in actuarial applications. For example, in most individual life insurances, periodic premiums form an annuity-due. In Section 3, we discuss continuous life annuities and in Section 4 we study discrete life annuities. Section 5 discusses the extension of Section 4 when payments form m-thly annuity.

Annuity	Present value	Accumulated value
Immediate	$a_{\overline{n}\rvert} = (1-v^n)/i$	$S_{\overline{n}\rvert} = ((1+i)^n - 1)/i$ $= (1+i)^n\, a_{\overline{n}\rvert}$
Due	$\ddot{a}_{\overline{n}\rvert} = (1-v^n)/d$ $= 1 + a_{\overline{n-1}\rvert}$	$\ddot{S}_{\overline{n}\rvert} = ((1+i)^n - 1)/d$ $= (1+i)^n\, \ddot{a}_{\overline{n}\rvert} = S_{\overline{n+1}\rvert} - 1$
m-thly immediate	$a_{\overline{n}\rvert}^{(m)} = (1-v^n)/i^{(m)}$ $= ia_{\overline{n}\rvert}/i^{(m)}$	$S_{\overline{n}\rvert}^{(m)} = ((1+i)^n - 1)/i^{(m)}$ $= iS_{\overline{n}\rvert}/i^{(m)}$
m-thly due	$\ddot{a}_{\overline{n}\rvert}^{(m)} = (1-v^n)/d^{(m)}$ $= d\,\ddot{a}_{\overline{n}\rvert}/d^{(m)}$	$\ddot{S}_{\overline{n}\rvert}^{(m)} = ((1+i)^n - 1)/d^{(m)}$ $= dS_{\overline{n}\rvert}/d^{(m)}$
Continuous	$\overline{a}_{\overline{n}\rvert} = (1-v^n)/\delta$ $= ia_{\overline{n}\rvert}/\delta = d\ddot{a}_{\overline{n}\rvert}/\delta$	$\overline{S}_{\overline{n}\rvert} = ((1+i)^n - 1)/\delta$ $= (1+i)^n\, \overline{a}_{\overline{n}\rvert}$ $= i\, S_{\overline{n}\rvert}/\delta = d\ddot{S}_{\overline{n}\rvert}/\delta$
m-Year deferred immediate	$_{m\rvert}a_{\overline{n}\rvert} = v^m\, a_{\overline{n}\rvert}$ $= a_{\overline{m+n}\rvert} - a_{\overline{m}\rvert}$	$_{m\rvert}S_{\overline{n}\rvert} = S_{\overline{n}\rvert}$
m-Year deferred due	$_{m\rvert}\ddot{a}_{\overline{n}\rvert} = v^m\, \ddot{a}_{\overline{n}\rvert}$ $= \ddot{a}_{\overline{m+n}\rvert} - \ddot{a}_{\overline{m}\rvert}$	$_{m\rvert}\ddot{S}_{\overline{n}\rvert} = \ddot{S}_{\overline{n}\rvert}$
m-Year deferred continuous	$_{m\rvert}\overline{a}_{\overline{n}\rvert} = v^m\, \overline{a}_{\overline{n}\rvert}$ $= \overline{a}_{\overline{m+n}\rvert} - \overline{a}_{\overline{m}\rvert}$	$_{m\rvert}\overline{S}_{\overline{n}\rvert} = \overline{S}_{\overline{n}\rvert}$

Table 6.6 Annuities certain: present values and accumulated values

6.3   Continuous Life Annuities

Life annuities play a major role in life insurance operations as life insurances are usually purchased by a life annuity of premiums rather than by a single premium. The amount payable at the time of claim may be converted through a settlement option into some form of life annuity for the beneficiary. For example, there may be a monthly income payable to a surviving spouse. A continuous life annuity can be any one of the following three types: (i) n-year temporary life annuity, (ii) whole life annuity and (iii) n-year certain and life annuity. Each of these is further classified depending upon the deferment or otherwise of payments. We discuss each of these annuities in detail in the present section.

n-Year Temporary Life Annuity: In this annuity, payment of one unit per annum is payable continuously for at most n years. Thus, the annuity payments cease when either the individual dies during the term of n-years or at the end of n years provided the individual survives the n-year term. Such a situation arises if a n-year term insurance or n-year endowment insurance, with benefit payable at the moment of death, is purchased by paying periodical premiums continuously at a certain rate per annum. Payments of premiums will stop if death occurs before the term of n-year is over or definitely at the end of n-years if the individual survives the term. To decide the amount of premium to be fixed at the issue of policy, we have to find the expected present value of premium payments which form a n-year temporary life annuity and balance it with the actuarial present value of the benefit.

In life annuity, the present value of the annuity becomes a random variable. The present value random variable Y for an n-year temporary life annuity of 1 per year, payable continuously while (x) survives during the next n years, is defined by noting the random period of annuity payments. It is $T(x)$ if death occurs within a span of n years after signing the contract at age x and is n years if the individual survives n years. Thus we get,

$$Y = \begin{cases} \bar{a}_{\overline{T}|} = (1 - v^T)/\delta & \text{if} & 0 \le T < n \\ \bar{a}_{\overline{n}|} = (1 - v^n)/\delta & \text{if} & T \ge n \end{cases} \tag{6.3.1}$$

The distribution of Y in this case is a mixed distribution as in the case of the present value random variable of benefit in n-year endowment insurance. With a probability $_nq_x$, it is distributed as $\bar{a}_{\overline{T}|}$ and with probability $_np_x = 1 - {}_nq_x$, it takes a value $\bar{a}_{\overline{n}|}$. The maximum value of Y is $\bar{a}_{\overline{n}|}$. Note that the random variable Y can be written as , $Y = (1 - Z)/\delta$ where,

$$Z = \begin{cases} v^T, & \text{if} & 0 \le T < n \\ v^n, & \text{if} & T \ge n \end{cases} \tag{6.3.2}$$

In Eq. 6.3.2, Z is the present value random variable of the unit benefit, to be paid at the moment of death, for an n-year endowment insurance. Hence,

$$E(Y) \quad = \quad E\left[(1 - Z)/\delta\right] \quad = \quad (1 - \bar{A}_{x:\overline{n}|})/\delta$$

$$\text{and} \quad Var(Y) \quad = \quad Var(Z)/\delta^2 \quad = \quad ({}^2\bar{A}_{x:\overline{n}|} - \bar{A}^2_{x:\overline{n}|})/\delta^2 \tag{6.3.3}$$

In international actuarial notation, the actuarial present value of the n-year temporary life annuity is denoted by $\bar{a}_{x:\overline{n}|}$. Thus,

$$\bar{a}_{x:\overline{n}|} = (1 - \bar{A}_{x:\overline{n}|})/\delta \tag{6.3.4}$$

From Chapter 5, we have

$$\bar{A}_{x:\overline{n}|} = \int_0^n v^t g(t)dt + v^n \, {}_np_x$$

Using the relation $g(t) = -\frac{d}{dt} \, {}_tp_x$ we solve the integral $\int_0^n v^t g(t)dt$ by parts and get $\bar{A}_{x:\overline{n}|} = 1 - \delta \int_0^n v^t \, {}_tp_x \, dt$. Substituting in Eq. 6.3.4, we get

$$\bar{a}_{x:\overline{n}|} = \int_0^n v^t \, {}_tp_x \, dt \tag{6.3.5}$$

This is known as the current payment integral for the actuarial present value for the n-year temporary annuity. This formula is similar to that for $\bar{a}_{\overline{n}|}$, except the factor $_tp_x$. This factor reflects the fact that payment is made at time t if (x) survives for t units, $t \le n$.

From Eq. 6.3.4, we get $1 = \delta \bar{a}_{x:\overline{n}|} + \bar{A}_{x:\overline{n}|}$. Hence, in terms of annuity values, $Var(Y)$ as given in Eq. 6.3.3 can be rewritten as,

$$Var(Y) \;=\; \frac{1 - 2\delta \; {}^2\bar{a}_{\overline{x:n}|} - (1 - \delta\bar{a}_{x:\overline{n}|})^2}{\delta^2} \;=\; \frac{2}{\delta}\left(\bar{a}_{x:\overline{n}|} - {}^2\,\bar{a}_{x:\overline{n}|}\right) - \left(\bar{a}_{x:\overline{n}|}\right)^2$$

In this expression, ${}^2\bar{a}_{x:\overline{n}|}$ denotes the present value of n-year temporary life annuity at the force of interest 2δ.

As we have defined the accumulated value $\overline{S}_{\overline{n}|}$ of n-year continuous annuity certain, we define the accumulated value of n-year temporary annuity as follows. It is denoted by $\overline{S}_{x:\overline{n}|}$ and is defined as $\overline{S}_{x:\overline{n}|} = \bar{a}_{x:\overline{n}|}/{}_nE_x$, where ${}_nE_x = v^n \; {}_np_x$. It denotes the accumulated value of 1 unit benefit per annum, paid continuously for at most n years. Observe that

$$\overline{S}_{\overline{n}|} \;=\; (1+i)^n \; \bar{a}_{\overline{n}|} \quad \text{and} \quad \overline{S}_{x:\overline{n}|} \;=\; \bar{a}_{x:\overline{n}|}/{}_nE_x \;=\; (1+i)^n \; \bar{a}_{x:\overline{n}|}/{}_np_x$$

Thus, $\overline{S}_{x:\overline{n}|}$ has a similar expression as that of $\overline{S}_{\overline{n}|}$ with one additional factor ${}_np_x$, as the annuity payments are conditional on the survival of (x). We need this expression of $\overline{S}_{x:\overline{n}|}$ in Chapter 8 while finding retrospective reserves.

The following examples illustrate the computation of $\bar{a}_{x:\overline{n}|}$.

Example 6.3.1. It is given that the force of interest is $\delta = 6\%$ and the actuarial present value at age 27 of a unit benefit to be paid at the moment of death in a 5-year endowment insurance with a force of interest δ is 0.7395 and with a force of interest 2δ is 0.6066. Find the actuarial present value at age 27 of a continuous 5-year temporary life annuity payable at the rate of 10000 per annum. Also find the standard deviation of the corresponding present value random variable.

Solution: We are given that $\bar{A}_{27:\overline{5}|} = 0.7395$ and ${}^2\bar{A}_{27:\overline{5}|} = 0.6066$. Using the formula derived in Eq. 6.3.4 we get

$$\bar{a}_{27:\overline{5}|} \;=\; (1 - \bar{A}_{27:\overline{5}|})/\delta \;=\; 4.341667 \quad \text{and} \quad 10000\,\bar{a}_{27:\overline{5}|} \;=\; 43416.67$$

Thus, 43416.67 is the actuarial present value of the annuity payable continuously, at the rate of 10000 per year, for at most 5 years. Further variance of the corresponding present value random variable is given by

$$({}^2\bar{A}_{27:\overline{5}|} - (\bar{A}_{27:\overline{5}|})^2)/\delta^2 \;=\; 14.92771$$

Hence the standard deviation corresponding to the annuity of 10000 is
$10000 \times 3.863639 \;=\; 38636.39$. Note that the standard deviation is very high, indicating a large spread in the present value random variable of the annuity. Thus, if we have a portfolio of 5-year temporary life annuity contracts, all issued to individuals of age 27, then there is large variation in the mortality behavior of these individuals and hence in annuity payments. ∎

Example 6.3.2. Suppose the force of mortality follows Gompertz' law given by $\mu_x = BC^x$. Find the values of $100\,\bar{a}_{x:\overline{n}|}$ for $n = 1, 2, \cdots, 10$, $x = 25, 30, 35$ and 40. Take $\delta = 0.05$, $B = 0.0001151$ and $C = 1.096$. Interpret the results.

Solution: In Example 5.3.7, we have obtained $\overline{A}_{x:\overline{n}|}$ for the specified mortality law, interest rate and values of n and x. We use the formula given in Eq. 6.3.4 to find $\overline{a}_{x:\overline{n}|}$ for various values of n and x. These are reported in Table 6.7. We compute these values using Part I of Code 6.6.4.

n	25	30	35	40
1	97.48	97.45	97.40	97.32
2	190.10	189.97	189.76	189.43
3	278.08	277.78	277.32	276.58
4	361.64	361.11	360.29	358.99
5	440.99	440.18	438.89	436.87
6	516.34	515.17	513.33	510.43
7	587.87	586.28	583.78	579.86
8	655.77	653.70	650.44	645.33
9	720.20	717.58	713.47	707.04
10	781.34	778.11	773.04	765.13

Table 6.7 Gompertz' law: $100\overline{a}_{x:\overline{n}|}$

The first entry $\overline{a}_{25:\overline{1}|} = 97.48$ is interpreted as follows: The present value of the payment made continuously at the rate of 100 per annum for at most 1 year is 97.48, that is if death of (25) occurs during the first year, then payment ceases and if (25) survives for 1 year, then payment ceases after one year. The expected present value of such cash flow is 97.48. In other words if premiums are paid continuously at the rate of 100 per annum for at most 1 year, then its present value, that is, the expected cash inflow to the insurance company corresponding to this policy is 97.48. Similarly, $\overline{a}_{40:\overline{10}|} = 765.13$ is the actuarial present value of the payment made continuously at the rate of 100 per annum to (40) for at most 10 years. Thus, if (40) purchases a 10-year temporary annuity which is payable continuously at a rate of 100 per annum, then its purchase price is at least 765.13. From the table we note that values of $100\,\overline{a}_{x:\overline{n}|}$ decrease marginally as age increases for the given mortality law with given parameters, however, they increase as the term n increases. ∎

Whole Life Annuity: This annuity, as the name implies, provides payments throughout the future life time of an individual, that is, until the death of the individual. Whole life insurance is usually purchased by whole life annuity of premiums, that is premiums are paid continuously at a certain rate per annum till the individual is alive.

The present value random variable corresponding to payments made continuously at the rate of 1 per year for future lifetime T of (x) is defined as

$$Y = \overline{a}_{\overline{T}|} = (1 - v^{T})/\delta$$

The support of this random variable is $(0, 1/\delta)$. The actuarial present value for a continuous whole life annuity is denoted by $\overline{a}_{x}$. We know from Chapter 4 that the probability density function of T is $_tp_x\,\mu_{x+t}$, so, the actuarial present value can be calculated by,

$$\overline{a}_{x} = E(Y) = \int_{0}^{\infty} \overline{a}_{\overline{t}|}\,_tp_x\mu_{x+t}\,dt \tag{6.3.6}$$

In Chapter 4, we have noted that, $\frac{d}{dt}\,{}_tp_x = -{}_tp_x\,\mu_{x+t}$. Hence using integration by parts, we obtain

$$\bar{a}_x = \int_0^\infty v^t\,{}_tp_x dt \tag{6.3.7}$$

This integral can be interpreted as involving a momentary payment of $1dt$ made at time t, discounted at interest back to time zero by multiplying by v^t and further multiplied by ${}_tp_x$ to reflect the probability that (x) has survived up to $x+t$ and hence payment is made at a time t. This is the current payment form of the actuarial present value for the whole life annuity. The formula for $\bar{a}_x$ as given in Eq. 6.3.7 can also be obtained by allowing n to tend to ∞ in the formula for $\bar{a}_{x:\overline{n}|}$ given in Eq 6.3.5.

The present value random variable corresponding to whole life annuity can be expressed as

$$Y = \bar{a}_{\overline{T}|} = (1 - v^T)/\delta = (1 - Z)/\delta$$

where $Z = v^T$ is the present value random variable corresponding to unit benefit to be payable at the moment of death in whole life insurance. Taking expectations on both sides we get a relation between the actuarial present values of benefit function and the annuity as given below.

$$\bar{a}_x = (1 - \overline{A}_x)/\delta \tag{6.3.8}$$

From Eq. 6.3.8, we obtain

$$1 = \delta\,\bar{a}_x + \overline{A}_x \tag{6.3.9}$$

This also has the similar interpretation as that for annuity certain due.

To measure the mortality risk in a continuous life annuity, we are interested in $Var(\bar{a}_{\overline{T}|})$. We find it as follows.

$$Var(\bar{a}_{\overline{T}|}) = Var\left((1 - v^T)/\delta\right) = Var(v^T)/\delta^2 = ({}^2\overline{A}_x - (\overline{A}_x)^2)/\delta^2$$

Further, we observe that since $1 = \delta\bar{a}_{\overline{T}|} + v^T$, $\quad Var(\delta\bar{a}_{\overline{T}|} + v^T) = 0$. Thus, there is no mortality risk for the combination of a continuous life annuity of δ per year and a life insurance of 1 payable at the moment of death.

The following examples illustrate the computations related to whole life annuity.

Example 6.3.3. It is given that the force of interest is $\delta = 6\%$ and the actuarial present value at age 27 of a unit benefit to be paid at the moment of death in a whole life insurance with a force of interest δ is 0.5321 and with a force of interest 2δ is 0.2935. Find the actuarial present value at age 27 of a continuous whole life annuity payable at the rate of 10000 per annum. Also find the standard deviation of the corresponding present value random variable.

Solution: We are given that $\overline{A}_{27} = 0.5321$ and ${}^2\overline{A}_{27} = 0.2935$. Using the formula derived in Eq. 6.3.8, we get

$$\bar{a}_{27} = (1 - \overline{A}_{27})/\delta = 7.798333 \quad \text{and} \quad 10000\,\bar{a}_{27} = 77983.33$$

77983.33 is the actuarial present value of an annuity, payable continuously at the rate of 10000 per annum, till (27) survives. Further variance is given by $Var(\bar{a}_{\overline{T}|}) = ({}^2\overline{A}_x - (\overline{A}_x)^2)/\delta^2 = 2.880442$. Hence, the standard deviation of the corresponding present value random variable of benefit of 10000 is $10000 \times (2.880442)^{1/2} = 10000 \times 1.697186 = 16971.86$ ∎

Example 6.3.4. Suppose the force of mortality follows Gompertz' law given by $\mu_x = BC^x$. Find the values of $100\,\bar{a}_x$ for $x = 25, 30, 35$ and 40. Take $\delta = 0.05$, $B = 0.0001151$ and $C = 1.096$.

Solution: In Example 5.3.7 we have obtained $\bar{A}_x$ for the specified values of x. We use the formula derived in Eq. 6.3.8 to find $\bar{a}_x$ for various values of x. These are computed using Part II of Code 6.6.4 and are reported in Table 6.8. From the table we have $100\,\bar{a}_{25} = 1694.93$.

Age	25	30	35	40
$100\bar{a}_x$	1694.93	1621.75	1534.60	1432.56

Table 6.8 Gompertz' law: $100\bar{a}_x$

This value is interpreted as follows. The expected present value of the periodical payments, paid continuously at the rate of 100 per annum, for the future life time of (25) is 1694.93. Thus, if the whole life insurance is purchased by whole life annuity of premium payments and if the premium is paid continuously at the rate of 100 per annum, then the expected present value of the premiums is 1694.93. Secondly the amount 1694.93 can also be interpreted as the purchase price of the whole life annuity which pays continuously at the rate of 100 per annum till (25) survives. As age increases, the actuarial present value decreases as the expected period of payments decreases. ∎

Deferred Whole Life Annuity: If the annuity payments are postponed for a certain period then we get deferred annuities. For example, suppose in the whole life annuity studied above, the payments are deferred for m years. The present value random variable Y for an m-year deferred whole life annuity is defined as

$$Y = \begin{cases} 0 & = \bar{a}_{\overline{T|}} - \bar{a}_{\overline{T|}} & \text{if} \quad 0 \leq T < m \\ v^m\,\bar{a}_{\overline{T-m|}} & = \bar{a}_{\overline{T|}} - \bar{a}_{\overline{m|}} & \text{if} \quad T \geq m \end{cases}$$

Here the maximum value of the random variable Y is v^m/δ and the probability that it takes a value 0 is $Pr(T \leq m) = {}_m q_x$. The actuarial present value of this annuity, denoted by ${}_{m|}\bar{a}_x$ is given by

$$\begin{aligned} {}_{m|}\bar{a}_x = E(Y) &= \int_m^\infty v^m \bar{a}_{\overline{t-m|}}\ {}_t p_x \mu_{x+t} dt = \int_0^\infty v^m \bar{a}_{\overline{s|}}\ {}_{m+s} p_x\ \mu_{x+m+s} ds \\ &= v^m\ {}_m p_x \int_0^\infty \bar{a}_{\overline{s|}}\ {}_s p_{x+m} \mu_{x+m+s} ds = {}_m E_x\ \bar{a}_{x+m} \end{aligned} \qquad (6.3.10)$$

To calculate the variance of Y for the deferred annuity, we proceed as follows:

$$
\begin{aligned}
Var(Y) &= \int_m^\infty v^{2m}(\overline{a}_{\overline{t-m|}})^2 \; {}_tp_x \; \mu_{x+t}dt \; - \; ({}_{m|}\overline{a}_x)^2 \\
&= v^{2m} \; {}_mp_x \int_0^\infty (\overline{a}_{\overline{s|}})^2 \; {}_sp_{x+m}\mu_{x+m+s}ds \; - \; ({}_{m|}\overline{a}_x)^2 \\
&= v^{2m} \; {}_mp_x \int_0^\infty 2\overline{a}_{\overline{s|}}v^s \; {}_sp_{x+m}ds - ({}_{m|}\overline{a}_x)^2, \quad \text{(integration by parts)} \\
&= (2/\delta)v^{2m} \; {}_mp_x \int_0^\infty (v^s - v^{2s}) \; {}_sp_{x+m}ds \; - \; ({}_{m|}\overline{a}_x)^2 \\
&= (2/\delta)v^{2m} \; {}_mp_x(\overline{a}_{x+m} - {}^2\overline{a}_{x+m}) - ({}_{m|}\overline{a}_x)^2 \qquad (6.3.11)
\end{aligned}
$$

Although the present value random variables of the whole life annuity and the n-year temporary annuity both satisfy the relationship $Y = (1 - Z)/\delta$, the m-year deferred annuity present value random variable does not satisfy such a relationship. An alternative way to find ${}_{m|}\overline{a}_x$ is from the definition of Y for an m-year deferred whole life annuity. It can be expressed as,

$$Y \text{ for a whole life annuity} - Y \text{ for an } m\text{-year temporary life annuity}$$

$$\text{Hence,} \quad {}_{m|}\overline{a}_x = \overline{a}_x - \overline{a}_{x:\overline{m|}} \qquad (6.3.12)$$

This expression of ${}_{m|}\overline{a}_x$ is useful in the calculation of deferred whole life annuity. Using Eqs 6.3.10 and 6.3.12, $\overline{a}_x$ can be expressed as

$$\overline{a}_x = \overline{a}_{x:\overline{m|}} + {}_{m|}\overline{a}_x = \overline{a}_{x:\overline{m|}} + {}_mE_x\overline{a}_{x+m} \qquad (6.3.13)$$

Example 6.3.5. For the setup of Examples 6.3.1 and 6.3.3, find the actuarial present value at age 27 of a 5-year deferred continuous life annuity payable at the rate of 5000 per annum.

Solution: By the formula derived in Eq. 6.3.12, we have

$$_{5|}\overline{a}_{27} = \overline{a}_{27} - \overline{a}_{27:\overline{5|}} = 3.4567$$

For 5000 per annum the actuarial present value is 17283.33. Thus, if the payments are deferred by 5 years, then the actuarial present value is 3.4567 for 1 unit payment of whole life annuity issued to (27). From Example 6.3.3 without deferment the value is 7.7983, larger than that for the deferred annuity. $\blacksquare$

Example 6.3.6. Suppose the force of mortality follows Gompertz' law given by $\mu_x = BC^x$. Find the values of $100_{m|}\overline{a}_x$ for $m = 1, 2, \cdots, 10$ and for $x = 25, 30, 35, 40$. Take $\delta = 0.05$, $B = 0.0001151$ and $C = 1.096$.

Solution: In Example 6.3.2 we have obtained $\overline{a}_{x:\overline{m|}}$ for the specified values of x and m. In Example 6.3.4 we have obtained $\overline{a}_x$ for the specified values of x. We use the formula derived in Eq. 6.3.12 to find ${}_{m|}\overline{a}_x$ for various values of x and m. These are computed using Part III of Code 6.6.4 and are reported in Table 6.9.

m	25	30	35	40
1	1597.44	1524.30	1437.20	1335.24
2	1504.83	1431.79	1344.85	1243.13
3	1416.85	1343.97	1257.29	1155.98
4	1333.29	1260.64	1174.31	1073.57
5	1253.94	1181.58	1095.71	995.69
6	1178.59	1106.58	1021.27	922.13
7	1107.05	1035.47	950.82	852.70
8	1039.16	968.06	884.16	787.22
9	974.73	904.17	821.13	725.52
10	913.59	843.65	761.56	667.43

Table 6.9 Gompertz' law: $100 {}_{m|}\bar{a}_x$

From Table 6.9, we note that if (40) purchases 10-year deferred whole life annuity, payable continuously at the rate of 100 per annum, till death, then its purchase price is 667.43. If (40) purchases 10-year whole life annuity, payable continuously at the rate of 100 per annum, till death, then from Table 6.8 its purchase price is 1432.56. Thus, deferred whole life annuity is cheaper than whole life annuity, as the insurer has no liability for the first ten years. Also note that as age increases, the actuarial present value decreases. It also decreases as the deferment period increases. ∎

n-**Year Certain and Whole Life Annuity:** As is clear from the name of the annuity, it is a combination of annuity certain and life annuity. Thus, n-year certain and whole life annuity is a whole life annuity with a guarantee of payments for the first n years and then continuation of payments till the individual survives. Thus, the period of annuity payments is $\max\{T(x), n\}$. For n-year temporary life annuity it is $\min\{T(x), n\}$. The present value random variable of n-year certain and whole life annuity is defined as,

$$Y = \begin{cases} \bar{a}_{\overline{n|}} & \text{if} \quad T \le n \\ \bar{a}_{\overline{T|}} & \text{if} \quad T > n \end{cases}$$

The minimum value and the maximum value of Y are $\bar{a}_{\overline{n|}}$ and $1/\delta$ respectively. The actuarial present value is denoted by $\bar{a}_{\overline{x:n|}}$. To find it, we express Y as follows.

$$Y = \begin{cases} \bar{a}_{\overline{n|}} + 0, & \text{if} \quad T \le n \\ \bar{a}_{\overline{n|}} + (\bar{a}_{\overline{T|}} - \bar{a}_{\overline{n|}}), & \text{if} \quad T > n \end{cases}$$

Thus, Y is the sum of a constant $\bar{a}_{\overline{n|}}$ and the present value random variable for the n-year deferred whole life annuity. Thus,

$$\begin{aligned} \bar{a}_{\overline{x:n|}} &= \bar{a}_{\overline{n|}} + {}_{n|}\bar{a}_x & (6.3.14) \\ &= \bar{a}_{\overline{n|}} + {}_nE_x\, \bar{a}_{x+n} & \text{by (6.3.10)} \\ &= \bar{a}_{\overline{n|}} + (\bar{a}_x - \bar{a}_{x:\overline{n|}}) & \text{by (6.3.12)} \end{aligned}$$

The expression of $\overline{a}_{\overline{x:n|}}$ as given in Eq. 6.3.14 shows that the actuarial present value of this annuity is larger than that for n-year certain annuity and the n-year deferred whole life annuity. It is of course obvious as in n-year certain and whole life annuity that there is a guarantee of payments for n years. Example 6.3.8 illustrates this comment for various values of n and various ages.

It is a well known result that variance is invariant to location change. Hence $Var(Y - \overline{a}_{\overline{n|}}) = Var(Y)$. Thus, the variance for the n-year certain and whole life annuity is the same as that of the n-year deferred whole life annuity given by Eq. 6.3.11. $\overline{a}_{\overline{x:n|}}$ can also be expressed as,

$$\overline{a}_{\overline{x:n|}} = \overline{a}_{\overline{n|}} + \int_n^\infty v^t \; {}_tp_x dt$$

This is the current payment form for the actuarial present value.

Example 6.3.7. For the setup of Example 6.3.1 and Example 6.3.3, find the actuarial present value at age 27 of a 5-year certain and continuous whole life annuity payable at the rate of 7000 per annum.

Solution: By Eq. 6.3.14, we have

$$\overline{a}_{\overline{27:5|}} = \overline{a}_{\overline{5|}} + {}_{5|}\overline{a}_{27} = 4.319696 + 3.4567 = 7.776396$$

For 7000 per annum the actuarial present value is $7000 * 7.776396 = 54434.77$.

∎

Example 6.3.8. Suppose the force of mortality follows Gompertz' law given by $\mu_x = BC^x$. Find the values of $100\overline{a}_{\overline{x:n|}}$ for $n = 1, 2, \cdots, 10$ and for $x = 25, 30, 35$ and 40. Take $\delta = 0.05$, $B = 0.0001151$ and $C = 1.096$.

Solution: In Example 6.3.6 we have obtained ${}_{n|}\overline{a}_x$ for the specified values of x and n. We use Eq. 6.3.14 to find $\overline{a}_{\overline{x:n|}}$ for various values of x and n. These are computed using Part IV of Code 6.6.4 and are reported in Table 6.10. From Table 6.10 we note that 1700.53 is the actuarial present value of 10-year certain and life annuity issued to (25). Thus, if payments are made continuously at the rate of 100 per annum, certainly for 10 years and then till (25) survives, then the actuarial present value of such payments is 1700.53. From Table 6.9, we note that it is larger than ${}_{10|}\overline{a}_x = 913.59$. The fourth column Table 6.5 reports $1000 \, \overline{a}_{\overline{10|}}$ with $i = 0.05$. Thus, the value of $100 \, \overline{a}_{\overline{10|}}$ for $\delta = 0.05$ will be approximately 791.32, which is smaller than 1700.60. Further we note that as age increases, the actuarial present values decrease as the total expected period of payments decreases. As n increases the actuarial present values also increase as the period for which payments are certain increases, however, the increase is at a very low rate. When the values in this table are compared to $100\overline{a}_x$ reported in Table 6.8, it is to be noted that $100\overline{a}_{\overline{x:1|}}$ are very close to $100 \, \overline{a}_x$ for all ages, as the period certain is just 1 year. ∎

n	25	30	35	40
1	1694.98	1621.84	1534.74	1432.78
2	1695.15	1622.11	1535.17	1433.45
3	1695.44	1622.56	1535.87	1434.56
4	1695.83	1623.18	1536.85	1436.11
5	1696.33	1623.98	1538.11	1438.08
6	1696.95	1624.95	1539.64	1440.49
7	1697.68	1626.09	1541.44	1443.33
8	1698.52	1627.42	1543.52	1446.58
9	1699.47	1628.91	1545.87	1450.26
10	1700.53	1630.59	1548.50	1454.37

Table 6.10 Gompertz' law: $100\overline{a}_{x:\overline{n}|}$

Table 6.11 displays the actuarial present values of n-year temporary, n-year deferred and n-year certain and whole life annuity, payable continuously with a rate of 100 per annum, when the mortality pattern of (30) is governed by Gompertz' law with parameters as in Example 6.3.1 and $\delta = 0.05$.

n	$100\,\overline{a}_{30:\overline{n}|}$	$100_{n|}\overline{a}_{30}$	$100\,\overline{a}_{\overline{30:n}|}$
1	97.45	1524.55	1622.09
2	189.97	1432.03	1622.36
3	277.78	1344.22	1622.80
4	361.11	1260.89	1623.42
5	440.18	1181.82	1624.22
6	515.17	1106.83	1625.19
7	586.28	1035.72	1626.34
8	653.70	968.30	1627.66
9	717.58	904.42	1629.16
10	778.11	843.89	1630.83

Table 6.11 Gompertz' law: comparison of actuarial present values

The actuarial present value of whole life annuity for the same setup is 1621.75. From Table 6.11, we note that the purchase price of n-year temporary annuity is less than any other annuity and that of n-year certain and life annuity is higher than any other annuity. This observation is reasonable in view of the nature of these annuities.

In the next section we discuss life annuities where payments are not made continuously but at discrete time points, may be at the beginning of payment intervals or at the end of the intervals. These are known as discrete life annuities.

6.4   Discrete Life Annuities

The theory of discrete life annuities, also known as annuities payable annually, is analogous to the theory of continuous life annuities. In continuous life annuities present value random variables are defined in terms of the future life time random variable $T(x)$, while in discrete life annuities present value random variables are defined in terms of the curtate future life time random variable $K(x)$. For continuous annuities, there was no distinction between payments at the beginning of payment intervals or at the end of the intervals, that is, between annuity due and annuity immediate. For discrete annuities, the distinction is meaningful. Discrete life annuity due is often used in applications. For example, most individual life insurances are purchased by an annuity due of periodic premiums. As in continuous life annuities, discrete life annuities can be any one of the following three types. (i) n-year temporary life annuity. (ii) Whole life annuity. (iii) n-year certain and life annuity. Each of these is further classified depending upon the deferment or otherwise of payments and as due and immediate. We discuss below each of these annuities in detail.

n-Year Temporary Life Annuity Due: In this annuity, one unit per annum is payable at the beginning of the year for at most n years. Thus, the annuity payments cease when an individual dies during the term of n-years or at the end of n years if individual survives the n-year term. Such a situation arises if n-year term insurance or n-year endowment insurance, with benefit payable at the end of year of death, is purchased by paying periodical premiums at the anniversary of the policy year. Payments of premiums will stop if the death occurs before the term of n-year is over or definitely at the end of n-year if individual survives the term. Thus, the total number of payments will be $1 + K(x)$, if death occurs before the end of n-year term and n if (x) survives the n-year term. Hence, the present value random variable of this annuity is defined as follows.

$$Y = \begin{cases} \ddot{a}_{\overline{K+1}|}, & \text{if} \quad K = 0, 1, \cdots, n-1 \\ \ddot{a}_{\overline{n}|}, & \text{if} \quad K = n, n+1, \cdots \end{cases}$$

Its actuarial present value is denoted by $\ddot{a}_{x:\overline{n}|}$ and is given by,

$$\ddot{a}_{x:\overline{n}|} \quad = \quad E(Y) = \sum_{k=0}^{n-1} \ddot{a}_{\overline{k+1}|} \; {}_kp_x q_{x+k} + \ddot{a}_{\overline{n}|} \; {}_np_x$$

Using the relation $\ddot{a}_{\overline{k+1}|} - \ddot{a}_{\overline{k}|} = v^k$ and expressing

$$P[K(x) = k] \quad = \quad {}_kp_x q_{x+k} = {}_kp_x(1 - p_{x+k}) = {}_kp_x - {}_kp_x p_{x+k} = {}_kp_x - {}_{k+1}p_x$$

the above sum can be simplified as follows.

$$\ddot{a}_{x:\overline{n}|} \quad = \quad \sum_{k=0}^{n-1} \ddot{a}_{\overline{k+1}|} \; {}_kp_x q_{x+k} + \ddot{a}_{\overline{n}|} \; {}_np_x$$

$$= \quad \sum_{k=0}^{n-1} \ddot{a}_{\overline{k+1}|} \left({}_kp_x - {}_{k+1}p_x \right) + \ddot{a}_{\overline{n}|} \; {}_np_x$$

$$= 1 + \sum_{k=1}^{n-1} \ddot{a}_{\overline{k+1}|}\, {}_kp_x - \sum_{k=0}^{n-2} \ddot{a}_{\overline{k+1}|}\, {}_{k+1}p_x - \ddot{a}_{\overline{n}|}\, {}_np_x + \ddot{a}_{\overline{n}|}\, {}_np_x$$

$$= 1 + \sum_{k=1}^{n-1} \ddot{a}_{\overline{k+1}|}\, {}_kp_x - \sum_{k=1}^{n-1} \ddot{a}_{\overline{k}|}\, {}_kp_x$$

$$= 1 + \sum_{k=1}^{n-1} {}_kp_x(\ddot{a}_{\overline{k+1}|} - \ddot{a}_{\overline{k}|}) = 1 + \sum_{k=1}^{n-1} {}_kp_x v^k$$

$$= \sum_{k=0}^{n-1} v^k\, {}_kp_x \tag{6.4.1}$$

It is the actuarial present value of Y in the current payment form. Observe that $\ddot{a}_{x:\overline{1}|} = 1$. Expressing Y in a different way, we get the relation of $\ddot{a}_{x:\overline{n}|}$ with the actuarial present value of the benefit function studied in Chapter 5. Note that Y can be written as $Y = (1 - Z)/d$, where,

$$Z = \begin{cases} v^{K+1}, & \text{if} \quad K = 0, 1, \cdots, n-1 \\ v^n, & \text{if} \quad K = n, n+1, \cdots \end{cases}$$

From Section 5.4 we recall that Z defined above is the present value random variable of a unit benefit, payable at the end of the year of death or at maturity, whichever is earlier in n-year endowment insurance. Taking expectations on both sides we have,

$$\ddot{a}_{x:\overline{n}|} = (1 - E(Z))/d = (1 - A_{x:\overline{n}|})/d. \tag{6.4.2}$$

Thus, the actuarial present value of n-year temporary discrete life annuity due, with unit payment and the actuarial present value of one unit benefit to be paid at the end of year of death in the n-year endowment insurance are related to each other. This relation is useful to find the values of one from the other and in the calculation of premiums, as will be seen in Chapter 7. Equation 6.4.2 can be rearranged as

$$1 = d\ddot{a}_{x:\overline{n}|} + A_{x:\overline{n}|} \tag{6.4.3}$$

Its interpretation is similar to that of annuity certain due in Section 2. We use the relation between Y and Z to calculate the variance of Y. Thus, $Var(Y) = Var(Z)/d^2 = ({}^2A_{x:\overline{n}|} - (A_{x:\overline{n}|})^2)/d^2$. The actuarial accumulated value at the end of the term of an n-year temporary life annuity due of 1 unit per year, payable till (x) survives, or up to n, whichever is earlier, is denoted by $\ddot{S}_{x:\overline{n}|}$ and is given by $\ddot{S}_{x:\overline{n}|} = \ddot{a}_{x:\overline{n}|}/ {}_nE_x = \ddot{a}_{x:\overline{n}|}/(v^n\, {}_np_x)$. This formula is analogous to the formula $\ddot{S}_{\overline{n}|} = \ddot{a}_{\overline{n}|}/v^n$ for accumulated values in annuities certain. The additional factor ${}_np_x$ comes in the denominator as we are concerned with life annuities and payments are made if (x) survives for the next n years.

n-Year Temporary Life Annuity Immediate: In this annuity, payment of one unit per annum is payable at the end of the payment intervals for at most n years. If death occurs in the first year after signing the contract then there is no payment. 1 unit will be paid at the end of the first policy year, if (x) survives the first year. If death occurs in the second year after signing the contract then there is only one payment of 1 unit at the end of the first year. Thus, the total number of payments will be $K(x)$, if death occurs during the n-year term and n if the (x) survives the n-year term. The present value random variable of n-year temporary life annuity immediate is defined as follows.

$$Y = \begin{cases} a_{\overline{K}|}, & \text{if} \quad K = 0, 1, \cdots, n-1 \\ a_{\overline{n}|}, & \text{if} \quad K = n, n+1, \cdots \end{cases}$$

Here we define $a_{\overline{0}|} = 0$. The actuarial present value of Y is denoted by $a_{x:\overline{n}|}$ and is given by,

$$\begin{aligned} a_{x:\overline{n}|} &= E(Y) = \sum_{k=0}^{n-1} a_{\overline{k}|} \, {}_kp_x q_{x+k} + a_{\overline{n}|} \, {}_np_x \\ &= \sum_{k=1}^{n-1} a_{\overline{k}|} \left({}_kp_x - {}_{k+1}p_x \right) + a_{\overline{n}|} \, {}_np_x \qquad \text{since} \quad a_{\overline{0}|} = 0 \\ &= \sum_{k=1}^{n-1} a_{\overline{k}|} \, {}_kp_x - \sum_{k=1}^{n-1} a_{\overline{k}|} \, {}_{k+1}p_x + a_{\overline{n}|} \, {}_np_x \\ &= v \, p_x + \sum_{k=2}^{n-1} a_{\overline{k}|} \, {}_kp_x - \sum_{k=2}^{n} a_{\overline{k-1}|} \, {}_kp_x + a_{\overline{n}|} \, {}_np_x \\ &= v \, p_x + \sum_{k=2}^{n} a_{\overline{k}|} \, {}_kp_x - \sum_{k=2}^{n} a_{\overline{k-1}|} \, {}_kp_x \\ &= v \, p_x + \sum_{k=2}^{n} {}_kp_x \left(a_{\overline{k}|} - a_{\overline{k-1}|} \right) \\ &= \sum_{k=1}^{n} v^k \, {}_kp_x \end{aligned}$$

$$(6.4.4)$$

It is the actuarial present value in the form of the current payment. The relation of $a_{x:\overline{n}|}$ with the actuarial present value of the benefit function is not obvious in this type of annuity. We write Y as $Y = (1 - Z_1 - Z_2)/i$, where,

$$Z_1 = \begin{cases} (1+i)v^{K+1}, & \text{if} \quad K = 0, 1, \cdots, n-1 \\ 0, & \text{if} \quad K = n, n+1, \cdots \end{cases}$$

and

$$Z_2 = \begin{cases} 0, & \text{if} \quad K = 0, 1, \cdots, n-1 \\ v^n, & \text{if} \quad K = n, n+1, \cdots \end{cases}$$

$Z_1/(i+1)$ is the present value random variable for a unit benefit in n-year term insurance, payable at the end of the year of death while Z_2 is the present value random variable for a unit benefit, n-year pure endowment insurance. Taking expectations on both sides and after some algebra, we have,

$$E(Y) = a_{x:\overline{n}|} = \frac{1 - (1+i)A^{\,1}_{x:\overline{n}|} - A_{x:\overline{n}|}^{\,1}}{i} = v \, \ddot{a}_{x:\overline{n}|} - A^{\,1}_{x:\overline{n}|} \qquad (6.4.5)$$

From Eqs 6.4.1 and 6.4.4, we derive one more relation between $a_{x:\overline{n}|}$ and $\ddot{a}_{x:\overline{n}|}$ as follows.

$$\ddot{a}_{x:\overline{n}|} = \sum_{k=0}^{n-1} v^k \, {}_kp_x = 1 + \sum_{k=1}^{n} v^k \, {}_kp_x - v^n \, {}_np_x = 1 + a_{x:\overline{n}|} - {}_nE_x \qquad (6.4.6)$$

Example 6.4.1. Prove that $a_{x:\overline{n}|} < \ddot{a}_{x:\overline{n}|}$.

Solution: It follows immediately from Eq. 6.4.5. We have,

$$a_{x:\overline{n}|} = v\,\ddot{a}_{x:\overline{n}|} - A^{\,1}_{x:\overline{n}|} < v\,\ddot{a}_{x:\overline{n}|} < \ddot{a}_{x:\overline{n}|}$$

as $A^{\,1}_{x:\overline{n}|}$ is positive and v is a positive fraction. Thus, the expected present value of n-year temporary life annuity immediate is less than that for due. It is logical in view of the nature of time points of the payment. ∎

Remark 6.4.1. This result is analogous to the result $a_{\overline{n}|} < \ddot{a}_{\overline{n}|}$ proved in Eq. 6.2.4.

Example 6.4.2. Prove that $a_{x:\overline{n}|} < \overline{a}_{x:\overline{n}|} < \ddot{a}_{x:\overline{n}|}$.

Solution: From Eq. 6.3.5, we have,

$$\overline{a}_{x:\overline{n}|} = \int_0^n v^t \, {}_tp_x \, dt = \sum_{k=0}^{n-1} \int_k^{k+1} v^t \, {}_tp_x \, dt = \sum_{k=0}^{n-1} \int_0^1 v^{k+s} \, {}_{k+s}p_x \, ds$$

$$= \sum_{k=0}^{n-1} v^k \, {}_kp_x \int_0^1 v^s \, {}_sp_{x+k} \, ds < \sum_{k=0}^{n-1} v^k \, {}_kp_x = \ddot{a}_{x:\overline{n}|}$$

as the integrand $v^s \, {}_sp_{x+k}$ in the integral $\int_0^1 v^s \, {}_sp_{x+k} \, ds$ is always less than 1. Thus we have $\overline{a}_{x:\overline{n}|} < \ddot{a}_{x:\overline{n}|}$. To prove $a_{x:\overline{n}|} < \overline{a}_{x:\overline{n}|}$, note that $0 < v < 1$ and $0 < s < 1$. Hence $v^s > v$. ${}_sp_{x+k}$ is a survival function, hence $s < 1$ implies that ${}_sp_{x+k} > {}_1p_{x+k} = p_{x+k}$. Hence, $\int_0^1 v^s \, {}_sp_{x+k} \, ds > v\,p_{x+k}$. Now

$$\overline{a}_{x:\overline{n}|} = \sum_{k=0}^{n-1} v^k \, {}_kp_x \int_0^1 v^s \, {}_sp_{x+k} \, ds > \sum_{k=0}^{n-1} v^k \, {}_kp_x \, v \, p_{x+k}$$

$$= \sum_{k=0}^{n-1} v^{k+1} \, {}_{k+1}p_x = \sum_{k=1}^{n} v^k \, {}_kp_x = a_{x:\overline{n}|}$$

Thus we have, $a_{x:\overline{n}|} < \overline{a}_{x:\overline{n}|} < \ddot{a}_{x:\overline{n}|}$ ∎

Remark 6.4.2. Result proved in Example 6.4.2 is analogous to the result in Eq. 6.2.17 which states that $a_{\overline{n}|} < \overline{a}_{\overline{n}|} < \ddot{a}_{\overline{n}|}$.

Example 6.4.3. Prove that $\overline{A}_{x:\overline{n}|} > A_{x:\overline{n}|}$.

Solution: From Chapter 5 we have,

$$\overline{A}_{x:\overline{n}|} \;=\; \int_0^n v^t g(t)dt \;+\; v^n \, {}_n p_x$$

Using the relation $g(t) \;=\; -\frac{d}{dt}\, {}_t p_x$, we solve the integral $\int_0^n v^t g(t)dt$ by parts and get $\overline{A}_{x:\overline{n}|} \;=\; 1 - \delta \int_0^n v^t \, {}_t p_x \, dt$. Thus we have,

$$\overline{A}_{x:\overline{n}|} \;=\; 1 - \delta \int_0^n v^t \, {}_t p_x \, dt = 1 - \delta \sum_{k=0}^{n-1} \int_k^{k+1} v^t \, {}_t p_x \, dt$$

$$=\; 1 - \delta \sum_{k=0}^{n-1} \int_0^1 v^{k+s} \, {}_{k+s} p_x \, ds = 1 - \delta \sum_{k=0}^{n-1} v^k \, {}_k p_x \int_0^1 v^s \, {}_s p_{x+k} \, ds$$

Observe that,

$$\int_0^1 v^s \, {}_s p_{x+k} \, ds \;<\; \int_0^1 v^s \, ds \;=\; (v-1)/\log v \;=\; d/\delta$$

With this inequality,

$$\overline{A}_{x:\overline{n}|} \;>\; 1 - \delta \, (d/\delta) \sum_{k=0}^{n-1} v^k \, {}_k p_x \;=\; 1 - d\ddot{a}_{x:\overline{n}|} \;=\; A_{x:\overline{n}|}$$

The inequality $\overline{A}_{x:\overline{n}|} \;>\; A_{x:\overline{n}|}$ is logically also appealing since $\overline{A}_{x:\overline{n}|}$ denotes the actuarial present value of the benefit to be paid at the moment of death while $A_{x:\overline{n}|}$ denotes the actuarial present value of the benefit to be paid at the end of year of death. ∎

Example 6.4.4. It is given that $q_{28} = 0.135$, $q_{29} = 0.146$, $q_{30} = 0.159$, $q_{31} = 0.173$, $q_{32} = 0.188$ and $i = 0.05$. Find the actuarial present value of the 5-year temporary annuity due and immediate issued to (28), at the rate of 5000 per annum. Also find the actuarial present value of 5-year temporary annuity payable continuously at the rate of 5000 per annum, under the assumption of uniformity of deaths in unit age interval. Compare the three values.

Solution: We first find $\ddot{a}_{28:\overline{5}|}$. By Eq. 6.4.1 we have,

$$\ddot{a}_{28:\overline{5}|} \;=\; \sum_{k=0}^{4} v^k \, {}_k p_{28} \;=\; 1 + v p_{28} + v^2 p_{28} p_{29} + v^3 p_{28} p_{29} p_{30} + v^4 p_{28} p_{29} p_{30} p_{31}$$

For the given values of q_x and i, we get $5000\,\ddot{a}_{28:\overline{5}|} \;=\; 5000 \times 3.453191 \;=\; 17265.96$. In Example 5.4.3, for the same mortality pattern and $i = 0.05$, we have obtained $A_{28:\overline{5}|} \;=\; 0.8355623$. Using Eq. 6.4.2, we can obtain $\ddot{a}_{28:\overline{5}|}$ from $A_{28:\overline{5}|}$. Both the approaches of course give the same answer. By Eq. 6.4.6 we have,

$$a_{28:\overline{5}|} \;=\; \ddot{a}_{28:\overline{5}|} + {}_5 E_{28} - 1 \;=\; 3.453191 + v^5 \, {}_5 p_{28} - 1$$

$$=\; 3.453191 + 0.3268775 - 1 = 2.780069 \;\Rightarrow\; 5000\, a_{28:\overline{5}|} = 13900.34$$

Now to find $\bar{a}_{28:\overline{5}|}$, from Eq. 6.3.4 we have, $\bar{a}_{28:\overline{5}|} = (1 - \bar{A}_{28:\overline{5}|})/\delta$, where $\bar{A}_{28:\overline{5}|} = \bar{A}^{1}_{28:\overline{5}|} + A_{28:\overline{5}|}^{1}$. Under the assumption of uniformity of deaths in unit age intervals, we have $\bar{A}^{1}_{28:\overline{5}|} = (i/\delta)A^{1}_{28:\overline{5}|}$. Note that

$$A_{28:\overline{5}|}^{1} = {}_5E_{28} = 0.3268775$$

$$\text{Further,} \quad \ddot{a}_{28:\overline{5}|} = (1 - A_{28:\overline{5}|})/d \Rightarrow A_{28:\overline{5}|} = 0.8355623$$

$$\Rightarrow A^{1}_{28:\overline{5}|} = A_{28:\overline{5}|} - A_{28:\overline{5}|}^{1} = 0.5086848$$

$$\Rightarrow \bar{A}^{1}_{28:\overline{5}|} = (i/\delta)A^{1}_{28:\overline{5}|} = 0.5212985$$

$$\Rightarrow \bar{A}_{28:\overline{5}|} = \bar{A}^{1}_{28:\overline{5}|} + A_{28:\overline{5}|}^{1} = 0.848176$$

$$\Rightarrow \bar{a}_{28:\overline{5}|} = (1 - \bar{A}_{28:\overline{5}|})/\delta = 3.111774 \text{ and } 5000\, \bar{a}_{28:\overline{5}|} = 15558.87$$

Thus we have,

$$5000\, a_{28:\overline{5}|} = 13900.34, \quad 5000\, \bar{a}_{28:\overline{5}|} = 15558.87, \quad 5000\, \ddot{a}_{28:\overline{5}|} = 17265.96.$$

Comparing the values we notice that these do satisfy the relation established in Example 6.4.2.

∎

Example 6.4.5. Suppose the force of mortality follows Gompertz' law given by $\mu_x = BC^x$. Find the values of (i) n-year temporary life annuity due and (ii) n-year temporary life annuity immediate for payment of 100 rupees per annum, for $n = 1, 2, \cdots, 10$ and for ages $25, 30, 35, 40$. Take $\delta = 0.05$, $B = 0.0001151$ and $C = 1.096$.

Solution: (i) From Eq. 6.4.2 we have, $\ddot{a}_{x:\overline{n}|} = (1 - A_{x:\overline{n}|})/d$. We have obtained in Example 5.4.2 the values of $A_{x:\overline{n}|}$ for the specified values of x and n for Gompertz' law with the given values of parameters. Using those values and Eq. 6.4.2, we find the values of $100\ddot{a}_{x:\overline{n}|}$. These are reported in Table 6.12. These can also be obtained directly from the formula in Eq. 6.4.1 and the values of ${}_kp_x$ for Gompertz' law. Code 6.6.6 is used to find these values. The first row of Table 6.12 is 100 for all ages, since by the formula for $\ddot{a}_{x:\overline{n}|}$, $\ddot{a}_{x:\overline{1}|} = 1$. Annuity due is paid at the beginning of the interval. Hence for a one year term, its present value is the same as the amount paid at the beginning of the interval. The entry $\ddot{a}_{25:\overline{10}|} = 801.73$ is interpreted as follows: the present value of the payments made at the beginning of the year at the rate of 100 per annum for at most 10 years is 801.73. In other words, if the mode of premium payments is such that the premium is paid at the beginning of the year at the rate of 100 per annum for at most 10 years, then its expected present value, that is, expected cash inflow to the insurance company corresponding to this policy, is 801.73. From the table we note that values of $100\ddot{a}_{x:\overline{n}|}$ decrease marginally as age increases for the given mortality law with the given parameters. However, these increase as the term n increases, because the period of annuity payments increases. The values in this table are marginally higher than those reported in Table 6.7 specifying the values of $100\bar{a}_{x:\overline{n}|}$ for the same mortality law. Here the annuity payments are at the beginning of the interval while in the setup of Table 6.7 the annuity payments are continuous at the same rate and hence such a marginal difference.

(ii) From Eq. 6.4.5 we have,

$$a_{x:\overline{n}|} = (1 - (1+i)A^{1}_{x:\overline{n}|} - A^{1}_{x:\overline{n}|})/i$$

n	25	30	35	40
1	100.00	100.00	100.00	100.00
2	195.01	194.94	194.84	194.68
3	285.27	285.07	284.76	284.27
4	371.00	370.61	369.99	369.01
5	452.42	451.77	450.74	449.12
6	529.74	528.77	527.23	524.81
7	603.15	601.79	599.64	596.27
8	672.84	671.02	668.17	663.69
9	738.98	736.65	732.99	727.26
10	801.73	798.82	794.26	787.13

Table 6.12 Gompertz' law: $100\ddot{a}_{x:\overline{n}|}$

We have obtained in Example 5.4.2 the values of $A^{\ 1}_{\ x:\overline{n}|}$ and $A_{x:\overline{n}|}^{\ \ 1}$ for the specified values of x and n for Gompertz' law with the given values of parameters. Using those values and Eq. 6.4.5, we find the values of $100a_{x:\overline{n}|}$. These are obtained using Code 6.6.6 and are reported in Table 6.13.

n	25	30	35	40
1	95.01	94.94	94.84	94.68
2	185.27	185.07	184.76	184.27
3	271.00	270.61	269.99	269.01
4	352.42	351.77	350.74	349.12
5	429.74	428.77	427.23	424.81
6	503.15	501.79	499.64	496.27
7	572.84	571.02	568.17	563.69
8	638.98	636.65	632.99	627.26
9	701.73	698.82	694.26	687.13
10	761.27	757.72	752.15	743.47

Table 6.13 Gompertz' law: $100a_{x:\overline{n}|}$

The entry $a_{25:\overline{10}|} = 761.27$ is interpreted as follows: The present value of the payments made at the end of the year at the rate of 100 per annum for at most 10 years is 761.27. From the table we note that values of $100a_{x:\overline{n}|}$ decrease marginally as age increases for the given mortality law with given parameters. However, these increase as the term n increases. The values in this table are marginally lower than those reported in Table 6.7 specifying the values of $100\overline{a}_{x:\overline{n}|}$ for the same mortality law. Here the annuity payments are at the end of the interval while in the setup of Table 6.7 annuity payments are continuous at the same rate and hence such a marginal difference. The values reported in Table 6.13 can also be obtained from those in Table 6.12, using the relation derived in Eq. 6.4.6 or the last part of Eq. 6.4.5.   ∎

For comparative study of the actuarial present values of annuity payments which are either continuous or at the beginning of the payment interval or at the end of the interval, we combine the results of Example 6.3.2 and Example 6.4.5 for age 25 and display in Table 6.14.

n	$100\,\ddot{a}_{25:\overline{n}\rvert}$	$100\,\overline{a}_{25:\overline{n}\rvert}$	$100\,a_{25:\overline{n}\rvert}$
1	100.00	97.48	95.01
2	195.01	190.10	185.27
3	285.27	278.08	271.00
4	371.00	361.64	352.42
5	452.42	440.99	429.74
6	529.74	516.34	503.15
7	603.15	587.87	572.84
8	672.84	655.77	638.98
9	738.98	720.20	701.73
10	801.73	781.34	761.27

Table 6.14 Gompertz' law: n-Year temporary life annuity

From Table 6.14, it is clear that

$$100\,\ddot{a}_{25:\overline{n}\rvert} > 100\,\overline{a}_{25:\overline{n}\rvert} > 100\,a_{25:\overline{n}\rvert} \quad \forall \ n$$

It illustrates the result proved in Example 6.4.2. Such inequality follows from the nature of time points of payments in these three cases. In $100\,\ddot{a}_{25:\overline{n}\rvert}$ it is at the beginning of the interval, in $100\overline{a}_{25:\overline{n}\rvert}$ it is paid continuously while in $100\,a_{25:\overline{n}\rvert}$ it is paid at the end of the interval.

Whole Life Annuity Due: In this annuity, a unit amount is paid at the beginning of each year till the annuitant survives. The present value random variable Y, for such an annuity is given by

$$Y = \ddot{a}_{\overline{K+1}\rvert}$$

The possible values of this random variable range from $\ddot{a}_{\overline{1}\rvert} = 1$ to $\ddot{a}_{\overline{w^*+1}\rvert}$. The actuarial present value of the whole life annuity due is denoted by $\ddot{a}_x$ and is given by,

$$\ddot{a}_x = E(Y) = E[\ddot{a}_{\overline{K+1}\rvert}] = \sum_{k=0}^{\infty} \ddot{a}_{\overline{k+1}\rvert}\ {}_k p_x\, q_{x+k} \tag{6.4.7}$$

Using the same procedure as in Eq. 6.4.1, we simplify the above sum as follows.

$$\ddot{a}_x = \sum_{k=0}^{\infty} \ddot{a}_{\overline{k+1|}} \, {}_kp_x q_{x+k} = \sum_{k=0}^{\infty} \ddot{a}_{\overline{k+1|}} ({}_kp_x - {}_{k+1}p_x)$$

$$= 1 + \sum_{k=1}^{\infty} \ddot{a}_{\overline{k+1|}} \, {}_kp_x - \sum_{k=1}^{\infty} \ddot{a}_{\overline{k|}} \, {}_kp_x = 1 + \sum_{k=1}^{\infty} {}_kp_x (\ddot{a}_{\overline{k+1|}} - \ddot{a}_{\overline{k|}})$$

$$= 1 + \sum_{k=1}^{\infty} {}_kp_x v^k = \sum_{k=0}^{\infty} v^k \, {}_kp_x \qquad (6.4.8)$$

The expression in Eq. 6.4.8 is the current payment form of the actuarial present value of a unit benefit in a whole life annuity due. Note that $\ddot{a}_x$ is the limit of $\ddot{a}_{x:\overline{n|}}$ as n tends to ∞. To find the relation of $\ddot{a}_x$ with the actuarial present value of benefit function we write the present value random variable Y of whole life annuity due as

$$Y = \ddot{a}_{\overline{K+1|}} = (1 - v^{K+1})/d = (1 - Z)/d \qquad (6.4.9)$$

where Z is the present value random variable of a unit benefit, payable at the end of year of death in whole life insurance. Taking expectations on both sides of Eq. 6.4.9, we get,

$$\ddot{a}_x = E\left[(1 - v^{K+1})/d\right] = (1 - A_x)/d \quad \Rightarrow \quad 1 = d\ddot{a}_x + A_x \qquad (6.4.10)$$

The variance of $\ddot{a}_{\overline{K+1|}}$ is

$$Var(\ddot{a}_{\overline{K+1|}}) = Var\left((1 - v^{K+1})/d\right) = Var(v^{K+1})/d^2 = ({}^2A_x - A_x^2)/d^2$$

The formula for $\ddot{a}_x$ given in Eq. 6.4.8 can be used to find a backward recurrence relation as follows.

$$\ddot{a}_x = 1 + \sum_{k=0}^{\infty} v^{k+1} \, {}_{k+1}p_x = 1 + v \, p_x \sum_{k=0}^{\infty} v^k \, {}_kp_{x+1} = 1 + vp_x\ddot{a}_{x+1} \qquad (6.4.11)$$

The initial value to use for the whole life annuity is $\ddot{a}_{w^*} = 0$, which follows from Eq. 6.4.8 since ${}_kp_{w^*} = 0 \; \forall \; k$.

Whole Life Annuity Immediate: In this annuity, a unit amount is paid at the end of each year till the annuitant survives. The present value random variable Y for such an annuity, is given by

$$Y = a_{\overline{K|}}$$

The actuarial present value of the whole life annuity immediate is denoted by a_x and is given by,

$$a_x = E(Y) = E(a_{\overline{K|}}) = \sum_{k=0}^{\infty} a_{\overline{k|}} \, {}_kp_x q_{x+k} = \sum_{k=1}^{\infty} v^k \, {}_kp_x \qquad (6.4.12)$$

In this setup, $Y = (1 - v^K)/i = [1 - (1 + i)v^{K+1}]/i$. Taking expectations we have,

$$a_x = E(Y) = (1 - (1 + i)A_x)/i = (1 - A_x)/i - A_x = (d/i)(1 - A_x)/d - A_x$$

$$= v \, \ddot{a}_x - A_x \qquad (6.4.13)$$

Note that a_x is the limit of $a_{x:\overline{n}|}$ as $n \to \infty$ and the above relation can be obtained from Eq. 6.4.5 by allowing n to tend to ∞ and recalling that A_x is the limit of $A^1_{x:\overline{n}|}$ as $n \to \infty$. From Eq. 6.4.12 we get a relation as follows.

$$\ddot{a}_x = \sum_{k=0}^{\infty} v^k \, {}_k p_x = 1 + \sum_{k=1}^{\infty} v^k \, {}_k p_x = 1 + a_x \qquad (6.4.14)$$

Example 6.4.6. Prove that $\ddot{a}_x > \overline{a}_x > a_x$.

Solution: The solution is similar to that of Example 6.4.2, where the upper limit in $\int_0^n$ and in summation is replaced by ∞. It can also be obtained from Example 6.4.2 by allowing n to tend to ∞.   ∎

Example 6.4.7. Suppose for the specific mortality law, q_x values are as given in Table 6.15. For the same setup and $i = 0.05$, $A_{50} = 0.251$. Find the actuarial present value of whole life annuity due and immediate when the yearly payment is 1000, for ages 40 to 49. Also find the actuarial present value of whole life annuity, payable continuously at the rate of 1000 per annum, under the assumption of uniformity in unit age intervals.

Age	40	41	42	43	44
q_x	0.0027	0.0029	0.0032	0.0034	0.0037
Age	45	46	47	48	49
q_x	0.0039	0.0043	0.0046	0.0050	0.0054

Table 6.15 q_x values

Solution: From Eq. 6.4.10 we have, $\ddot{a}_{50} = (1 - A_{50})/d = 15.729$. Then we use the recurrence relation given in Eq. 6.4.11 to find $\ddot{a}_x$ for $x = 40$ to 49. From Eq. 6.4.14, we have $a_x = \ddot{a}_x - 1$. We find $\overline{a}_x$ using the relation $\overline{a}_x = (1 - \overline{A}_x)/\delta$ as given in Eq. 6.4.8. Under the assumption of uniformity in unit age intervals, $\overline{A}_x = (i/\delta)A_x$. We are given A_{50}. We use the backward recurrence relation $A_x = vq_x + vp_x A_{x+1}$ given in Eq. 5.4.3 to find A_x values for $x = 40$ to 49. A_x values for $x = 40$ to 49 can also be obtained from the values of $\ddot{a}_x$. We adopt the above mentioned procedure to write a Code 6.6.5 to find the actuarial present values of all these annuities for payment of 1000 per annum. These are obtained using Code 6.6.5 and are reported in Table 6.16.

Note that for all ages the actuarial present values satisfy the relation as proved in Example 6.4.6 and in Eq. 6.4.14.   ∎

Example 6.4.8. Suppose the force of mortality follows Gompertz' law given by $\mu_x = BC^x$. Find the values of whole life annuity due and whole life annuity immediate for payment of 100 for ages $25, 30, 35, 40$. Take $\delta = 0.05, B = 0.0001151$ and $C = 1.096$. Compare these values with $100\,\overline{a}_x$.

Solution: From Eq. 6.4.10 and Eq. 6.4.14 we have,

$$\ddot{a}_x = (1 - A_x)/d \quad \text{and} \quad a_x = \ddot{a}_x - 1$$

Age x	1000 a_x	1000 $\overline{a}_x$	1000 $\ddot{a}_x$
40	16283.26	16778.45	17283.26
41	16143.71	16638.88	17143.71
42	16000.19	16495.34	17000.19
43	15854.14	16349.25	16854.14
44	15703.64	16198.72	16703.64
45	15550.05	16045.11	16550.05
46	15391.48	15886.50	16391.48
47	15230.85	15725.84	16230.85
48	15066.30	15561.25	16066.30
49	14899.11	15394.03	15899.11
50	14729.00	15223.89	15729.00

Table 6.16 Actuarial present value of whole life annuity

We have obtained in Example 5.6.6 the values of A_x under Gompertz' law for the same values of parameters. Using those values and the above relation we get $100\ddot{a}_x$ and $100a_x$. Values of $100\,\overline{a}_x$ are obtained in Example 6.3.4. Table 6.17 reports all these values. These are obtained using Code 6.6.6.

Age	100 $\ddot{a}_x$	100 $\overline{a}_x$	100 a_x
25	1745.38	1694.93	1645.38
30	1672.20	1621.75	1572.20
35	1585.05	1534.60	1485.05
40	1483.02	1432.56	1383.02

Table 6.17 Gompertz' law: Actuarial present value of whole life annuity

$100\ddot{a}_{25} = 1745.38$ is the actuarial present value of the payments of 100 at the beginning of the year till (25) survives while $100a_{25} = 1645.38$ is the actuarial present value of the payments of 100 at the end of the year till (25) survives. If the payment is made continuously at the rate of 100 per annum then the actuarial present value of such payments is 1694.93. From the table it is clear that, $100\ddot{a}_x > 100\,\overline{a}_x > 100\,a_x$ for all ages, verifying the results of Example 6.4.6. Such an inequality is again the consequence of the nature of time points of payments in these three cases. In $100\ddot{a}_x$, it is at the beginning of the year, in $100\,\overline{a}_x$, it is paid continuously while in $100a_x$, it is paid at the end of the year. Note that $100\ddot{a}_{25} = 100a_{25}+100$. ∎

n-year deferred whole life annuity due, n-year deferred whole life annuity immediate, n-year certain and whole life annuity due and immediate can be defined analogously. Their present value random variables, formulae and international actuarial notation for actuarial present values are summarized in Table 6.18.

Monthly pension that retired employees receive is an illustration of deferred whole life annuity immediate, since the contribution to the pension fund begins when the person is employed but the benefit payments are deferred till he retires. In n-year certain and whole life discrete annuity due, the present value random variable is $\ddot{a}_{\overline{n|}}$ + present value random

Annuity	Y	$E(Y)$
n-Year deferred		
• Due	$\begin{cases} 0, \\ K = 0, \cdots, n-1 \\ \ddot{a}_{\overline{K+1}} - \ddot{a}_{\overline{n}}, \\ K = n, n+1, \cdots \end{cases}$	$\begin{aligned} {}_{n\|}\ddot{a}_x &= \sum_{k=n}^{\infty} v^k\ {}_kp_x \\[2mm] &= \ddot{a}_x - \ddot{a}_{x:\overline{n}} = {}_nE_x\ \ddot{a}_{x+n} \end{aligned}$
• Immediate	$\begin{cases} 0, \\ K = 0, \cdots, n-1 \\ a_{\overline{K}} - a_{\overline{n}}, \\ K = n, n+1, \cdots \end{cases}$	$\begin{aligned} {}_{n\|}a_x &= \sum_{k=n+1}^{\infty} v^k\ {}_kp_x \\[2mm] &= a_x - a_{x:\overline{n}} = {}_{(n+1)\|}\ddot{a}_x \\ &= {}_nE_x\ a_{x+n} \end{aligned}$
n-Year certain and whole life		
• Due	$\begin{cases} \ddot{a}_{\overline{n}}, \\ K = 0, \cdots, n-1 \\ \ddot{a}_{\overline{K+1}}, \\ K = n, n+1, \cdots \end{cases}$	$\ddot{a}_{\overline{x:\overline{n}}} = \ddot{a}_{\overline{n}} + {}_{n\|}\ddot{a}_x$
• Immediate	$\begin{cases} a_{\overline{n}}, \\ K = 0, \cdots, n-1 \\ a_{\overline{K}}, \\ K = n, n+1, \cdots \end{cases}$	$a_{\overline{x:\overline{n}}} = a_{\overline{n}} + {}_{n\|}a_x$

Table 6.18 Summary of discrete deferred n-year term and n-year certain and whole life annuities

variable for n-year deferred whole life annuity. This relation is reflected in the formula for actuarial present value. Same is the case with annuity immediate.

Example 6.4.9. Prove that ${}_{m\|}\ddot{a}_x > {}_{m\|}\overline{a}_x > {}_{m\|}a_x$.

Solution: Using the formulae as given in Table 6.18, the solution is similar to that of Example 6.4.2, where the lower and upper limit in $\int_0^n$ and in summation are replaced by m and ∞ respectively. Alternatively we have from Example 6.4.6, $\ddot{a}_{x+m} > \overline{a}_{x+m} > a_{x+m}$. Multiplying each term by ${}_mE_x$ in this inequality we get, ${}_{m\|}\ddot{a}_x > {}_{m\|}\overline{a}_x > {}_{m\|}a_x$. ∎

Example 6.4.10. Suppose the force of mortality follows Gompertz' law given by $\mu_x = BC^x$. Find the actuarial present values of n-year deferred whole life annuity due and whole life annuity immediate for payment of 100 issued to (25) for $n = 1, 2, \cdots, 10$. Take $\delta = 0.05$, $B = 0.0001151$ and $C = 1.096$. Compare these values with $100\ {}_{n\|}\overline{a}_{25}$.

Solution: From Table 6.18, we have

$$_{n|}\ddot{a}_x \;=\; \ddot{a}_x - \ddot{a}_{x:\overline{n|}} \quad \text{and} \quad _{n|}a_x = a_x - a_{x:\overline{n|}} = \;_{(n+1)|}\ddot{a}_x$$

For the specified Gompertz' law, we have calculated $\ddot{a}_x$ and a_x in Example 6.4.8 and values of $a_{x:\overline{n|}}$ and $\ddot{a}_{x:\overline{n|}}$ in Example 6.4.5. Hence, using the above formulae we get the values of $100\;_{n|}\ddot{a}_{25}$ and $100\;_{n|}a_{25}$. Values of $_{n|}\overline{a}_{25}$ are already computed in Example 6.3.6. All these are reported in Table 6.19. These are obtained using Code 6.6.6.

n	$100\;_{n|}\ddot{a}_{25}$	$100\;_{n|}\overline{a}_{25}$	$100_{n|}a_{25}$
1	1645.38	1597.52	1550.37
2	1550.37	1504.90	1460.11
3	1460.11	1416.92	1374.38
4	1374.38	1333.36	1292.96
5	1292.96	1254.01	1215.64
6	1215.64	1178.66	1142.23
7	1142.23	1107.13	1072.54
8	1072.54	1039.23	1006.40
9	1006.40	974.80	943.65
10	943.65	913.66	884.11

Table 6.19 Gompertz' law: APV of n-year deferred whole life annuity

$100_{1|}\ddot{a}_{25} = 1645.38$ is the actuarial present value of the payments of 100, deferred by 1 year, at the beginning of the year till (25) survives while $100_{1|}a_{25} = 1550.37$ is the actuarial present value of the payments of 100, deferred by 1 year, at the end of the year till (25) survives. If the payment is made continuously at the rate of 100 per annum, then the actuarial present value of such 1 year deferred payments is 1597.52. From the table it is clear that $100\;_{n|}\ddot{a}_{25} > 100\;_{n|}\overline{a}_{25} > 100\;_{n|}a_{25}$ for all n. We have proved it algebraically in Example 6.4.9. Such an inequality is again the consequence of the nature of time points of payments in these three cases. Further note that $_{(n+1)|}\ddot{a}_{25} = \;_{n|}a_{25}$ for all n. From Table 6.17, we have $100a_{25} = 1645.38$. Thus, we note that $100_{1|}\ddot{a}_{25} = 1645.38 = 100\;a_{25}$. It is a particular case of the relation $_{(n+1)|}\ddot{a}_{25} = \;_{n|}a_{25}$, with $n = 0$. It can also be derived from the formula of $_{1|}\ddot{a}_x$ as follows.

$$_{1|}\ddot{a}_x \;=\; \ddot{a}_x - \ddot{a}_{x:\overline{1|}} \;=\; \ddot{a}_x - 1 \;=\; a_x$$

∎

Example 6.4.11. Prove that (i) $\ddot{a}_{x:\overline{n+1|}} = 1 + a_{x:\overline{n|}}$. (ii) $\ddot{a}_{x:\overline{1|}} = \ddot{a}_x$
(iii) $\ddot{a}_{x:\overline{n|}} > \overline{a}_{x:\overline{n|}} > a_{x:\overline{n|}} \;\; \forall \; n$

Solution: (i) From Table 6.18 we have,

$$\ddot{a}_{x:\overline{n+1|}} \;=\; \ddot{a}_{\overline{n+1|}} + \;_{n+1|}\ddot{a}_x \;=\; 1 + a_{\overline{n|}} + \;_{n|}a_x \;=\; 1 + a_{x:\overline{n|}}$$

(ii) $\quad \ddot{a}_{x:\overline{1|}} \;=\; \ddot{a}_{\overline{1|}} + \;_{1|}\ddot{a}_x \;=\; 1 + \ddot{a}_x - \ddot{a}_{x:\overline{1|}} \;=\; 1 + \ddot{a}_x - 1 \;=\; \ddot{a}_x$

(iii) It follows from Example 6.4.9 and the order relation $\ddot{a}_{\overline{n|}} > \overline{a}_{\overline{n|}} > a_{\overline{n|}}$. ∎

Example 6.4.12. Suppose the force of mortality follows Gompertz' law given by $\mu_x = BC^x$. Find the values of n-year certain and whole life annuity due and n-year certain and whole life annuity immediate for payment of 100, issued to (25) for $n = 1, 2, \cdots, 10$. Take $\delta = 0.05$, $B = 0.0001151$ and $C = 1.096$. Compare these values with $100\overline{a}_{\overline{25:n|}}$, as reported in Table 6.10.

Solution: From Table 6.18, we have

$$\ddot{a}_{\overline{x:n|}} = \ddot{a}_{\overline{n|}} + {}_{n|}\ddot{a}_x \quad \text{and} \quad a_{\overline{x:n|}} = a_{\overline{n|}} + {}_{n|}a_x = v\,\ddot{a}_{\overline{n|}} + {}_{n|}a_x$$

For the specified Gompertz' law, we have calculated ${}_{n|}\ddot{a}_x$ and ${}_{n|}a_x$ in Example 6.4.10. We have $a_{\overline{n|}} = (1 - v^n)/i$ and $\ddot{a}_{\overline{n|}} = (1 - v^n)/d$. Hence using the above formulae, we get the values of $100\,\ddot{a}_{\overline{x:n|}}$ and $100\,a_{\overline{x:n|}}$, for $x = 25$. Values of $100\,\overline{a}_{\overline{25:n|}}$ are already computed in Table 6.10. All these are reported in Table 6.20. $100\ddot{a}_{\overline{25:1|}} = 1745.38$ is the actuarial present value of the

n	$100\,\ddot{a}_{\overline{25:n|}}$	$100\,\overline{a}_{\overline{25:n|}}$	$100\,a_{\overline{25:n|}}$
1	1745.38	1695.06	1645.49
2	1745.49	1695.23	1645.71
3	1745.71	1695.51	1646.05
4	1746.05	1695.90	1646.51
5	1746.50	1696.41	1647.07
6	1747.06	1697.02	1647.74
7	1747.73	1697.75	1648.52
8	1748.51	1698.59	1649.41
9	1749.40	1699.54	1650.42
10	1750.41	1700.60	1651.53

Table 6.20 Gompertz' law: APV of n-year certain and whole life annuity

payments of 100 of 1 year certain and whole life annuity, paid at the beginning of the interval for (25) while $100a_{\overline{25:1|}} = 1645.49$ is the actuarial present value of the payments of 100, of 1 year certain and whole life annuity, paid at the end of the interval for (25). If the payment is made continuously at the rate of 100 per annum then the actuarial present value is 1695.06. From the table it is clear that $\ddot{a}_{\overline{25:n|}} > \overline{a}_{\overline{25:n|}} > a_{\overline{25:n|}}$ for all n, verifying the result (iii) in Example 6.4.11. Such inequality is again the consequence of the nature of time points of payments in these three cases. It is to be noted that $100\,\ddot{a}_{\overline{25:1|}} = 1745.38 = 100\,\ddot{a}_{25}$ from Table 6.16, verifying the result (ii) in Example 6.4.11. Rest of the values in the first column of this table are close to $100\,\ddot{a}_{25}$, which indicates that the impact of n-year certain, part of the annuity is marginal and it is mainly governed by whole life annuity, for the mortality under consideration. Further, note that $\ddot{a}_{\overline{x:n+1|}} = 1 + a_{\overline{x:n|}}$, as proved in (i) of Example 6.4.11. ∎

Example 6.4.13. Suppose for the specific mortality law, q_x values are the same as in Example 6.4.7 and are given Table 6.15. For the same setup and $i = 0.05$, $A_{50} = 0.251$. (i) Find the actuarial present value of 5-year deferred whole life annuity due and immediate issued to (40), when the yearly payment is 1000. (ii) Also find the actuarial present value of 5-year certain and whole life annuity due and immediate issued to (40), when the yearly payment is 1000.

Solution: We have to find $1000 \; _{5|}\ddot{a}_{40:\overline{5}|}$, $1000 \; _{5|}a_{40:\overline{5}|}$ and $1000 \; \ddot{a}_{\overline{40:\overline{5}|}}$, $1000 \; a_{\overline{40:\overline{5}|}}$. From Table 6.18, we have $_{5|}\ddot{a}_{40:\overline{5}|} = \ddot{a}_{40} - \ddot{a}_{40:\overline{5}|}$. From Example 6.4.7, we have $\ddot{a}_{40} = 17.28326$. We find $\ddot{a}_{40:\overline{5}|}$ as follows. By Eq. 6.4.1, we have

$$\ddot{a}_{40:\overline{5}|} = \sum_{k=0}^{4} v^k \; _kp_{40} = 1 + vp_{40} + v^2 p_{40}p_{41} + v^3 p_{40}p_{41}p_{42} + v^4 p_{40}p_{41}p_{42}p_{43}$$

For the given values of q_x and i we get $\ddot{a}_{40:\overline{5}|} = 4.520736$. Hence,

$$_{5|}\ddot{a}_{40:\overline{5}|} = 12.76252 \quad \text{and} \quad 1000 \; _{5|}\ddot{a}_{40:\overline{5}|} = 12762.52$$

From Table 6.18, we have $_{5|}a_{40:\overline{5}|} = a_{40} - a_{40:\overline{5}|}$. Further, $a_{40} = \ddot{a}_{40} - 1$. We find $a_{40:\overline{5}|}$ as follows. By Eq. 6.4.6,

$$\begin{aligned} a_{40:\overline{5}|} &= \ddot{a}_{40:\overline{5}|} + \; _5E_{40} - 1 = 4.520736 + v^5 \; _5p_{40} - 1 \\ &= 4.520736 + 0.7711468 - 1 = 4.291883 \\ \Rightarrow \quad _{5|}a_{40:\overline{5}|} &= 11.99138 \quad \text{and} \quad 1000 \; _{5|}a_{40:\overline{5}|} = 11991.38 \end{aligned}$$

From Table 6.18,

$$\ddot{a}_{\overline{40:\overline{5}|}} = \ddot{a}_{\overline{5}|} + _{5|}\ddot{a}_{40:\overline{5}|} = 17.30847 \quad \text{and} \quad a_{\overline{40:\overline{5}|}} = a_{\overline{5}|} + _{5|}a_{40:\overline{5}|} = 16.32085$$

$$\text{Hence,} \quad 1000 \; \ddot{a}_{\overline{40:\overline{5}|}} = 17308.47 \quad \text{and} \quad 1000 \; a_{\overline{40:\overline{5}|}} = 16320.85$$

We have used Code 6.6.7 to compute these values. ∎

Example 6.4.14. It is given that $_{10}E_x = 0.40$, $_{10|}\ddot{a}_x = 7$ and $\ddot{S}_{x:\overline{10}|} = 15$. Calculate $\ddot{a}_x$.

Solution: From Table 6.18,

$$\ddot{a}_x = \ddot{a}_{x:\overline{10}|} + _{10|}\ddot{a}_x = \ddot{S}_{x:\overline{10}|} \; _{10}E_x + _{10|}\ddot{a}_x = 15(0.40) + 7 = 13$$

∎

Example 6.4.15. Given the following information, calculate $A^{1}_{x:\overline{10}|}$.
(i) $_{10|}\ddot{a}_x = 4.0$, (ii) $\ddot{a}_x = 10.0$, (iii) $\ddot{S}_{x:\overline{10}|} = 15.0$ and (iv) $v = 0.94$.

Solution: $A^{1}_{x:\overline{10}|} = A_{x:\overline{10}|} - v^{10} \; _{10}p_x = 1 - d\ddot{a}_{x:\overline{10}|} - v^{10} \; _{10}p_x$. From the information given, we have $\ddot{a}_{x:\overline{10}|} = \ddot{a}_x - _{10|}\ddot{a}_x = 6.0$, $d = 1 - v = 0.06$. Now, from the relationship $\ddot{a}_{x:\overline{10}|} = v^{10} \; _{10}p_x \ddot{S}_{x:\overline{10}|}$, we get, $v^{10} \; _{10}p_x = 6/15 = 0.4$. Hence, $A^{1}_{x:\overline{10}|} = 1 - (0.06)(6) - 0.4 = 0.24$. ∎

Table 6.21 displays the international actuarial notation for present values of continuous and discrete life annuities studied in this section and the previous section, for annuity of 1 unit per annum.

Example 6.4.16. Upon payment of a death benefit, a beneficiary of age 40 is given the following options: (i) a lump sum payment of 10,000 or (ii) an annual payment of c at the beginning of each year guaranteed for 10 years and continuing as long as the beneficiary is alive. The two options are actuarially equivalent. It is given that (i) $i = 0.04$, (ii) $A_{40} = 0.30$, (iii) $A_{50} = 0.35$ and (iv) $A_{40:\overline{10|}}^{\ 1} = 0.09$. Calculate the value of c.

Solution: The two options are actuarially equivalent.   Hence we get the equation $10000 = c[\ddot{a}_{\overline{10|}} + {}_{10|}\ddot{a}_{40}]$. To compute the right hand side of the above equation, note that

$$
\begin{aligned}
\ddot{a}_{\overline{10|}} &= (1 - v^{10})/d = 8.435, \qquad \ddot{a}_{50} = (1 - A_{50})/d = 16.9 \\
A_{40} &= A_{40:\overline{10|}}^{\ 1} + v^{10} \, {}_{10}p_{40} A_{50} \ \Rightarrow \ v^{10} \, {}_{10}p_{40} = (0.3 - 0.09)/0.35 = 0.6 \\
&\Rightarrow \ {}_{10|}\ddot{a}_{40} = v^{10} \, {}_{10}p_{40}\ddot{a}_{50} = 10.14 \ \Rightarrow \ c = 10000/(8.435 + 10.14) = 538 \qquad \blacksquare
\end{aligned}
$$

The next section extends the results of this section to m-thly annuities.

6.5   Life Annuities with m-thly Payments

In practice, life annuities are often payable on a monthly, quarterly or semi-annual basis. In this section we discuss how to obtain actuarial present values of such annuities. Such annuities are of course discrete annuities where payment may be at the beginning of the mth part or at the end of the mth part of the annum. We begin with m-thly whole life annuity due.

m-**thly Whole Life Annuity Due:** In this annuity 1 unit per year is payable in instalments of $1/m$ at the beginning of each mth part of the year till (x) survives. In international actuarial notation, the actuarial present value of this annuity is denoted by $\ddot{a}_x^{(m)}$. To find its expression we proceed on similar lines as those for $\ddot{a}_{\overline{n|}}^{(m)}$ derived in Section 2. When period n of payments is fixed, there are mn parts for m-thly payments. In life annuity we find how many m parts the individual has survived. (x) has definitely survived for $m\,K(x)$ parts. Suppose $U(x) = T(x) - K(x)$ is the fractional part of the year of death, that (x) survives. We have to find out how many m parts of $U(x)$ the individual has survived. Hence, we define a random variable $J = J(x)$ as the greatest integer in $U(x)m$. Thus, J is the number of complete m-parts of a year lived in the year of death. For example if $U(x) = 0.8$ and $m = 4$, then $J = 3$, if $U(x) = 0.5$ and $m = 4$, then $J = 2$. If $U(x) = 0.8$ and $m = 4$, then (x) survives 3 complete quarters and dies in the fourth quarter. In annuity due, payments are at the beginning of the quarter, thus in this case there would be $4 = J + 1$ payments in the year of death as well.

Annuity contract	Continuous	Discrete
Whole life annuity	$\overline{a}_x = \int_0^\infty v^t\, {}_tp_x dt$	
• Due		$\ddot{a}_x = \sum_{k=0}^\infty v^k\, {}_kp_x$
• Immediate		$a_x = \sum_{k=1}^\infty v^k\, {}_kp_x$ $= \ddot{a}_x - 1$
n-Year temporary life annuity	$\overline{a}_{x:\overline{n}|} = \int_0^n v^t\, {}_tp_x dt$	
• Due		$\ddot{a}_{x:\overline{n}|} = \sum_{k=0}^{n-1} v^k\, {}_kp_x$
• Immediate		$a_{x:\overline{n}|} = \sum_{k=1}^{n} v^k\, {}_kp_x$ $= \ddot{a}_{x:\overline{n}|} + {}_nE_x - 1$
n-Year deferred whole life annuity	${}_{n|}\overline{a}_x = \int_n^\infty v^t\, {}_tp_x dt$ $= \overline{a}_x - \overline{a}_{x:\overline{n}|}$ $= {}_nE_x\, \overline{a}_{x+n}$	
• Due		${}_{n|}\ddot{a}_x = \sum_{k=n}^\infty v^k\, {}_kp_x$ $= \ddot{a}_x - \ddot{a}_{x:\overline{n}|}$ $= {}_nE_x\, \ddot{a}_{x+n}$
• Immediate		${}_{n|}a_x = \sum_{k=n+1}^\infty v^k\, {}_kp_x$ $= a_x - a_{x:\overline{n}|} = {}_{(n+1)|}\ddot{a}_x$ $= {}_nE_x\, a_{x+n}$
n-Year certain and whole life annuity	$\overline{a}_{\overline{x:n}|} = \overline{a}_{\overline{n}|} + \int_n^\infty v^t\, {}_tp_x dt$ $= \overline{a}_{\overline{n}|} + {}_{n|}\overline{a}_x$	
• Due		$\ddot{a}_{\overline{x:n}|} = \ddot{a}_{\overline{n}|} + \sum_{k=n}^\infty v^k\, {}_kp_x$ $= \ddot{a}_{\overline{n}|} + {}_{n|}\ddot{a}_x$
• Immediate		$a_{\overline{x:n}|} = a_{\overline{n}|} + \sum_{k=n+1}^\infty v^k\, {}_kp_x$ $a_{\overline{n}|} + {}_{n|}a_x$

Table 6.21 Actuarial notation and formulae for APV of life annuities

In general, for an annuity due there would be m payments for each of the K complete years and then $J+1$ payments of $1/m$ in the year of death. Hence the present value random variable corresponding to the m-thly annuity due is given by,

$$Y = \frac{1}{m} \sum_{j=0}^{mK+J} (v^{1/m})^j = \frac{1 - (v^{1/m})^{mK+J+1}}{m(1 - v^{1/m})}$$

$$= (1 - v^{K+(J+1)/m})/d^{(m)} = \ddot{a}_{\overline{K+(J+1)/m|}}^{(m)} \tag{6.5.1}$$

$$\Rightarrow \quad E(Y) = (1 - A_x^{(m)})/d^{(m)} = \ddot{a}_x^{(m)} \tag{6.5.2}$$

where $A_x^{(m)} = E(v^{K+(J+1)/m})$. This expectation can be interpreted as the actuarial present value of the unit benefit paid at the end of the mth part of year of death. Similarly variance of Y is given by,

$$Var(Y) = \frac{Var(v^{K+(J+1)/m})}{(d^{(m)})^2} = \frac{{}^2A_x^{(m)} - (A_x^{(m)})^2}{(d^{(m)})^2}$$

with the usual interpretation for ${}^2A_x^{(m)}$. To find $A_x^{(m)}$, we note that it involves the fractional part of the year of death. Hence to find it, we have to make some assumption regarding the mortality pattern in fractional ages, as discussed in Section 5.5. If we assume that deaths occur according to a uniform distribution in unit age interval, then we know that $U = U(x)$ has a uniform distribution on $(0, 1)$. It can be shown that J has discrete uniform distribution on the integers $\{0, 1, \cdots, m - 1\}$. Further, J and K are independent random variables. Hence,

$$A_x^{(m)} = E(v^{K+1+(J+1)/m \ -1}) = v^{-1} E(v^{K+1}) \ E(v^{(J+1)/m})$$

$$= v^{-1} A_x \left\{ \frac{1}{m} \sum_{j=0}^{m-1} v^{(j+1)/m} \right\} = v^{-1} A_x \left\{ \frac{1}{m} v^{1/m} \sum_{j=0}^{m-1} (v^{1/m})^j \right\}$$

$$= v^{-1} A_x \left\{ \frac{1}{m} v^{1/m} (1 - v)/(1 - v^{1/m}) \right\}$$

$$= \left\{ \frac{1-v}{v} \right\} \left\{ \frac{A_x}{m(v^{-1/m} - 1)} \right\} = \frac{i}{i^{(m)}} A_x \tag{6.5.3}$$

In the relation, $A_x^{(m)} = (i/i^{(m)})A_x$, if we allow m to tend to ∞, then we get a relation $\overline{A}_x = (i/\delta)A_x$ as derived in Eq. 5.5.3. Substituting the expression of $A_x^{(m)}$ in Eq. 6.5.2 and then replacing A_x by $1 - d\,\ddot{a}_x$, we get

$$\ddot{a}_x^{(m)} = \frac{1 - \frac{i}{i^{(m)}} A_x}{d^{(m)}} = \frac{i^{(m)} - i(1 - d\ddot{a}_x)}{i^{(m)} d^{(m)}} = \frac{i\,d}{i^{(m)}\,d^{(m)}} \ddot{a}_x - \frac{i - i^{(m)}}{i^{(m)}\,d^{(m)}}$$

$$= \alpha(m)\ddot{a}_x - \beta(m) \tag{6.5.4}$$

$$\text{where,} \quad \alpha(m) = id/i^{(m)}d^{(m)} \quad \text{and} \quad \beta(m) = (i - i^{(m)})/i^{(m)}d^{(m)} \tag{6.5.5}$$

Thus, $\ddot{a}_x^{(m)} = \alpha(m)\ddot{a}_x - \beta(m)$ is the actuarial present value of m-thly whole life annuity due. This relation is useful to find $\ddot{a}_x^{(m)}$ when $\ddot{a}_x$ are given. It is to be noted that $\alpha(m)$ and $\beta(m)$ depend only on m and the rate of interest, and are independent of the age. Note that for $m = 1$, $i^{(m)} = i$ and $d^{(m)} = d$. As a consequence, $\alpha(1) = 1$ and $\beta(1) = 0$. Further with $m = 1$, the expression

$\ddot{a}_x^{(m)} = (1 - (i/i^{(m)})A_x)/d^{(m)}$ reduces to $\ddot{a}_x = (1 - A_x)/d$, as in Eq. 6.4.10. $\ddot{a}_x^{(m)}$ can also be expressed as follows.

$$\ddot{a}_x^{(m)} = \frac{1 - \frac{i}{i^{(m)}}A_x}{d^{(m)}} = \frac{d\ddot{a}_x + A_x - \frac{i}{i^{(m)}}A_x}{d^{(m)}}$$

$$= \frac{d}{d^{(m)}}\ddot{a}_x - \frac{A_x(i - i^{(m)})}{i^{(m)}\ d^{(m)}} \doteq \frac{d}{d^{(m)}}\ddot{a}_x - \beta(m)A_x \qquad (6.5.6)$$

An alternative widely used formula for $\ddot{a}_x^{(m)}$ is $\ddot{a}_x^{(m)} = \ddot{a}_x - (m-1)/2m$. This result is obtained by assuming that the function $v^{k+(j/m)}\ _{k+(j/m)}p_x$ is linear in j for $j = 0, 1, 2, \cdots, m - 1$. However, it has been observed that formula of $\ddot{a}_x^{(m)}$, derived from this expression can produce distorted annuity values, for high rates of interest and low rates of mortality.

The following example establishes the order relation among the actuarial present values of the whole life annuity with different mode of payments.

Example 6.5.1. Prove that $\overline{a}_x < \ddot{a}_x^{(m)} < \ddot{a}_x$, under the assumption of uniformity in unit age interval.

Solution: In Eq. 6.2.20, we have proved that $i > i^{(m)} > \delta > d^{(m)} > d$. Hence, $(i/i^{(m)}) < (i/\delta)$ and $\delta > d^{(m)} \Rightarrow 1/d^{(m)} > 1/\delta$. From Eq. 6.5.3, under the assumption of uniformity we have,

$$A_x^{(m)} = \frac{i}{i^{(m)}}A_x < \frac{i}{\delta}A_x = \overline{A}_x$$

$$\Rightarrow \ddot{a}_x^{(m)} = \frac{1 - A_x^{(m)}}{d^{(m)}} > \frac{1 - \overline{A}_x}{d^{(m)}} > \frac{1 - \overline{A}_x}{\delta} = \overline{a}_x$$

$$\text{Further, } \ddot{a}_x^{(m)} = \frac{d}{d^{(m)}}\ddot{a}_x - \beta(m)A_x < \frac{d}{d^{(m)}}\ddot{a}_x < \ddot{a}_x, \text{ since } \frac{d}{d^{(m)}} < 1.$$

Thus, we have obtained a relation $\overline{a}_x < \ddot{a}_x^{(m)} < \ddot{a}_x$, similar to that for annuities certain and logically appealing. ∎

m-thly Whole Life Annuity Immediate: In this annuity 1 unit per year is payable in instalments of $1/m$ at the end of each mth part of the year while (x) survives. The actuarial present value for the annuity immediate with m-thly payments at a rate of 1 per annum, is denoted by $a_x^{(m)}$. As in m-thly annuity due, there will be $mK(x)$ payments in $K(x)$ complete years that (x) has survived but there will be J payments in the year of death, as the payments are at the end of mth part. For example, if $U(x) = 0.8$ and $m = 4$, then (x) survives 3 complete quarters and dies in the fourth quarter. In annuity immediate, payments are at the end of the quarter, thus in this case there would be three payments in the year of death. In general, for an annuity immediate there would be m payments for each of the K complete years and then J payments of $1/m$ in the year of death. Hence, the present value random variable corresponding to m-thly whole life annuity immediate is given by,

$$Y = \frac{1}{m} \sum_{j=1}^{mK+J} (v^{1/m})^j = \frac{1}{m} \left(\sum_{j=0}^{mK+J} (v^{1/m})^j - 1 \right)$$

$$\Rightarrow E(Y) = a_x^{(m)} = \ddot{a}_x^{(m)} - 1/m \tag{6.5.7}$$

To prove a result similar to that in Example 6.5.1 for m-thly whole life annuity immediate, we rewrite the present value random variable Y as follows.

$$Y = \frac{1}{m} \sum_{j=1}^{mK+J} (v^{1/m})^j = \frac{v^{1/m} \{1 - (v^{1/m})^{mK+J}\}}{m(1 - v^{1/m})}$$

$$= \frac{1 - v^{K+J/m}}{m(v^{-1/m} - 1)} = \frac{1 - v^{-1/m} v^{K+(J+1)/m}}{i^{(m)}}$$

Taking expectations on both sides, using the relation in Eq. 6.5.3 and the fact that $v^{1/m} i^{(m)} = d^{(m)}$ we get,

$$a_x^{(m)} = E(Y) = \frac{1 - v^{-1/m} A_x^{(m)}}{i^{(m)}} = \frac{1 - v^{-1/m} \frac{i}{i^{(m)}} A_x}{i^{(m)}}$$

$$= \{1 - (i/d^{(m)}) A_x\}/i^{(m)} \tag{6.5.8}$$

This expression of $a_x^{(m)}$ is similar to that of $\ddot{a}_x^{(m)}$ in the first step of Eq. 6.5.4, with interchange of $i^{(m)}$ and $d^{(m)}$. Further, the expression $a_x^{(m)} = (1 - (i/d^{(m)}) A_x)/i^{(m)}$ with $m = 1$ reduces to $a_x = (1 - (1+i) A_x)/i$, as in Eq. 6.4.13. From the expression of $a_x^{(m)}$ as derived in Eq. 6.5.8, replacing A_x by $1 - d \, \ddot{a}_x$, and with some algebra we get $a_x^{(m)} = \ddot{a}_x^{(m)} - 1/m$ as in Eq. 6.5.7. In the same expression, replacing A_x by $1 - d \, (a_x + 1)$ and with some algebra we get $a_x^{(m)} = \alpha(m) a_x + \{\alpha(m) - \beta(m) - 1/m\}$. We now establish an order relation among the actuarial present values for the whole life annuities immediate, similar to that in Example 6.5.1.

Example 6.5.2. Prove that $a_x < a_x^{(m)} < \bar{a}_x$, under the assumption of uniformity in each unit age interval.

Solution: We have,

$$a_x = \frac{1 - (1+i) A_x}{i}, \quad a_x^{(m)} = \frac{1 - \frac{i}{d^{(m)}} A_x}{i^{(m)}} \quad \& \quad \bar{a}_x = \frac{1 - \bar{A}_x}{\delta}$$

Observe that,

$$d^{(m)} > d = \frac{i}{1+i} \quad \Rightarrow \quad (1+i) > \frac{i}{d^{(m)}} \quad \Rightarrow \quad 1 - (1+i) A_x < 1 - \frac{i}{d^{(m)}} A_x$$

From $i > i^{(m)}$ we get,

$$a_x = \frac{1 - (1+i) A_x}{i} < \frac{1 - (1+i) A_x}{i^{(m)}} < \frac{1 - \frac{i}{d^{(m)}} A_x}{i^{(m)}} = a_x^{(m)}$$

To prove the second part of the inequality, note that,

$$\delta > d^{(m)} \quad \Rightarrow \quad \frac{i\, A_x}{d^{(m)}} > \frac{i\, A_x}{\delta} = \overline{A}_x \quad \Rightarrow \quad 1 - \frac{i\, A_x}{d^{(m)}} < 1 - \overline{A}_x$$

Further, using the relation $i^{(m)} > \delta$, we get,

$$a_x^{(m)} = \frac{1 - \frac{i\, A_x}{d^{(m)}}}{i^{(m)}} < \frac{1 - \frac{i\, A_x}{d^{(m)}}}{\delta} < \frac{1 - \overline{A}_x}{\delta} = \overline{a}_x$$

■

Combining the results in Examples 6.5.1 and 6.5.2 we get the ordering as,

$$a_x \;<\; a_x^{(m)} \;<\; \overline{a}_x \;<\; \ddot{a}_x^{(m)} \;<\; \ddot{a}_x \tag{6.5.9}$$

under the assumption of uniformity in each unit age interval. This result is similar to that in Eq. 6.2.18 for annuities certain.

Example 6.5.3. Suppose the force of mortality follows Gompertz' law given by $\mu_x = BC^x$. Find the actuarial present values of m-thly whole life annuity due and immediate for payment of 100 issued to (25) and (30) for $m = 2, 4$ and 12. Take $\delta = 0.05, B = 0.0001151$ and $C = 1.096$. Use uniformity assumption for fractional ages.

Solution: For given $\delta = 0.05$ we find $i = 0.0512711$ and $d = 0.04877058$. Using these we find $i^{(m)}$ and $d^{(m)}$. From Eq. 6.5.5 we find $\alpha(m)$ and $\beta(m)$. These are given in Table 6.22.

m	$i^{(m)}$	$d^{(m)}$	$\alpha(m)$	$\beta(m)$
2	0.05063	0.04938	1.00016	0.25633
4	0.05031	0.04969	1.00020	0.38291
12	0.05010	0.04990	1.00021	0.46671

Table 6.22 Values $i^{(m)}$, $d^{(m)}$, $\alpha(m)$ and $\beta(m)$ for $\delta = 0.05$

From Table 6.22, we note that $i^{(m)}$ is a decreasing function of m while $d^{(m)}$ is an increasing function of m. Using Eqs 6.5.5, 6.5.7 and the values of $\ddot{a}_{25} = 17.4535$ and $\ddot{a}_{30} = 16.7219$ reported in Table 6.17, we find the values of $\ddot{a}_x^{(m)}$ and $a_x^{(m)}$ for $x = 25$ and 30. These are obtained using Code 6.6.8 and are reported in Table 6.23.

m	$100\ddot{a}_{25}^{(m)}$	$100a_{25}^{(m)}$	$100\ddot{a}_{30}^{(m)}$	$100a_{30}^{(m)}$
2	1719.99	1669.99	1646.81	1596.81
4	1707.40	1682.40	1634.22	1609.22
12	1699.04	1690.71	1625.86	1617.53

Table 6.23 Gompertz' law: m-thly whole life annuity due and immediate

From Table 6.23 we note that as age increases, the actuarial present values decrease for all m, because expected future life time decreases. Further as m increases, the actuarial present values decrease for m-thly annuity due but increase for m-thly annuity immediate. It is again the result of the nature of time points of payments of annuity. ∎

Remark 6.5.1. (i) In Table 6.17 we have obtained the values of $100\ddot{a}_x$, $100\overline{a}_x$ and $100a_x$ for $x = 25$ and 30 and the same mortality pattern and interest rate. For comparison, we reproduce part of the Table 6.17 in Table 6.24. We note that $a_x < a_x^{(m)} < \overline{a}_x < \ddot{a}_x^{(m)} < \ddot{a}_x$, for

Age	$100\ \ddot{a}_x$	$100\overline{a}_x$	$100\ a_x$
25	1745.38	1694.93	1645.38
30	1672.20	1621.75	1572.20

Table 6.24 Gompertz' law: Whole life annuity for ages 25 and 30

$x = 25$ and 30, verifying the results proved in Examples 6.5.1 and 6.5.2.
(ii) From the values of $i^{(m)}$ and $d^{(m)}$ in Table 6.22, we note that $i^{(2)} > i^{(4)} > i^{(12)}$ and $d^{(2)} < d^{(4)} < d^{(12)}$, as proved in Example 6.2.8.
(iii) From Table 6.23, we observe that $\ddot{a}_x^{(12)} < \ddot{a}_x^{(4)} < \ddot{a}_x^{(2)}$ and $a_x^{(2)} < a_x^{(4)} < a_x^{(12)}$.

In the following example we prove the results noted in (iii) of the above remark. In Example 6.2.8 it is proved that $i^{(2)} > i^{(4)} > i^{(12)}$ and hence, $a_{\overline{n}|}^{(2)} < a_{\overline{n}|}^{(4)} < a_{\overline{n}|}^{(12)}$. Similarly, it is shown that $d^{(2)} < d^{(4)} < d^{(12)}$ and hence, $\ddot{a}_{\overline{n}|}^{(2)} > \ddot{a}_{\overline{n}|}^{(4)} > \ddot{a}_{\overline{n}|}^{(12)}$. In the next example, we show that such inequalities are valid for life annuities as well.

Example 6.5.4. Under the assumption of uniformity in unit age interval, prove that (i) $\ddot{a}_x^{(12)} < \ddot{a}_x^{(4)} < \ddot{a}_x^{(2)}$ and (ii) $a_x^{(2)} < a_x^{(4)} < a_x^{(12)}$.

Solution: From the first step of derivation of $\ddot{a}_x^{(m)}$ in Eq. 6.5.4, we have $\ddot{a}_x^{(m)} = (1 - (i/i^{(m)})A_x)/d^{(m)}$. Note that

$$i^{(2)} > i^{(4)} \implies (1 - (i/i^{(2)})A_x) > (1 - (i/i^{(4)})A_x)$$
$$d^{(2)} < d^{(4)} \implies \ddot{a}_x^{(2)} = (1 - (i/i^{(2)})A_x)/d^{(2)} > (1 - (i/i^{(2)})A_x)/d^{(4)}$$
$$> (1 - (i/i^{(4)})A_x)/d^{(4)} = \ddot{a}_x^{(4)}$$

Similarly we get $\ddot{a}_x^{(4)} > \ddot{a}_x^{(12)}$.
(ii) We have $a_x^{(m)} = (1 - (i/d^{(m)})A_x)/i^{(m)}$. As in (i),

$$d^{(2)} < d^{(4)} \implies (1 - (i/d^{(2)})A_x) < (1 - (i/d^{(4)})A_x)$$
$$i^{(2)} > i^{(4)} \implies a_x^{(2)} = (1 - (i/d^{(2)})A_x)/i^{(2)} < (1 - (i/d^{(2)})A_x)/i^{(4)}$$
$$< (1 - (i/d^{(4)})A_x)/i^{(4)} = a_x^{(4)}$$

Similarly we get $a_x^{(4)} < a_x^{(12)}$. ∎

Example 6.5.5. Suppose for the specific mortality law, q_x values are as given in Table 6.15. For the same setup and $i = 0.05$ and $A_{50} = 0.251$. Find the actuarial present value of whole life annuity due and immediate paid monthly with the monthly payment of 1000, for ages 40 to 49.

Solution: From Eq. 6.5.5, we have $\ddot{a}_x^{(12)} = \alpha(12)\ddot{a}_x - \beta(12)$. Note that

$$i = 0.05 \quad \Rightarrow \quad d = 0.04761905, \quad i^{(12)} = 0.04888949 \ \text{ and } \ d^{(12)} = 0.04869111$$

$$\Rightarrow \quad \alpha(12) = 1.000197 \quad \text{and} \quad \beta(12) = 0.466508$$

For the given q_x values, $i = 0.05$ and $A_{50} = 0.251$ we have obtained the values of $\ddot{a}_x$ for $x = 40$ to 50 in Example 6.4.7. Using these we get $\ddot{a}_x^{(12)}$. Further using the relation given in Eq. 6.5.7, we have $a_x^{(12)} = \ddot{a}_x^{(12)} - 1/12$. Note that the monthly instalment is 1000, hence the yearly instalment is 12000. So we multiply $\ddot{a}_x^{(12)}$ and $a_x^{(12)}$ by 12000 to get the required values. These are found by using Code 6.6.9 and are reported in Table 6.25. $\blacksquare$

Age x	$12000\ddot{a}_x^{(12)}$	$12000a_x^{(12)}$
40	201841.90	200841.90
41	200166.90	199166.90
42	198444.40	197444.40
43	196691.40	195691.40
44	194885.00	193885.00
45	193041.70	192041.70
46	191138.50	190138.50
47	189210.50	188210.50
48	187235.50	186235.50
49	185228.80	184228.80
50	183187.10	182187.10

Table 6.25 Monthly whole life annuity immediate

The formulae for actuarial present values for other annuities such as n-year temporary life annuity and n-year deferred whole life annuity, with m-thly payments can be derived on similar lines. The actuarial present values for n-year temporary life annuity due and n-year temporary life annuity immediate, with m-thly payments at a rate of 1 unit per annum, are denoted by $\ddot{a}_{x:\overline{n}|}^{(m)}$ and $a_{x:\overline{n}|}^{(m)}$ respectively. Their formulae are derived below.

m-thly n-Year Temporary Annuity Due: Proceeding on similar lines as those for m-thly whole life annuity due, the present value random variable Y for n-year m-thly temporary life annuity due, with m-thly payments at a rate of 1 unit per annum, is given by,

$$Y = \begin{cases} \ddot{a}_{\overline{K+(J+1)/m}|}^{(m)}, & \text{if} \quad K = 0, 1, \cdots, n-1 \\ \ddot{a}_{\overline{n}|}^{(m)}, & \text{if} \quad K = n, n+1, \cdots \end{cases}$$

It can be expressed as

$$Y = \begin{cases} \{1 - v^{K+(J+1)/m}\}/d^{(m)}, & \text{if} \quad K = 0, 1, \cdots, n-1 \\ (1 - v^n)/d^{(m)}, & \text{if} \quad K = n, n+1, \cdots \end{cases}$$

We write Y as $Y = (1 - Z_1)/d^{(m)}$ where $Z_1 = Z_2 + Z_3$ and Z_2 and Z_3 are defined as,

$$
Z_2 = \begin{cases} v^{K+(J+1)/m}, & \text{if} \quad K = 0, 1, \cdots, n-1 \\ 0, & \text{if} \quad K = n, n+1, \cdots \end{cases}
$$

$$
Z_3 = \begin{cases} 0, & \text{if} \quad K = 0, 1, \cdots, n-1 \\ v^n, & \text{if} \quad K = n, n+1, \cdots \end{cases}
$$

Note that $E(Z_3) = {}_nE_x$. Using the arguments as in the derivation of Eq. 6.5.3 we get,

$$
E(Z_2) = (i/i^{(m)})A^1_{x:\overline{n}|} \quad \Rightarrow \quad E(Z_1) = (i/i^{(m)})A^1_{x:\overline{n}|} + {}_nE_x
$$

With this expression for $E(Z_1)$, $E(Y) = \ddot{a}^{(m)}_{x:\overline{n}|}$ is obtained as follows.

$$
\begin{aligned}
\ddot{a}^{(m)}_{x:\overline{n}|} &= E(Y) = \frac{1 - \frac{i}{i^{(m)}}A^1_{x:\overline{n}|} - {}_nE_x}{d^{(m)}} \\
&= \frac{1 - \{\frac{i}{i^{(m)}}(A_{x:\overline{n}|} - {}_nE_x)\} - {}_nE_x}{d^{(m)}} \\
&= \frac{i^{(m)} - i(1 - d\ddot{a}_{x:\overline{n}|} - {}_nE_x) - i^{(m)}\,{}_nE_x}{i^{(m)}d^{(m)}} \\
&= \frac{i\,d}{i^{(m)}\,d^{(m)}}\ddot{a}_{x:\overline{n}|} - \frac{i - i^{(m)}}{i^{(m)}\,d^{(m)}} + \frac{(i - i^{(m)})\,{}_nE_x}{i^{(m)}\,d^{(m)}} \\
&= \alpha(m)\ddot{a}_{x:\overline{n}|} - \beta(m)(1 - {}_nE_x) \qquad (6.5.10)
\end{aligned}
$$

This expression can also be obtained from the actuarial present value of m-thly whole life annuity due as follows.

$$
\begin{aligned}
\ddot{a}^{(m)}_{x:\overline{n}|} &= \ddot{a}^{(m)}_x - {}_nE_x\ddot{a}^{(m)}_{x+n} \\
&= \alpha(m)\ddot{a}_x - \beta(m) - {}_nE_x(\alpha(m)\ddot{a}_{x+n} - \beta(m)) \\
&= \alpha(m)\ddot{a}_{x:\overline{n}|} - \beta(m)(1 - {}_nE_x)
\end{aligned}
$$

Using Equation (6.5.6), $\ddot{a}^{(m)}_{x:\overline{n}|}$ can also be expressed as follows.

$$
\begin{aligned}
\ddot{a}^{(m)}_{x:\overline{n}|} &= \ddot{a}^{(m)}_x - {}_nE_x\ddot{a}^{(m)}_{x+n} \\
&= \frac{d}{d^{(m)}}\ddot{a}_x - \beta(m)A_x - {}_nE_x\left(\frac{d}{d^{(m)}}\ddot{a}_{x+n} - \beta(m)A_{x+n}\right) \\
&= \frac{d}{d^{(m)}}\ddot{a}_{x:\overline{n}|} - \beta(m)(A_x - {}_nE_xA_{x+n}) \\
&= \frac{d}{d^{(m)}}\ddot{a}_{x:\overline{n}|} - \beta(m)A^{\,1}_{x:\overline{n}|} \qquad (6.5.11)
\end{aligned}
$$

In all these expressions, if we allow $n \to \infty$, we get the expression for actuarial present value of m-thly whole life annuity due.

Example 6.5.6. Prove that $\overline{a}_{x:\overline{n}|} \; < \; \ddot{a}^{(m)}_{x:\overline{n}|} \; < \; \ddot{a}_{x:\overline{n}|}$, under the assumption of uniformity in unit age interval.

Solution: Observe that under the assumption of uniformity

$$i \; > \; i^{(m)} \; > \; \delta \; > \; d^{(m)} \; > \; d \; \Rightarrow \; i/i^{(m)} \; < \; i/\delta$$

$$\Rightarrow \; \frac{i}{i^{(m)}} A^1_{x:\overline{n}|} + \; _nE_x \; < \; \frac{i}{\delta} A^1_{x:\overline{n}|} + \; _nE_x = \overline{A}^1_{x:\overline{n}|} + \; _nE_x = \overline{A}_{x:\overline{n}|}$$

Further, $\; \delta \; > \; d^{(m)} \; \Rightarrow \; 1/d^{(m)} \; > \; 1/\delta$

$$\Rightarrow \; \ddot{a}^{(m)}_{x:\overline{n}|} = \frac{1 - \frac{i}{i^{(m)}} A^1_{x:\overline{n}|} - \; _nE_x}{d^{(m)}} > \frac{1 - \overline{A}_{x:\overline{n}|}}{d^{(m)}} > \frac{1 - \overline{A}_{x:\overline{n}|}}{\delta} = \overline{a}_{x:\overline{n}|}$$

Now by Equation (6.5.11),

$$\ddot{a}^{(m)}_{x:\overline{n}|} \; = \; \frac{d}{d^{(m)}} \ddot{a}_{x:\overline{n}|} - \beta(m) A^{\;\;1}_{x:\overline{n}|} < \frac{d}{d^{(m)}} \ddot{a}_{x:\overline{n}|} < \ddot{a}_{x:\overline{n}|}, \;\; \text{since} \;\; \frac{d}{d^{(m)}} < 1$$

Thus, we have $\overline{a}_{x:\overline{n}|} \; < \; \ddot{a}^{(m)}_{x:\overline{n}|} \; < \; \ddot{a}_{x:\overline{n}|}$.   ∎

In Chapter 7, we apply the result proved in Example 6.5.6 to derive an ordering among premiums. As in part (iii) of Example 6.2.8, we have the following order relation among the actuarial present values of m-thly n-year temporary annuities due.

Example 6.5.7. Prove that $\ddot{a}^{(12)}_{x:\overline{n}|} \; < \; \ddot{a}^{(4)}_{x:\overline{n}|} \; < \; \ddot{a}^{(2)}_{x:\overline{n}|}$, under the assumption of uniformity in unit age interval.

Solution: We use following formula to prove the inequality.

$$\ddot{a}^{(m)}_{x:\overline{n}|} \; = \; \frac{1 - \frac{i}{i^{(m)}} A^1_{x:\overline{n}|} - \; _nE_x}{d^{(m)}}$$

Adopting the same steps as in part (iii) of Example 6.2.8 we get the result.   ∎

m-thly n-Year Temporary Annuity Immediate: The formula for actuarial present values in this set up is derived on similar lines as those for annuity due. Thus, using Eq. 6.5.7 we get,

$$a^{(m)}_{x:\overline{n}|} \; = \; a^{(m)}_x - \; _nE_x \, a^{(m)}_{x+n} \; = \; \ddot{a}^{(m)}_x - 1/m - \; nE_x(\ddot{a}^{(m)}_{x+n} - 1/m)$$

$$= \; \ddot{a}^{(m)}_{x:\overline{n}|} - (1 - \; _nE_x)/m \tag{6.5.12}$$

However, to derive the result parallel to that in Example 6.5.2, we define the present value random variable in this case and derive an alternate expression for the actuarial present value. Proceeding on similar lines as those for annuity due, the present value random variable Y for n-year m-thly temporary life annuity immediate, with m-thly payments at a rate of 1 unit per annum, is given by,

$$Y = \begin{cases} a^{(m)}_{\overline{K+J/m}|} \,, & \text{if} \quad K = 0, 1, \cdots, n-1 \\ a^{(m)}_{\overline{n}|} \,, & \text{if} \quad K = n, n+1, \cdots \end{cases}$$

It can be expressed as

$$Y = \begin{cases} (1 - v^{K+J/m})/i^{(m)}, & \text{if} \quad K = 0, 1, \cdots, n-1 \\ (1 - v^n)/i^{(m)}, & \text{if} \quad K = n, n+1, \cdots \end{cases}$$

We write Y as $Y = (1 - Z_1)/i^{(m)}$ where $Z_1 = Z_2 + Z_3$ and Z_2 and Z_3 are defined as,

$$Z_2 = \begin{cases} v^{K+J/m}, & \text{if} \quad K = 0, 1, \cdots, n-1 \\ 0, & \text{if} \quad K = n, n+1, \cdots \end{cases}$$

$$Z_3 = \begin{cases} 0, & \text{if} \quad K = 0, 1, \cdots, n-1 \\ v^n, & \text{if} \quad K = n, n+1, \cdots \end{cases}$$

We rewrite Z_2 as $Z_2 = v^{-1/m} \, v^{K+J/m+1/m}$, for $K = 0, 1, \cdots, n-1$, to find its expectation using the arguments as in the derivation of Eq. 6.5.3. Thus,

$$E(Z_2) = \frac{i}{i^{(m)}} A^1_{x:\overline{n}|} \, v^{-1/m} = \frac{i}{d^{(m)}} A^1_{x:\overline{n}|}$$

$$E(Z_3) = {}_nE_x \quad \Rightarrow \quad E(Z_1) = \frac{i}{d^{(m)}} A^1_{x:\overline{n}|} + {}_nE_x$$

$$\text{Hence,} \quad E(Y) = a^{(m)}_{x:\overline{n}|} = \frac{1 - \frac{i}{d^{(m)}} A^1_{x:\overline{n}|} - {}_nE_x}{i^{(m)}} \tag{6.5.13}$$

Example 6.5.8. Prove that $a_{x:\overline{n}|} < \ddot{a}^{(m)}_{x:\overline{n}|} < \overline{a}_{x:\overline{n}|}$, under the assumption of uniformity in unit age interval.

Solution: We use the formula $a^{(m)}_{x:\overline{n}|} = (1 - (i/d^{(m)})A^1_{x:\overline{n}|} - {}_nE_x)/i^{(m)}$ as derived in Eq. 6.5.13 and the steps analogous to those in Example 6.5.2, to get the required inequality. ∎

Combining the results in Examples 6.5.6 and 6.5.8 we get the ordering as,

$$a_{x:\overline{n}|} < \ddot{a}^{(m)}_{x:\overline{n}|} < \overline{a}_{x:\overline{n}|} < \ddot{a}^{(m)}_{x:\overline{n}|} < \ddot{a}_{x:\overline{n}|} \tag{6.5.14}$$

This result is similar to that for annuities certain and whole life annuities.

Example 6.5.9. Prove that $a^{(2)}_{x:\overline{n}|} < a^{(4)}_{x:\overline{n}|} < a^{(12)}_{x:\overline{n}|}$, under the assumption of uniformity in unit age interval.

Solution: We use the following formula $a^{(m)}_{x:\overline{n}|} = (1 - (i/d^{(m)})A^1_{x:\overline{n}|} - {}_nE_x)/i^{(m)}$ as derived in Eq. 6.5.13 and the steps analogous to those in part (iv) of Example 6.2.8 to get the required inequality. ∎

The actuarial present values for n-year deferred whole life annuity due and n-year deferred whole life annuity immediate, with m-thly payments at a rate of 1 per annum, are denoted by ${}_{n|}\ddot{a}^{(m)}_x$ and ${}_{n|}a^{(m)}_x$ respectively. These are given by,

$$_{n|}\ddot{a}^{(m)}_x = \ddot{a}^{(m)}_x - \ddot{a}^{(m)}_{x:\overline{n}|} = \alpha(m) \, {}_{n|}\ddot{a}_x - \beta(m) \, {}_nE_x \tag{6.5.15}$$

$$\text{and} \quad _{n|}a^{(m)}_x = a^{(m)}_x - a^{(m)}_{x:\overline{n}|} = {}_{n|}\ddot{a}^{(m)}_x - {}_nE_x/m \tag{6.5.16}$$

Example 6.5.10. Suppose the force of mortality follows Gompertz' law given by $\mu_x = BC^x$. Find the actuarial present values of m-thly n-year temporary life annuity due and immediate for payment of 100 issued to (25) for $m = 2, 4$ and 12 and $n = 1, 2, \cdots, 10$. Take $\delta = 0.05$, $B = 0.0001151$ and $C = 1.096$. Use uniformity assumption for fractional ages.

Solution: For given $\delta = 0.05$ we find i, d, $i^{(m)}$ and $d^{(m)}$. These are the same as given in Table 6.22. Using Eqs 6.5.10, 6.5.12 and the values of $\ddot{a}_{25:\overline{n}|}$ reported in Table 6.12, we find the values of $\ddot{a}^{(m)}_{25:\overline{n}|}$ and $a^{(m)}_{25:\overline{n}|}$. These are obtained using Code 6.6.10 and are reported in Table 6.26 and Table 6.27 respectively.

n	$100\ddot{a}^{(2)}_{25:\overline{n}|}$	$100\ddot{a}^{(4)}_{25:\overline{n}|}$	$100\ddot{a}^{(12)}_{25:\overline{n}|}$
1	98.74	98.11	97.69
2	192.54	191.32	190.50
3	281.66	279.86	278.67
4	366.30	363.96	362.41
5	446.68	443.83	441.93
6	523.01	519.67	517.44
7	595.48	591.66	589.13
8	664.27	660.01	657.18
9	729.55	724.86	721.75
10	791.49	786.39	783.01

Table 6.26 Gompertz' law: m-thly n-year temporary life annuity due

n	$100a^{(2)}_{25:\overline{n}|}$	$100a^{(4)}_{25:\overline{n}|}$	$100a^{(12)}_{25:\overline{n}|}$
1	96.24	96.86	97.28
2	187.67	188.88	189.69
3	274.52	276.29	277.48
4	357.01	359.32	360.86
5	435.34	438.16	440.04
6	509.72	513.02	515.23
7	580.32	584.09	586.60
8	647.33	651.54	654.35
9	710.92	715.55	718.64
10	771.25	776.28	779.64

Table 6.27 Gompertz' law: m-thly n-year temporary life annuity immediate

Comparing the values reported in Table 6.26 with those of $100\, \ddot{a}_{25:\overline{n}|}$ reported in Table 6.12 and with those of $100\, \overline{a}_{25:\overline{n}|}$ reported in Table 6.7, we notice that $100\, \overline{a}_{25:\overline{n}|} < 100\, \ddot{a}^{(m)}_{25:\overline{n}|} < 100\, \ddot{a}_{25:\overline{n}|}$ for all the values of m and n, verifying the result proved in Example 6.5.6. It is again a consequence of the fact that in $100\, \ddot{a}_{25:\overline{n}|}$ payments are made at the beginning of the year while in $100\, \ddot{a}^{(m)}_{25:\overline{n}|}$ payments are made at the beginning of every m^{th} part of the year and in $100\, \overline{a}_{25:\overline{n}|}$

payments are made continuously. Further as m increases, the actuarial present values decrease for all n, verifying the result proved in Example 6.5.7. However as n increases, actuarial present values also increase as the total period of payments increases.

Comparing the values reported in Table 6.27 with those of $100\ a_{25:\overline{n}|}$ reported in Table 6.13, we notice that $100\ a^{(m)}_{25:\overline{n}|} > 100\ a_{25:\overline{n}|}$ for all the values of m and n, verifying the result of Example 6.5.8. It is again a consequence of the fact that in $100\ a_{25:\overline{n}|}$ payments are made at the end of the year while in $100\ a^{(m)}_{25:\overline{n}|}$ payments are made at the end of every m^{th} part of the year. Further as m increases, in this case actuarial present values also increase for all n, verifying the result of Example 6.5.9. As n increases, actuarial present values also increase, as the total period of payments increases. ∎

Example 6.5.11. It is given that $q_{28} = 0.135$, $q_{29} = 0.146$, $q_{30} = 0.159$, $q_{31} = 0.173$, $q_{32} = 0.188$ and $i = 0.05$. Find the actuarial present value of a 5-year temporary annuity due and immediate payable six monthly, at the rate of 5000 per six months issued to (28).

Solution: From Eq. 6.5.10 we have $\ddot{a}^{(2)}_{28:\overline{5}|} = \alpha(2)\ddot{a}_{28:\overline{5}|} - \beta(2)(1 - {}_5E_{28})$. Note that

$$i = 0.05 \quad \Rightarrow \quad d = 0.04761905, \quad i^{(2)} = 0.04939015 \ \text{ and } \ d^{(2)} = 0.04819985$$
$$\Rightarrow \quad \alpha(2) = 1.000149 \ \text{ and } \ \beta(2) = 0.2561738$$

For the given q_x values and $i = 0.05$ we have obtained the value of $\ddot{a}_{28:\overline{5}|} = 3.453191$ in Example 6.4.4. Using these we get, $\ddot{a}^{(2)}_{28:\overline{5}|} = 3.281269$. Further using the relation given in Eq. 6.5.12, we have

$$a^{(2)}_{28:\overline{5}|} = \ddot{a}^{(2)}_{28:\overline{5}|} - (1 - {}_5E_{28})/2 = 2.944707$$

Note that six monthly instalment is 5000, hence yearly instalment is 10000. So we multiply $\ddot{a}^{(2)}_{28:\overline{5}|}$ and $a^{(2)}_{28:\overline{5}|}$ by 10000 to get the values as

$$10000\ \ddot{a}^{(2)}_{28:\overline{5}|} = 32812.69 \quad \text{and} \quad 10000\ a^{(2)}_{28:\overline{5}|} = 29447.07 \qquad ∎$$

Example 6.5.12. It is given that $A^{\ 1}_{x:\overline{n}|} = 0.01419$, ${}_nE_x = 0.54733$, $i = 0.05$, $\alpha(4) = 1.00019$ and $\beta(4) = 0.38272$. Calculate $\ddot{a}^{(4)}_{x:\overline{n}|}$ under the assumption of uniformity over each year of age.

Solution: From the given data, we have

$$\ddot{a}_{x:\overline{n}|} = (1 - A^{\ 1}_{x:\overline{n}|} - {}_nE_x)/d = 9.20808$$
$$\Rightarrow \ddot{a}^{(4)}_{x:\overline{n}|} = \alpha(4)\ddot{a}_{x:\overline{n}|} - \beta(4)(1 - {}_nE_x) = 9.03658 \qquad ∎$$

Example 6.5.13. It is given that (i) $\ddot{a}^{(4)}_{\overline{\infty}|} = 17.287$, (ii) $A_x = 0.1025$ and (iii) deaths are uniformly distributed over each year of age. Calculate $\ddot{a}^{(4)}_x$.

Solution: We have $\ddot{a}_x^{(4)} = (1 - A_x^{(4)})/d^{(4)}$ and under uniformity assumption, $A_x^{(4)} = (i/i^{(4)})A_x$. Observe that

$$17.287 = \ddot{a}_{\overline{\infty}|}^{(4)} = 1/d^{(4)} \;\Rightarrow\; d^{(4)} = 0.05785$$

$$\Rightarrow\; i^{(4)} = \frac{d^{(4)}}{1 - \frac{d^{(4)}}{4}} = 0.05870 \;\text{ and }\; i = \left[1 + \frac{i^{(4)}}{4}\right]^4 - 1 = 0.06$$

$$\Rightarrow\; A_x^{(4)} = \frac{0.06}{0.05870}(0.1025) = 0.1048 \;\text{ and }\; \ddot{a}_x^{(4)} = \frac{1 - 0.1048}{0.05785} = 15.5 \qquad \blacksquare$$

As in Section 5.6, we discuss computations of various annuity functions when the mortality pattern is specified by a set of q_x values for integer values of x. In Section 5.6, R code is given to find actuarial present values involved in term insurance, pure endowment insurance, endowment insurance and whole life insurance corresponding to a set of q_x values. In this code, we add some functions to get the actuarial present values of the annuities. We use the relations between the actuarial present values of benefit and the actuarial present values of annuities, as stated in Eqs 6.3.4, 6.3.8 and 6.4.2, to add these new functions. In the next example we compute various annuity functions when the mortality pattern is specified by the q_x values is as given in Table 4.8. We use Code 6.6.11.

Example 6.5.14. Suppose a set of q_x values is as given in Table 4.8 and the effective rate of interest is $i = 0.06$. Find the actuarial present values of n-year temporary annuities due, immediate, continuous and payable m-thly, for a benefit of 1000. Take $m = 2, 4, 12$ and $n = 1$ to 10. Also find the actuarial present values of whole life annuities due, immediate, continuous and payable m-thly, with $m = 2, 4, 12$, for a benefit of 1000. Use uniformity assumption for fractional ages.

Solution: In Example 5.5.5 we have computed the actuarial present values of the benefit involved in the term insurance, the endowment insurance and whole life insurance issued to (25), when the benefit of 1000 is paid (i) at the end of the year of death and (ii) at the moment of death, under the assumption of uniformity in unit age interval. These are corresponding to the mortality law specified by values of q_x as given in Table 4.8. These are reported in Table 5.23. Values of $1000A_{25}$ and $1000\overline{A}_{25}$, are $1000A_{25} = 73.88$, $1000\overline{A}_{25} = 76.08$. We use these values to compute the actuarial present values of the various types of annuities discussed in this chapter, using the following links.

$$\ddot{a}_{x:\overline{n}|} = (1 - A_{x:\overline{n}|})/d \quad \text{by Eq. 6.4.2}$$

$$a_{x:\overline{n}|} = \{1 - (1 + i)A_{\;x:\overline{n}|}^{\,1} - A_{x:\overline{n}|}^{\,1}\}/i \quad \text{by Eq. 6.4.5}$$

$$\overline{a}_{x:\overline{n}|} = (1 - \overline{A}_{x:\overline{n}|})/\delta \quad \text{by Eq. 6.3.4}$$

These values are obtained using Code 6.6.11 and are presented in Table 6.28.

From $a_{25:\overline{n}|}$ and $\ddot{a}_{25:\overline{n}|}$ obtained above, we compute the actuarial present values of the m-thly annuities due and immediate using the following formulae.

$$\ddot{a}(m)_{25:\overline{n}|} = \alpha(m)\ddot{a}_{25:\overline{n}|} - \beta(m)(1 - {}_nE_x) \quad \text{by Eq. 6.5.10}$$

$$a(m)_{25:\overline{n}|} = \ddot{a}_{x:\overline{n}|}^{(m)} - (1 - {}_nE_x)/m \quad \text{by Eq. 6.5.12}$$

n	$a_{25:\overline{n}|}$	$\overline{a}_{25:\overline{n}|}$	$\ddot{a}_{25:\overline{n}|}$
1	942.29	970.86	1000.00
2	1830.19	1885.69	1942.29
3	2666.82	2747.69	2830.19
4	3455.09	3559.90	3666.82
5	4197.76	4325.15	4455.09
6	4897.40	5046.09	5197.76
7	5556.48	5725.25	5897.40
8	6177.29	6365.00	6556.48
9	6762.01	6967.59	7177.29
10	7312.70	7535.12	7762.01

Table 6.28 n-Year temporary annuity immediate, due and continuous

n	$1000\ \ddot{a}^{(2)}_{25:\overline{n}|}$	$1000\ \ddot{a}^{(4)}_{25:\overline{n}|}$	$1000\ \ddot{a}^{(12)}_{25:\overline{n}|}$
1	985.36	978.09	973.27
2	1913.85	1899.73	1890.36
3	2788.74	2768.17	2754.51
4	3613.10	3586.43	3568.73
5	4389.80	4357.39	4335.88
6	5121.55	5083.73	5058.62
7	5810.90	5767.97	5739.47
8	6460.27	6412.52	6380.82
9	7071.92	7019.63	6984.91
10	7648.01	7591.43	7553.86

Table 6.29 n-Year temporary m-thly annuity due

These are obtained using Code 6.6.11 and are reported in Table 6.29 and Table 6.30 respectively.

The actuarial present values of whole life annuities for a benefit of 1000 are $1000\ a_{25} = 15361.39$, $1000\ \overline{a}_{25} = 15856.17$ and $1000\ \ddot{a}_{25} = 16361.39$. The actuarial present values of m-thly whole life annuities due and immediate for a benefit of 1000 are presented in Table 6.31. ∎

n-year deferred annuities and n-year certain and whole life annuities with m-thly payments are defined on similar lines. Table 6.32 displays the actuarial notation for present values of life annuities with m-thly payments studied in this section for annuity of 1 unit per annum. For ready reference, the formulae of $\alpha(m)$, $\beta(m)$, $i^{(m)}$ and $d^{(m)}$ are also given at the end of the table.

With discrete annuities, payable either annually or m-thly, in practice some adjustment is needed in the year of death. For example, suppose that a life insurance contract is purchased by annuity due, that is payments are made at the beginning of each contract year. If the insured dies 3 months after making an annual payment, the premium for the remaining months should be refunded. On the other hand if premium payments form annuity immediate and if annuitant

n	$1000\ a^{(2)}_{25:\overline{n}|}$	$1000\ a^{(4)}_{25:\overline{n}|}$	$1000\ a^{(12)}_{25:\overline{n}|}$
1	956.51	963.66	968.46
2	1857.80	1871.71	1881.02
3	2707.05	2727.32	2740.89
4	3507.23	3533.50	3551.09
5	4261.13	4293.06	4314.43
6	4971.38	5008.64	5033.59
7	5640.44	5682.74	5711.06
8	6270.68	6317.72	6349.22
9	6864.29	6915.81	6950.30
10	7423.35	7479.10	7516.42

Table 6.30 n-Year temporary m-thly annuity immediate

m	$1000\ \ddot{a}^{(m)}_{25}$	$1000\ a^{(m)}_{25}$
2	16107.48	15607.48
4	15981.50	15731.50
12	15897.87	15814.54

Table 6.31 m-thly whole life annuity due and immediate

dies 3 months before the due date of the next payment, there has to be a final payment for the 9 months. Thus, one has to consider first the appropriate size for the adjustment payment in such cases. It gives rise to the type of annuities, referred to as apportionable annuities due and complete annuities immediate. We do not discuss these in this book. Interested reader may refer to Section 5.5 of the book by Bowers et al [1].

The next section presents the codes used to compute the variety of monetary functions discussed in this chapter.

6.6   R Codes

Monetary functions studied in this chapter are the present values and actuarial present values of the annuities of various types, such as certain and life, continuous and discrete, discrete being further classified as due, immediate and m-thly payments. In this section we present R codes used to compute these functions for a specified mortality and interest pattern. We begin with the codes for annuities certain.

Code 6.6.1. In Section 2, we discussed how to compute present values of annuity certain with its types as due, immediate and continuous. We have presented these values for specified rates of interest and for $n = 1$ to 10, in Table 6.1, Table 6.2 and Table 6.4. This code generates these tables.

```
# Annuity Certain:due,immediate and continuous
n=1:10; int=c(.05,.06,.07,.08,.09,.10); v=1/(1+int)
```

Annuity	Notation	Formula
Whole life annuity due	$\ddot{a}_x^{(m)}$	$\ddot{a}_x^{(m)} = \alpha(m)\ddot{a}_x - \beta(m)$
Whole life annuity immediate	$a_x^{(m)}$	$a_x^{(m)} = \ddot{a}_x^{(m)} - 1/m$
n-Year temporary life annuity due	$\ddot{a}_{x:\overline{n}|}^{(m)}$	$\ddot{a}_{x:\overline{n}|}^{(m)} = \alpha(m)\ddot{a}_{x:\overline{n}|} - \beta(m)(1 - {}_nE_x)$
n-Year temporary life annuity immediate	$a_{x:\overline{n}|}^{(m)}$	$a_{x:\overline{n}|}^{(m)} = \ddot{a}_{x:\overline{n}|}^{(m)} - (1/m)(1 - {}_nE_x)$
n-Year deferred whole life annuity due	${}_{n|}\ddot{a}_x^{(m)}$	${}_{n|}\ddot{a}_x^{(m)} = \alpha(m)\,{}_{n|}\ddot{a}_x - \beta(m)\,{}_nE_x$ $= {}_nE_x\ddot{a}_{x+n}^{(m)}$
n-Year deferred whole life annuity immediate	${}_{n|}a_x^{(m)}$	${}_{n|}a_x^{(m)} = {}_{n|}\ddot{a}_x^{(m)} - (1/m)\,{}_nE_x$ $= {}_nE_x a_{x+n}^{(m)}$
p-Year deferred n-year temporary life annuity due	${}_{p|}\ddot{a}_{x:\overline{n}|}^{(m)}$	${}_{p|}\ddot{a}_{x:\overline{n}|}^{(m)} = {}_{p|}\ddot{a}_x^{(m)} - {}_{p+n|}\ddot{a}_x^{(m)}$
p-Year deferred n-year temporary life annuity immediate	${}_{p|}a_{x:\overline{n}|}^{(m)}$	${}_{p|}a_{x:\overline{n}|}^{(m)} = {}_{p|}a_x^{(m)} - {}_{p+n|}a_x^{(m)}$
n-Year certain and whole life annuity due	$\ddot{a}_{\overline{x:n|}}^{(m)}$	$\ddot{a}_{\overline{x:n|}}^{(m)} = \ddot{a}_{\overline{n}|}^{(m)} + {}_{n|}\ddot{a}_x^{(m)}$
n-Year certain and whole life annuity immediate	$a_{\overline{x:n|}}^{(m)}$	$a_{\overline{x:n|}}^{(m)} = a_{\overline{n}|}^{(m)} + {}_{n|}a_x^{(m)}$
		$\alpha(m) = \dfrac{id}{i^{(m)}d^{(m)}}$ $\beta(m) = \dfrac{i - i^{(m)}}{i^{(m)}d^{(m)}}$ $i^{(m)} = m\{(1 + i)^{1/m} - 1\}$ $d^{(m)} = m(1 - v^{1/m})$

Table 6.32 Actuarial notation and formulae for life annuities with m-thly payments

```
d=1-v; del=-log(v)
x1=x2=x3=matrix(nrow=length(n),ncol=length(int))
for(i in 1:length(n))
{
for(j in 1:length(int))
{
x1[i,j]=round(1000*(1-v[j]^n[i])/int[j],2)
x2[i,j]=round(1000*(1-v[j]^n[i])/d[j],2)
x3[i,j]=round(1000*(1-v[j]^n[i])/del[j],2)
}
}
x1;x2;x3 # Tables 6.1, 6.2 and 6.4 respectively
```

∎

Code 6.6.2. This code generates Table 6.3, which displays the present values of m-thly annuity certain due and immediate, for $m = 2, 4, 12$ and $n = 1$ to 5.

```
# Annuity Certain:m-thly
n=1:5; m=c(2,4,12); int=.05; v=1/(1+int)
im=m*((1+int)^(1/m) -1); dm=m*(1-v^(1/m)); im;dm
x1=x2=matrix(nrow=length(n),ncol=length(m))
for(i in 1:length(n))
{
for(j in 1:length(m))
{
x1[i,j]=round(1000*(1-v^n[i])/im[j],2)
x2[i,j]=round(1000*(1-v^n[i])/dm[j],2)
}
}
x=cbind(n,x1,x2);x # Table 6.3
```

∎

Code 6.6.3. This code generates Table 6.5, which presents the values of $1000\, a_{\overline{n}|}$, $1000\, a_{\overline{n}|}^{(m)}$, $1000\, \overline{a}_{\overline{n}|}$, $1000\, \ddot{a}_{\overline{n}|}^{(m)}$ and $1000\, \ddot{a}_{\overline{n}|}$ for $m = 12$, for $n = 1$ to 10 and $i = 0.05$. This table is useful to compare the present values of annuity certain of different types.

```
m=12; n=1:10; int=.05; v=1/(1+int); d=1-v;
del=-log(v); im=m*((1+int)^(1/m)-1);dm=m*(1-v^(1/m))
x1=1000*(1-v^n)/int; x2=1000*(1-v^n)/im
x3=1000*(1-v^n)/del; x4=1000*(1-v^n)/dm;
```

```
x5=1000*(1-v^n)/d;
d1=round(data.frame(n,x1,x2,x3,x4,x5),2); d1
```

■

Code 6.6.4. In Example 6.3.2, when the mortality law is Gompertz' law, we have obtained the values of $100\ \bar{a}_{x:\overline{n}|}$ for $n = 1, 2, \cdots, 10$ and are reported in Table 6.7. In Example 6.3.4, under the same mortality law, we have obtained the values of $100\ \bar{a}_x$, which are displayed in Table 6.8. In Example 6.3.6, Table 6.9 presents the values of $100\ {}_{m|}\bar{a}_x$ for $m = 1, 2, \cdots, 10$ under Gompertz' law. In Example 6.3.8, we have found the values of $100\ \bar{a}_{\overline{x:n}|}$ for $n = 1, 2, \cdots, 10$, under the same mortality law. These values are displayed in Table 6.10. In all these examples $x = 25, 30, 35$ and 40, $\delta = 0.05$ and the parameters of the Gompertz' law are the same, $C = 1.096$ and $B = 0.0001151$. Computation of all these actuarial present values is combined in this code. It generates Table 6.7, Table 6.8, Table 6.9 and Table 6.10.

```
# Part I: Mortality and interest pattern
B=0.0001151; C=1.096; m=B/log(C); del=0.05; n=1:10;
la=(-del/log(C))+1; x=c(25,30,35,40); alx=m*C^x
# Part II: n-year temporary life annuity
T=P=U=DW=WC=matrix(nrow=length(n),ncol=length(x))
for(i in 1:length(n))
{
for(j in 1:length(x))
{
U[i,j]=pgamma(C^n[i],la,alx[j])-pgamma(1,la,alx[j])
T[i,j]=exp(alx[j])*gamma(la)*alx[j]^(1-la)*U[i,j]
P[i,j]=exp(-del*n[i]-alx[j]*(C^n[i]-1))
}
}
E=T+P; # Endowment insurance
A=(1-E)/del; A1=round(100*A,2); A1 # Table 6.7
# Part III: Whole life annuity
w=c()
for(j in 1:length(x))
{
w[j]=exp(alx[j])*gamma(la)*alx[j]^(1-la)*(1-pgamma(1,la,alx[j]))
}
WA=(1-w)/del
WA1=round(100*WA,2)
WA1 # Table 6.8 # Whole life annuity
# Part IV: n-year deferred whole life annuity
for(i in 1:length(n))
```

```
{
DW[i,]=WA-A[i,]
}
DW1=round(100*DW,2)
DW1 # Table 6.9 # Deferred whole life annuity
# Part V: n-year certain and whole life annuity
v=exp(-del);v; ACim=(1-v^n)/del; ACim
for(j in 1:length(x))
{
WC[,j]=ACim+DW[,j]
}
WC1=round(100*WC,2)
WC1 # Table 6.10 # n-year certain and whole life annuity
```

■

Code 6.6.5. In Example 6.4.7 the mortality law is specified by q_x values as given in Table 6.15. When $A_{50} = 0.251$, we have found the actuarial present value of whole life annuity due and immediate when the yearly payment is 1000, for ages 40 to 49. We have also obtained the actuarial present value of whole life annuity, payable continuously at the rate of 1000 per annum, under the assumption of uniformity for fractional ages. We used the backward recurrence formulae to obtain the actuarial present value of A_x and also for annuities. These values are presented in Table 6.16, which are obtained using this code.

```
# Part I: Mortality law and interest rates
int=0.05; v=(1+int)^(-1); d=int/(1+int);del=log(1+int)
q=c(.0027,.0029,.0032,.0034,.0037,.0039,.0043,.0046,.0050,.0054)
# Part II : APVs of annuities
p=1-q; l=length(p); A50=0.251; ad50=(1-A50)/d; ad=c(1:l,ad50)
for (i in 1:l)
{
ad[l+1-i]=1+v*p[l+1-i]*ad[l+1-i+1]
}
ai=ad-1; A=c(1:l,A50)
for (i in 1:l)
{
A[l+1-i]<-v*q[l+1-i] + v*p[l+1-i]*A[l+1-i+1]
}
Abar=(int/del)*A; ac=(1-Abar)/del; age=40:50
d=round(1000*data.frame(ai,ac,ad),2)
d1=data.frame(age,d);d1 # Table 6.16
```

■

Code 6.6.6. In Section 4, when the mortality pattern is specified by Gompertz' law, we have obtained the actuarial present values of the n-year temporary annuity due and immediate, whole life annuity, deferred whole life annuity and n-year certain and whole life annuity. The parameters of Gompertz' law, rate of interest, values of n and x are the same for all the annuities. These values are reported in Table 6.12, Table 6.13, Table 6.17 and Table 6.19 and Table 6.20. This code generates all these tables.

```r
# Part I: Gompertz' Mortality law and interest rate
B=0.0001151; C=1.096; m=B/log(C); del=0.05; v=exp(-del)
d=1-v; int=1/v-1; x=c(25,30,35,40); k=0:(100-min(x))
y=matrix(nrow=length(k),ncol=length(x))# pmf of K(x)
for(i in 1:length(k))
{
for(j in 1:length(x))
{
y[i,j]=exp(-m*C^x[j]*(C^k[i]-1))-exp(-m*C^x[j]*(C^(k[i]+1)-1))
}
}
b=v^(k+1);w=c(); t=matrix(nrow=length(k),ncol=length(x))
for(j in 1:length(x))
{
w[j]=sum(y[,j]*b)
t[,j]=cumsum(y[,j]*b)
}
n=1:10; T=t[1:length(n),] # Term insurance
P=matrix(nrow=length(n),ncol=length(x))
for(i in 1:length(n))
{
for(j in 1:length(x))
{
P[i,j]=v^(n[i])*exp(-m*C^x[j]*(C^n[i]-1))
}
}
E=T+P; E # Endowment insurance
# Part II: Temporary life annuity due and immediate
adue=(1-E)/d; adue1=round(100*adue,2);adue1 # Table 6.12
aimm=(1-(1+int)*T-P)/int; aimm1=round(100*aimm,2)
aimm1 # Table 6.13
# Part III: Whole life annuity due and immediate
wadue=(1-w)/d
```

```
waduel=round(100*wadue,2);wadue1 # Table 6.17
waimm=wadue-1
waimm1=round(100*waimm,2);waimm1 # Table 6.17
# Part IV: Deferred whole life annuity due and immediate
defwadue=wadue[1]-adue[,1]
defwadue1=round(100*defwadue,2);defwadue1 # Table 6.19
defwaimm=waimm[1]-aimm[,1]
defwaimm1=round(100*defwaimm,2);defwaimm1 # Table 6.19
# Part V: Whole life and certain annuity due and immediate
ncwadue=(1-v^n)/d+defwadue
ncwadue1=round(100*ncwadue,2);ncwadue1 # Table 6.20
ncwaimm=(1-v^n)/int+defwaimm
ncwaimm1=round(100*ncwaimm,2);ncwaimm1 # Table 6.20
```

■

Code 6.6.7. In Example 6.4.13 the mortality law is specified by q_x values, which are the same as in Example 6.4.7. For $i = 0.05$, $A_{50} = 0.251$, we have obtained the actuarial present value of 5-year deferred whole life annuity due and immediate, and 5-year certain and whole life annuity due and immediate issued to (40), when the yearly payment is 1000. We use this code to compute these values.

```
# Part I: Mortality law and interest rates
int=0.05; v=(1+int)^(-1); d=int/(1+int)
q=c(.0027,.0029,.0032,.0034,.0037,.0039,.0043,.0046,.0050,.0054)
# Part II: APVs of annuities
p=1-q; l=length(p); A50=0.251; ad50=(1-A50)/d; ad=c(1:l,ad50)
for (i in 1:l)
{
ad[l+1-i]=1+v*p[l+1-i]*ad[l+1-i+1]
}
p1=c(1,p[1], p[1]*p[2], p[1]*p[2]*p[3], p[1]*p[2]*p[3]*p[4]);
v1=c(1,v,v^2,v^3,v^4); ad540=sum(p1*v1); ad540
dad=ad[1]-ad540; dad1<-1000*dad; dad1
p2=p[1]*p[2]*p[3]*p[4]*p[5]; v2=v^5*p2; v2
ai=ad540+v2-1; dai=ad[1]-1-ai; dai1=1000*dai;dai1
a2d=(1-v^5)/d+dad; a2d1=1000*a2d;a2d1
a2i=(1-v^5)/int+dai; a2i1=1000*a2i;a2i1
```

■

Code 6.6.8. In Example 6.5.3, we have obtained the actuarial present values of m-thly whole life annuity due and immediate for payment of 100 issued to (25) and (30) for $m = 2, 4$ and

12, when the force of mortality follows Gompertz' law. This code obtains the actuarial present values required in Example 6.5.3 and reported in Table 6.23. Values $i^{(m)}$, $d^{(m)}$, $\alpha(m)$ and $\beta(m)$ for $\delta = 0.05$ are given in Table 6.22.

```
# Part I: Mortality law and interest rate
m=c(2,4,12); n=1:10; del=.05; v=exp(-del); int=exp(del)-1
d=1-v; im=m*((1+int)^(1/m)-1); dm=m*(1-v^(1/m))
alm=(int*d)/(im*dm); bm=(int-im)/(im*dm)
d1=round(data.frame(m,im,dm,alm,bm),5); d1 # Table 6.22
B=0.0001151; C=1.096; m1=B/log(C); x=c(25,30);k=0:(100-min(x))
# Part II: APVs of annuities
y=matrix(nrow=length(k),ncol=length(x))# pmf of K(x)
for(i in 1:length(k))
{
for(j in 1:length(x))
{
y[i,j]=exp(-m1*C^x[j]*(C^k[i]-1))-exp(-m1*C^x[j]*(C^(k[i]+1)-1))
}
}
b=v^(k+1);w=wadue=u=z=c()
for(j in 1:length(x))
{
w[j]=sum(y[,j]*b)
wadue[j]=(1-w[j])/d
}
w;wadue
u1=alm*wadue[1]-bm; z1=u1-1/m
u2=alm*wadue[2]-bm; z2=u2-1/m
d2=round(100*data.frame(u1,z1,u2,z2),2)
d3=data.frame(m,d2); d3 # Table 6.23
```

■

Code 6.6.9. In Example 6.5.5, we have obtained the actuarial present value of whole life annuity due and immediate paid monthly with the monthly payment of 1000, for ages 40 to 49 for the mortality law, specified by q_x values as given in Table 6.15. This code obtains these actuarial present values, which are reported in Table 6.25.

```
# Part I: Mortality law and interest rate
int=0.05; v=(1+int)^(-1); d=int/(1+int); m=12
im=m*((1+int)^(1/m)-1); dm=m*(1-v^(1/m))
alm=(int*d)/(im*dm); bm=(int-im)/(im*dm)
```

```
q=c(.0027,.0029,.0032,.0034,.0037,.0039,.0043,.0046,.0050,.0054)
# Part II: APVs of annuities
p=1-q; l=length(p); A50=0.251; ad50=(1-A50)/d; ad=c(1:l,ad50)
for (i in 1:l)
{
ad[l+1-i]=1+v*p[l+1-i]*ad[l+1-i+1]
}
adue12=alm*ad-bm; aimm12=adue12-1/12
d1=round(12000*data.frame(adue12,aimm12),3)
age=c(40:50); d2=data.frame(age,d1);d2 # Table 6.25
```

■

Code 6.6.10. In Example 6.5.10, we have obtained the actuarial present values of m-thly n-year temporary life annuity due and immediate for payment of 100 issued to (25) for $m = 2, 4$ and 12 and $n = 1, 2, \cdots, 10$, when the force of mortality follows Gompertz' law. This code is used to obtain these actuarial present values, which are reported in Table 6.26 and Table 6.27.

```
# Part I: Mortality law and interest rate
m=c(2,4,12); n=1:10; del=.05; v=exp(-del); int=exp(del)-1
d=1-v; im=m*((1+int)^(1/m)-1); dm=m*(1-v^(1/m))
alm=(int*d)/(im*dm); bm=(int-im)/(im*dm)
B=0.0001151; C=1.096; m1=B/log(C); x=25;k=0:(100-x)
# Part II: APVs of benefit
y=c() # pmf of K(x)
for(i in 1:length(k))
{
y[i]=exp(-m1*C^x*(C^k[i]-1))-exp(-m1*C^x*(C^(k[i]+1)-1))
}
b=v^(k+1); t=cumsum(y*b)
T=t[1:length(n)] # Term insurance
P=v^n*exp(-m1*C^x*(C^n-1)) # nEx
E=T+P # Endowment insurance
adue=(1-E)/d # n-year annuity due
# Part III: APVs of annuities
aduem=aimem=matrix(nrow=length(n),ncol=length(m))
for(i in 1:length(n))
{
for(j in 1:length(m))
{
aduem[i,j]=alm[j]*adue[i]-bm[j]*(1-P[i])
aimem[i,j]=aduem[i,j]-(1-P[i])/m[j]
```

```
}
}
aduem1=data.frame(n,round(100*aduem,2)); aduem1 # Table 6.26
aimem1=data.frame(n,round(100*aimem,2)); aimem1 # Table 6.27
```

∎

Code 6.6.11. In Example 6.5.14 the mortality law is specified by the set of q_x values is as given in Table 4.8. We have found the actuarial present values of n-year temporary annuities due, immediate, continuous and payable m-thly, for the benefit of 1000. We have also obtained the actuarial present values of whole life annuities due, immediate, continuous and payable m-thly, with $m = 2, 4, 12$, for the benefit of 1000. We obtain these values using this code. These actuarial present values are reported in Table 6.28, Table 6.29, Table 6.30 and Table 6.31.

```
# Part I
z=read.table("E://qx.txt", header=T)
x=z[,1] # column of x values running from 0 to 110
q=z[,2] # column of qx values
pr=1-q # column of px values
p=pr[26:111] # column of px values for x = 25 to 110
p1=c(p[1],2:85) # vector to store values of kpx for x=25
# and k=1 to 85,first element being p25
for (i in 2:85)
{
p1[i]=p1[i-1]*p[i]
}
q=1-p; # column of qx values for x = 25 to 110
p3=c(q[1],2:84)
for(i in 2:85)
{
p3[i]=p1[i-1]*q[i]
}
# Part II: APVs of benefits
int=0.06; v=(1 + int)^(-1); del=log((1+int));d=1-v
k=0:84; n=1:10; b=v^(k+1); v1=v^n
w25=cumsum(p3*b); t25=w25[1:10] # Term insurance
tbar25=(int/del)*t25; pu25=p1[1:10]*v1
e25=t25+pu25; ebar25=tbar25+pu25
d1=data.frame(t25,tbar25, pu25,e25,ebar25)
d2=round(data.frame(n,1000*d1),2);d2
wh25=sum(p3*b);wbar25=(int/del)*wh25
```

```
wh25; wbar25
# Part III: APVs of temporary annuities
abar25=(1-ebar25)/del
ad25=(1-e25)/d; ai25=(1-(1+int)*t25-pu25)/int
d3=data.frame(ai25,abar25,ad25);d3
d4=round(data.frame(n,1000*d3),2);d4 # Table 6.28
# Part IV: APVs of six monthly temporary annuities
m=c(2,4,12); im=m*((1+int)^(1/m)-1); dm=m*(1-v^(1/m))
alm=(int*d)/(im*dm); bm=(int-im)/(im*dm)
aduem=aimem=matrix(nrow=length(n),ncol=length(m))
for(i in 1:length(n))
{
for(j in 1:length(m))
{
aduem[i,j]=alm[j]*ad25[i]-bm[j]*(1-pu25[i])
aimem[i,j]=aduem[i,j]-(1-pu25[i])/m[j]
}
}
aduem1=data.frame(n,round(1000*aduem,2)); aduem1 # Table 6.29
aimem1=data.frame(n,round(1000*aimem,2)); aimem1 # Table 6.30
# Part V: APVs of whole annuities
abar25=(1-wbar25)/del; ad25=(1-wh25)/d; ai25=ad25-1
wha=round(1000*c(ai25,abar25, ad25),2);wha
# Part VI: APVs of m-thly whole life annuities
adm25=alm*ad25-bm; aim25=adm25-1/m
d5=round(1000*data.frame(adm25,aim25),2)
d6=data.frame(m,d5);d6 # Table 6.31
```

■

6.7   Conceptual Exercises

6.7.1 Find the present value and the accumulated value of a 15-year annuity immediate and due of Rs 15000 per annum if the effective rate of interest is 4%.

6.7.2 An annuity of Rs 10000 per year is payable for 10 years. What is the price of this annuity purchased on 1 January 2020 and if the first payment is made on the same day? Suppose the effective annual rate of interest is 4%.

6.7.3 A housing loan of Rs 25 lakhs is taken on 1 July 2020. It has to be repaid by equal monthly instalments payable at the beginning of the month for 10 years. Based on an

5% annual rate of interest, determine the amount of the instalment. In the same setup, determine the amount of the instalment if the period of repayment is 15 years and 20 years. Compare the monthly instalments in three cases.

6.7.4 An annuity of Rs 12000 per quarter, payable quarterly in arrear is purchased for 10 years. Find its price if the annual rate of interest is 4%. Find its accumulated value at the end of 10 years.

6.7.5 An annuity of Rs 20000 per annum payable continuously is purchased for 7 years. Find its price if the annual rate of interest is 4%. Find its accumulated value at the end of 7 years.

6.7.6 An amount of Rs 20000 per annum is payable on 1 January for 10 years. The first payment is due in 2027. Find the purchase price of this annuity on 1 January 2023 with 5% annual effective rate of interest.

6.7.7 Calculate the present value at 4% per annum of a continuous annuity for 4 years under which the payment at time t is at the rate of 100 e^t per annum.

6.7.8 An annuity of Rs 10000 per annum is payable for 8 years. Suppose the annual rate of interest is 5%. Find the present value of the annuity at the beginning of the first year if the payments are made (a) yearly in advance, (b) yearly in arrears, (c) monthly in advance, (d) monthly in arrears and (e) continuously. Comment on the results.

6.7.9 Prepare a table showing the values of $1000\, a_{\overline{n}|}$, $1000\, a_{\overline{n}|}^{(m)}$, $1000\, \ddot{a}_{\overline{n}|}$, $1000\, \ddot{a}_{\overline{n}|}^{(m)}$ and $1000\, \overline{a}_{\overline{n}|}$ for $m = 4$, for values of n from 1 to 10 and $i = 0.06$. Prepare the table of corresponding accumulated values. Interpret the results obtained in two tables.

6.7.10 It is given that the force of interest is $\delta = 6\%$ and the actuarial present value at age 32 of a unit benefit to be paid at the moment of death in a 5-year endowment insurance with force of interest δ is 0.80219 and with a force of interest 2δ is 0.6506. Find the actuarial present value at age 32 of a continuous 5-year temporary life annuity payable at the rate of 8000 per annum. Also find the standard deviation of the present value random variable of this annuity.

6.7.11 It is given that the force of interest is $\delta = 6\%$ and the actuarial present value at age 32 of a unit benefit to be paid at the moment of death in a whole life insurance with a force of interest δ is 0.47632 and with a force of interest 2δ is 0.2365. Find the actuarial present value at age 32 of a continuous whole life annuity payable at the rate of 8000 per annum. Also find the standard deviation of the present value random variable of this annuity.

6.7.12 For the set up of Exercises 6.7.10 and 6.7.11, find the actuarial present value at age 32 of a 5-year deferred continuous whole life annuity payable at the rate of 8000 per annum and the actuarial present value at age 32 of a 5-year certain and continuous whole life annuity payable at the rate of 8000 per annum.

6.7.13 Using the assumption of a uniform distribution of deaths in each year of age and the information that, $A_{45} = 0.1875$, $^2A_{45} = 0.09412$, with an interest at the effective annual rate of 6%, calculate (a) $\overline{a}_{45}$ and
(b) $Var(\overline{a}_{\overline{T(45)}|})$.

6.7.14 It is given that, $q_{25} = 0.145, q_{26} = 0.141, q_{27} = 0.156, q_{28} = 0.174, q_{29} = 0.185$, $q_{30} = 0.194$ and $i = 0.06$. Find the actuarial present value of the 6-year temporary annuity due and immediate issued to (25), at the rate of 10000 per annum. Also find the actuarial present value of the 6-year temporary annuity payable continuously at the rate of 10000 per annum, under the assumption of uniformity of deaths in unit age intervals. Compare the three values.

6.7.15 Suppose for a specific mortality law, q_x values are as given in Table 6.15. For the same setup and $\delta = 0.06$, $A_{50} = 0.251$. (a) Find the actuarial present value of 5-year deferred whole life annuity due and immediate issued to (40), when the yearly payment is 1000. (b) Find the actuarial present value of 5-year certain and whole life annuity due and immediate issued to (40), when the yearly payment is 1000. (c) Find the actuarial present value of quarterly whole life annuity due and immediate when the quarterly payment is 250, for ages 40 to 49.

6.7.16 For the setup of Exercise 6.7.15, find the actuarial present value of 6-year temporary annuity due and immediate issued to (25), if the payments are made monthly, at the rate of 10000 per annum. Compare these values with those corresponding to those in Exercise 6.7.15.

6.7.17 It is given that $\mu_{x+t} = \mu$, $t \geq 0$ and $\delta_t = \delta$, $t \geq 0$. Which of the following is a correct expression for $\bar{a}_{x:\overline{1}|}$?

(a) $(1 - e^{-(\mu+\delta)})/(\mu + \delta)^2$. (b) $(\ddot{a}_{x:\overline{1}|} + a_{x:\overline{1}|})/2$. (c) $(\ddot{a}_{x:\overline{1}|} - a_{x:\overline{1}|})/(-\log_e a_{x:\overline{1}|})$.

6.7.18 (a) Show that $a_x = a_{x:\overline{n}|} + {}_nE_x a_{x+n}$ and $\ddot{a}_{x:\overline{n}|} = a_{x:\overline{n}|} - {}_nE_x + 1$.

(b) Given $a_x = 20$, $a_{x:\overline{n}|} = 18$ and $a_{x+n} = 8$, find ${}_nE_x$ and $\ddot{a}_{x:\overline{n}|}$.

6.7.19 Obtain a backward recurrence relation for a_x.

6.7.20 It is given that $\ddot{a}_x = 7$ for all integers x and $i = 0.05$. Calculate ${}_7q_{40}$.

6.7.21 Express in international actuarial notation, the actuarial present values of the following annuities:
(a) An annuity of Rs 5000 per annum payable for 15 years, is purchased on 1 January 2023, in which the first payment is made on 31 December 2023.
(b) An annuity of Rs 10000 per annum payable for 20 years, is purchased on 1 January 2022, in which the first payment is to be made on 1 January 2022.
(c) An annuity of Rs 1000 per month, payable monthly in advance for 10 years.
(d) A life annuity of Rs 20000 per annum payable in arrears for a term of 20 years, is purchased by (50).
(e) A whole life annuity of Rs 5000 per month, payable monthly in advance to (40).
(f) A life annuity issued to (50) pays continuously, Rs 8000 per annum, between ages 60 and 80.
(g) An annuity of 5000 per annum, issued to (55), makes guaranteed payments, at the beginning of the year, for the first 5 years and then till (55) survives.

6.7.22 For an annuity payable semiannually find $\ddot{a}_{74}^{(2)}$, given the following information: (i) Deaths are uniformly distributed over each year of age, (ii) $q_{74} = 0.03$, (iii) $i = 0.06$ and $1000\,\overline{A}_{75} = 530$.

6.7.23 Calculate $\overline{a}_x$ when $\delta = 0.05$, $\mu_{x+t} = 0.03, 0 \le t < 5$ and $\mu_{x+t} = 0.04, t \ge 5$.

6.7.24 For a continuous whole life annuity of 1 issued to (x), the force of interest and force of mortality are constant and equal. If $\overline{a}_x = 12.5$, find $Var(\overline{a}_{\overline{T}|})$.

6.7.25 Prove that $\ddot{a}_{\overline{x:n|}} > \overline{a}_{\overline{x:n|}} > a_{\overline{x:n|}}$

6.7.26 Prove the following: (a) $a_{\overline{x:n|}} = {}_1E_x \ddot{a}_{\overline{x+1:n|}}$ (b) ${}_{n|}a_x = \dfrac{A_{\overline{x:n|}} - A_x}{d} - {}_nE_x$ (c) $A_{\overline{x:n|}} = v\ddot{a}_{\overline{x:n|}} - a_{\overline{x:n-1|}}$.

6.7.27 (a) Prove that $\lim\limits_{m\to\infty} \ddot{a}_x^{(m)} = \overline{a}_x$. (b) Use the result in (a) to show that $\overline{a}_x \cong a_x + 1/2$.

6.8 Computational Exercises

6.8.1 (i) Find $a_{\overline{n|}}$, $a_{\overline{n|}}^{(12)}$, $\overline{a}_{\overline{n|}}$, $\ddot{a}_{\overline{n|}}^{(12)}$, $\ddot{a}_{\overline{n|}}$, ${}_{5|}\ddot{a}_{\overline{n|}}$, ${}_{5|}\overline{a}_{\overline{n|}}$ and ${}_{5|}a_{\overline{n|}}$ for values of $n = 1, 2, \cdots, 10$ and $i = 0.05, 0.06, 0.07, 0.08, 0.09, 0.10$, when the benefit amount is Rs 1000 per annum. (ii) Find the corresponding accumulated values.

Note: For the following exercises, use the mortality pattern and interest pattern that you have adopted in the computational Exercise 5.8.2. State the assumptions, if needed.

6.8.2 For $x = 25, 30, 35, 40$, find $1000\,a_x$, $1000\,a_x^{(m)}$, $1000\,\overline{a}_x$, $1000\,\ddot{a}_x^{(m)}$, $1000\,\ddot{a}_x$ and $1000\,{}_{5|}\ddot{a}_x$ for $m = 2, 4, 12$.

6.8.3 For $x = 25$, find $1000\,a_{x:\overline{n|}}$, $1000\,a_{x:\overline{n|}}^{(m)}$, $1000\,\overline{a}_{x:\overline{n|}}$, $1000\,\ddot{a}_{x:\overline{n|}}^{(m)}$, $1000\,\ddot{a}_{x:\overline{n|}}$, $1000\,a_{\overline{x:n|}}$, $1000\,\ddot{a}_{\overline{x:n|}}$ and $1000\,\overline{a}_{\overline{x:n|}}$ for $m = 2, 4, 12$ and for values of $n = 1, 2, \cdots, 10$.

6.9 Multiple Choice Questions

Note: Unless specified otherwise, you have to identify which of the options is correct. Answers are given in the solutions of conceptual exercises.

6.9.1 The present value of a 10-year annuity immediate of Rs 1000 per annum, with the effective rate of interest 5% is

(a) 7107.82
(b) 7721.74
(c) 7.7212
(d) 8107.82

6.9.2 A loan of Rs 50,000 is taken on 1 January 2023. It has to be repaid by 5 equal instalments payable yearly at the beginning of the year. The amount of the instalment based on an 6% annual rate of interest is

(a) 10000.00
(b) 10235.78
(c) 11197.94
(d) 13423.56

6.9.3 The following are three statements. (I) $a_{\overline{n}|}^{(m)} = (1 - v^n)/i^{(m)}$. (II) $a_{\overline{n}|}^{(m)} = (i/i^{(m)})\, a_{\overline{n}|}$. (III) $a_{\overline{n}|}^{(m)} = (d/d^{(m)})\, a_{\overline{n}|}$.

(a) Only (I) is true.
(b) Only (I) and (II) are true.
(c) Only (I) and (III) are true.
(d) All three are true.

6.9.4 The following are three statements. (I) $\ddot{a}_{\overline{n}|}^{(m)} = (1 - v^n)/d^{(m)}$. (II) $\ddot{a}_{\overline{n}|}^{(m)} = (i/d^{(m)})\, a_{\overline{n}|}$. (III) $\ddot{a}_{\overline{n}|}^{(m)} = (d/d^{(m)})\, \ddot{a}_{\overline{n}|}$.

(a) Only (I) is true.
(b) Only (I) and (II) are true.
(c) Only (I) and (III) are true.
(d) All three are true.

6.9.5 Which of the following is a correct relationship among the present values of annuities?

(a) $a_{\overline{n}|} < a_{\overline{n}|}^{(m)} < \overline{a}_{\overline{n}|} < \ddot{a}_{\overline{n}|}^{(m)} < \ddot{a}_{\overline{n}|}$

(b) $\ddot{a}_{\overline{n}|} < \ddot{a}_{\overline{n}|}^{(m)} < \overline{a}_{\overline{n}|} < a_{\overline{n}|}^{(m)} < a_{\overline{n}|}$

(c) $a_{\overline{n}|} < \overline{a}_{\overline{n}|} < a_{\overline{n}|}^{(m)} < \ddot{a}_{\overline{n}|}^{(m)} < \ddot{a}_{\overline{n}|}$

(d) $\overline{a}_{\overline{n}|} < a_{\overline{n}|}^{(m)} < a_{\overline{n}|} < \ddot{a}_{\overline{n}|}^{(m)} < \ddot{a}_{\overline{n}|}$

6.9.6 Which of the following is a correct relationship ?

(a) $i > i^{(m)} > d^{(m)} > d$
(b) $i < i^{(m)} < d^{(m)} < d$
(c) $i^{(m)} > i > d^{(m)} > d$
(d) $i^{(m)} > i > d > d^{(m)}$

6.9.7 The following are two inequalities. (I) $a_{\overline{n}|}^{(12)} > a_{\overline{n}|}^{(4)} > a_{\overline{n}|}^{(2)}$. (II) $\ddot{a}_{\overline{n}|}^{(12)} < \ddot{a}_{\overline{n}|}^{(4)} < \ddot{a}_{\overline{n}|}^{(2)}$.

(a) (I) is true but (II) is false.
(b) (I) is false but (II) is true.
(c) Both (I) and (II) are true.
(d) Both (I) and (II) are false.

6.9.8 The following are four statements. (I) $\overline{a}_{\overline{n}|} = (i/\delta)\, a_{\overline{n}|}$. (II) $\overline{a}_{\overline{n}|} = (d/\delta)\, \ddot{a}_{\overline{n}|}$. (III) $\overline{a}_{\overline{n}|} = \lim_{m\to\infty} a_{\overline{n}|}^{(m)}$. (IV) $\overline{a}_{\overline{n}|} = \lim_{m\to\infty} \ddot{a}_{\overline{n}|}^{(m)}$.

    (a) Only (I) and (II) are true.
    (b) Only (III) and (IV) are true.
    (c) Only (I) and (III) are true.
    (d) All four are true.

6.9.9 The following are four statements. (I) $\overline{S}_{\overline{n}|} = (1+i)^n\, \overline{a}_{\overline{n}|}$. (II) $\overline{S}_{\overline{n}|} = (i/\delta)S_{\overline{n}|}$. (III) $\overline{S}_{\overline{n}|} = (d/\delta)\ddot{S}_{\overline{n}|}$. (IV) $\overline{S}_{\overline{n}|} = (1-i)^n\, \overline{a}_{\overline{n}|}$.

    (a) Only (I) and (II) are true.
    (b) Only (III) and (IV) are true.
    (c) Only (I), (II) and (III) are true.
    (d) All four are true.

6.9.10 The following are two statements. (I) $a_{\overline{n}|} < \overline{a}_{\overline{n}|} < \ddot{a}_{\overline{n}|}$. (II) $S_{\overline{n}|} > \overline{S}_{\overline{n}|} > \ddot{S}_{\overline{n}|}$.

    (a) (I) is true but (II) is false.
    (b) (I) is false but (II) is true.
    (c) Both (I) and (II) are true.
    (d) Both (I) and (II) are false.

6.9.11 The following are three statements. (I) $a_{\overline{n}|}^{(m)} < \overline{a}_{\overline{n}|}$. (II) $\overline{a}_{\overline{n}|} < \ddot{a}_{\overline{n}|}^{(m)}$. (III) $\ddot{a}_{\overline{n}|}^{(m)} < \overline{a}_{\overline{n}|}$.

    (a) Only (I) and (II) are true.
    (b) Only (I) and (III) are true.
    (c) Only (II) is true.
    (d) Only (I) is true.

6.9.12 The following are three statements. (I) $i > i^{(m)} > \delta$. (II) $\delta > d^{(m)} > d$. (III) $i^{(m)} > \delta > d^{(m)}$.

    (a) Only (I) and (II) are true.
    (b) Only (I) and (III) are true.
    (c) Only (II) and (III) are true.
    (d) All three are true.

6.9.13 The following are two statements. (I) $S_{\overline{n}|} < S_{\overline{n}|}^{(m)} < \overline{S}_{\overline{n}|}$. (II) $\overline{S}_{\overline{n}|} < \ddot{S}_{\overline{n}|}^{(m)} < \ddot{S}_{\overline{n}|}$.

    (a) (I) is true but (II) is false.
    (b) (I) is false but (II) is true.
    (c) Both (I) and (II) are true.
    (d) Both (I) and (II) are false.

6.9.14 The following are three statements. (I) $_{m|}S_{\overline{n}|} = S_{\overline{n}|}$. (II) $_{m|}\overline{S}_{\overline{n}|} = \overline{S}_{\overline{n}|}$. (III) $_{m|}\ddot{S}_{\overline{n}|} = \ddot{S}_{\overline{n}|}$.

    (a) All three are false.
    (b) All three are true.

(c) Only (II) and (III) are true.
(d) Only (I) and (II) are true.

6.9.15 The present value random variable Y for an n-year temporary life annuity of 1 per year, payable continuously while (x) survives during the next n years, is defined by

(a)
$$Y = \begin{cases} (1 - v^T)/\delta & \text{if} & 0 \leq T < n \\ (1 - v^n)/\delta & \text{if} & T \geq n \end{cases}$$

(b)
$$Y = \begin{cases} v^T/\delta & \text{if} & 0 \leq T < n \\ v^n/\delta & \text{if} & T \geq n \end{cases}$$

(c)
$$Y = \begin{cases} (1 - v^T)/d & \text{if} & 0 \leq T < n \\ (1 - v^n)/d & \text{if} & T \geq n \end{cases}$$

(d)
$$Y = \begin{cases} (1 - v^T)/i & \text{if} & 0 \leq T < n \\ (1 - v^n)/i & \text{if} & T \geq n \end{cases}$$

6.9.16 The present value random variable Y for an n-year temporary life annuity of 1 per year, payable continuously while (x) survives during the next n years, is defined by

(a)
$$Y = \begin{cases} \bar{a}_{\overline{T}|} & \text{if} & 0 \leq T < n \\ 0 & \text{if} & T \geq n \end{cases}$$

(b)
$$Y = \begin{cases} \bar{a}_{\overline{T}|} & \text{if} & 0 \leq T < n \\ \bar{a}_{\overline{n}|} & \text{if} & T \geq n \end{cases}$$

(c)
$$Y = \begin{cases} \bar{a}_{\overline{n}|} & \text{if} & 0 \leq T < n \\ \bar{a}_{\overline{T}|} & \text{if} & T \geq n \end{cases}$$

(d) $Y = \bar{a}_{\overline{T}|}$,   if $T \geq 0$

6.9.17 The actuarial present value $\bar{a}_{\overline{x:n}|}$ in n-year certain and whole life annuity is given by
(I) $\bar{a}_{\overline{n}|} + {}_{n|}\bar{a}_x$, (II) $\bar{a}_{\overline{n}|} + {}_nE_x\,\bar{a}_{x+n}$ and (III) $\bar{a}_{\overline{n}|} + (\bar{a}_x - \bar{a}_{\overline{x:n}|})$.
(a) All three are false.
(b) All three are true.
(c) Only (II) and (III) are true.
(d) Only (I) and (II) are true.

6.9.18 The following are three identities. (I) $\ddot{a}_{x:\overline{n}|} = (1 - A_{x:\overline{n}|})/d$. (II) $\ddot{a}_{x:\overline{n}|} = \sum_{k=0}^{n-1} v^k\,{}_kp_x$.
(III) $\ddot{a}_{x:\overline{n}|} = 1 + a_{x:\overline{n}|} - {}_nE_x$.

(a) Only (II) is true.
(b) Only (I) and (II) are true.

(c) Only (II) and (III) are true.
(d) All three are true.

6.9.19 The following are four statements. (I) $a_{x:\overline{n}|} < \overline{a}_{x:\overline{n}|} < \ddot{a}_{x:\overline{n}|}$.
(II) $a_{\overline{n}|} < \overline{a}_{\overline{n}|} < \ddot{a}_{\overline{n}|}$. (III) $a_{x:\overline{n}|} > \overline{a}_{x:\overline{n}|} > \ddot{a}_{x:\overline{n}|}$. (IV) $a_{\overline{n}|} > \overline{a}_{\overline{n}|} > \ddot{a}_{\overline{n}|}$.

(a) Only (II) is true.
(b) Only (I) and (II) are true.
(c) Only (III) are true.
(d) Only (III) and (IV) are true.

6.9.20 The following are two statements. (I) $a_{x:\overline{n}|} < \overline{a}_{x:\overline{n}|}$. (II) $\overline{A}_{x:\overline{n}|} > A_{x:\overline{n}|}$.

(a) (I) is true but (II) is false.
(b) (I) is false but (II) is true.
(c) Both (I) and (II) are true.
(d) Both (I) and (II) are false.

6.9.21 The following are three statements. (I) $\ddot{a}_x = (1 - A_x)/d$. (II) $\ddot{a}_x = \lim_{n\to\infty} \ddot{a}_{x:\overline{n}|}$.
(III) $\ddot{a}_x = \lim_{n\to\infty} a_{x:\overline{n}|}$.

(a) Only (II) is true.
(b) Only (I) and (II) are true.
(c) Only (III) is true.
(d) Only (I) and (III) are true.

6.9.22 Which of the following identities is correct?

(a) $1 = \delta\,\overline{a}_x + \overline{A}_x$
(b) $1 = \delta\,\overline{A}_x + \overline{a}_x$
(c) $1 = \delta + \overline{A}_x$
(d) $1 = \overline{a}_x + \overline{A}_x$

6.9.23 Which of the following identities is correct?

(a) $\ddot{a}_{x+1} = 1 + vp_x\ddot{a}_x$
(b) $\ddot{a}_x = 1 + vq_x\ddot{a}_{x+1}$
(c) $\ddot{a}_x = 1 + vp_x\ddot{a}_{x+1}$
(d) $\ddot{a}_x = v + p_x\ddot{a}_{x+1}$

6.9.24 The following are three statements. (I) $a_x = v\,\ddot{a}_x - A_x$. (II) $a_x = \lim_{n\to\infty} a_{x:\overline{n}|}$.
(III) $a_x = \ddot{a}_x - 1$.

(a) Only (I) is true.
(b) Only (I) and (II) are true.
(c) Only (III) is true.
(d) All three are true.

6.9.25 The following are three statements. (I) $\ddot{a}_x > \overline{a}_x$. (II) $\overline{a}_x > a_x$. (III) $a_x < \ddot{a}_x$.

(a) Only (I) is true.
(b) Only (I) and (II) are true.
(c) Only (II) and (III) is true.

(d) All three are true.

6.9.26 The following are three statements. $(I)\ddot{a}_{\overline{x:n+1|}} = 1 + a_{\overline{x:n|}}$. (II) $\ddot{a}_{\overline{x:1|}} = \ddot{a}_x$.
    (III) $\ddot{a}_{\overline{x:n|}} > \overline{a}_{\overline{x:n|}} > a_{\overline{x:n|}}$ $\forall$ n.

(a) Only (I) is true.
(b) Only (I) and (II) are true.
(c) Only (II) and (III) is true.
(d) All three are true.

6.9.27 The following are three statements. $(I)\ddot{a}_{\overline{x:n+1|}} = \ddot{a}_{\overline{n+1|}} + {}_{n+1|}\ddot{a}_x$.
    (II) $\ddot{a}_{\overline{x:n+1|}} = 1 + a_{\overline{n|}} + {}_{n|}a_x$. (III) $\ddot{a}_{\overline{x:n+1|}} = 1 + a_{\overline{x:n|}}$.

(a) Only (I) is true.
(b) Only (I) and (II) are true.
(c) Only (II) and (III) is true.
(d) All three are true.

6.9.28 If it is given that ${}_{10}E_x = 0.5$, ${}_{10|}\ddot{a}_x = 8$ and $\ddot{S}_{x:\overline{10|}} = 14$, what is the value of $\ddot{a}_x$?

(a) 22
(b) 15
(c) 18
(d) cannot be calculated in view of insufficient information

6.9.29 The actuarial present value at age 27 of a unit benefit to be paid at the moment of death in a whole life insurance with a force of interest $\delta = 5\%$, is 0.5 and with a force of interest 2δ is 0.3. The variance of the present value random variable corresponding to the continuous life annuity payable at the rate of 10 per annum,

(a) is 100
(b) is 20
(c) is 2000
(d) cannot be calculated in view of insufficient information

6.9.30 In international actuarial notation, the actuarial present values of a life annuity of Rs 20000 per annum payable in arrears for a term of 20 years, purchased by (50) is expressed as

(a) $20000\ \ddot{a}_{50:\overline{20|}}$
(b) $20000\ \overline{a}_{50:\overline{20|}}$
(c) $a_{50:\overline{20|}}$
(d) $20000\ a_{50:\overline{20|}}$

6.9.31 Which of the following identities is correct?

(a) $\ddot{a}_x^{(m)} = \beta(m)\ddot{a}_x - \alpha(m)$
(b) $\ddot{a}_x^{(m)} = (1/m)\ \ddot{a}_x$
(c) $\ddot{a}_x^{(m)} = \alpha(m)\ddot{a}_x - \beta(m)$
(d) $\ddot{a}_x^{(m)} = \alpha(m)\ddot{a}_x + \beta(m)$

6.9.32 The following are two statements. (I) $\ddot{a}_x^{(m)} = \alpha(m)\ddot{a}_x - \beta(m)$
(II) $(d/d^{(m)})\ddot{a}_x - \beta(m)A_x$.

    (a) (I) is true but (II) is false.
    (b) (I) is false but (II) is true.
    (c) Both (I) and (II) are true.
    (d) Both (I) and (II) are false.

6.9.33 The following are two statements. (I) $\overline{a}_{\overline{n}|} < \ddot{a}_{\overline{n}|}^{(m)} < \ddot{a}_{\overline{n}|}$.

(II) $\overline{a}_x < \ddot{a}_x^{(m)} < \ddot{a}_x$, under the assumption of uniformity in unit age interval.

    (a) (I) is true but (II) is false.
    (b) (I) is false but (II) is true.
    (c) Both (I) and (II) are false.
    (d) Both (I) and (II) are true.

6.9.34 The following are three statements. (I) $a_x^{(m)} = \ddot{a}_x^{(m)} - 1/m$
(II) $a_x^{(m)} = \alpha(m)a_x + \{\alpha(m) - \beta(m)\}$.
(III) $a_x^{(m)} = \alpha(m)a_x + \{\alpha(m) - \beta(m) - 1/m\}$.

    (a) Only (I) is true.
    (b) Only (II) and (III) are true.
    (c) Only (I) and (III) are true.
    (d) All three are true

6.9.35 The following are two statements. (I) $a_x < a_x^{(m)} < \overline{a}_x < \ddot{a}_x^{(m)} < \ddot{a}_x$, under the assumption of uniformity in each unit age interval.
(II) $a_{\overline{n}|} < a_{\overline{n}|}^{(m)} < \overline{a}_{\overline{n}|} < \ddot{a}_{\overline{n}|}^{(m)} < \ddot{a}_{\overline{n}|}$.

    (a) (I) is true but (II) is false.
    (b) (I) is false but (II) is true.
    (c) Both (I) and (II) are false.
    (d) Both (I) and (II) are true.

6.9.36 The following are two statements. (I) $i^{(m)}$ is an increasing function of m. (II) $d^{(m)}$ is a decreasing function of m.

    (a) (I) is true but (II) is false.
    (b) (I) is false but (II) is true.
    (c) Both (I) and (II) are false.
    (d) Both (I) and (II) are true.

6.9.37 The following are two statements. Under the assumption of uniformity in each unit age interval, (I) $\ddot{a}_x^{(m)}$ is a decreasing function of m, (II) $a_x^{(m)}$ is an increasing function of m.

    (a) (I) is true but (II) is false.
    (b) (I) is false but (II) is true.
    (c) Both (I) and (II) are false.
    (d) Both (I) and (II) are true.

6.9.38 The following are two statements. (I) $\bar{a}_{x:\overline{n}|} \;<\; \ddot{a}^{(m)}_{x:\overline{n}|} \;<\; \ddot{a}_{x:\overline{n}|}$. (II) $\ddot{a}^{(m)}_{x:\overline{n}|}$ is a decreasing function of m, under the assumption of uniformity in each unit age interval.

(a) (I) is true but (II) is false.
(b) (I) is false but (II) is true.
(c) Both (I) and (II) are false.
(d) Both (I) and (II) are true.

Chapter 7

Premiums

Key Terms: Benefit premiums, Contract premiums, Expense loaded premiums, Equivalence principle premiums, Fully continuous premiums, Fully discrete premiums, Loss at issue random variable, True fractional premiums.

7.1 Introduction

The theory developed in all the previous chapters is mainly motivated by the computation of premiums for a variety of insurance products. In this chapter we proceed with this main aim of determination of premiums corresponding to various modes of premium payments and for various types of insurance products. The premiums charged by an insurance company for an insurance policy serve two purposes. Firstly, these generate a fund needed to meet the benefit paying liabilities and the expenses of running the business. Secondly, premium calculations also have to take into account the profit margin for the company, as after all, it is a business.

An important issue in the determination of the premiums is that these are to be fixed at the time of signing the contract. As a consequence, the premium calculations depend heavily on the actuarial present values of the benefit to be paid, whenever claim arises and the actuarial present values of the premium payments for various modes of payment of the premiums. Chapters 5 and 6 discuss the actuarial present values of the benefit payments of some life insurance products and of annuities respectively. These two aspects are combined in this chapter to determine the premiums corresponding to different insurance products.

There are a number of approaches to determine the premium rates, which may result in different numerical values for the premium. However, all the principles are based on the impact of the insurance on the wealth of the insurance company. One approach requires that the chance of loss to the insurance company should be as small as possible while the second approach requires that the expected loss to the company should be 0. One more approach brings into the picture, the utility function, as defined in Chapter 3, of the wealth of the insurer. In this book we use the second approach, known as the equivalence principle, and touch upon the other two by giving one illustration of each. In any of these approaches, we need to define a random variable that gives the present value at issue of the insurer's loss, when the insurance contract specifies the benefit function, the time of its payment, premium level and its mode of payment. Hence we begin the next section by defining a loss at issue random variable.

Insurance policies are mainly classified as fully continuous and fully discrete policies, depending on the time of payment of benefit and the mode of premium payment. If the death benefit is payable at the moment of death and the premiums are payable as a continuous annuity, the policy is referred to as a fully continuous policy. If the death benefit is payable at the end of the year of death and the premiums are payable as a discrete life annuity due, the policy is referred to as a fully discrete policy. In a fully discrete policy, the premium payments may be made annually, semi-annually, quarterly or monthly. If the death benefit is payable at the moment of death and the premiums are payable as m-thly annuity due, $m = 1, 2, 4, 12$, then the policy is referred to as a semi-continuous policy. Section 3 discusses the determination of premiums for a fully continuous policy while Section 4 is devoted to the determination of premiums in a fully discrete policy and semi-continuous policy. Section 5 specifies the modifications needed in these two types, when the premium payments form m-thly annuity.

The premiums studied in Sections 3 to 5 are determined only by the pattern of benefits and the mode of premium payments and do not consider expenses, profit or contingency margins. Section 6 illustrates how the expenses are incorporated into the model for premiums. Premiums calculated without an allowance for expenses are called net premiums, while premiums which take into account expenses are called gross premiums or expense loaded premiums. Premiums which take into consideration expenses and profit and which are actually charged, are known as office or contract premiums. It is important to note that the net premiums are uniquely determined by the mortality pattern of the group under study and the interest patterns dictated by the financial market conditions. Both the factors are beyond the control of the insurance company.

7.2   Loss at Issue Random Variable

In Chapters 5 and 6 we have noted that to decide the actuarial present value of benefit and annuity, the first step was to define the present value random variable. On similar lines, in this chapter we begin with the definition of a present value of loss to the insurance company related to a specific portfolio of policies, as loss to the company is the main governing principle in the determination of the premium. An insurance policy usually involves a contract in which a life annuity of premiums is used to pay the policy's benefits. If death occurs t units after the policy issue, then the loss-at-issue function $l(t)$ is defined as,

$$l(t) = \text{ present value of benefits paid} - \text{present value of premiums received},$$

where the present values are calculated as per the interest rate at the policy issue date. For all insurance policies $l(t)$ is a decreasing function of t, since the later the death occurs, the smaller will be the present value of the death benefit and the more premiums will have been received. In practice, the time from signing the contract to death is a future life time random variable T. Hence, in a fully continuous policy the loss at issue random variable $L = L(T)$ for a unit benefit is defined as

$$L(T) = Z_T - PY$$

where the policy has level premiums at annual rate P, Z_T as defined in Chapter 5 represents the present value random variable for 1 unit death benefit to be paid at the moment of death and Y represents the present value random variable of the continuous annuity of premiums.

In the case of a fully discrete policy, wherein the death benefit is payable at the end of the year of death and the premiums are payable as a discrete life annuity due, the loss at issue random variable for a unit benefit is of the form

$$L(K) = Z_K - PY$$

where K is a curtate future life time random variable, the policy has level premiums at an annual rate P, Z_K as defined in Chapter 5 represents the present value random variable for 1 unit death benefit to be paid at the end of the year of death and Y represents the present value random variable of the discrete annuity of premiums. For a policy with discrete premiums, unless specified otherwise, it is assumed that premiums are payable as annuity due.

In a semi-continuous policy loss at issue, the random variable is defined similarly, where Z is a function of T while Y is a function of K, which in turn is a function of T.

The following two examples illustrate the concept of a loss function.

Example 7.2.1. On 6 May 2006, (65) bought a Rs 1 lakh whole life insurance policy, with the death benefit payable at the end of the year of death. The policy is purchased by means of annual premiums, payable at the start of each year the policy remains in force. The policy holder died on 16 August 2013 and the loss to the insurer was Rs 30000. If $i = 0.06$, what was the annual premium paid?

Solution: The benefit is payable on 5 May 2014, which is the end of the policy year in which death occurs. This period is 8 years after the policy is issued. Suppose the annual premium is P, which is paid for 8 years at the beginning of the year. Thus, premium payments form 8-year annuity certain due. Hence the loss to the insurer corresponding to benefit of 1 lakh is given by $\;l(8) \;=\; 10^5 v^8 \;-\; P\ddot{a}_{\overline{8|}} \;=\; 62741 - 6.58P$. Equating it to 30000, we have $P = (62741 - 30000)/6.58 = 5000$. ∎

Example 7.2.2. A fully continuous 10-year term insurance of face amount Rs $10,000$ has an annual premium rate of Rs 100 and the force of interest is 0.05. Find the value of the loss at issue-random variable (i) if death occurs exactly 5 years after the issue and (ii) if death occurs exactly 15 years after the issue.

Solution: (i) The annual premium of 100 corresponds to the benefit of 10000, hence for 1 unit benefit it will be 0.01. For 1 unit benefit, the value of loss at issue random variable is given by,

$$l(5) \;=\; e^{-5(0.05)} \;-\; (0.01)\overline{a}_{\overline{5|}} \;=\; 0.735$$

Hence, for the benefit of 10000, it is Rs 7350. In the second case, the policy ends after 10 years and death occurs after 15 years, hence there is no death benefit or the corresponding Z has value 0, with probability 1. Hence for the 1 unit benefit, value of loss at issue random variable is given by,

$$l(15) \;=\; 0 - (0.01)\overline{a}_{\overline{10|}} \;=\; -0.079$$

For the benefit of 10000, the value is -790 rupees. The negative loss is of course the gain to the insurer. $\blacksquare$

The following three principles are generally used for premium calculation.

(i) **Percentile Premium:** In this approach, the value of the annual premium rate P is found so that probability statement $P[L > 0] = c$ is satisfied for some specified c. For instance, if $c = 0.05$, then P is found so that there is only a 5% probability that the loss is positive. The premium calculated with this approach is referred to as the $100\,c$-th percentile the premium.

(ii) **Exponential Premium:** Using an exponential utility function, P is found so that the expected utility of wealth with insurance is equal to the utility of wealth without insurance, that is, by solving the equation $U(w) = E(U(w - L))$, where $U(w)$ is the exponential utility function, as defined in Section 3.3.

(iii) **Equivalence Principle Premium:** In equivalence principle premium, P is found so that the expected loss $E(L)$ to the company is 0, that is by solving the equation $E(Z) = P\,E(Y)$. It leads to the following principle to determine the premium.

$$\text{Actuarial present value of benefit} \quad = \quad \text{Actuarial present value of premiums}$$

This result also follows from the equation $U(w) = E(U(w - L))$ with $U(w) = w$. Insurance companies usually have a large number of contracts. Hence, by the law of large numbers, the expected loss describes the average actual loss quite accurately. In this approach, it is not necessary to derive the distribution of L. It is enough to have knowledge of $E(Z)$ and $E(Y)$.

Premiums found on the basis of the third principle are called benefit premiums or net premiums. As a special case, if the policy is purchased by a single premium on the issue date, the annuity random variable is the constant 1, since only one premium is to be paid. From the equivalence principle we get,

$$E(Z) = PE(Y) = P$$

For this reason $E(Z)$ is called the net single premium in Chapter 5. It is also known as the single benefit premium. Benefit premiums are thus obtained by solving an expression of the form $A = P\ddot{a}$, in the fully discrete case, or $\overline{A} = P\overline{a}$ in the fully continuous case, or $\overline{A} = P\ddot{a}$ in the semi-continuous case, where post or pre subscripts to A and a are as per the policy under consideration. Premiums are usually calculated using this approach.

The following two examples illustrate the application of the first two approaches. In rest of the chapter, we decide premiums using the equivalence principle.

Example 7.2.3. For a fully continuous whole life insurance of 1000 issued to (30), it is given that (i) $\delta = 0.05$ and (ii) mortality follows Gompertz' law with $B = 0.0012$ and $C = 1.1$. Calculate the 5-th percentile premium P for this insurance.

Solution: We wish to find the premium such that

$$P[L(T) > 0] = 0.05 \quad \text{where} \quad L(T) \;=\; e^{-\delta T} - P\overline{a}_{\overline{T}|}$$

To find the premium for which $P[L(T) > 0] = 0.05$, we need the knowledge of the distribution of L. In general, it is very difficult to find the distribution of L. We can avoid this derivation by noting the fact that $L(T)$ is a decreasing function of T. Hence, $P[L(T) > 0] = P[T < t_0]$, for some t_0. Hence, we find a point t_0 such that $_{t_0}q_{30} = P[T(30) \leq t_0] = 0.05$. With Gompertz' law,

$$_{t_0}q_{30} = 1 - {}_{t_0}p_{30} = 1 - \exp\{-mC^{30}(C^{t_0} - 1)\} = 0.05$$

where $m = B/log_e C$. With the given values of B and C, we get $t_0 = 44.88392$. Since the loss-at-issue decreases as time of death increases, we choose P so that $L = 0$ if death is at age 75, or in other words, if $l(45) = 0$. Thus,

$$l(45) = 1000e^{-.05(45)} - P\,\overline{a}_{\overline{45}|} = 0 \Rightarrow P = 1000e^{-.05(45)}/\overline{a}_{\overline{45}|} = 5.89$$

It is the 5-th percentile premium paid continuously at a rate of 5.89 per annum till individual survives. ∎

Example 7.2.4. A general insurance company sells a 1-year insurance of benefit of 1000, to be paid at the end of the year, to cover damage to the car due to accident. Suppose the chance of accident in a year is 0.01. Determine the premium P, to be paid at the issue of the policy, so that the insurer is indifferent to two options: accepting or not accepting the risk under a utility of wealth function, $u(w) = -\exp\{-0.007w\}$.

Solution: With the expected utility principle to find P, the utility of wealth of the insurer assuming that the insurance is not issued and the expected utility of wealth of the insurer assuming that the insurance is issued, are equal. The premium is found from such an indifference equation. The insurer's utility of wealth assuming that the insurance is not issued is $u(w) = -\exp\{-0.007w\}$. If the insurance is issued and if the accident does not occur, the insurer's wealth will be $w + P$, with utility $-\exp\{-0.007(w + P)\}$. If the accident occurs, the insurer's wealth will be $w + P - 1000$, with utility $-\exp\{-0.007(w + P - 1000)\}$. We are given that the probability of an accident is 0.01, so that if insurance is issued, the insurer's expected utility of wealth is given by $0.99[-\exp\{-0.007(w + P)\}] - 0.01[\exp\{-0.007(w + P - 1000)\}]$. The equivalence equation is therefore given by,

$$e^{-0.007w} = 0.99[e^{-0.007(w+P)}] + 0.01[e^{-0.007(w+P-1000)}]$$
$$\Rightarrow P = 1000 \, log_e(0.99 + 0.01e^7)/7 = 354.47$$

Thus, 354.47 is the annual premium for a benefit of Rs 1000. ∎

In the following section we discuss the computation of fully continuous benefit premiums.

7.3   Fully Continuous Premiums

As outlined in the previous section, fully continuous premiums using equivalence principle are obtained by solving the equation,

$$E(Z_T) = PE(Y)$$

Expressions for $E(Z_T)$ and $E(Y)$ will change as insurance product changes and mode of premium payments changes, respectively. We begin with the first insurance product discussed in Section 5.3 and it is an n-year term insurance with level, one unit benefit payable immediately on the death of (x). Suppose the premiums are paid continuously till the claim is made in n years after the issue of the policy or for n years if claim is not made in n years. Thus, the premium payments form a n-year temporary continuous life annuity. The corresponding loss at issue random variable is defined as,

$$L(T) = \begin{cases} v^T - P\overline{a}_{\overline{T}|} & \text{if} \quad T \le n \\ 0 - P\,\overline{a}_{\overline{n}|} & \text{if} \quad T > n \end{cases}$$

Recall that in term insurance, if the claim is not made in the term of n years, the insurer has no liability to pay the benefit. But the insured has paid the premiums as a n-year temporary continuous life annuity. Thus, when $T > n$ the loss random variable takes value $-P\,\overline{a}_{\overline{n}|}$, which is negative, resulting in gain for the insurer. In this setup,

$$E(Z_T) = \overline{A}\,{}^{1}_{x:\overline{n}|} \quad \& \quad E(Y) = \overline{a}_{x:\overline{n}|} \quad \text{hence,} \quad E(L(T)) = 0 \;\Rightarrow\; P = \overline{A}\,{}^{1}_{x:\overline{n}|}/\overline{a}_{x:\overline{n}|}$$

In international actuarial notation, benefit premium for 1 unit benefit, in this case is denoted by $\overline{P}(\overline{A}\,{}^{1}_{x:\overline{n}|})$. Thus,

$$\overline{P}(\overline{A}\,{}^{1}_{x:\overline{n}|}) = \overline{A}\,{}^{1}_{x:\overline{n}|}/\overline{a}_{x:\overline{n}|} \tag{7.3.1}$$

For whole life insurance with unit benefit, the loss random variable is given by,

$$L(T) = v^T - P\,\overline{a}_{\overline{T}|}$$

The benefit premium corresponding to unit benefit is denoted by $\overline{P}(\overline{A}_x)$. Hence,

$$E(L) = 0 \;\Rightarrow\; \overline{P}(\overline{A}_x) = \overline{A}_x/\overline{a}_x \tag{7.3.2}$$

It is to be noted that $\overline{P}(\overline{A}_x)$ is the limit of $\overline{P}(\overline{A}\,{}^{1}_{x:\overline{n}|})$ as $n \to \infty$. The symbols $\overline{P}(\overline{A}\,{}^{1}_{x:\overline{n}|})$ and $\overline{P}(\overline{A}_x)$ refer to the premium corresponding to unit benefit. The premium for benefit b is simply obtained by multiplying premium for a unit benefit by b.

L is a random variable due to the random nature of time-until-death. To measure the variability of losses on an individual whole life insurance, the variance of L is used as a measure of the variability. When $E(L) = 0$, $Var(L) = E(L^2)$. However, instead of obtaining the second moment of L, it is simple to find the variance using the following approach. Note that for the whole life insurance, with $\overline{P} \equiv \overline{P}(\overline{A}_x)$,

$$\begin{aligned} Var(v^T - \overline{P}\,\overline{a}_{\overline{T}|}) &= Var\left[v^T - \overline{P}(1 - v^T)/\delta\right] = Var\left[v^T\left(1 + \overline{P}/\delta\right) - \overline{P}/\delta\right] \\ &= Var\left[v^T\left(1 + \overline{P}/\delta\right)\right] = Var(v^T)\left(1 + \overline{P}/\delta\right)^2 \\ &= \left[{}^{2}\overline{A}_x - (\overline{A}_x)^2\right]\left(1 + \overline{P}/\delta\right)^2 \end{aligned} \tag{7.3.3}$$

For the premium determined by the equivalence principle, the identity,

$$\begin{aligned} \delta\overline{a}_x + \overline{A}_x &= 1 \;\Rightarrow\; 1 + \overline{P}/\delta = 1 + \overline{A}_x/\overline{a}_x\delta = (\overline{a}_x\delta + \overline{A}_x)/\overline{a}_x\delta = 1/\overline{a}_x\delta \\ \Rightarrow\; Var(L) &= ({}^{2}\overline{A}_x - (\overline{A}_x)^2)/(\delta\overline{a}_x)^2 \end{aligned} \tag{7.3.4}$$

From Eq. 7.3.3, we note that

$$Var(v^T - \overline{P}\overline{a}_{\overline{T}|}) \; > \; {}^2\overline{A}_x - (\overline{A}_x)^2$$

Thus, variance of the loss associated with a single premium whole life insurance is less than the variance of the loss associated with an annual premium whole life insurance. The result is reasonable as in single premium whole life insurance, the entire premium is already paid at the issue of the policy.

Example 7.3.1. For a whole life insurance with unit benefit, calculate $\overline{P}(\overline{A}_x)$ when the force of mortality is constant μ and the force of interest is δ.

Solution: Under the constant force of mortality assumption, in Chapter 5, we have shown that $\overline{A}_x = \mu/(\mu + \delta)$. Further $\delta\overline{a}_x + \overline{A}_x = 1$ implies $\overline{a}_x = 1/(\mu + \delta)$. Hence, $\overline{P}(\overline{A}_x) = \overline{A}_x/\overline{a}_x = \mu$, which does not depend on the force of interest or the age at issue. It is the same as the force of mortality. It again shows that constant force of mortality is not at all a realistic mortality pattern for human life. $\blacksquare$

Formulae for annual premiums of other fully continuous life insurances can be derived analogously. $\overline{P}$ is a general symbol for a fully continuous net annual premium, followed by a symbol of actuarial present value of a unit benefit in brackets. Table 7.1 summarizes the international actuarial notation and the formula for fully continuous net annual premium for a unit benefit, for a variety of insurance products studied in Section 5.3.

In some cases using the relation between actuarial present value of a unit benefit and the corresponding annuity, premium formulae can be expressed either in terms of the annuity symbol a or the net single premium symbol A. In Chapter 6, we have derived the following two identities.

$$\delta\overline{a}_x + \overline{A}_x = 1 \quad \text{and} \quad \delta\overline{a}_{x:\overline{n}|} + \overline{A}_{x:\overline{n}|} = 1$$

These are used to express the formula for premium either in actuarial present value symbol of insurance or of annuity, for whole life insurance and for n-year endowment insurance. From the formulae of premiums for n-year term insurance, n-year pure endowment and n-year endowment, it is clear that

$$\overline{P}(\overline{A}_{x:\overline{n}|}) \; = \; \overline{P}(\overline{A}\,{}^{1}_{x:\overline{n}|}) \; + \; \overline{P}(A_{x:\overline{n}|}^{\;\;1})$$

In the fourth insurance product in Table 7.1, whole life insurance is purchased by whole life annuity while in the fifth product, it is purchased by h-year temporary continuous life annuity. In the sixth product also n-year endowment is purchased by h-year temporary continuous life annuity, $h \, < \, n$. The last row gives the annual premium for deferred whole life annuity, paid for at most n years. It can be interpreted as an annual contribution to the pension fund if the contribution is done for at most n years. For example, an individual is employed at age 22 and retires from service at age 60. Suppose he is entitled for annual pension benefit after age 60 and for that he has to contribute to the pension fund for the period of 38 years, from age 22 to 60, provided the individual survives this period. Then the pension scheme can be described as 38-year deferred whole life annuity. If death occurs before 60, the contribution to the pension fund stops and of course there is no pension benefit.

Insurance product	Notation and formula
n-Year term insurance	$\overline{P}(\overline{A}\,{}^{1}_{x:\overline{n}|}) = \overline{A}\,{}^{1}_{x:\overline{n}|}/\overline{a}_{x:\overline{n}|}$
n-Year pure endowment	$\overline{P}(A_{x:\overline{n}|}^{\;\;1}) = A_{x:\overline{n}|}^{\;\;1}/\overline{a}_{x:\overline{n}|}$
n-Year endowment insurance	$\overline{P}(\overline{A}_{x:\overline{n}|}) = \dfrac{\overline{A}_{x:\overline{n}|}}{\overline{a}_{x:\overline{n}|}} = \dfrac{\delta\overline{A}_{x:\overline{n}|}}{1-\overline{A}_{x:\overline{n}|}} = \dfrac{1-\delta\overline{a}_{x:\overline{n}|}}{\overline{a}_{x:\overline{n}|}}$
Whole life insurance	$\overline{P}(\overline{A}_x) = \dfrac{\overline{A}_x}{\overline{a}_x} = \dfrac{\delta\overline{A}_x}{1-\overline{A}_x} = \dfrac{1-\delta\overline{a}_x}{\overline{a}_x}$
h-Payment, whole life insurance	$_h\overline{P}(\overline{A}_x) = \overline{A}_x/\overline{a}_{x:\overline{h}|}$
h-Payment, n-year endowment insurance $h < n$	$_h\overline{P}(\overline{A}_{x:\overline{n}|}) = \overline{A}_{x:\overline{n}|}/\overline{a}_{x:\overline{h}|}$
n-Year Deferred whole life annuity	$\overline{P}(\,_{n|}\overline{a}_x) = A_{x:\overline{n}|}^{\;\;1}\overline{a}_{x+n}/\overline{a}_{x:\overline{n}|}$

Table 7.1 Fully continuous premiums: actuarial notation and formulae

Example 7.3.2. Suppose the effective rate of interest is 0. Show that $\overline{P}(\overline{A}_x)$ is the reciprocal of complete expectation, that is, expectation of $T(x)$.

Solution: Observe that

$$i \;=\; 0 \;\Rightarrow\; v = 1 \;\Rightarrow\; \overline{A}_x \;=\; E(v^{T(x)}) \;=\; 1$$

$$\&\;\; \overline{a}_x \;=\; \int_0^\infty v^t \,_tp_x dt \;=\; \int_0^\infty \,_tp_x dt = E(T(x)) = e_x^0$$

$$\Rightarrow\;\; \overline{P}(\overline{A}_x) \;=\; \overline{A}_x/\overline{a}_x = 1/e_x^0$$

This relation is logically appealing since as age increases, the expected remaining life, that is, complete expectation decreases and the premium increases. ∎

Example 7.3.3. The actuarial present value at age 27 of a unit benefit to be paid at the moment of death in a 5-year endowment insurance with force of interest $\delta = 0.06$ is 0.7395 and with force of interest 2δ is 0.6066. The actuarial present value at age 27 of a unit benefit to be paid at the moment of death in a whole life insurance with force of interest δ is 0.5321. (i) Write the expression for loss at issue random variable to find the premium per annum, payable as a continuous 5-year temporary life annuity, when the benefit amount is Rs 1000, payable at the moment of death, in a 5-year endowment insurance. Find the variance of the corresponding loss at issue random variable. (ii) Find the premium per annum, payable as a continuous whole life annuity when the benefit amount is Rs 1000, payable at the moment of death in a whole life insurance. (iii) Find the annual premium, payable continuously for at most 5 years, for 5-year deferred continuous whole life annuity payable to (27) at the rate of Rs 7000 per annum. (iv) Find the premium per annum, payable as a 5-year continuous annuity certain, for 5-year certain and continuous whole life annuity payable at the rate of Rs 7000 per annum, issued to (27). Compare the premiums in (iii) and (iv).

Solution: (i) The expression for loss at issue random variable when the premiums are payable as a continuous 5-year temporary life annuity and when the benefit amount is 1 unit, payable at the moment of death, in a 5-year endowment insurance is as follows.

$$L(T) = \begin{cases} v^T - P\bar{a}_{\overline{T}|} & \text{if} \quad T \leq 5 \\ v^5 - P\bar{a}_{\overline{5}|} & \text{if} \quad T > 5 \end{cases}$$

From the given information

$$\bar{A}_{27:\overline{5}|} = 0.7395, \quad {}^2\bar{A}_{27:\overline{5}|} = 0.6066 \quad \& \quad \bar{A}_{27} = 0.5321$$

From Table 7.1, for the unit benefit the premium is,

$$\overline{P}(\bar{A}_{27:\overline{5}|}) = \frac{\bar{A}_{27:\overline{5}|}}{\bar{a}_{27:\overline{5}|}} = \frac{\delta \bar{A}_{27:\overline{5}|}}{1 - \bar{A}_{27:\overline{5}|}} = 0.141938 \quad \Rightarrow \quad 1000\overline{P}(\bar{A}_{27:\overline{5}|}) = 141.94$$

Premium will be paid continuously for at most 5 years at the rate of 141.94 per annum. Further, variance of loss at issue random variable can be derived as in Eq. 7.3.3. Suppose the premium $\overline{P}(\bar{A}_{27:\overline{5}|})$ is denoted by $\overline{P}$. Then

$$\begin{aligned} Var(L(T)) &= [{}^2\bar{A}_{27:\overline{5}|} - (\bar{A}_{27:\overline{5}|})^2]\left(1 + \overline{P}/\delta\right)^2 \\ &= 0.6767013 \quad \text{for 1 unit benefit} \\ \Rightarrow \quad Var(L(T)) &= 10^6 \times 0.6767013 = 676701 \quad \text{for benefit of 1000} \end{aligned}$$

(ii) From Table 7.1, for unit benefit, the premium in whole life insurance is given by,

$$\overline{P}(\bar{A}_{27}) = \bar{A}_{27}/\bar{a}_{27} = \delta\bar{A}_{27}/(1 - \bar{A}_{27}) = 0.05686 \quad \Rightarrow \quad 1000\overline{P}(\bar{A}_{27}) = 56.86$$

Thus, the annual premium will be paid continuously till (27) is alive, at the rate of 56.86 per annum. Note that premium for whole life insurance is less than that for 5-year endowment insurance, since in the latter case the period for accumulation is comparatively less.

(iii) For the given information, in Example 6.3.5 we have calculated the actuarial present value at age 27 of a 5-year deferred continuous life annuity. It is 3.4567 for 1 unit benefit. Hence, for the payment of Rs 7000 per annum, it is 24196.90. This is the expected present value of the outflow of the company. Now if the premium P per annum is paid as 5-year temporary continuous life annuity, then the actuarial present value of the inflow to the company is

$$P\bar{a}_{27:\overline{5}|} = P(1 - \bar{A}_{27:\overline{5}|})/\delta = 4.341667\,P$$

Equating the expected present value of the outflow to the expected present value of the inflow, we get

$$7000 \; _{5|}\bar{a}_{27} \; = \; P\bar{a}_{27:\overline{5}|} \quad \Rightarrow \quad P = 5573.18$$

Thus, $P = 5573.18$ per annum is paid continuously for at most 5 years and in return (27) receives Rs 7000 per annum continuously, after 5 years, till (27) survives.

(iv) For the given information, in Example 6.3.7 we have calculated the actuarial present value at age 27 of a 5-year certain and continuous whole life annuity payable at the rate of Rs 7000 per annum. It is 54434.77. This is the expected present value of the outflow of the company and can be interpreted as the net single premium for the said annuity. Now if the premium P per annum is paid as 5-year continuous annuity certain, then the present value of the inflow to the company is $P\bar{a}_{\overline{5}|} \; = \; 4.319696 \; P$. Equating expected present value of the outflow to the expected present value of the inflow, we get

$$7000 \; \overline{a}_{\overline{27:\overline{5}|}} \; = \; P\bar{a}_{\overline{5}|} \quad \Rightarrow \quad P = 12601.53$$

Thus premium $P = 12601.53$ per annum, will be paid continuously for 5 years and in return (27) will receive from the insurer guaranteed payment of Rs 7000 per annum for the first five years and then payment of Rs 7000 per annum till (27) is alive. Observe that the premium for 5-year certain and continuous whole life annuity is more than double of that for 5-year deferred whole life annuity. It seems reasonable as in 5-year certain and continuous whole life annuity, payment of Rs 7000 per annum is guaranteed while in 5-year deferred whole life annuity there is no payment for the first 5 years after signing the contract. ∎

Example 7.3.4. Suppose the force of mortality follows Gompertz' law given by $\mu_x = BC^x$ with $B = 0.0001151$ and $C = 1.096$. For the following fully continuous policies issued to (x), find the premium for the benefit of 1000. Take $\delta = 0.05$. (i) n-year endowment insurance, $n = 1, 2, \cdots, 10$ and $x = 25, 30$, premiums paid for at most n years. (ii) 5-year endowment insurance, $x = 25$, premiums paid for at most 3 years. (iii) Whole life insurance, $x = 25, 30$, premiums paid as a whole life annuity. (iv) Whole life insurance, $x = 25$, premiums paid for at most 20 years. (v) 30-year deferred whole life annuity, $x = 30$, premiums paid for at most 30 years. Compare the premium as age changes in (i) and (iii). Compare premiums in (i) and (ii). Compare premiums in (iii) and (iv).

Solution: (i) From Table 7.1, for the 1 unit benefit the premium for n-year endowment insurance is given by,

$$P(\overline{A}_{x:\overline{n}|}) = \overline{A}_{x:\overline{n}|}/\bar{a}_{x:\overline{n}|} \; = \; \delta\overline{A}_{x:\overline{n}|}/(1 - \overline{A}_{x:\overline{n}|})$$

In Example 5.3.7 we have derived the formula for $\overline{A}_{x:\overline{n}|}$, when the mortality is governed by Gompertz' law. Thus we have,

$$\overline{A}_{x:\overline{n}|} \; = \; \alpha_x \, e^{\alpha_x}\Gamma(\lambda) \, \alpha_x^{-\lambda} \, P[1 \leq W \leq C^n] + e^{-\delta n - \alpha_x(C^n - 1)}$$

where W follows gamma distribution with shape parameter λ and scale parameter α_x and $\alpha_x \; = \; mC^x$. We obtain the probability $P[1 \leq W \leq C^n]$ using R. Values of the annual premium for benefit of 1000 are reported in columns 3 and 5 of Table 7.2 for $x = 25$ and 30

Term	Single	Annual	Single	Annual
n	25	25	30	30
1	951.26	975.81	951.27	976.15
2	904.95	476.04	905.02	476.41
3	860.96	309.61	861.11	310.00
4	819.18	226.52	819.44	226.92
5	779.50	176.76	779.91	177.18
6	741.83	143.67	742.41	144.11
7	706.06	120.10	706.86	120.57
8	672.11	102.49	673.15	102.98
9	639.90	88.85	641.21	89.36
10	609.33	77.99	610.95	78.52

Table 7.2 Fully continuous premiums for endowment insurance

respectively. For the sake of comparison, columns 2 and 4 report values of net single premium for $x = 25$ and 30, which are nothing but $\overline{A}_{25:\overline{n}|}$ and $\overline{A}_{30:\overline{n}|}$ respectively. These values are obtained using Part II of Code 7.7.1. From Table 7.2, we note that as the term increases, the annual premium decreases for both the ages. For this mortality pattern, the difference between premiums for two ages is marginal for any value of n. For $n = 1$, the single premium is less than the annual premium, it is in view of the fact that a single premium is paid at the issue while annual premium is paid continuously throughout the year. For small values of n, it seems that, it is better to purchase the insurance cover by one-time premium than the periodical premiums.
(ii) When the 5-year endowment insurance is purchased by (25) with at most 3 instalments, the annual premium is given by,

$$1000 \,_3\overline{P}(\overline{A}_{25:\overline{5}|}) = 1000 \frac{\overline{A}_{25:\overline{5}|}}{\overline{a}_{25:\overline{3}|}} = 1000 \frac{\delta \overline{A}_{25:\overline{5}|}}{1 - \overline{A}_{25:\overline{3}|}} = \frac{0.05 \times 779.50}{1 - 0.86096} = 280.32$$

It is obtained using Part III of Code 7.7.1. Note that it is larger than the annual premium 176.76, when premiums are paid for at most 5 years.
(iii) From Table 7.1, for unit benefit whole life insurance, the premium is given by, $\overline{P}(\overline{A}_x) = \overline{A}_x/\overline{a}_x = \delta \overline{A}_x/(1 - \overline{A}_x)$. In Example 5.3.7, we have derived the formula for $\overline{A}_x$, when the mortality is governed by Gompertz' law. It is given by,

$$\overline{A}_x = \alpha_x \, e^{\alpha_x} \Gamma(\lambda) \, \alpha_x^{-\lambda} \, P[W \geq 1]$$

where W follows gamma distribution as in (i) above. We obtain the probability $P[W \geq 1]$ using R. The second and third columns of Table 7.3 report the values of the one-time premium and the annual premium which is paid continuously till (x) survives, for the benefit amount 1000.
(iv) When the whole life insurance is purchased by 20-year temporary continuous annuity then the annual premium is given by,

$$_{20}\overline{P}(\overline{A}_x) = \overline{A}_x/\overline{a}_{x:\overline{20}|} = \delta \overline{A}_x/(1 - \overline{A}_{x:\overline{20}|})$$

Using the approach in (i) above we find $\overline{A}_{x:\overline{20}|}$. The last column of Table 7.3 reports the annual premium for benefit of 1000, when premiums are paid as 20-year temporary continuous annuity. Values in Table 7.3 are obtained by Part IV and Part V of Code 7.7.1.

Age	Single	Annual	20 Years
25	152.54	9.00	12.29
30	189.12	11.66	15.39

Table 7.3 Fully continuous premiums for whole life insurance

From Table 7.3 we note that as age increases, the premium of any type increases. If the premium payment period is restricted then naturally the annual premiums are higher than those paid as whole life annuity.

(v) For 30-year deferred whole life annuity, issued to (30) with premiums paid for at most 30 years, the annual premium is given by the formula

$$\overline{P}(\,_{30|}\overline{a}_{30}) = A_{30:\overline{30|}}^{\,1}\,\overline{a}_{60}/\overline{a}_{30:\overline{30|}} \;=\; v^{30}\,_{30}p_{30}(1 - \overline{A}_{60})/(1 - \overline{A}_{30:\overline{30|}})$$

For the given mortality pattern and $\delta = 0.05$, we find $\overline{A}_{30:\overline{30|}}$ and $\overline{A}_{60}$ as in (i) and (iii) above. Thus we get,

$$1000\overline{P}(\,_{30|}\overline{a}_{30}) = \frac{1000 \times 0.1673596 \times (1 - 0.5534345)}{1 - 0.2638599} = 101.53$$

These are obtained by Part VI of Code 7.7.1. The annual premium 101.53 for 30-year deferred whole life annuity, issued to (30), can be interpreted as follows. If (30) survives up to 60 and pays continuously at the rate of 101.53 per annum for 30 years in a pension fund, then after his age 60, he will get the benefit of 1000 per annum paid continuously till his death.   ■

In the next section, we proceed to the discussion on fully discrete premiums, where the future life time random variable T is replaced by the curtate future life random variable K.

7.4   Fully Discrete Premiums

In this section, we discuss the derivation of annual benefit premium in fully discrete policies, that is, when the sum insured is payable at the end of the policy year in which death occurs and the first premium is payable when the insurance is issued. Subsequent premiums are payable on anniversaries of the policy issue date while the insured survives during the contractual premium payment period. The payment of annual premiums thus forms a life annuity due. We consider various insurance products as discussed in Section 5.4 and annuities as discussed in Section 6.4. The procedure to find fully discrete premiums is similar to that adopted for fully continuous policies in the previous section. The loss at issue random variable L is now a function of K, the curtate future life time random variable. As an illustration we consider whole life insurance when the unit benefit is payable at the end of the year of death and premiums form whole life discrete annuity due. In this case, the international actuarial notation for the level annual premium for a unit benefit is P_x, where the absence of $(\overline{A}_x)$ means that the insurance is payable at the end of the policy year in which death occurs. The loss random variable for this insurance is

$$L \;=\; v^{K+1} - P_x\,\ddot{a}_{\overline{K+1|}}$$

where the curtate random variable K takes values $0, 1, 2, \cdots$. The equivalence principle requires that $E(L) = 0$. Thus,

$$E(v^{K+1}) - P_x\, E(\ddot{a}_{\overline{K+1|}}) = 0, \quad \Rightarrow \quad P_x = A_x/\ddot{a}_x \tag{7.4.1}$$

This is the discrete analogue of Eq. 7.3.2. Using steps parallel to those in the derivation of Eq. 7.3.3, we obtain,

$$Var(L) \;=\; (^2A_x - A_x^2)/(d\ddot{a}_x)^2 \tag{7.4.2}$$

Formulae for annual benefit premium in other fully discrete policies and semi-continuous policies are derived similar to that in Eq. 7.4.1. Table 7.4 summarizes the international actuarial notation and the formulae for net annual premiums for fully discrete policy and semi-continuous policy, for a unit benefit for insurance products studied in Section 5.4.

Insurance product	Fully discrete	Semi-continuous
n-Year term insurance	$P^1_{x:\overline{n|}} = \dfrac{A^1_{x:\overline{n|}}}{\ddot{a}_{x:\overline{n|}}}$	$P(\overline{A}^1_{x:\overline{n|}}) = \dfrac{\overline{A}^1_{x:\overline{n|}}}{\ddot{a}_{x:\overline{n|}}}$
n-Year endowment	$P_{x:\overline{n|}} = \dfrac{A_{x:\overline{n|}}}{\ddot{a}_{x:\overline{n|}}} = \dfrac{dA_{x:\overline{n|}}}{1 - A_{x:\overline{n|}}}$ $= \dfrac{1 - d\ddot{a}_{x:\overline{n|}}}{\ddot{a}_{x:\overline{n|}}}$	$P(\overline{A}_{x:\overline{n|}}) = \dfrac{\overline{A}_{x:\overline{n|}}}{\ddot{a}_{x:\overline{n|}}}$
Whole life insurance	$P_x = \dfrac{A_x}{\ddot{a}_x}$ $= \dfrac{dA_x}{1 - A_x} = \dfrac{1 - d\ddot{a}_x}{\ddot{a}_x}$	$P(\overline{A}_x) = \dfrac{\overline{A}_x}{\ddot{a}_x}$
h-Payment, whole life insurance	$_hP_x = \dfrac{A_x}{\ddot{a}_{x:\overline{h|}}}$	$_hP(\overline{A}_x) = \dfrac{\overline{A}_x}{\ddot{a}_{x:\overline{h|}}}$
h-Payment, n-Year endowment	$_hP_{x:\overline{n|}} = \dfrac{A_{x:\overline{n|}}}{\ddot{a}_{x:\overline{h|}}}$	$_hP(\overline{A}_{x:\overline{n|}}) = \dfrac{\overline{A}_{x:\overline{n|}}}{\ddot{a}_{x:\overline{h|}}}$
n-Year deferred whole life annuity	$P(\,_{n|}\ddot{a}_x) = \dfrac{A^{\;\;1}_{x:\overline{n|}}\,\ddot{a}_{x+n}}{\ddot{a}_{x:\overline{n|}}}$	$P(\,_{n|}\overline{a}_x) = \dfrac{A^{\;\;1}_{x:\overline{n|}}\,\overline{a}_{x+n}}{\ddot{a}_{x:\overline{n|}}}$

Table 7.4 Notation and formulae for fully discrete and semi-continuous annual benefit premiums

P is a general symbol for a fully discrete net annual premium. It is followed by a subscript x or $x : \overline{n|}$ depending on the insurance product. In Chapter 6 we have derived following two identities.

$$d\ddot{a}_x + A_x = 1 \quad \text{and} \quad d\ddot{a}_{x:\overline{n|}} + A_{x:\overline{n|}} = 1$$

These can be used to express the formula for premium either in actuarial present value symbol of insurance or of annuity, for whole life insurance and for n-year endowment insurance.

The following examples illustrate the computation of fully discrete premiums.

Example 7.4.1. It is given that $q_{28} = 0.135$, $q_{29} = 0.146$, $q_{30} = 0.159$, $q_{31} = 0.173$, $q_{32} = 0.188$ and $i = 0.05$. (i) Find the annual premium payable as 5-year temporary discrete annuity due, for the benefit of 1000, payable at the end of the year of death, in a 5-year term insurance, issued to (28). (ii) Find the annual premium payable as 5-year temporary discrete annuity due, for the benefit of 1000, payable at the end of year of death, in a 5-year endowment insurance, issued to (28). (iii) Find the annual premium payable as 2-year temporary discrete annuity due for the setup of (i) and (ii). Compare the premiums. (iv) Find the annual premium payable as 5-year temporary discrete annuity due for the benefit of 1000, payable at the moment of death, in a 5-year endowment insurance, issued to (28). State the assumptions, if needed, to calculate the premium.

Solution: (i) From Table 7.4, the annual premium payable as 5-year temporary discrete annuity due for benefit of 1 unit in a 5-year term insurance is given by, $P^1_{28:\overline{5}|} = A^1_{28:\overline{5}|}/\ddot{a}_{28:\overline{5}|}$. In Example 5.4.3, we have obtained $A^1_{28:\overline{5}|} = 0.5086848$ for the given values of q_x and $i = 0.05$. Further for the same setup in Example 6.4.4, we have obtained $\ddot{a}_{28:\overline{5}|} = 3.453191$. Hence,

$$P^1_{28:\overline{5}|} = 0.1473086 \quad \Rightarrow \quad 1000 P^1_{28:\overline{5}|} = 147.31$$

(ii) The annual premium payable as 5-year temporary discrete annuity due for benefit of 1 unit in a 5-year endowment insurance is given by, $P_{28:\overline{5}|} = A_{28:\overline{5}|}/\ddot{a}_{28:\overline{5}|}$. In Example 5.4.3, we have obtained $A_{28:\overline{5}|} = 0.8355623$ for the given values of q_x and $i = 0.05$. Hence,

$$P_{28:\overline{5}|} = 0.2419682 \quad \Rightarrow \quad 1000 P_{28:\overline{5}|} = 241.97$$

This premium is larger than that for the term insurance as in the endowment insurance there is a guarantee of payment of 1000 at the end of 5-year term, provided (28) survives the term while in term insurance there is no benefit payment if (28) survives the term.

(iii) To find the premiums paid as 2-year annuity due, we first find

$$\ddot{a}_{28:\overline{2}|} = 1 + p_{28}v = 1 + 0.865/(1.05) = 1.823810$$

Hence the annual premium for 5-year term insurance with benefit 1000 and at most two premium payments, is given by,

$$1000 \; {}_2P^1_{28:\overline{5}|} = 1000 A^1_{28:\overline{5}|}/\ddot{a}_{28:\overline{2}|} = 278.91$$

The annual premium for 5-year endowment insurance with benefit of 1000 and at most two premium payments, is given by,

$$1000 \; {}_2P_{28:\overline{5}|} = 1000 \; A_{28:\overline{5}|}/\ddot{a}_{28:\overline{2}|} = 458.14$$

We again note that the annual premium is larger for endowment insurance than for term insurance. Further, if the number of premium payments is less, the annual premium is high, as expected.

(iv) In this case the policy is semi-continuous. Hence from Table 7.4, the annual premium is given by, $P(\overline{A}_{28:\overline{5}|}) = \overline{A}_{28:\overline{5}|}/\ddot{a}_{28:\overline{5}|}$. Here we are given values of q_x for some x and not the distribution of $T(28)$. Hence we cannot find $\overline{A}_{28:\overline{5}|}$ directly but it can be obtained under the assumption of uniformity in unit age intervals. For the given setup, under the assumption of uniformity in unit age intervals, we have obtained in Example 6.4.4, $\overline{A}_{28:\overline{5}|} = 0.848176$. Hence, the annual premium for benefit of 1000 is given by,

$$1000P(\overline{A}_{28:\overline{5}|}) = 1000\overline{A}_{28:\overline{5}|}/\ddot{a}_{28:\overline{5}|} = 245.62$$

In a fully discrete policy, the annual premium is 241.97. Thus, in a semi-continuous policy, the annual premium is slightly higher, which is logically acceptable.   ∎

Example 7.4.2. Suppose for the specific mortality law, q_x values are as given in Table 6.15. For the same mortality pattern and $i = 0.05$, $A_{50} = 0.251$.
(i) Find the annual premium for ages 40 to 50, for the whole life insurance of benefit of 1000, payable at the end of the year of death, when the premiums are payable as the whole life annuity due. (ii) Find the premium per annum, if the benefit is payable at the moment of death, under the assumption of uniformity in unit age intervals. (iii) Find the annual premium for the whole life insurance to (40), when the benefit is payable either at the end of the year of death or at the moment of death and if the premium payments form a 5-year temporary annuity due. Make the assumption of uniformity in unit age interval wherever needed. (iv) Find the annual premium for the whole life annuity due of benefit of 1000 per annum, issued to (40), if it is purchased by a 5-year temporary annuity due.

Solution: (i) From Table 7.4 we have, $P_x = A_x/\ddot{a}_x = (1 - d\ddot{a}_x)/\ddot{a}_x$. For the given mortality pattern and $i = 0.05$, we have obtained the values of $\ddot{a}_x$ for $x = 40$ to 50 in Example 6.4.7. Hence we use the formula of P_x in terms of annuity symbols to find the premiums. These are reported in the second column of Table 7.5.
(ii) In semi-continuous policy, the premium is given by $P(\overline{A}_x) = \overline{A}_x/\ddot{a}_x$. To find $\overline{A}_x$, we use the backward recurrence relation $A_x = vq_x + vp_xA_{x+1}$ given in Eq. 5.4.3 to find A_x values, for $x = 40$ to 49 from the knowledge of $A_{50} = 0.251$. Under the assumption of uniformity in unit age intervals, $\overline{A}_x = (i/\delta)A_x$. Thus, we find the premiums in a semi-continuous policy. These are obtained using Part II of Code 7.7.2 and are reported in the third column of Table 7.5. Note that annual premiums in semi-continuous policy are marginally higher than those for fully discrete policy.
(iii) We have to find $1000 \, _5P_{40} = A_{40}/\ddot{a}_{40:\overline{5}|}$ and $1000 \, _5P(\overline{A}_{40}) = \overline{A}_{40}/\ddot{a}_{40:\overline{5}|}$

Thus, we first find $\ddot{a}_{40:\overline{5}|}$. It is given by,

$$\ddot{a}_{40:\overline{5}|} = \sum_{k=0}^{4} v^k \, _kp_{40} = 1 + vp_{40} + v^2p_{40}p_{41} + v^3p_{40}p_{41}p_{42}$$

$$+ \; v^4p_{40}p_{41}p_{42}p_{43} = 4.520736$$

$$\Rightarrow \; 1000 \, _5P_{40} = (1 - d\ddot{a}_{40})/\ddot{a}_{40:\overline{5}|} = 39.15$$

$$\text{and } 1000 \, _5P(\overline{A}_{40}) = \overline{A}_{40}/\ddot{a}_{40:\overline{5}|} = 40.12$$

Age x	1000 P_x	1000 $P(\overline{A}_x)$
40	10.24	10.49
41	10.71	10.98
42	11.20	11.48
43	11.71	12.00
44	12.25	12.55
45	12.80	13.12
46	13.39	13.72
47	13.99	14.34
48	14.62	14.99
49	15.28	15.66
50	15.96	16.35

Table 7.5 Annual discrete premiums for whole life insurance

Premiums in (iii) are larger than those in (i) and (ii) for age (40). It is reasonable since the premium paying term is less in (iii).These premiums are obtained using Part III of Code 7.7.2. (iv) In Example 6.4.7, we have calculated the actuarial present value of the whole life annuity due issued to (40), with a benefit of 1000 per annum. It is 17283.26. The actuarial present value of the premiums which form a 5-year temporary annuity due is $4.520736P$. Hence, $P = 17283.26/4.520736 = 3823.11$. Thus, the annual premium of 3823.11 is to be paid for at most 5 years to get the survival benefit of 1000 per annum till (40) survives. These premiums are obtained using Part IV of Code 7.7.2. ∎

Example 7.4.3. Suppose L is the loss-at-issue random variable for a fully discrete whole life insurance of 1 issued to (49). Calculate P and $E(L)$ on the basis of the following information. (i) $A_{49} = 0.23882$, $\ddot{a}_{49} = 13.4475$, $^2A_{49} = 0.08873$, (ii) $i = 0.06$ and (iii) $Var(L) = 0.10$.

Solution: Suppose the annual premium is P. Since the policy is whole life, the loss-at-issue random variable is given by,

$$L = Z - PY = Z - P(1-Z)/d = (1 + P/d)Z - P/d$$

$$\Rightarrow E(L) = E(Z) - PE(Y) = A_{49} - P\ddot{a}_{49} = 0.23882 - 13.4475P$$

$$\text{Now, } 0.10 = Var(L) = (1 + P/d)^2 Var(Z)$$

$$= (1 + P/d)^2(^2A_{49} - (A_{49})^2)$$

$$= (1 + P/0.0556)^2(0.08873 - (0.23882)^2)$$

$$\Rightarrow P = 0.044 \Rightarrow E(L) = -0.35$$

∎

It is to be noted that P is not obtained using equivalence principle.

Example 7.4.4. The following information is given about the premiums paid by (x). (i) The premium for a 20-year endowment insurance of 1 is 0.0349. (ii) The premium for a 20-year deferred whole life annuity due of 1 per year is 0.2087 and is paid for 20 years. (iii) $_{20}E_x/\ddot{a}_{x:\overline{20|}} = 0.0230$. (It is the premium for a 20-year pure-endowment of benefit 1 unit, denoted by $P_{x:\overline{20|}}^{\;\;1}$.) (iv) All premiums are fully discrete annual benefit premiums. (v) $i = 0.05$. Calculate the premium for a 20-payment whole life insurance of 1.

Solution: We are given $P_{x:\overline{20|}} = 0.0349$, $P_{x:\overline{20|}}^{\;\;1} = 0.0230$, $P(\,_{20|}\ddot{a}_x) = 0.2087$. We have to find $_{20}P_x$. Note that the identity

$$A_x = A_{x:\overline{20|}}^{\;\;1} + \,_{20}E_x\,A_{x+20} \Rightarrow \frac{A_x}{\ddot{a}_{x:\overline{20|}}} = \frac{A_{x:\overline{20|}}^{\;\;1}}{\ddot{a}_{x:\overline{20|}}} + \frac{_{20}E_x}{\ddot{a}_{x:\overline{20|}}} A_{x+20}$$

$$\Rightarrow\ _{20}P_x = P_{x:\overline{20|}}^{\;\;1} + P_{x:\overline{20|}}^{\;\;1}\,A_{x+20} = P_{x:\overline{20|}} - P_{x:\overline{20|}}^{\;\;1} + P_{x:\overline{20|}}^{\;\;1}\,A_{x+20}$$

$$= 0.0119 + 0.0230 A_{x+20}$$

To find A_{x+20}, we proceed as follows. Observe that,

$$0.0349 = P_{x:\overline{20|}} = (1 - d\ddot{a}_{x:\overline{20|}})/\ddot{a}_{x:\overline{20|}}$$

$$\Rightarrow \ddot{a}_{x:\overline{20|}} = 1/(P_{x:\overline{20|}} + d) = 1/(0.0349 + 0.04762) = 12.12$$

$$\text{and}\ _{20}E_x = P_{x:\overline{20|}}^{\;\;1}\,\ddot{a}_{x:\overline{20|}} = 0.279$$

$$0.2087 = P(\,_{20|}\ddot{a}_x) = \,_{20|}\ddot{a}_x/\ddot{a}_{x:\overline{20|}} \Rightarrow \,_{20|}\ddot{a}_x = 2.53$$

$$\text{and}\ \ddot{a}_{x+20} = \,_{20|}\ddot{a}_x/\,_{20}E_x = 2.53/0.279 = 9.07$$

$$\Rightarrow A_{x+20} = 1 - d\ddot{a}_{x+20} = 0.57$$

$$\Rightarrow\ _{20}P_x = 0.0119 + 0.0230 A_{x+20} = 0.025$$

■

Example 7.4.5. Express in international actuarial notation the premiums corresponding to following insurances. (i) If (40) dies within 25 years, a death benefit of Rs 50000 is payable at the end of year of death, otherwise a sum of Rs 2000 is payable yearly in advance from age 65. Premiums are payable as 25-year temporary life annuity due. (ii) An insurance issued to (50) provides guaranteed benefit of 20000, at the end of term of 10 years. Annual premiums are payable as 10-year temporary annuity due. (iii) (30) signs an insurance contract, in which annual payment of 15 year annuity immediate is 12000 for first 5 years and 24000 for the remaining term. Annual premiums are payable as 5-year temporary annuity due.

Solution: In international actuarial notation the premiums can be expressed as follows.

(i) $(50000A_{40:\overline{25|}}^{\;\;1} + 2000\,_{25|}\ddot{a}_{40})/\ddot{a}_{40:\overline{25|}}$

(ii) $20000v^{10}/\ddot{a}_{50:\overline{10|}}$

(iii) $(12000a_{30:\overline{5|}} + \,_5E_{30}24000a_{35:\overline{10|}})/\ddot{a}_{30:\overline{5|}}$

■

Example 7.4.6. Suppose the mortality law is Gompertz' law with force of mortality $\mu_x = BC^x$, $x > 0$ and $S(x) = \exp\{-m(c^x - 1)\}$, where $B = 0.0001151$ and $C = 1.096$. Suppose $\delta = 0.05$. (i) Find the premiums for a n-year endowment insurance for a benefit of Rs 1000, when the premiums are payable as n-year temporary annuity due, for $n = 1$ to 10 and $x = 25, 30, 35, 40$. (ii) Find premiums for the whole life insurance, when the benefit of Rs 1000 is payable at the end of the year of death or at the moment of death, when the premiums are payable as whole life annuity due and as 10-year temporary life annuity due.

Solution: In Example 4.3.4, we have obtained the distribution of $K(x)$ and $_np_x$, when the life length random variable is modelled by Gompertz' law. These are given by,

$$P[K(x) = k] = \exp[-mC^x(C^k - 1)] - \exp[-mC^x(C^{k+1} - 1)]$$

and $_np_x = \exp[-mC^x(C^n - 1)]$. From Table 7.4, the annual premium payable as n-year temporary discrete annuity due for benefit of 1 unit in a n-year endowment insurance is given by, $P_{x:\overline{n}|} = A_{x:\overline{n}|}/\ddot{a}_{x:\overline{n}|}$, where

$$A_{x:\overline{n}|} = \sum_{k=0}^{n-1} v^{k+1}\, {}_kp_x q_{x+k} + v^n\, {}_np_x \quad \text{and} \quad \ddot{a}_{x:\overline{n}|} = (1 - A_{x:\overline{n}|})/d$$

We use Part II of Code 7.7.3 to obtain these monetary functions and hence the required premiums. These are reported in Table 7.6. From Table 7.6, we note that there is no difference,

n	$P_{25:\overline{n}|}$	$P_{30:\overline{n}|}$	$P_{35:\overline{n}|}$	$P_{40:\overline{n}|}$
1	951.23	951.23	951.23	951.23
2	464.02	464.20	464.47	464.90
3	301.78	302.02	302.40	303.01
4	220.77	221.06	221.51	222.22
5	172.26	172.58	173.09	173.89
6	140.00	140.35	140.90	141.77
7	117.02	117.40	118.00	118.94
8	99.85	100.26	100.89	101.90
9	86.55	86.98	87.66	88.73
10	75.96	76.41	77.13	78.27

Table 7.6 Fully discrete premiums $P_{x:\overline{n}|}$ for n-year endowment insurance

up to two decimal places, among values of $P_{x:\overline{1}|}$ for $x = 25, 30, 35$ and 40. As n increases, there are slight differences. As age increases, the premiums also increase, but again the difference is marginal for the given mortality pattern and the values of the parameters.

(ii) To find the premiums for whole life insurance, we note that the fully discrete premium is given by $P_x = A_x/\ddot{a}_x$, where $\ddot{a}_x = (1 - A_x)/d$. The semi-continuous premium $P(\overline{A}_x)$, when the benefit is payable at the moment of death, is given by $P(\overline{A}_x) = \overline{A}_x/\ddot{a}_x$. Under the assumption of uniform distribution of deaths between integral ages, we know that $\overline{A}_x = (i/\delta)A_x$. Hence, we can find $\overline{A}_x$, so that $P(\overline{A}_x)$ can be computed. These are obtained using Part III of Code 7.7.3 and are reported in the second and third columns of Table 7.7. Suppose now the premiums are payable as 10-year temporary life annuity due. In this mode, the premium for the whole life insurance, when the benefit of 1 unit is payable at the end of year of death or at the moment of death, are given by,

$$_{10}P_x = A_x/\ddot{a}_{x:\overline{10|}} \quad \& \quad _{10}P(\overline{A}_x) = \overline{A}_x/\ddot{a}_{x:\overline{10|}}$$

We have already computed these functions. We use Part IV of Code 7.7.3 to compute the premiums for the whole life insurance, when premiums are payable as 10-year temporary life annuity due. These are reported in the last two columns of Table 7.7. From the table we observe

x	P_x	$P(\overline{A}_x)$	$_{10}P_x$	$_{10}P(\overline{A}_x)$
25	8.52	8.74	18.56	19.03
30	11.03	11.31	23.01	23.59
35	14.32	14.68	28.31	29.03
40	18.66	19.13	34.52	35.39

Table 7.7 Fully discrete and semi-continuous premiums for whole life insurance

that premiums do not differ much in discrete and semi-continuous setup. As age increases, the premiums increase, as expected. We have a similar observation for premiums payable as 10-year temporary life annuity due. Note that premiums payable as 10-year life annuity due are higher than those payable as whole life annuity, as expected.   ∎

In practice, sometimes in fully discrete policies and semi-continuous policies, premiums are payable m times in a policy year rather than annually. We discuss such a setup in the following section.

7.5   True m-thly Payment Premiums

If in fully discrete policies and semi-continuous policies, premiums are payable m times a policy year, rather than annually, with no adjustment in the death benefit, the resulting premiums are called true fractional premiums. Typically, $m = 2, 4$ or 12 corresponding to semi-annual, quarterly and monthly premiums respectively. The case $m = 1$ corresponding to annual premiums is already covered in Section 4. These fractional premiums are also determined using the equivalence principle.

As an illustration, consider a unit benefit whole life insurance payable at the end of year of death and premiums are payable in m-thly instalments at the beginning of each m-thly period at the rate of 1 unit per annum. In international actuarial notation, $P_x^{(m)}$ denotes the true net level annual benefit premium in this setup. The actual amount of each fractional premium, payable m times each policy year, during the premium paying period and the survival of (x), is $P_x^{(m)}/m$. The symbol $P^{(m)}(\overline{A}_x)$ is used when a unit benefit in a whole life insurance is payable at the moment of death and premiums are payable in m-thly instalments at the beginning of each m-thly period at the rate of 1 unit per annum. Table 7.8 specifies the symbols and formulae for true fractional premiums for common life insurances. The premium formulae are obtained by applying the equivalence principle.

Insurance product	Fully discrete	Semi-continuous
n-Year term insurance	$P^{1(m)}_{x:\overline{n}|} = \dfrac{A^{\ 1}_{x:\overline{n}|}}{\ddot{a}^{(m)}_{x:\overline{n}|}}$	$P^{(m)}(\overline{A}^{\ 1}_{x:\overline{n}|}) = \dfrac{\overline{A}^{\ 1}_{x:\overline{n}|}}{\ddot{a}^{(m)}_{x:\overline{n}|}}$
n-Year endowment insurance	$P^{(m)}_{x:\overline{n}|} = \dfrac{A_{x:\overline{n}|}}{\ddot{a}^{(m)}_{x:\overline{n}|}}$	$P^{(m)}(\overline{A}_{x:\overline{n}|}) = \dfrac{\overline{A}_{x:\overline{n}|}}{\ddot{a}^{(m)}_{x:\overline{n}|}}$
Whole life insurance	$P^{(m)}_x = \dfrac{A_x}{\ddot{a}^{(m)}_x}$	$P^{(m)}(\overline{A}_x) = \dfrac{\overline{A}_x}{\ddot{a}^{(m)}_x}$
h-Payment years, whole life insurance	${}_hP^{(m)}_x = \dfrac{A_x}{\ddot{a}^{(m)}_{x:\overline{h}|}}$	${}_hP^{(m)}(\overline{A}_x) = \dfrac{\overline{A}_x}{\ddot{a}^{(m)}_{x:\overline{h}|}}$
h-Payment years, n-year endowment insurance	${}_hP^{(m)}_{x:\overline{n}|} = \dfrac{A_{x:\overline{n}|}}{\ddot{a}^{(m)}_{x:\overline{h}|}}$	${}_hP^{(m)}(\overline{A}_{x:\overline{n}|}) = \dfrac{\overline{A}_{x:\overline{n}|}}{\ddot{a}^{(m)}_{x:\overline{h}|}}$
n-Year deferred whole life annuity due	$P^{(m)}({}_{n|}\ddot{a}_x) = \dfrac{A^{\ 1}_{x:\overline{n}|}\,\ddot{a}_{x+n}}{\ddot{a}^{(m)}_{x:\overline{n}|}}$	$P^{(m)}({}_{n|}\overline{a}_x) = \dfrac{A^{\ 1}_{x:\overline{n}|}\,\overline{a}_{x+n}}{\ddot{a}^{(m)}_{x:\overline{n}|}}$

Table 7.8 Notation and formulae for true fractional benefit premiums

Note that formulae for premiums in Table 7.8 are similar to those given in Table 7.4, with the only change that annual annuity functions are replaced by m-thly annuity functions. In the table h refers to the number of payment years, not to the number of payments. $P^{(12)}({}_{n|}\ddot{a}_x)/12$ denotes the monthly contribution to the pension fund.

Example 7.5.1. Derive a relation between ${}_hP^{(m)}_{x:\overline{n}|}$ and ${}_hP_{x:\overline{n}|}$.

Solution: From Table 7.8 we have

$$ {}_hP^{(m)}_{x:\overline{n}|} \;=\; \frac{A_{x:\overline{n}|}}{\ddot{a}^{(m)}_{x:\overline{h}|}} = \frac{{}_hP_{x:\overline{n}|}\,\ddot{a}_{x:\overline{h}|}}{\ddot{a}^{(m)}_{x:\overline{h}|}} = \; {}_hP_{x:\overline{n}|}\;\frac{\ddot{a}_{x:\overline{h}|}}{\ddot{a}^{(m)}_{x:\overline{h}|}} \tag{7.5.1} $$

The formula in Eq. 7.5.1 expresses the m-thly payment premium as equal to the corresponding annual payment premium times a ratio of annuity values. ∎

In the following example we continue with the mortality pattern as specified in Example 7.4.2 to find m-thly premiums.

Example 7.5.2. Suppose for the specific mortality law, q_x values are as given in Table 6.15. For the same mortality law and $i = 0.05$, $A_{50} = 0.251$. (i) Find the monthly premium for ages 40 to 50, for whole life insurance with benefit of 1000, payable at the end of year of death, when the premiums are paid as monthly whole life annuity due. (ii) For the setup in (i) find the monthly premium, if the benefit is payable at the moment of death, under the assumption of uniformity in unit age intervals. (iii) Find the quarterly premium for whole life insurance to (40), when the benefit is payable either at the end of the year of death or at the moment of death, if the premium payments form 5-year temporary quarterly annuity due. Assume uniformity in unit age interval, wherever needed. (iv) Find the six monthly premium for whole life annuity due of benefit of 1000 per annum, issued to (40), if it is purchased by 5-year temporary six monthly annuity due. (v) Find the monthly premium for 5-year endowment insurance issued to (40), when the benefit of 1000 is payable at the end of the year of death and premiums form 5-year temporary monthly annuity due. (vi) Find the monthly premium for 5-year term insurance issued to (40), when the benefit of 1000 is payable at the end of year of death and premiums form 5-year temporary monthly annuity due.(vii) Repeat (v) and (vi) if the benefit of 1000 is payable at the moment of death.(viii) Find $1000\overline{P}(\overline{A}_{40:\overline{5}|})$, $1000P^{(m)}(\overline{A}_{40:\overline{5}|})$ for $m = 2, 4$ and 12, and $1000P(\overline{A}_{40:\overline{5}|})$. Order these premiums in decreasing order and comment on the result.

Solution: (i) From Table 7.8 we have $P_x^{(12)} = A_x/\ddot{a}_x^{(12)}$. We are given A_{50}. Using the backward recurrence relation $A_x = vq_x + vp_x A_{x+1}$ given in Eq. 5.4.3, we find A_x values for $x = 40$ to 49 from the knowledge of $A_{50} = 0.251$. Further $\ddot{a}_x^{(12)}$ is given by $\ddot{a}_x^{(12)} = \alpha(12)\ddot{a}_x - \beta(12)$. For $i = 0.05$, $d = 0.04761905$, $i^{(12)} = 0.04888949$, $d^{(12)} = 0.04869111$, $\alpha(12) = 1.000197$ and $\beta(12) = 0.466508$. With $\ddot{a}_x = (1-A_x)/d$, we find the values of $P_x^{(12)}$ for $x = 40$ to 50. Note that the monthly premium is given by $P_x^{(12)}/12$. $P_x^{(12)}$ and the corresponding monthly premiums are reported in the second and third columns respectively of Table 7.9.

(ii) For a semi-continuous policy we have $P^{(12)}(\overline{A}_x) = \overline{A}_x/\ddot{a}_x^{(12)}$. Under the assumption of uniformity in unit age intervals $\overline{A}_x = (i/\delta)A_x$. We have already calculated $\ddot{a}_x^{(12)}$ in (i). Thus, we find the premiums in a semi-continuous policy. $P^{(12)}(\overline{A}_x)$ and corresponding monthly premiums are reported in the fourth and fifth column respectively of Table 7.9. Comparing

Age x	1000 $P_x^{(12)}$	1000 $P_x^{(12)}/12$	1000 $P^{(12)}(\overline{A}_x)$	1000 $P^{(12)}(\overline{A}_x)/12$
40	10.52	0.88	10.78	0.90
41	11.01	0.92	11.28	0.94
42	11.52	0.96	11.80	0.98
43	12.04	1.00	12.34	1.03
44	12.60	1.05	12.91	1.08
45	13.17	1.10	13.50	1.12
46	13.78	1.15	14.12	1.18
47	14.40	1.20	14.76	1.23
48	15.06	1.25	15.43	1.29
49	15.74	1.31	16.13	1.34
50	16.44	1.37	16.85	1.40

Table 7.9 Monthly premiums for whole life insurance

column 2 with the second column of Table 7.5, specifying $1000\, P_x$, we note that

$$1000 P_x^{(12)} \quad > \quad 1000 P_x$$

A similar relation is noted for a semi-continuous policy. We use Part I and Part II of Code 7.7.4 to compute all the monetary functions and premiums reported in Table 7.9.
(iii) We have to find

$$_5P_{40}^{(4)} = A_{40}/\ddot{a}_{40:\overline{5|}}^{(4)} \quad \text{and} \quad _5P^{(4)}(\overline{A}_{40}) = \overline{A}_{40}/\ddot{a}_{40:\overline{5|}}^{(4)}$$

From (i) and (ii) we have $A_{40} = 0.1769877$ & $\overline{A}_{40} = 0.1813764$. For $i = 0.05$, $d = 0.04761905$, $i^{(4)} = 0.04908894$ $d^{(4)} = 0.04849381$, $\alpha(4) = 1.000186$ and $\beta(4) = 0.3827173$. Further, $_5E_{40} = v^5 \, _5p_{40} = 0.7711468$. From Example 7.4.2, $\ddot{a}_{40:\overline{5|}} = 4.520736$. From Table 6.32,

$$\ddot{a}_{40:\overline{5|}}^{(4)} = \alpha(4)\ddot{a}_{40:\overline{5|}} - \beta(4)(1 - \,_5E_{40}) = 4.433991$$

$$\Rightarrow \quad 1000 \,_5P_{40}^{(4)} = A_{40}/\ddot{a}_{40:\overline{5|}}^{(4)} = 39.92 \;\; \text{and} \;\; 1000 \,_5P_{40}^{(4)}/4 = 9.98$$

$$1000 \,_5P^{(4)}(\overline{A}_{40}) = \overline{A}_{40}/\ddot{a}_{40:\overline{5|}}^{(4)} = 40.91 \;\; \text{and} \;\; 1000 \,_5P^{(4)}(\overline{A}_{40})/4 = 10.23$$

We use Part III of Code 7.7.4 to compute these values.
(iv) We have to find the six monthly premium for whole life annuity due of benefit of 1000 per annum, issued to (40), if it is purchased by 5-year temporary six monthly annuity due. The formula for the required premium is
$1000_5P^{(2)}(\ddot{a}_{40}) = 1000\ddot{a}_{40}/\ddot{a}_{40:\overline{5|}}^{(2)}$, where $\ddot{a}_{40} = (1 - A_{40})/d = 17.28326$. For $i = 0.05$ and $d = 0.04761905$,

$$i^{(2)} = 0.04939015 \;\; \text{and} \;\; d^{(2)} = 0.04819985$$

$$\alpha(2) = 1.000149 \;\; \text{and} \;\; \beta(2) = 0.2561738$$

$$\Rightarrow \quad \ddot{a}_{40:\overline{5|}}^{(2)} = \alpha(2)\ddot{a}_{40:\overline{5|}} - \beta(2)(1 - \,_5E_{40}) = 4.462782$$

$$\Rightarrow \quad 1000P^{(2)}(\ddot{a}_{40}) = 3872.75 \;\; \text{and} \;\; 1000P^{(2)}(\ddot{a}_{40})/2 = 1936.38$$

Thus the six-monthly premium of 1936.38, per six months, is to be paid for at most 5 years to get the survival benefit of 1000 per annum till (40) survives. We use Part IV of Code 7.7.4 to compute these values. From (i), (ii), (iii) and (iv) of this example we note that,

$$i > i^{(2)} > i^{(4)} > i^{(12)} > d^{(12)} > d^{(4)} > d^{(2)} > d$$

(v) It is required to find the monthly premium for 5-year endowment insurance issued to (40), when the benefit of 1000 is payable at the end of year of death and premiums form 5-year temporary monthly annuity due. Such a premium is given by the formula,

$$P_{40:\overline{5|}}^{(12)} = \frac{A_{40:\overline{5|}}}{\ddot{a}_{40:\overline{5|}}^{(12)}} = \frac{1 - d\ddot{a}_{40:\overline{5|}}}{\alpha(12)\ddot{a}_{40:\overline{5|}} - \beta(12)(1 - \,_5E_{40})}$$

We have calculated all the relevant quantities above. Hence substituting we get,

$$1000\, P^{(12)}_{40:\overline{5}|} = 177.75 \quad \text{and} \quad 1000\, P^{(12)}_{40:\overline{5}|}/12 = 14.81$$

Part V of Code 7.7.4 is related to the computation of these values.

(vi) We need to find the monthly premium for a 5-year term insurance issued to (40), when the benefit of 1000 is payable at the end of year of death and premiums form 5-year temporary monthly annuity due. This premium is given by,

$$1000 P^{1(12)}_{40:\overline{5}|} = 1000 \frac{A^{1}_{40:\overline{5}|}}{\ddot{a}^{(12)}_{40:\overline{5}|}} = 1000 \frac{1 - d\ddot{a}_{40:\overline{5}|} - {}_5E_{40}}{\ddot{a}^{(12)}_{40:\overline{5}|}} = 3.08$$

and $1000 P^{1(12)}_{40:\overline{5}|}/12 = 0.2563$, where $A^{1}_{40:\overline{5}|} = 0.01358002$ and $\ddot{a}^{(12)}_{40:\overline{5}|} = 4.414865$

(vii) We have to find the premiums as computed in (v) and (vi), when the benefit of 1000 is payable at the moment of death. When the benefit of 1000 is payable at the moment of death in term insurance, $A^{1}_{40:\overline{5}|}$ in the formula in (vi) is replaced by

$$\overline{A}^{1}_{40:\overline{5}|} = (i/\delta)A^{1}_{40:\overline{5}|} = (i/\delta)0.01358002 = 0.01391676$$

Hence, the premium is given by

$$1000\, P^{(12)}(\overline{A}^{1}_{40:\overline{5}|}) = 1000\, \frac{\overline{A}^{1}_{40:\overline{5}|}}{\ddot{a}^{(12)}_{40:\overline{5}|}} = 3.15 \quad \text{and} \quad 1000\, P^{(12)}(\overline{A}^{1}_{40:\overline{5}|})/12 = 0.2626$$

When the benefit of 1000 is payable at the moment of death in endowment insurance, $A_{40:\overline{5}|}$ in the formula in (v) is replaced by $\overline{A}_{40:\overline{5}|} = (i/\delta)A^{1}_{40:\overline{5}|} + {}_5E_{40}$ which comes out to be 0.7850636. Hence, the premium is given by

$$1000 P^{(12)}(\overline{A}_{40:\overline{5}|}) = 1000 \frac{\overline{A}_{40:\overline{5}|}}{\ddot{a}^{(12)}_{40:\overline{5}|}} = 177.82 \quad \text{and} \quad 1000\, P^{(12)}(\overline{A}^{1}_{40:\overline{5}|})/12 = 14.82$$

We use Parts VI and VII of Code 7.7.4 to compute these premiums. From the solutions of (v), (vi) and (vii), we note that the premiums for a benefit of 1000 do not change much whether the benefit is paid at the moment of death or at the end of year of death.

(viii) From $\overline{A}_{40:\overline{5}|} = 0.7850636$, we get $\overline{a}_{40:\overline{5}|} = 4.405322$ and hence

$$1000\overline{P}(\overline{A}_{40:\overline{5}|}) = 1000\overline{A}_{40:\overline{5}|}/\overline{a}_{40:\overline{5}|} = 178.21$$

In this solution we have computed the following annuity values.

$$\overline{a}_{40:\overline{5}|} = 4.405322 \quad \ddot{a}^{(12)}_{40:\overline{5}|} = 4.414865 \quad \ddot{a}^{(4)}_{40:\overline{5}|} = 4.433991$$

$$\ddot{a}^{(2)}_{40:\overline{5}|} = 4.462782 \quad \ddot{a}_{40:\overline{5}|} = 4.520736$$

It is to be noted that

$$\overline{a}_{40:\overline{5}|} \;<\; \ddot{a}^{(12)}_{40:\overline{5}|} \;<\; \ddot{a}^{(4)}_{40:\overline{5}|} \;<\; \ddot{a}^{(2)}_{40:\overline{5}|} \;<\; \ddot{a}_{40:\overline{5}|}$$

verifying the result proved in Example 6.5.6 and Example 6.5.7. From the formulae of premiums we note that we have computed all the required actuarial present values. Hence the values of premium are as follows.

$$
\begin{aligned}
1000\overline{P}(\overline{A}_{40:\overline{5}|}) &= 1000\overline{A}_{40:\overline{5}|}/\overline{a}_{40:\overline{5}|} = 178.21 \\
1000P^{(12)}(\overline{A}_{40:\overline{5}|}) &= 1000\overline{A}_{40:\overline{5}|}/\ddot{a}^{(12)}_{40:\overline{5}|} = 177.82 \\
1000P^{(4)}(\overline{A}_{40:\overline{5}|}) &= 1000\overline{A}_{40:\overline{5}|}/\ddot{a}^{(4)}_{40:\overline{5}|} = 177.06 \\
1000P^{(2)}(\overline{A}_{40:\overline{5}|}) &= 1000\overline{A}_{40:\overline{5}|}/\ddot{a}^{(2)}_{40:\overline{5}|} = 175.91 \\
1000P(\overline{A}_{40:\overline{5}|}) &= 1000\overline{A}_{40:\overline{5}|}/\ddot{a}_{40:\overline{5}|} = 173.66
\end{aligned}
$$

We use Part VIII of Code 7.7.4 to find these premiums. We note that,

$$\overline{P}(\overline{A}_{40:\overline{5}|}) \;>\; P^{(12)}(\overline{A}_{40:\overline{5}|}) \;>\; P^{(4)}(\overline{A}_{40:\overline{5}|}) \;>\; P^{(2)}(\overline{A}_{40:\overline{5}|}) \;>\; P(\overline{A}_{40:\overline{5}|})$$

This ordering follows from the ordering of the actuarial present values of annuities as proved in Example 6.5.6 and Example 6.5.7. In the next example we prove these inequalities for any n. $\blacksquare$

Example 7.5.3. Under the assumption of uniformity in unit age interval, prove that

$$\overline{P}(\overline{A}_{x:\overline{n}|}) \;>\; P^{(12)}(\overline{A}_{x:\overline{n}|}) \;>\; P^{(4)}(\overline{A}_{x:\overline{n}|}) \;>\; P^{(2)}(\overline{A}_{x:\overline{n}|}) \;>\; P(\overline{A}_{x:\overline{n}|}).$$

Solution: From the formulae of premiums we have,

$$\overline{P}(\overline{A}_{x:\overline{n}|}) \;=\; \frac{\overline{A}_{x:\overline{n}|}}{\overline{a}_{x:\overline{n}|}}, \qquad P^{(m)}(\overline{A}_{x:\overline{n}|}) \;=\; \frac{\overline{A}_{x:\overline{n}|}}{\ddot{a}^{(m)}_{x:\overline{n}|}}, \qquad P(\overline{A}_{x:\overline{n}|}) \;=\; \frac{\overline{A}_{x:\overline{n}|}}{\ddot{a}_{x:\overline{n}|}}$$

Note that numerator of all the premiums is the same. In Example 6.5.6 we have proved that

$$\overline{a}_{x:\overline{n}|} \;<\; \ddot{a}^{(m)}_{x:\overline{n}|} \;<\; \ddot{a}_{x:\overline{n}|}$$

and in Example 6.5.7 we have proved that

$$\ddot{a}^{(12)}_{x:\overline{n}|} \;<\; \ddot{a}^{(4)}_{x:\overline{n}|} \;<\; \ddot{a}^{(2)}_{x:\overline{n}|}$$

Hence the result follows. Thus, a premium with continuous payment is larger than that with m-thly payment which itself is larger than that for annual payment at the beginning of the year. $\blacksquare$

In the following example we find the net single premium, annual and monthly premiums for the benefit of 1000, payable either at the moment of death or at the end of year of death when the mortality pattern is governed by Makeham's law, which is a good model for human life length.

Example 7.5.4. Suppose the force of mortality follows Makeham's law given by $\mu_x = A + BC^x$. (i) Find the values of net single premium, annual premium and monthly premium for a benefit of 1000 to be paid at the end of year of death, for whole life insurance for ages 30 to 50. (ii) Find the values of net single premium, annual premium and monthly premium, paid as whole life annuity due, for the benefit of 1000 to be paid at the moment of death, for whole life insurance for ages 30 to 50. Take $A_{50} = 0.40356, i = 0.05, A = 0.0007, B = 0.0001151$ and $C = 1.096$.

Solution: Values of net single premium A_x are derived in Example 5.4.2 for Makeham's law with the given set of parameters. From these we get $\overline{A}_x$ by multiplying it by i/δ. Then using the formulae for premiums as illustrated in Example 7.5.2, we find the values of the annual and monthly premiums. These are reported in Table 7.10. We use Code 7.7.5 for these computations.

x	Single $1000\,A_x$	Annual $1000\,P_x$	Monthly $\dfrac{1000P_x^{(12)}}{12}$	Single $1000\,\overline{A}_x$	Annual $1000\,P(\overline{A}_x)$	Monthly $\dfrac{1000P^{(12)}(\overline{A}_x)}{12}$
30	180.30	10.73	0.92	184.88	11.00	0.94
31	188.58	11.33	0.97	193.38	11.62	1.00
32	197.15	11.98	1.03	202.17	12.28	1.05
33	206.02	12.65	1.09	211.26	12.98	1.11
34	215.19	13.37	1.15	220.66	13.71	1.18
35	224.65	14.13	1.21	230.37	14.49	1.24
36	234.43	14.93	1.28	240.39	15.31	1.31
37	244.51	15.78	1.36	250.72	16.19	1.39
38	254.90	16.68	1.43	261.38	17.11	1.47
39	265.60	17.64	1.52	272.36	18.09	1.56
40	276.62	18.65	1.60	283.65	19.12	1.65
41	287.95	19.72	1.70	295.27	20.22	1.74
42	299.59	20.86	1.80	307.21	21.39	1.84
43	311.54	22.07	1.90	319.46	22.63	1.95
44	323.80	23.35	2.01	332.03	23.95	2.06
45	336.36	24.72	2.13	344.91	25.35	2.19
46	349.23	26.17	2.26	358.10	26.84	2.32
47	362.38	27.72	2.39	371.60	28.42	2.46
48	375.83	29.37	2.54	385.38	30.11	2.60
49	389.56	31.12	2.69	399.46	31.91	2.76
50	403.56	33.00	2.86	413.82	33.84	2.93

Table 7.10 Makeham's law: Premiums for whole life insurance

From the table we note that as age increases, the premium of any type increases, as expected. Premiums corresponding to benefit payable at the moment of death are higher than corresponding to those when the benefit is payable at the end of year of death.   ∎

Remark 7.5.1. Note that the premiums computed in Example 7.5.4 are the net premiums. The premiums actually charged are higher than these to account for expenses incurred by the insurance company. Hence, Table 7.10 may not depict the real picture of premium payments, however this example illustrates the procedure to prepare the ready reckoner for the net single premium, annual and monthly premium for the benefit of 1000, payable either at the moment of death or at the end of year of death if the mortality pattern of the group under study is governed by Makeham's law and the effective rate of interest is 5%. On similar lines we can prepare the ready reckoner for the term insurance and the endowment insurance.

As mentioned in previous chapters, the mortality pattern is usually specified by a set of q_x values for integer values of x starting from 0 to some limiting age w. In the following example we find premiums for the insurance products discussed in Chapter 5, when the premiums are payable in various modes of annuity as discussed in Chapter 6. Given a set of q_x values, we cannot find the exact values of actuarial present values of the benefit to be paid at the moment of death. We find these values from the discrete setup, under the assumption of uniformity in unit age interval. We use Code 7.7.6 to compute the premiums.

Example 7.5.5. Suppose the mortality law is specified by values of q_x as specified in Table 4.8. Compute the fully discrete, fully continuous and semi-continuous premiums when the benefit of 1000 is paid to (25) (a) at the end of the year of death and (b) at the moment of death, under the assumption of uniformity in unit age interval, for the following insurance products and corresponding modes of premium payments. (i) n-year term insurance and n-year endowment insurance, when the premiums are payable as n-year temporary life annuity. (ii) n-year term insurance and n-year endowment insurance, when the premiums are payable as m-thly n-year temporary life annuity for $m = 2, 4, 12$. (iii) Whole life insurance when the premiums are payable as m-thly whole life annuity due for $m = 1, 2, 4, 12$. (iv) Whole life insurance when the benefit is paid at the moment of death and the premiums are payable as continuous whole life annuity. Suppose the effective rate of interest is $i = 0.06$.

Solution: As discussed in Chapter 4, when the mortality pattern is specified by a set of q_x values for integer values of x starting from 0 to some limiting age w, Code 4.7.2 computes the probability distribution of the curtate future life time random variable $K(25)$. Once we know the probability distribution of $K(25)$, we can find the actuarial present values of the benefit involved in term insurance, endowment insurance, and whole life insurance issued to (25). From these actuarial present values we find the values of actuarial present values of the benefit to be paid at the moment of death, under the assumption of uniformity in unit age interval. We have obtained these values in Example 5.5.5. We use the relations between the actuarial present values of benefit and the actuarial present values of annuities, as stated in Eqs 6.3.4, 6.3.8 and 6.4.2, to find the actuarial present values of the annuities. We have adopted this approach in Example 6.5.14, using Code 6.6.11. Once we have these two components required in the equivalence principle, we find the premiums using the formulae developed in this chapter. We use Code 7.7.6 to compute the premiums. Part I of the code specifies the mortality and interest pattern and related functions. (i) Parts II, III and IV of the code computes premiums for the n-year term insurance and n-year endowment insurance, when the premiums are payable as n-year temporary life annuity, for $n = 1$ to 10. These are reported in Table 7.11.

From Table 7.11, we observe that for the term and endowment insurance, the difference between the fully discrete and semi-continuous premiums is marginal for all values of n. Further, fully continuous premiums are larger than these two for all values of n. We note

n	$P^1_{25:\overline{n}|}$	$P(\overline{A}^{\,1}_{25:\overline{n}|})$	$\overline{P}(\overline{A}^{\,1}_{25:\overline{n}|})$	$P_{25:\overline{n}|}$	$P(\overline{A}_{25:\overline{n}|})$	$\overline{P}(\overline{A}_{25:\overline{n}|})$
1	1.10	1.14	1.17	943.40	943.43	971.75
2	1.11	1.15	1.18	458.25	458.28	472.04
3	1.12	1.16	1.19	296.73	296.76	305.67
4	1.14	1.17	1.21	216.11	216.15	222.64
5	1.16	1.19	1.23	167.86	167.89	172.94
6	1.18	1.22	1.25	135.79	135.82	139.90
7	1.21	1.24	1.28	112.96	113.00	116.40
8	1.23	1.27	1.31	95.92	95.95	98.84
9	1.26	1.29	1.33	82.72	82.76	85.25
10	1.28	1.32	1.36	72.23	72.27	74.44

Table 7.11 q_x Values: Annual premiums for term and endowment insurance

that the premiums for the endowment insurance are substantially higher than those for the term insurance, clearly reflecting the different nature of these two products.

(ii) We use Part V of Code 7.7.6 to compute the m-thly premiums, $m = 2, 4, 12$ for n-year term insurance and n-year endowment insurance. Table 7.12 displays monthly premiums, which divided by 12 give premiums to be paid per month.

n	$P^{1(12)}_{25:\overline{n}|}$	$P^{(12)}(\overline{A}^{\,1}_{25:\overline{n}|})$	$P^{(12)}_{25:\overline{n}|}$	$P^{(12)}(\overline{A}_{25:\overline{n}|})$
1	1.13	1.17	969.31	969.34
2	1.14	1.18	470.84	470.87
3	1.15	1.19	304.88	304.92
4	1.17	1.20	222.05	222.09
5	1.19	1.23	172.47	172.51
6	1.21	1.25	139.52	139.56
7	1.24	1.28	116.07	116.11
8	1.26	1.30	98.56	98.60
9	1.29	1.33	85.00	85.04
10	1.32	1.36	74.22	74.26

Table 7.12 q_x Values: Monthly premiums for term and endowment insurance

As observed in Table 7.11 for the term and endowment insurance, in Table 7.12 we note that the difference between the fully discrete and semi-continuous premiums is marginal for all values of n and the premiums for the endowment insurance are substantially higher than those for the term insurance, as expected. Table 7.13 displays quarterly premiums, which divided by 4 give premiums to be paid per quarter.

We note the similar features of the values of quarterly premium as those for monthly premiums. Table 7.14 displays six-monthly premiums, which divided by 2 give premiums to be paid for each half part of the year.

n	$P^{1(4)}_{25:\overline{n}\rvert}$	$P^{(4)}(\overline{A}^{\,1}_{25:\overline{n}\rvert})$	$P^{(4)}_{25:\overline{n}\rvert}$	$P^{(4)}(\overline{A}_{25:\overline{n}\rvert})$
1	1.13	1.16	964.53	964.56
2	1.14	1.17	468.52	468.55
3	1.15	1.18	303.38	303.41
4	1.16	1.20	220.96	222.99
5	1.18	1.22	171.62	171.66
6	1.21	1.24	138.83	138.87
7	1.23	1.27	115.50	115.53
8	1.26	1.30	98.07	98.11
9	1.28	1.32	84.58	84.62
10	1.31	1.35	73.85	73.89

Table 7.13 q_x Values: Quarterly premiums for term and endowment insurance

n	$P^{1(2)}_{25:\overline{n}\rvert}$	$P^{(2)}(\overline{A}^{\,1}_{25:\overline{n}\rvert})$	$P^{(2)}_{25:\overline{n}\rvert}$	$P^{(2)}(\overline{A}_{25:\overline{n}\rvert})$
1	1.12	1.15	957.41	957.45
2	1.13	1.16	465.06	465.10
3	1.14	1.17	301.14	301.17
4	1.15	1.19	219.33	219.36
5	1.18	1.21	170.36	170.39
6	1.20	1.23	137.81	137.84
7	1.22	1.26	114.64	114.68
8	1.25	1.29	97.35	97.38
9	1.27	1.31	83.96	83.99
10	1.30	1.34	73.31	73.34

Table 7.14 q_x Values: Six-monthly premiums for term and endowment insurance

For six-monthly premiums also we note the similar patterns as those for monthly premiums. (iii) Using Part VI of the Code 7.7.6 we compute premiums for whole life insurance when the premiums are payable as m-thly whole life annuity due for $m = 1, 2, 4, 12$, where $m = 1$ corresponds to annual premiums. These are reported in Table 7.15.

m	$P^{(m)}_{25}$	$P^{(m)}(\overline{A}_{25})$
1	4.52	4.65
2	4.59	4.72
4	4.62	4.76
12	4.65	4.79

Table 7.15 q_x Values: Premiums for whole life insurance

From Table 7.15, we observe that $P^{(m)}(\overline{A}_{25}) > P_{25}^{(m)}$ for all m. Further, annual premiums are the smallest while the monthly premiums are the largest.

(iv) Premium for the whole life insurance, when the benefit is paid at the moment of death and the premiums are payable as continuous whole life annuity, is given by $\overline{P}(\overline{A}_{25}) = 4.80$. In the fully continuous policy $\overline{P}(\overline{A}_{25})$ is the highest if we consider all the five modes of premium payment. Premiums for other ages can be found analogously. Thus, with appropriate changes in the Code 7.7.6, one can prepare a ready reckoner for annul or m-thly premiums for the benefit of 1000 in term, endowment and whole life insurances.   ■

Premiums for a variety of insurance and annuity contracts are thus determined by writing an equivalence relation between the actuarial present value of the outflow of the company and the actuarial present value of the inflow to the company.

The following example illustrates some variations in examples discussed earlier.

Example 7.5.6. A whole life insurance of 1 lakh is issued to (40), with the benefit payable at the end of the year of death. The annual premium, payable as a whole life annuity due is reduced by half after ten years, provided the policy is in force. Write the equivalence relation to determine the premium.

Solution: The actuarial present value of the outflow of the company, that is, of benefit is $100000A_{40}$. Suppose P is the annual premium to be determined. For first 10 years it is paid as 10-year temporary life annuity due. Hence its actuarial present value is $P\ddot{a}_{40:\overline{10}|}$. After 10 years if the policy is in force, the annual premium is $P/2$, paid as annuity due till (40) survives. Hence its actuarial present value is $(0.5P)(\,_{10|}\ddot{a}_{40})$. Thus, the equivalence relation to determine the premium is given by,

$$100000A_{40} = P\ddot{a}_{40:\overline{10}|} + 0.5P\,_{10|}\ddot{a}_{40}$$

■

The entire discussion so far in this section is for the true fractional premiums. A second type of fractional premium is the apportionable premium. Here, at death, a refund is made of a portion of the premium related to the length of time between the time of death and the time of the next scheduled premium payment. We will not discuss it in this book. Interested readers may refer to Section 6.5 in the book by Bowers et al [1]. We now proceed to discuss how to obtain gross premiums.

7.6   Gross Premiums

Insurance models discussed so far have not incorporated several aspects of insurance practice. For example, an insurer has cash outflows other than claim payments. These are expenses to run the business. Expenditures of various types include salaries paid to the staff, commissions paid to the agents, taxes and license fees, rent paid for the offices, cost of establishing and maintaining the offices, as well as those for selling and providing service for active insurance policies. Most expenses are classified as, (a) initial, that is, incurred at the issue of the policy and (b) renewal, that is, incurred regularly while the policy is in force. There are settlement expenses, expenses of payment of benefits and the expenses of maintaining records of policies with continuing benefits after premiums have ceased. Some expenses are related to the premium and benefit amount and some are not related to either of them. For example commissions are related to the premium or the benefit amount while the cost of purchasing the office or rent of the office does not depend on the benefit or premium. When expenses are related to the benefit, these are specified as a percentage of premium or benefit.

Some insurance companies employ a policy fee system, whereby a fixed number is added to the annual premium. For example, the gross annual premium for a policy with a benefit of Rs 20,000 may be quoted as 'Rs 50 per Rs 1000 plus a policy fee of Rs 50'. Thus the premium for a benefit of Rs 20,000 will be $20 \times 50 + 50 = 1050$ rupees. This system reflects the fact that certain administrative costs do not depend on the size of the benefit.

The expenses of running the business must be met from premiums and investment income. Consequently, the insurance company has to add an amount to the net premium to cover all these operating costs and to provide itself with some profit. The total amount added to the net premium to cover all of the insurer's cost of doing business is called loading. The net premium with the loading added is called the gross premium, which is the premium amount the insurer charges the policy owner to keep the policy in force. The difference between the gross premium and the net premium is known as expense loading. We discuss below how to incorporate expenses into the equivalence principle for net premiums to find the gross premiums.

The equivalence principle to determine net premiums is

$$\text{Expected present value of inflow} \ = \ \text{Expected present value of outflow}$$

where the inflow to the company is via net premiums and the outflow of the company is via benefit claims. This equivalence principle is extended to take into account the administrative expenses. Thus, the expected present value of the outflow of the insurance company considers output corresponding to death benefits as well as administrative expenses. We assume that administrative expenses are not random variables but are fixed by the company at the time of issue of the policy. The expense augmented equivalence principle is as follows.

$$\text{Expected present value of gross premiums}$$
$$= \text{Expected present value of benefits and expenses}$$

To elaborate on such an expense augmented principle, as an illustration, consider the 15-year endowment insurance issued to (30), with a benefit of 1000, payable at the end of year of death. Expenses are Rs 50 per renewal premium payment (including the first), plus Rs 100 at the time of signing the contract. Then the expected present value of benefit and expenses is $1000A_{30:\overline{15|}} + 50\ddot{a}_{30:\overline{15|}} + 100$. The expected present value of annual premiums, with a gross premium G, is $G\,\ddot{a}_{30:\overline{15|}}$. Hence the expenses augmented equivalence principle gives the equation

$$1000A_{30:\overline{15|}} + 50\ddot{a}_{30:\overline{15|}} + 100 = G\ddot{a}_{30:\overline{15|}}$$

Solving the equation we get the gross premium G. Thus, to determine the gross premium G along with the interest and mortality pattern, it is essential to have information regarding the expenses of the insurance company. In the following example we take a hypothetical situation describing the expenses, in one of the examples of determining the net premium and discuss the computation of gross premium.

Example 7.6.1. It is given that $q_{28} = 0.135$, $q_{29} = 0.146$, $q_{30} = 0.159$, $q_{31} = 0.173$, $q_{32} = 0.188$ and $i = 0.05$. Find the annual premium paid as 3-year temporary discrete annuity due, for a benefit of 50,000, payable at the end of year of death, in a 5-year term insurance to (28). Expenses, paid at the beginning of the policy year, are as shown in Table 7.16. The settlement expenses, paid at the end of the year of death, are 50 per policy plus 10 per 1000 of insurance.

Year	Per policy	Per 1000 of insurance	Fraction of premium
First	50	6	0.25
Second	25	3	0.15
Third	25	3	0.15

Table 7.16 Expenses per policy year

Using the equivalence principle find the gross premium G for an insurance of Rs 50,000.

Solution: Note that the insurance policy is a fully discrete 5-year term insurance. For the benefit of Rs 50,000, the actuarial present value of the benefit is $50000A^{1}_{28:\overline{5}|}$. In Example 5.4.3, we have obtained $A^{1}_{28:\overline{5}|} = 0.5086848$ for the given values of q_x and $i = 0.05$. Hence, $50,000A^{1}_{28:\overline{5}|} = 25434.24$. Now we compute the expected present values of expenses. Settlement expenses are Rs 50 per policy plus 10 per 1000, hence, for a policy of Rs 50,000, settlement expenses are $50 + 10 \times 50 = 550$ rupees. These are paid at the end of the year of death. If death occurs in the first year after signing the contract, then 550 is the expense incurred at the end of first year to settle the policy, hence its present value is $550 \times v \times q_{28}$. Death may or may not occur in the 5-year term. If it occurs after 5 years then no settlement is needed. Thus for the given data, the actuarial present value E of settlement expenses is given by,

$$
\begin{aligned}
E &= 550 \left(v\, q_{28} + v^2 p_{28} q_{29} + v^3 p_{28} p_{29} q_{30} + v^4 p_{28} p_{29} p_{30} q_{31} \right. \\
&+ \left. v^5 p_{28} p_{29} p_{30} p_{31} q_{32} \right) = 550 A^{1}_{28:\overline{5}|} \\
&= 550 \times 0.5086848 = 279.7766
\end{aligned}
$$

Other expenses given in Table 7.16 are paid at the beginning of the year. Expenses given in the second column are per 1000 of insurance, hence for the insurance of Rs 50,000, these numbers are to be multiplied by 50. Hence the actuarial present value E_1 of these expenses is,

$$
\begin{aligned}
E_1 &= (50 + 300 + 0.25G) + p_{28} v (25 + 150 + 0.15G) \\
&+ p_{28} p_{29} v^2 (25 + 150 + 0.15G) \\
&= 350 + 0.25G + (175 + 0.15G) \left(\ddot{a}_{28:\overline{3}|} - 1 \right) = 611.4223 + 0.4740762G
\end{aligned}
$$

as for the given data $\ddot{a}_{28:\overline{3}|} = 2.493841$. There are at most 3 premiums, paid at the beginning of the year. Hence, its actuarial present value is

$$
G + p_{28} v G + p_{28} p_{29} v^2 G = G\ddot{a}_{28:\overline{3}|} = 2.493841G
$$

With the extended equivalence principle,

actuarial present value of benefits $+$ actuarial present value of expenses

$=$ actuarial present value of expense loaded premiums

$\Rightarrow\quad 50550A\,^{1}_{28:\overline{5}|} + 350 + 0.25G + (175 + 0.15G)\,(\ddot{a}_{28:\overline{3}|} - 1) = G\ddot{a}_{28:\overline{3}|}$

$\Rightarrow\quad 25434.24 \;+\; 279.7766 \;+\; 611.4223 + 0.4740762G = 2.493841G$

$\Rightarrow\quad G = 13033.91$

Thus, $G = 13033.91$ is the annual premium to be paid at the beginning of the year for at most 3 years. Without expenses the annual premium is $25434.24/2.493841 = 10198.82$ for the cover of 50000. Thus, the loading for expenses is $13033.91 - 10198.82 = 2835.09$. ∎

Example 7.6.2. A 10-year endowment insurance is sold to (40). A benefit of Rs 10,000 is payable at the end of year of death or at the end of the term of 10 years, whichever is earlier. Suppose there are initial expenses of Rs 200 and renewal expenses of 0.1% of the benefit insured, incurred at the beginning of the year, including the first year. There are settlement expenses of Rs 50. Find the purchase price of the insurance, if the force of interest is 5% and the mortality of (40) is modelled by Gompertz' law given by, $\mu_x = BC^x$, with $B = 0.0001151$ and $C = 1.096$.

Solution: Note that the purchase price of the 10-year endowment insurance is the actuarial present value of the benefit and the expenses. The initial expense is Rs 200, the renewal expenses of 0.1% of the benefit insured, incurred at the beginning of the year form a 10-year temporary life annuity due, with an actuarial present value $0.001 \times 10000 \times \ddot{a}_{40:\overline{10}|}$. The settlement expense may be at the end of the first year or the second year or at the end of 10th year, depending on when the death occurs. Thus the actuarial present value of the settlement expense is given by $50A_{40:\overline{10}|}$. For the specified mortality law and the force of interest we have from Example 5.4.4, $A_{40:\overline{10}|} = 0.61611$ and from Example 6.4.5, $\ddot{a}_{40:\overline{10}|} = 7.8713$. From this information we write the actuarial present value of the benefit and expenses as follows, which is the purchase price of the policy. We denote it by G. Thus,

$$\begin{aligned} G &= 10000A_{40:\overline{10}|} + 200 + 0.001 \times 10000 \times \ddot{a}_{40:\overline{10}|} + 50A_{40:\overline{10}|} \\ &= 6161.10 + 200 + 78.71 + 50 \times 0.6161 = 6470.62 \end{aligned}$$

Without expenses the purchase price is 6161.10. Thus, the loading for expenses is 309.52. ∎

Example 7.6.3. An insurance company offers a whole life annuity product that pays the annuitant Rs 12,000 at the beginning of each year and a death benefit of 10000, that will be paid at the end of the year of death. Expenses, payable at the beginning of the year are as follows:
(a) Taxes are 0.5% of the expense-loaded premium. (b) Commissions are 2% of the expense-loaded premium. (c) Other expenses are 100 in the first year and 50 in following years till the contract is in force.

Suppose such a contract is signed by (25) by paying the premium at the beginning of each year, including the first, for at most 10 years. Assume that the life length of (25) is modelled by Gompertz' law with a force of mortality $\mu_x = BC^x$, $x > 0$ with $B = 0.0001151$ and $C = 1.096$. Find the gross premium, if the force of interest is 0.05.

Solution: In this contract, there is a survival benefit of Rs 12,000 per annum and a death benefit of Rs 10,000. Hence for this insurance contract, the actuarial present value of the benefit is given by

$$12000\ddot{a}_{25} + 10000A_{25} = 12000\ddot{a}_{25} + 10000(1 - d\ddot{a}_{25})$$

The expenses payable at the beginning of the year form annuity due. Suppose G denotes the gross premium. Thus, the actuarial present value of expenses is given by,

$$0.005G\ddot{a}_{25} + 0.02G\ddot{a}_{25} + 100 + 50\ddot{a}_{25} - 50$$

The first term corresponds to expenses due to taxes, the second corresponds to expenses due to commission, the third term specifies the initial cost and the last two correspond to renewal expenses, Rs 50 is subtracted as the renewal expenses are from the second year onwards. For the given mortality and interest pattern we have calculated $\ddot{a}_{25} = 17.4538$ in Example 6.4.8. Thus the actuarial present value of the benefit and the expenses is given by $210933.3 + 0.436345\ G + 922.69$. A premium G is paid as 10-year temporary life annuity due. Hence, the actuarial present value of the premium payments is given by, $G\ddot{a}_{25:\overline{10}|}$. For the given mortality and interest pattern we have calculated $\ddot{a}_{25:\overline{10}|} = 8.0173$ in Example 6.4.5. Hence, the equivalence relation to determine G is obtained by equating the actuarial present value of the premium payments to the actuarial present value of the benefit and the expenses. Thus, we get the equation,

$$\ddot{a}_{25:\overline{10}|}G = 8.0173G = 210933.3 + 0.436345\ G + 922.69$$
$$\Rightarrow\ G = 27945.82$$

By paying the premium G for at most 10 years, (25) gets the annual benefit of Rs 12,000 per annum till he survives and a death benefit of Rs 10,000 at the end of year of death. ■

Example 7.6.4. An insurance company offers a whole life insurance product with the specified benefit amount to be paid at the end of year of death. If death occurs during a term of n-years then along with the death benefit, the bonus is also paid. If death occurs after the end of the term, no bonus is paid. The amount of bonus is not fixed a priori. At the end of each year, depending on the reserve calculations and after accounting for the company's expenses and profits, the bonus is determined. A loading on premium is charged to pay for the bonus. The premium loading is 2.25 per thousand benefit if the term is 5 to 9 years, 1.5 per thousand benefit if the term is 10 to 14 years, Rs 1.25 per thousand if the term is 15 to 19 years, 1.15 per thousand benefit if the term is 20 to 24 years and 1 per thousand benefit if the term is 25 years or more. There is a rebate on premium, with rate of rebate depending on the sum assured. The rebate is Rs 1 per thousand, if the benefit amount is more than or equal to 1.5 lakhs but less than 5 lakhs, Rs 1.5 per thousand, if the benefit amount is more than or equal to 5 lakhs but less than 10 lakhs, Rs 1.75 per thousand, if the benefit amount is more than or equal to 10 lakhs. Suppose (40) signs this contract with premiums payable as n-year temporary annuity due. Determine the annual premium in the following two setups. (i) Benefit amount Rs 2 lakhs and premium paying period of 5 years and (ii) benefit amount Rs 5 lakhs and premium paying period of 10 years. Assume that life length of (40) is modelled by Gompertz' law with a force of mortality $\mu_x = BC^x$, $x > 0$ with $B = 0.0001151$ and $C = 1.096$ and the force of interest is 0.05.

Solution: In this contract, the actuarial present value of the premiums is given by $G\ddot{a}_{40:\overline{n}|}$, where G denotes the annual premium. This premium has to take into account the bonus payments and the rebate given on the premiums. The bonus amount is available only through the premium loading. Thus, the actuarial present value of the premiums paid should equal the actuarial present value of the benefit, allowance for the bonus and rebate. The actuarial present value of the benefit of b amount is bA_{40}. If c denotes the loading for bonus per 1000, then the loading for bonus for benefit b will be $c(b/1000)$. It is payable at the beginning of each year for at most n years. Hence, its actuarial present value is $c(b/1000)\ddot{a}_{40:\overline{n}|}$. Suppose R denotes the rebate on the premium, when benefit is of 1000. For b benefit, the rebate will be $R(b/1000)$, which again forms a n-year temporary life annuity due. Hence, its actuarial present value is $R(b/1000)\ddot{a}_{40:\overline{n}|}$. Hence, the equivalence relation to determine G is given by,

$$\begin{aligned}
G\ddot{a}_{40:\overline{n}|} &= bA_{40} + c(b/1000)\ddot{a}_{40:\overline{n}|} + R(b/1000)\ddot{a}_{40:\overline{n}|}\\
\Rightarrow G &= bA_{40}/\ddot{a}_{40:\overline{n}|} + c(b/1000) + R(b/1000)
\end{aligned}$$

For Gompertz' law with a force of mortality $\mu_x = BC^x$, with $B = 0.0001151$ and $C = 1.096$ and the force of interest 0.05, we have computed in Example 5.4.4 and in Example 6.4.5

$$A_{40} = 0.27673, \quad \ddot{a}_{40:\overline{5}|} = 4.4912 \quad \text{and} \quad \ddot{a}_{40:\overline{10}|} = 7.8713$$

Using these values we find premium in (i) and (ii).Thus,

$$\begin{aligned}
b &= 2,00,000 \quad \text{and} \quad n = 5, \quad \Rightarrow \quad G = 12973.21\\
\text{and} \quad b &= 5,00,000 \quad \text{and} \quad n = 10, \quad \Rightarrow \quad G = 19078.42
\end{aligned}$$

∎

Examples 7.6.1 to 7.6.4 illustrate the procedure to find the gross premiums. Using a similar procedure, in all the examples in Sections 7.3, 7.4 and 7.5, we can find gross premiums by incorporating the specified expenses in the equivalence principle.

In the first two case studies described in Section 1.5, the premiums reported are office premiums, which take into account information about the interest rate, mortality pattern, expense patterns and also the profit. We do not have information on any of these four components. Hence, we cannot compute these although we are now familiar with the technique to determine such premiums. In the following illustration, we compute the net premiums for the products in the first two case studies under the assumption of some mortality law and rate of interest. Similarly, for case studies 3 and 4 also we do not have information on any of the four components needed to compute the annual and quarterly benefit amounts. We compute these benefit annuity amounts under the assumption of some mortality law and rate of interest, in the absence of the information related to expense and profit and compare with the reported ones in Section 1.5. We use Code 7.7.7 for these computations.

Case study 1: It is concerned with LIC's Jeevan Umang plan signed by Vidula, a lady of 45 years, for a cover of Rs 5 lakhs with a survival benefit equal to 8% of basic sum assured, payable each year after the premium paying term of 15 years is over. From the calculation on the LIC website, she will have to pay an annual premium of Rs 38772 for 15 years, including taxes, provided she survives till 60. We assume that the policy is fully discrete and the underlying force of mortality follows Gompertz' law given by $\mu_x = BC^x$ with $B = 0.0001151$ and $C = 1.096$. Suppose the effective rate of interest is $i = 0.04$. We do not know the mortality and interest pattern that LIC has adopted. It may be different from what we have adopted. The equivalence principle to determine premium P for a unit benefit leads to the following equation.

$$\begin{aligned}
P\ddot{a}_{45:\overline{15|}} &= A_{45} + 0.08\ _{15|}\ddot{a}_{45} = A_{45} + 0.08(\ddot{a}_{45} - \ddot{a}_{45:\overline{15|}}) \\
\Rightarrow P &= (A_{45} + 0.08\ddot{a}_{45})/\ddot{a}_{45:\overline{15|}} - 0.08 = 0.07152043 \\
\text{and } 5 \times 10^5 P &= 35760.21
\end{aligned}$$

For the assumed mortality and interest pattern, $A_{45} = 0.4129972, \ddot{a}_{45} = 15.26207$ and $\ddot{a}_{45:\overline{15|}} = 10.78378$. The premium reported in the case study is Rs 38772. Thus, $38772 - 35760.21 = 3011.79$ could be a loading for expenses and profit. If the effective rate of interest is taken as $i = 0.05$, then the premium comes out to be 31047.36.

Case study 2: In this case study Mr Ketan, a healthy 50 year old person, opts for a 10 year term Saral Jeevan Bima policy for sum assured Rs 20 lakhs. From the SBI Life website, Mr Ketan is required to pay a premium of Rs 22140 for 10 years or till the policy is active. We assume that the policy is fully discrete and underlying force of mortality follows Gompertz' law given by $\mu_x = BC^x$ with $B = 0.0001151$ and $C = 1.087$. Suppose the effective rate of interest is $i = 0.06$. In this case also the mortality and interest pattern that SBI Life has adopted may be different from what we have adopted. Using the equivalence principle, the premium $P^1_{50:\overline{10|}}$ for a unit benefit is given by,

$$\begin{aligned}
P^1_{50:\overline{10|}} &= A^1_{50:\overline{10|}}/\ddot{a}_{50:\overline{10|}} = 0.07818631/7.505004 = 0.01041789 \\
\Rightarrow 20 \times 10^5 P &= 20835.78
\end{aligned}$$

Thus, $22140 - 20835.78 = 1304.22$ could be a loading for expenses and profit.

Case study 3: It is related to the annuity product by HDFC Life in which Mr Bapat of age 57 purchased the immediate annuity policy with purchase price Rs 10,58,652. The annual benefit of Rs 68503 is payable annually in arrears. We assume that the annuity contract is fully discrete and underlying force of mortality follows Gompertz' law given by $\mu_x = BC^x$ with $B = 0.0001151$ and $C = 1.096$. Suppose the effective rate of interest is 6.5%. The equivalence principle to determine the benefit amount B leads to the following equation.

$$\begin{aligned}
1058652 &= Ba_{57} + 1058652A_{57} = 8.443088B + 1058652 \times 0.4236613 \\
\Rightarrow B &= 72265.28
\end{aligned}$$

Note that this benefit amount does not take into account the expenses and profit, hence the actual benefit amount will be smaller than this. It is reported as 68503, under different mortality and interest pattern. Thus, $72265.28 - 68503 = 3762.28$ could be a loading for expenses and profit.

Case study 4: It is also an annuity product by HDFC Life in which Mr Singh of age 65 purchased the policy with a single premium of Rs 4911591. The quarterly benefit payment is

Rs 67708.43, payable in arrears. We assume that the annuity contract is fully discrete and the underlying force of mortality follows Gompertz' law given by $\mu_x = BC^x$ with $B = 0.0001151$ and $C = 1.096$. Suppose the effective rate of interest is 5.4%. The equivalence principle to determine the benefit amount B leads to the following equation.

$$4911591 = 4Ba_{65}^{(4)} + 4911591A_{65} = 7.192109B + 4911591 \times 0.599163$$
$$\Rightarrow B = 68434.29$$

Note that this quarterly benefit amount does not take into account expenses and profit, hence the actual benefit amount will be smaller than this. It is reported as 67708.43. Thus, $4(768434.29 - 67708.43) = 2902.57$ could be the annual loading for expenses and profit.

The next section presents the codes used to compute the premiums discussed in this chapter.

7.7   R Codes

Chapter 5 and Chapter 6 provide the two components necessary in the computation of benefit premiums for various insurance products and various modes of premium payment. Consequently R codes for the calculation of premiums can be easily written from R codes in Section 5.6 and 6.6, with some additional functions. In this section we present some codes used to solve the examples in Sections $3, 4, 5$ and 6.

Code 7.7.1. In Example 7.3.4 we have obtained fully continuous premiums for some insurance products with different modes of premium payments. The underlying force of mortality follows Gompertz' law given by $\mu_x = BC^x$ with $B = 0.0001151$ and $C = 1.096$. It is given that $\delta = 0.05$. Part I of the code specifies the parameters of mortality law, value of δ and other related parameters. (i) We use Part II of this code to obtain fully continuous premiums for the benefit of 1000 for n-year endowment insurance, $n = 1, 2, \cdots, 10$ and issued to (x) where $x = 25, 30$, when premiums are paid for at most n years. These are reported in Table 7.2. (ii) Part III of this code computes the premium for the 5-year endowment insurance issued to (25), when premiums are paid for at most 3 years. (iii) Part IV of this code computes the premium for the whole life insurance for (x) with $x = 25, 30$, when premiums are paid as a whole life annuity. Premiums for whole life insurance are displayed in Table 7.3. (iv) Part V of this code computes the premium for the whole life insurance for (25), when premiums are paid for at most 20 years. (v) Part VI of this code computes the premium for the 30-year deferred whole life annuity for (30), when premiums are paid for at most 30 years.

```
# Part I: parameters of Gompertz' law and interest pattern
B=0.0001151; C=1.096; m=B/log(C); del=0.05; n=1:10;v=exp(-del)
la=(-del/log(C))+1; x=c(25,30); alx=m*C^x
T=P=S=matrix(nrow=length(n),ncol=length(x))
for(i in 1:length(n))
{
for(j in 1:length(x))
```

```r
{
S[i,j]=pgamma(C^n[i],la,alx[j])-pgamma(1,la,alx[j])
T[i,j]=exp(alx[j])*gamma(la)*alx[j]^(1-la)*S[i,j]
P[i,j]=exp(-del*n[i]-alx[j]*(C^n[i]-1))
}
}
E=T+P # Endowment insurance
# Part II: Fully continuous premiums for endowment insurance
# Premiums paid as n-year temporary continuous life annuity
E1=del*E/(1-E)
d1=round(1000*data.frame(E[,1],E1[,1],E[,2],E1[,2]),2)
PbarEn=data.frame(n,d1); PbarEn # Table 7.2
# Part III: Fully continuous premiums for 5-year
# endowment insurance for (25), premiums paid as 3-year
# temporary continuous life annuity
Pbar25En3=round(1000*del*E[5,1]/(1-E[3,1]),2); Pbar25En3 # (ii)
# Part IV: Whole life insurance, premiums as whole life annuity
w=c()
for(j in 1:length(x))
{
w[j]=exp(alx[j])*gamma(la)*alx[j]^(1-la)*(1-pgamma(1,la,alx[j]))
}
w1=round(1000*w,2); w1
PbarW=round(1000*del*w/(1-w),2); PbarW # (iii)
# Part V: Whole life insurance, premiums at most 20 years
n=20; T=P=S=c()
for(j in 1:length(x))
{
S[j]=pgamma(C^n,la,alx[j])-pgamma(1,la,alx[j])
T[j]=exp(alx[j])*gamma(la)*alx[j]^(1-la)*S[j]
P[j]=exp(-del*n-alx[j]*(C^n-1))
}
E20=T+P ; E20 # Endowment insurance n=20
Pbar20W=round(1000*del*w/(1-E20),2); Pbar20W # (iv)
d1=data.frame(x,w1,PbarW,Pbar20W );d1 # Table 7.3
# Part VI: 30-year deferred whole life annity
# premiums at most 30 years
n=30; Pu=exp(-del*n-alx[2]*(C^n-1));Pu
x=60;alx60=m*C^x
```

```
w60=exp(alx60)*gamma(la)*alx60^(1-la)*(1-pgamma(1,la,alx60));w60
S1=pgamma(C^n,la,alx[2])-pgamma(1,la,alx[2])
T1=exp(alx[2])*gamma(la)*alx[2]^(1-la)*S1
P1=exp(-del*n-alx[2]*(C^n-1))
E30=T1+P1 ; E30 # Endowment insurance n=30
Pbar30defwa=round(1000*Pu*(1-w60)/(1-E30),2)
Pbar30defwa # (v)
```

∎

Code 7.7.2. In Example 7.4.2, the mortality law is specified by q_x values as given in Table 6.15, $i = 0.05$ and $A_{50} = 0.251$. Part I of the code specifies the mortality pattern and interest pattern. (i) We have obtained the fully discrete annual premiums for the whole life insurance of benefit of 1000, issued to (x) for $x = 40$ to 50, payable at the end of the year of death, when the premiums are paid as the whole life annuity due. (ii) We have also obtained semi-continuous annual premiums for the same product, when benefit is payable at the moment of death, under the assumption of uniformity in unit age intervals. We use Part II of this code to obtain the premiums in (i) and (ii), which are reported in Table 7.5. (iii) We have obtained the annual premium for the whole life insurance to (40), when the benefit is payable either at the end of year of death or at the moment of death and when the premium payments form a 5-year temporary annuity due. We use Part III of this code to obtain the values. (iv) Using Part IV of the code, we obtain the annual premium for the whole life annuity due of benefit of 1000 per annum, issued to (40), when it is purchased by a 5-year temporary annuity due.

```
# Part I: Mortality law and interest rates
int=0.05; v=(1+int)^(-1); d=int/(1+int);del=log(1+int)
q=c(.0027,.0029,.0032,.0034,.0037,.0039,.0043,.0046,.0050,.0054)
# Part II: Discrete and semi-continuous premiums
p=1-q; l=length(p); A50=0.251; ad50=(1-A50)/d; ad=c(1:l,ad50)
for (i in 1:l)
{
ad[l+1-i]=1+v*p[l+1-i]*ad[l+1-i+1]
}
ad
Px=(1-d*ad)/ad # discrete premium
A=c(1:l,A50)
for (i in 1:l)
{
A[l+1-i]<-v*q[l+1-i] + v*p[l+1-i]*A[l+1-i+1]
}
Abar=(int/del)*A; age=40:50
Pbar=Abar/ad # semi-continuous premium
d1=round(1000*data.frame(Px,Pbar),2)
```

```
d2=data.frame(age,d1);d2 # Table 7.5
# Part III: # Premium payments as 5-year temporary annuity due
p1=c(1,p[1], p[1]*p[2], p[1]*p[2]*p[3], p[1]*p[2]*p[3]*p[4])
v1=c(1,v,v^2,v^3,v^4); ad540=sum(p1*v1); ad540
P540=round(1000*A[1]/ad540,2); P540
Pbar540=round(1000*Abar[1]/ad540,2); Pbar540
# Part IV: # Premium for whole life annuity
# as 5-year temporary annuity due
wa40=1000*(1-A[1])/d; wa40
Pwa40=round(wa40/ad540,2); Pwa40
```

∎

Code 7.7.3. In Example 7.4.6, the mortality law is Gompertz' law. Part I of this code specifies the parameters of Gompertz' law and the interest rate.
(i) Using Part II of this code, we have obtained the fully discrete premiums for an n-year endowment insurance, where the premiums are payable as n-year temporary annuity due, for $n = 1$ to 10 and $x = 25, 30, 35, 40$. These are reported in Table 7.6. (ii) Using Part III of this code, we have found the fully discrete premiums for the whole life insurance, where the premiums are payable as whole life annuity due. We use Part IV to compute the fully discrete premiums for the whole life insurance, when premiums are payable as 10-year temporary life annuity due. These are reported in Table 7.7.

```
# Part I: Gompertz' Mortality law and interest rate
B=0.0001151; C=1.096; m=B/log(C); del=0.05; v=exp(-del)
d=1-v; int=1/v-1; x=c(25,30,35,40); k=0:(100-min(x))
y=t=matrix(nrow=length(k),ncol=length(x))# pmf of K(x)
for(i in 1:length(k))
{
for(j in 1:length(x))
{
y[i,j]=exp(-m*C^x[j]*(C^k[i]-1))-exp(-m*C^x[j]*(C^(k[i]+1)-1))
}
}
b=v^(k+1);w=c()
for(j in 1:length(x))
{
w[j]=sum(y[,j]*b)
t[,j]=cumsum(y[,j]*b)
}
n=1:10; T=t[1:length(n),] # Term insurance
P=matrix(nrow=length(n),ncol=length(x))
```

```
for(i in 1:length(n))
{
for(j in 1:length(x))
{
P[i,j]=v^(n[i])*exp(-m*C^x[j]*(C^n[i]-1))
}
}
E=T+P; # Endowment insurance
# Part II: Fully discrete premiums for n-year endowment insurance
# premium payable as n-year temporary life annuity due
adue=(1-E)/d; PrEn=round(1000*E/adue,2);PrEn # Table 7.6
# Part III: Premiums for whole life insurance,
# payable as whole life annuity due
wadue=(1-w)/d; wbar=(int/del)*w
Pwadue=round(1000*(w/wadue),2)
Pbarwadue=round(1000*(wbar/wadue),2)
# Part IV: Premiums for whole life insurance,
# payable as 10-year temporary life annuity due
P10adue=round(1000*(w/adue[10]),2)
Pbar10adue=round(1000*(wbar/adue[10]),2)
d1=data.frame(x,Pwadue,Pbarwadue,P10adue,Pbar10adue)
d1 # Table 7.7
```

∎

Code 7.7.4. In Example 7.5.2 the mortality law is specified by q_x values as given in Table 6.15. Further, $i = 0.05$, $A_{50} = 0.251$. We have obtained the monthly premium for ages 40 to 50, for whole life insurance with a benefit of 1000, payable at the end of the year of death, when the premiums are paid as monthly whole life annuity due. (ii) For the setup in (i) we have also found the monthly premium, if the benefit is payable at the moment of death. These are displayed in Table 7.9. We have also obtained quarterly, six monthly premiums for a variety of insurance products as described in Example 7.5.2. We use eight parts of this code to compute these premiums in various cases.

```
# Part I: Mortality law and interest rate
int=0.05; v=(1+int)^(-1); d=int/(1+int); m=c(2,4,12)
im=m*((1+int)^(1/m)-1); dm=m*(1-v^(1/m))
alm=(int*d)/(im*dm); bm=(int-im)/(im*dm)
alm;bm
q=c(.0027,.0029,.0032,.0034,.0037,.0039,.0043,.0046,.0050,.0054)
# Part II: Monthly premiums
p=1-q; l=length(p); A50=0.251; A=c(1:l,A50)
```

```
for (i in 1:l)
{
A[l+1-i]<-v*q[l+1-i] + v*p[l+1-i]*A[l+1-i+1]
}
A; Abar=(int/del)*A
ad50=(1-A50)/d; ad=c(1:l,ad50)
for (i in 1:l)
{
ad[l+1-i]=1+v*p[l+1-i]*ad[l+1-i+1]
}
adue12=alm[3]*ad-bm[3];
Px12=A/adue12; Pbarx12=Abar/adue12
d1=round(1000*data.frame(Px12,Px12/12,Pbarx12,Pbarx12/12),2)
age=c(40:50); d2=data.frame(age,d1);d2 # Table 7.9
# Part III: quarterly premiums,5-year annuity due
# Premium payments as 5-year temporary quarterly annuity due
p1=c(1,p[1], p[1]*p[2], p[1]*p[2]*p[3], p[1]*p[2]*p[3]*p[4])
v1=c(1,v,v^2,v^3,v^4); ad540=sum(p1*v1); ad540
E540=v^5*p[1]*p[2]*p[3]*p[4]*p[5]; E540
ad4540=alm[2]*ad540-bm[2]*(1-E540); ad4540
P404=round(1000*A[1]/ad4540,2); P404; P404/4 # (iii)
Pbar404=round(1000*Abar[1]/ad4540,2); Pbar404; Pbar404/4 #(iii)
# Part IV: six monthly premium,5-year six monthly annuity due
# whole life annuity
ad2540=alm[1]*ad540-bm[1]*(1-E540); ad2540
wa40=(1-A[1])/d; wa40
P2wa40=round(1000*wa40/ad2540,2);P2wa40; round(P2wa40/2,2) #(iv)
# Part V: monthly premium for 5-year endowment insurance
# 5-year temporary monthly annuity due
ad12540=alm[3]*ad540-bm[3]*(1-E540); ad12540
A540=1-d*ad540
P12540=round(1000*A540/ad12540,2); P12540
round(P12540/12,2) # (v)
# Part VI: monthly premium for 5-year term insurance
# 5-year temporary monthly annuity due
TA540=1-d*ad540-E540; TA540
TP12540=round(1000*TA540/ad12540,2); TP12540
round(TP12540/12,2) # (vi)
# Part VII: monthly semi-continuous premium for 5-year term
```

```
# insurance, 5-year temporary monthly annuity due
TAbar540=(int/del)*(1-d*ad540-E540); TAbar540
TPbar12540=round(1000*TAbar540/ad12540,2); TPbar12540
round(TPbar12540/12,2) # (vii)
Abar540=TAbar540 + E540
Pbar12540=round(1000*Abar540/ad12540,2); Pbar12540
round(Pbar12540/12,2) # (vii)
# Part VIII: continuous and m-thly premiums for 5-year endowment
# insurence, 5-year temporary continuous and m-thly annuity due
abar540=(1-Abar540)/del; abar540
Pbar540=round(1000*Abar540/abar540,2); Pbar540 # (viii)
P12bar540=round(1000*Abar540/ad12540,2); P12bar540
P4bar540=round(1000*Abar540/ad4540,2); P4bar540
P2bar540=round(1000*Abar540/ad2540,2); P2bar540
P1bar540=round(1000*Abar540/ad540,2); P1bar540
```

∎

Code 7.7.5. In Example 7.5.4 the force of mortality follows Makeham's law. We use this code to obtain the values of net single premium, annual premium and monthly premium for benefit of 1000 to be paid at the end of the year of death, for whole life insurance for ages 30 to 50. Using this code we have also found the values of the net single premium, annual premium and monthly premium, paid as whole life annuity due, for the benefit of 1000 to be paid at the moment of death. These are reported in Table 7.10.

```
# Part I: Makeham's law,parameters and interest parameters
A=0.0007; B=0.0001151; C=1.096; m=B/log(C); del=0.05; v=exp(-del)
d=1-v; int=d/(1-d)
A50=0.40356; x=30:50; y=exp(A*x -m*(C^x-1)); l=length(x)
p=1:(l-1)
for (i in 1:(l-1))
{
p[i]=y[i+1]/y[i]
}
p1=1-p; A1=c(1:(l-1),A50)
for(i in 1:20)
{
A1[l-i]=v*p1[l-i] + v*p[l-i]*A1[l-i+1]
}
# Part II: Premiums
A1 # NSP in whole life insurance
ad=(1-A1)/d # discrete whole life annuity
```

308   Actuarial Statistics: An Introduction Using R

```
PW=A1/ad # annual premium as discrete whole life annuity
Abar=A1*(int/del); PbarW=Abar/ad # semi-continuous
# annual premium as discrete whole life annuity
m1=12; im=m1*((1+int)^(1/m1)-1); dm=m1*(1-v^(1/m1))
alm=(int*d)/(im*dm); bm=(int-im)/(im*dm); alm; bm
ad12=alm*ad-bm; P12=A1/ad12; Pbar12=Abar/ad12
mP12=P12/12; mPbar12=Pbar12/12
d1=round(1000*data.frame(A1,PW,mP12,Abar,PbarW,mPbar12),2)
d2=data.frame(x,d1);d2 # Table 7.10
```

■

Code 7.7.6. In Example 7.5.5, the mortality law is specified by values of q_x as specified in Table 4.8. We use this code to compute the fully discrete, fully continuous and semi-continuous premiums when the benefit of 1000 is paid (a) at the end of the year of death and (b) at the moment of death, for different insurance products and various modes of premium payments. These include n-year term insurance and n-year endowment insurance, when the premiums are payable as n-year temporary life annuity, n-year term insurance and n-year endowment insurance, when the premiums are payable as m-thly n-year temporary life annuity for $m = 2, 4, 12$, whole life insurance when the premiums are payable as m-thly whole life annuity due for $m = 1, 2, 4, 12$ and whole life insurance when the benefit is paid the at the moment of death and the premiums are payable as continuous whole life annuity. These are reported in Table 7.11, Table 7.12, Table 7.13, Table 7.14 and Table 7.15.

```
# Part I: Mortality pattern in terms of qx values
z=read.table("F://qx.txt", header=T)
x=z[,1] # column of x values running from 0 to 110
q=z[,2] # column of qx values
pr=1-q # column of px values
p=pr[26:111] # column of px values for x = 25 to 110
p1=c(p[1],2:85) # vector to store values of kpx for x=25
# and k=1 to 85,first element being p25
for (i in 2:85)
{
p1[i]=p1[i-1]*p[i]
}
q25=1-p; # column of qx values for x = 25 to 110
p3=c(q25[1],2:84)
for(i in 2:85)
{
p3[i]=p1[i-1]*q25[i]
}
sum(p3)
```

```
#Part II:APVs of benefits in term,endowment and whole life insurance
int=0.06; v=(1 + int)^(-1); del=log((1+int));d=1-v
k=0:84; n=1:10; b=v^(k+1); v1=v^n
w25=cumsum(p3*b); t25=w25[1:10] # Term insurance
tbar25=(int/del)*t25; pu25=p1[1:10]*v1
e25=t25+pu25; ebar25=tbar25+pu25
wh25=sum(p3*b);whbar25=(int/del)*wh25
wh25; whbar25
# Part III: APVs of temporary annuities
abar25=(1-ebar25)/del; ad25=(1-e25)/d; wa25=(1-wh25)/d
wabar25=(1-whbar25)/del
# Part IV: Annual premiums for n-year term and endowment insurance
Pt=t25/ad25; cPt=tbar25/abar25; sePt=tbar25/ad25
d1=data.frame(Pt,sePt,cPt)
d2=round(1000*d1,2); d2
PEn=e25/ad25; cPEn=ebar25/abar25; sePEn=ebar25/ad25
d3=round(1000*data.frame(PEn,sePEn,cPEn),2)
d4=data.frame(n,d2,d3);d4 # Table 7.11
# Part V: m-thly premiums
m=c(2,4,12); im=m*((1+int)^(1/m)-1); dm=m*(1-v^(1/m))
alm=(int*d)/(im*dm); bm=(int-im)/(im*dm)
aduem=matrix(nrow=length(n),ncol=length(m))
for(i in 1:length(n))
{
for(j in 1:length(m))
{
aduem[i,j]=alm[j]*ad25[i]-bm[j]*(1-pu25[i])
}
}
# monthly premiums for term and endowment insurance
Pt12=t25/aduem[,3]; sePt12=tbar25/aduem[,3]; PEn12=e25/aduem[,3]
sePEn12=ebar25/aduem[,3]
d5=round(1000*data.frame(Pt12,sePt12,PEn12,sePEn12),2)
d6=data.frame(n,d5); d6 # Table 7.12
# quarterly premiums for term and endowment insurance
Pt4=t25/aduem[,2]; sePt4=tbar25/aduem[,2]; PEn4=e25/aduem[,2]
sePEn4=ebar25/aduem[,2]
d7=round(1000*data.frame(Pt4,sePt4,PEn4,sePEn4),2)
d8=data.frame(n,d7); d8 # Table 7.13
```

```
# six monthly premiums for term and endowment insurance
Pt2=t25/aduem[,1]; sePt2=tbar25/aduem[,1]; PEn2=e25/aduem[,1]
sePEn2=ebar25/aduem[,1]
d9=round(1000*data.frame(Pt2,sePt2,PEn2,sePEn2),2)
d10=data.frame(n,d9); d10 # Table 7.14
# Part VI: whole life insurance
wa25=(1-wh25)/d; PWh25=wh25/wa25; PbarWh25=whbar25/wa25
cPbarWh25=round(1000*whbar25/wabar25,2);cPbarWh25
wam25=alm*wa25-bm
PWhm25=wh25/wam25; PbarWhm25=whbar25/wam25
d11=data.frame(PWhm25,PbarWhm25)
d12=round(1000*rbind(c(PWh25,PbarWh25),d11),2)
m1=c(1,2,4,12)
d13= data.frame(m1,d12); d13 # Table 7.15
```

■

Code 7.7.7. At the end of Section 7.6, we discussed the computation of premiums in the first two case studies and computation of benefit amounts for the last two case studies. We use this code for these computations.

```
#Case Study 1: LIC Jeevan Umang
B=0.0001151; C=1.096; m1=B/log(C); x=45;k=0:(100-x)
y=c() # pmf of K(x)
for(i in 1:length(k))
{
y[i]=exp(-m1*C^x*(C^k[i]-1))-exp(-m1*C^x*(C^(k[i]+1)-1))
}
n=15; int=.04; v=(1+int)^(-1); d=1-v;
b=v^(k+1); A=sum(y*b); A
t=cumsum(y*b)
T=t[n] # Term insurance
pu=v^n*exp(-m1*C^x*(C^n-1)) # nEx
E=T+pu # Endowment insurance
adue=(1-E)/d; adue # n-year annuity due
wadue=(1-A)/d; wadue # whole life annuity due
P=(A+.08*(wadue-adue))/adue; b=5*10^5
Pr=b*P; Pr; ExL1=38772-Pr; ExL1

#Case Study 2:SBI Term insurance
B=0.0001151; C=1.087; m1=B/log(C); x=50;k=0:(110-x)
```

```
y=c() # pmf of K(x)
for(i in 1:length(k))
{
y[i]=exp(-m1*C^x*(C^k[i]-1))-exp(-m1*C^x*(C^(k[i]+1)-1))
}
sum(y); n=10; int=.06; v=(1+int)^(-1); d=1-v
b=v^(k+1); t=cumsum(y*b)
T=t[n]; T # Term insurance
pu=v^n*exp(-m1*C^x*(C^n-1)) # nEx
E=T+pu # Endowment insurance
adue=(1-E)/d; adue # n-year annuity due
P=T/adue; P; b=20*10^5; Pr=b*P; Pr
ExL2=22140-Pr;ExL2


#Case Study 3:HDFC yearly Immediate annuity
B=0.0001151; C=1.096; m1=B/log(C); x=57;k=0:(100-x)
y=c() # pmf of K(x)
for(i in 1:length(k))
{
y[i]=exp(-m1*C^x*(C^k[i]-1))-exp(-m1*C^x*(C^(k[i]+1)-1))
}
sum(y); int=.065; v=(1+int)^(-1); d=1-v
b=v^(k+1); A=sum(y*b); A
wadue=(1-A)/d; wadue # whole life annuity due
waimm=wadue-1; waimm # whole life annuity immediate
b=1058652; B=(b*(1-A))/waimm; B
ExL3=B-68503; ExL3


#Case Study 4:HDFC quarterly Immediate annuity
B=0.0001151; C=1.096; m1=B/log(C); x=65;k=0:(100-x)
y=c() # pmf of K(x)
for(i in 1:length(k))
{
y[i]=exp(-m1*C^x*(C^k[i]-1))-exp(-m1*C^x*(C^(k[i]+1)-1))
}
sum(y); int=.054; v=(1+int)^(-1); int4=4*((1+int)^.25-1)
d4=4*(1-v^.25); b=v^(k+1); A=sum(y*b); A
waimm=(1-(int/d4)*A)/int4; waimm # whole life quarterly
# annuity immediate
```

```
b=4911591; B=(b*(1-A))/(4*waimm); B
ExL4=B-67708.65; ExL5=4*ExL4; ExL5
```

■

In summary, using some simple functions from R and some simple loops, it is possible to write codes to prepare a ready reckoner of premiums for a variety of insurance and annuity products and for various modes of premium payments. Further, from Examples 7.4.1, 7.4.2 and 7.4.6, we note that premiums do depend on the underlying mortality patterns, as is expected.

7.8   Conceptual Exercises

7.8.1 For a fully discrete whole life insurance of 1000 issued to (x), it is given that (i) $^2A_x = 0.08$ (ii) $A_x = 0.2$ and (iii) the annual premium is determined using the equivalence principle and to be payable as whole life annuity due. S is the sum of the loss at issue random variable for 100 such independent policies. Calculate the standard deviation of S.

7.8.2 L is the loss at issue random variable for a fully discrete whole life insurance with a benefit of 1 unit, issued to (x). The annual premium charged for this insurance is 0.044 and is payable at the beginning of the year. It is given that $A_x = 0.40$, $\ddot{a}_x = 10$ and $Var(L) = 0.12$. An insurer has a portfolio of 100 independent lives. Eighty of these insurances have a death benefit of 4 units and 20 have a death benefit of 1 unit. Assuming that the total loss for this portfolio is distributed normally, calculate the probability that the present value of the gain for this portfolio is greater than 22 units.

7.8.3 For a fully discrete whole life insurance issued to (x), it is given that $d = 0.05$, $A_x = 0.4$, $^2A_x = 0.2$ and the annual premium is 0.05 times the amount of insurance. (a) Write an expression for L, the loss at issue random variable for a policy of this type, with a benefit of 1 unit. (b) Which of the following is a correct expression for $E(L)$? (i) $(P_x - 0.05)\ddot{a}_x$, (ii) $P_x\ddot{a}_x - A_x$, (iii) 0, (iv) $(0.05 - P_x)\ddot{a}_x$, where $P_x = A_x/\ddot{a}_x$. (c) Which of the following is a correct expression for $Var(L)$? (i) $E(L^2)$, (ii) $^2A_x - A_x^2$, (iii) $4(\,^2A_x - A_x^2)$, (iv) $(\,^2A_x - A_x^2)(1 + 20P_x)$. (d) Find $E(L)$ and $Var(L)$. (e) An insurance company issues 145 policies of this type with the amounts of insurance as given in Table 7.17. Assume the losses are independent. Use the normal approximation to

Amount of insurance	Number of policies
1	135
3	10

Table 7.17 Number of policies

calculate the probability that the present value at issue of the insurer's total gain on these policies will exceed 45 units.

7.8.4 It is given that the force of interest is $\delta = 6\%$ and the actuarial present value at age 32 of a unit benefit to be paid at the moment of death in a 5-year endowment insurance with a force of interest δ is 0.80219 and the same with a force of interest 2δ is 0.6506. (32) purchases this insurance by paying net annual premiums as a continuous 5-year temporary life annuity. (a) Write the expression for loss random variable at policy issue and its variance. (b) Find the annual benefit premium for benefit of 1000, using equivalence principle and the value of the variance. (c) Suppose expenses are 4% of each annual premium including the first, with additional initial expenses of 2% of the sum insured and settlement expense of Rs 100. Calculate the annual gross premium and the loading for expenses. (d) Suppose the actuarial present value at age 32 of a unit benefit to be paid at the moment of death in a whole life insurance with a force of interest δ is 0.47632 and with a force of interest 2δ is 0.2365. Whole life insurance is purchased by paying net annual premiums as a continuous whole life annuity. Write the expression for the loss random variable at policy issue and its variance. Find the annual benefit premium for a benefit of Rs 1000, using the equivalence principle and the value of the variance. Also find the gross premium if the expenses are as specified in (c) and the corresponding loading. (e) Find the annual net premium payable as 5-year certain annuity for 5-year deferred continuous whole life annuity payable at the rate of 1000 per annum and also for the 5-year certain and continuous whole life annuity payable at the rate of 1000 per annum, issued to (32).

7.8.5 It is given that $q_{25} = 0.145$, $q_{26} = 0.141$, $q_{27} = 0.156$, $q_{28} = 0.174$, $q_{29} = 0.185$, $q_{30} = 0.194$ and $i = 0.06$. (a) Find the annual premium paid as 6-year temporary discrete annuity due for benefit of Rs 1000, payable at the end of year of death, in a 6-year term insurance, 6-year pure endowment insurance and 6-year endowment insurance, issued to (25). (b) Find the annual premium paid as 6-year temporary discrete annuity due for a benefit of Rs 1000, payable at the moment of death, in a 6-year endowment insurance, issued to (25). State the assumption, if needed to calculate the premium. (c) Find the annual premium paid as 3-year temporary discrete annuity due for the set up of (a). Compare the premiums. (d) Suppose expenses are 5% of each annual premium including the first, with additional initial expenses of 3% of the sum insured and settlement expense of Rs 150. Calculate the annual gross premium and the loading for expenses in each case.

7.8.6 Suppose the force of mortality follows Gompertz' law given by $\mu_x = BC^x$ with $B = 0.0001151$ and $C = 1.096$. Suppose the interest pattern is specified by $\delta = 0.05$. (a) Find the monthly premium for a whole life insurance issued to (30), of a benefit of 1000, payable at the end of the year of death, when the premiums are paid as monthly whole life annuity due. (b) Find the monthly premium, if a benefit of Rs 1000 is payable at the moment of death, under the assumption of uniformity in unit age intervals, in the set up of (a). (c) Find the quarterly premium for whole life insurance to (30), when a benefit of Rs 1000 is payable either at the end of year of death or at the moment of death, if the premium payments form a 6-year temporary quarterly annuity due. Make the assumption of uniformity in unit age interval wherever needed. (d) Find the six-monthly premium for whole life annuity due of a benefit of Rs 1000 per annum, issued to (30), if it is purchased by a 6-year temporary six-monthly annuity due. (e) Find the monthly premium for a 6-year endowment insurance issued to (30), when the benefit of Rs 1000 is payable at the end of the year of death and premiums form the 6-year temporary monthly annuity due. (f) Find the monthly premium for a 6-year term insurance issued to (30), when the benefit of Rs 1000 is payable at the end of year of death and premiums form the 6-year temporary monthly annuity due. (g) Repeat (e) and (f) if a benefit of Rs 1000 is payable at the moment of death.

7.8.7 If $P^{1(12)}_{x:\overline{20}|} \big/ P^{\ 1}_{x:\overline{20}|} = 1.032$ and $P_{x:\overline{20}|} = 0.040$, what is the value of $P^{(12)}_{x:\overline{20}|}$?

7.8.8 Arrange the following in the ascending order.

$$P^{(2)}(\overline{A}_{40:\overline{25}|}), \ \overline{P}(\overline{A}_{40:\overline{25}|}), \ \ P(\overline{A}_{40:\overline{25}|})$$

7.8.9 If $_{15}P_{45} = 0.038$, $P_{45:\overline{15}|} = 0.056$ and $A_{60} = 0.625$, calculate $P^{\ 1}_{45:\overline{15}|}$.

7.8.10 Express in international actuarial notation, the premium corresponding to the following insurances. (a) A 20-year pure endowment insurance issued to (40), with a death benefit of 20000, is purchased by a 20-year temporary life annuity, payable at the beginning of each year. (b) A whole life annuity of Rs 8000 per month, payable at the end of the month, starting at age 60, is purchased by (30), by paying a monthly premium, at the beginning of the month, between ages 30 and 60, provided (30) survives this period. (c) An insurance to (40) provides a guaranteed benefit of Rs 25,000 per annum for 20 years, payable at the beginning of the year, by paying an annual premium at the beginning of the year for at most 20 years. (d) A 10-year deferred continuous life annuity of Rs 20,000 per annum is purchased by (50) by monthly payments payable at the beginning of the month, for at most 5 years. (e) A 25-year endowment insurance of 1 lakh, payable at the end of the year of death is purchased by (25) by paying quarterly premiums for at most 10 years. (f) A monthly annuity of Rs 5000 per month, payable monthly in advance from age 60 is issued to (40). Premiums are payable monthly in advance for at most 20 years. The initial expense is 20% of the monthly benefit and there are renewal expenses of 5% of each monthly premium, excluding the first.

7.8.11 A fully continuous level annual premium whole life insurance issued to (x) pays a benefit $b_t = (1+i)^t$ at death. The premium is determined by the equivalence principle using an interest rate i. Show that the expression for L, the loss-at-issue random variable for this insurance, is given by $(v^T - \overline{A}_x)/(1 - \overline{A}_x)$.

7.8.12 Show that

$$_{20}P^{\ 1}_{x:\overline{30}|} - P^{\ 1}_{x:\overline{20}|} = \ _{20}P(\ _{20|}A^1_{x:\overline{10}|}).$$

7.9   Computational Exercises

7.9.1 Collect the information on any one insurance product or annuity scheme from any insurance company and write a brief report specifying premium instalments for a benefit of Rs 1000.

Note: For the following questions, use the mortality pattern and interest pattern that you have adopted in the computational exercise 5.8.2. State the assumptions, if needed.

7.9.2 Suppose (30) signs a whole life insurance contract for a benefit of Rs 1000. Find (i) net single premium, (ii) annual premium paid as a whole life annuity continuously at a rate of P per annum, (iii) annual premium paid as whole life annuity due and (iv) monthly premium paid as monthly whole life annuity due. Calculate the premiums for both the cases—(a) benefit payable at the moment of death and (b) benefit payable at the end of the year of death.

7.9.3 Calculate the premiums in (ii), (iii) and (iv) of Exercise 7.9.2 when premiums form a 10-year temporary annuity due.

7.9.4 Suppose (30) signs a 10-year term insurance contract for a benefit of Rs 1000. Find the (i) one time premium, (ii) annual premium paid continuously, (iii) annual premium paid at the beginning of the year, (iv) quarterly premium paid at the beginning of the quarter, when premiums form (a) 10-year temporary life annuity and (b) 5-year temporary life annuity. Calculate these premiums when the benefit is payable at the moment of death or at the end of the year of death.

7.9.5 Suppose (30) signs a 10-year endowment insurance contract for a benefit of Rs 1000. Find the (i) one time premium, (ii) annual premium paid continuously, (iii) annual premium paid at the beginning of the year, (iv) quarterly premium paid at the beginning of the quarter, when premiums form (a) 10-year temporary life annuity and (b) 5-year temporary life annuity. Calculate these premiums when the benefit is payable at the moment of death or at the end of the year of death.

7.9.6 Suppose (30) signs a 15-year deferred whole life annuity contract with a benefit of Rs 1000 to be received at the beginning of (a) a year or (b) a month. Find the premiums paid as (i) 15-year temporary life annuity due and (ii) 15-year temporary monthly life annuity due.

7.9.7 Suppose (30) signs a 10-year certain and whole life annuity contract with a benefit of Rs 1000 per annum. Find the premium, if these form (i) 10-year certain six-monthly annuity due, (ii) 10-year certain annuity due, (iii) 10-year temporary six-monthly life annuity due and (iv) 10-year temporary annuity due.

7.9.8 On the basis of the Life Table constructed in computational exercise 4.9.2, uniformity assumption and an interest rate of 6%, calculate the values for the annual premiums as stated in Table 7.18. In semi-continuous and discrete cases, find the monthly premiums.

Fully continuous	Semi-continuous	Fully discrete
$\overline{P}(\overline{A}_{35:\overline{10|}})$	$P(\overline{A}_{35:\overline{10|}})$	$P_{35:\overline{10|}}$
$\overline{P}(\overline{A}_{35})$	$P(\overline{A}_{35})$	P_{35}
$\overline{P}(\overline{A}^{\,1}_{35:\overline{30|}})$	$P(\overline{A}^{\,1}_{35:\overline{30|}})$	$P^{\,1}_{35:\overline{30|}}$

Table 7.18 Annual premiums

7.9.9 In any one of the above exercises, find the gross premium if the pattern of expenses is as specified below: (i) initial expense of Rs 100, (ii) renewal expense of 1% of the benefit insured, incurred at the beginning of the year, including the first and (iii) settlement expenses of Rs 100. Find the loading for the expenses.

7.10   Multiple Choice Questions

Note: Unless specified otherwise, you have to identify which of the options is correct. Answers are given in the solutions of conceptual exercises.

7.10.1 In a 10-year term insurance with level, one unit benefit payable immediately on the death of (25) and when the premiums are payable as a 10-year temporary continuous life annuity, the loss at issue random variable is

(a) always positive
(b) always negative
(c) negative with probability $_{10}p_{25}$
(d) negative with probability $_{10}q_{25}$

7.10.2 In a 5-year endowment insurance with one unit level benefit, payable immediately on the death of (25) and when the premiums P are payable as a 5-year temporary continuous life annuity, the loss at issue random variable

(a) is always positive
(b) is positive at maturity if $P < ((1+i)^5 - 1)/\log(1+i)$
(c) is negative at maturity if $P > \log(1+i)/((1+i)^5 - 1)$
(d) is negative with probability $_5p_{25}$

7.10.3 In a whole life insurance with one unit level benefit payable immediately on the death of (25) and when the premiums P are payable as whole life continuous annuity, the loss at issue random variable is positive with a probability

(a) $_aq_{25}$
(b) $_bq_{25}$
(c) $_ap_{25}$
(d) $_bp_{25}$

where $a = (\log(1 + P/\delta) - \log(P/\delta))/\log(1+i)$ and $b = -a$.

7.10.4 Suppose L_1 is the loss at issue random variable associated with a single premium whole life insurance and L_2 is the loss at issue random variable associated with an annual premium whole life, payable as whole life continuous annuity. Then

(a) $Var(L_1) > Var(L_2)$
(b) $Var(L_1) < Var(L_2)$
(c) $Var(L_1) = Var(L_2)$
(d) The relation between $Var(L_1)$ and $Var(L_2)$ depends on the force of interest.

7.10.5 The following are three statements about $\overline{P}(\overline{A}_x)$. (I) $\overline{P}(\overline{A}_x) = \overline{A}_x/\overline{a}_x$. (II) $\overline{P}(\overline{A}_x) = \lim_{n\to\infty} \overline{P}(\overline{A}\,{}^{1}_{x:\overline{n}|})$. (III) $\overline{P}(\overline{A}_x) = \lim_{n\to\infty} \overline{P}(\overline{A}_{x:\overline{n}|})$.

    (a) Only (I) is true
    (b) Only (I) and (II) are true
    (c) Only (I) and (III) are true
    (d) All three are true

7.10.6 Suppose the effective rate of interest is 0. Then

    (a) $\overline{P}(\overline{A}_x) = e^{0}_{x}$
    (b) $\overline{P}(\overline{A}_x) = e_x$
    (c) $\overline{P}(\overline{A}_x) = 1/e^{0}_{x}$
    (d) $\overline{P}(\overline{A}_x) = 1/e_x$

7.10.7 Suppose the effective rate of interest is 0. Then

    (a) $\overline{P}(\overline{A}_x) = e^{0}_{x}$
    (b) $\overline{a}_x = e^{0}_{x}$
    (c) $\overline{P}(\overline{A}_x) = 1/e_x$
    (d) $\overline{a}_x = 0$

7.10.8 Suppose the effective rate of interest is 0. Then

    (a) $\overline{P}(\overline{A}_x) = e^{0}_{x}$
    (b) $\overline{A}_x = e^{0}_{x}$
    (c) $\overline{A}_x = e_x$
    (d) $\overline{A}_x = 1$

7.10.9 Suppose the force of mortality is constant 0.04. Then the premium for 1 unit benefit for whole life insurance

    (a) is 0.04
    (b) is > 0.04
    (c) is < 0.04
    (d) cannot be computed in view of insufficient information

7.10.10 For some specific mortality law and effective rate of interest $i = 0.06$, $A_{35} = 0.35$. Then $1000P_{35}$

    (a) is 32.31
    (b) is 30.48
    (c) is 31.37
    (d) cannot be computed in view of insufficient information

7.10.11 For some specific mortality law and effective rate of interest $i = 0.05$, $P_{45:\overline{20}|} = 0.04$. Then $\ddot{a}_{45:\overline{20}|}$

    (a) is 11.2613
    (b) is 11.1111
    (c) is 11.4130
    (d) cannot be computed in view of insufficient information

7.10.12 For some specific mortality law and effective rate of interest $i = 0.05$, $A_{40} = 0.5$. Then $1000P(\overline{A}_{40})$

    (a) is 51.26
    (b) is 46.46
    (c) is 48.80
    (d) cannot be computed in view of insufficient information

7.10.13 For some specific mortality law $p_{30} = 0.87$ and effective rate of interest is $i = 0.05$. Then $1000P_{30:\overline{2}|}$

    (a) is 588.84
    (b) is 499.26
    (c) is 491.24
    (d) cannot be computed in view of insufficient information

7.10.14 The annual premium $\overline{P}(\overline{A}_{30:\overline{5}|})$, payable as a 5-year temporary life annuity for a 5-year endowment insurance, issued to (30) is given by.
(I) $\overline{A}_{30:\overline{5}|}/\overline{a}_{30:\overline{5}|}$. (II) $\delta\overline{A}_{30:\overline{5}|}/(1 - \overline{A}_{30:\overline{5}|})$. (III) $(1 - \delta\overline{a}_{30:\overline{5}|})/\overline{a}_{30:\overline{5}|}$.

    (a) Only (I) is true
    (b) Only (II) and (III) are true
    (c) Only (I) and (III) are true
    (d) All three are true

7.10.15 The annual premium $\overline{P}(\overline{A}\,{}^{1}_{30:\overline{5}|})$ payable as 5-year temporary life annuity for a 5-year term insurance, issued to (30) is given by (I) $\overline{A}\,{}^{1}_{30:\overline{5}|}/\overline{a}_{30:\overline{5}|}$.
(II) $\delta\overline{A}_{30:\overline{5}|}/(1 - \overline{A}_{30:\overline{5}|})$. (III) $(1 - \delta\overline{a}_{30:\overline{5}|})/\overline{a}_{30:\overline{5}|}$.

    (a) Only (I) is true
    (b) Only (II) and (III) are true
    (c) Only (I) and (III) are true
    (d) All three are true

7.10.16 The annual premium $\overline{P}(\overline{A}_{32})$ payable as whole life annuity for a whole insurance, issued to (32) is given by (I) $\overline{A}_{32}/\overline{a}_{32}$. (II) $\delta\overline{A}_{32}/(1 - \overline{A}_{32})$. (III) $(1 - \delta\overline{a}_{32})/\overline{a}_{32}$.

    (a) Only (I) is true
    (b) Only (II) and (III) are true
    (c) Only (I) and (III) are true
    (d) All three are true

7.10.17 The annual premium P_{28} payable as whole life annuity for a whole insurance, issued to (28) is given by (I) $A_{28}/\ddot{a}_{28}$. (II) $\delta A_{28}/(1 - A_{28})$. (III) $(1 - \delta\ddot{a}_{28})/\ddot{a}_{28}$.

    (a) Only (I) is true
    (b) Only (II) and (III) are true
    (c) Only (I) and (III) are true

(d) All three are true

7.10.18 The annual premium P_{30} payable as whole life annuity for a whole insurance, issued to (30) is given by (I) $A_{30}/\ddot{a}_{30}$. (II) $dA_{30}/(1 - A_{30})$.
(III) $(1 - d\ddot{a}_{30})/\ddot{a}_{30}$.

(a) Only (I) is true
(b) Only (II) and (III) are true
(c) Only (I) and (III) are true
(d) All three are true

7.10.19 The annual premium $P(\overline{A}_{25:\overline{5}|})$ payable as 5-year temporary life annuity for a 5-year endowment insurance, issued to (25) is given by
(I) $\overline{A}_{25:\overline{5}|}/\ddot{a}_{25:\overline{5}|}$. (II) $A_{25:\overline{5}|}/\overline{a}_{25:\overline{5}|}$. (III) $A_{25:\overline{5}|}/\ddot{a}_{25:\overline{5}|}$.

(a) Only (I) is true
(b) Only (II) and (III) are true
(c) Only (I) and (III) are true
(d) All three are true

7.10.20 The following are three statements. (I) $\overline{P}(\overline{A}_{40:\overline{5}|}) > P^{(12)}(\overline{A}_{40:\overline{5}|})$.
(II) $P^{(4)}(\overline{A}_{40:\overline{5}|}) > P^{(2)}(\overline{A}_{40:\overline{5}|})$. (III) $P^{(2)}(\overline{A}_{40:\overline{5}|}) > P(\overline{A}_{40:\overline{5}|})$.

(a) Only (I) is true
(b) Only (II) and (III) are true
(c) Only (I) and (III) are true
(d) All three are true

7.10.21 The premium $_4P^{(2)}_{25:\overline{5}|}$ is given by

(a) $_4P_{25:\overline{5}|} \times (\ddot{a}_{25:\overline{4}|}/\ddot{a}^{(2)}_{25:\overline{5}|})$

(b) $_4P_{25:\overline{5}|} \times (\ddot{a}_{25:\overline{5}|}/\ddot{a}^{(2)}_{25:\overline{5}|})$

(c) $_4P_{25:\overline{5}|} \times (\ddot{a}_{25:\overline{4}|}/\ddot{a}^{(2)}_{25:\overline{4}|})$

(d) $_4P_{25:\overline{5}|} \times (\ddot{a}_{25:\overline{5}|}/\ddot{a}^{(2)}_{25:\overline{4}|})$

7.10.22 If $P^{1(12)}_{37:\overline{20}|} = 1.1 \times P^{\,1}_{37:\overline{20}|}$ and $P_{37:\overline{20}|} = 0.04$. Then $P^{(12)}_{37:\overline{20}|}$

(a) is > 0.044
(b) is 0.044
(c) is < 0.044
(d) cannot be computed in view of insufficient information

7.10.23 Under the assumption of uniformity of deaths in a unit interval, we have the following inequalities. (I) $_{20}\overline{P}(\overline{A}_{30}) > _{20}P^{(2)}(\overline{A}_{30}) > _{20}P(\overline{A}_{30})$.
(II) $_{20}\overline{P}(\overline{A}_{30}) > _{20}P^{(12)}(\overline{A}_{30}) > _{20}P(\overline{A}_{30})$.
(III) $_{20}\overline{P}(\overline{A}_{30}) > _{20}P^{(4)}(\overline{A}_{30}) > _{20}P(\overline{A}_{30})$.

(a) Only (I) is true

(b) Only (II) and (III) are true
(c) Only (I) and (III) are true
(d) All three are true

7.10.24 If $_{10}P_{40} = 0.04$, $P_{40:\overline{10|}} = 0.06$ and $A_{50} = 0.6$, then $P^1_{40:\overline{10|}}$

(a) is 0.014
(b) is 0.001
(c) is 0.010
(d) cannot be computed in view of insufficient information

Chapter 8

Reserves

Key Terms: Annuity function formula, Paid up insurance formula, Premium difference formula, Policy values, Prospective loss random variable, Prospective reserve, Reserve for fractional premiums, Retrospective reserve.

8.1 Introduction

In Chapter 7 we have discussed how an equivalence principle is used to determine the benefit premiums for a variety of insurance products and annuity contracts. With the equivalence principle, an equivalence is established between the actuarial present values of inflow to the insurance company and the outflow of the insurance company.

Another important computation in the insurance industry is the reserve calculation. In comparison to premium calculation, the concept of reserve and its calculation are rather difficult to understand. In our every day life, we use the term reserve to mean something extra, something that is available in addition to the usual supply. For example, in a broad financial sense, people use the term reserve to refer to a fund of additional money that is available in case of some special need. In the insurance industry, however, reserve is liability representing the amount of money an insurer estimates for the fulfillment of future obligations. More precisely, for an insurance or annuity contract, reserve at time t, that is, after t units of time the contract is signed, is the amount that insurer needs at time t to ensure that obligations under the insurance or annuity contract are met.

Calculation of reserve for each policy is an important task of an actuary. It is known as the valuation of the policy. There are two commonly used approaches to calculate reserves at time t. These are the retrospective approach and the prospective approach. Retrospective approach deals with the cash flow, including inflow via premiums and interest, and outflow via benefit payments of the company before a time t, while the prospective approach deals with the cash flow after a time t. Thus, the reserve $\mathrm{RV}(t)$ at time t, using the retrospective approach is defined as,

$$\mathrm{RV}(t) = \text{Accumulated value of premiums at } t$$

$-$ Accumulated value of death benefits paid by time t

By convention, if the premium payment is due at a time t, then it is not included in the calculation of the accumulated value of premiums. The reserve $\mathrm{PV}(t)$ at a time t, using the prospective approach is defined as,

$$\mathrm{PV}(t) = \text{Actuarial present value at time } t \text{ of death benefits to be paid}$$
$$- \text{ Actuarial present value at time } t \text{ of premiums yet to be received}$$

We elaborate on these two approaches in Sections 2 and 3. The following example illustrates the concept of reserve.

Example 8.1.1. Suppose there is a portfolio of 1000 policies all issued at exactly the same time, to individuals of exact age 30. The policies are five year annual premium endowment insurances with a benefit of 1000 units to be paid at the end of the year of death and premiums are payable as 5-year temporary life annuity due. Suppose the mortality pattern of this group of 1000 individuals is according to Gompertz' law given by $\mu_x = BC^x$ with $B = 0.0001151$ and $C = 1.096$. Suppose the force of interest is $\delta = 0.05$. Find the expected cash flow of this group of 1000 policies over the next five years.

Solution: For the given mortality pattern and force of interest, we have from Example 5.4.4, $1000A_{30:\overline{5}|} = 779.67$. Hence, the annual premium, for the benefit of 1000, payable as a 5-year temporary life annuity due is given by,

$$1000\,P_{30:\overline{5}|} = 1000A_{30:\overline{5}|}/\ddot{a}_{30:\overline{5}|} = 779.67 \times d/(1 - 0.77967) = 172.58$$

where $d = 0.04877058$ corresponding to $\delta = 0.05$. Using the survival function of Gompertz' law with specified parameters, we obtain p_x and q_x values. We have $l_{30} = 1000$. The expected number d_{30} of deaths in the age group $(30 - 31)$ is given by $d_{30} = 1000q_{30}$ and the expected number l_{31} of survivors at age 31 is given by $l_{31} = 1000p_{30}$. Further,

$$d_{31} = q_{31}\,l_{31} = 0.002064 \times 998.1162 = 2.0606$$
$$\text{and } l_{32} = p_{31}\,l_{31} = 0.997936 \times 998.1160 = 996.0556$$

The remaining values are obtained along similar lines. We use the first part of Code 8.4.1 to find these values. The values of p_x, q_x, expected number d_x of deaths in age group $(x, x + 1)$ and the expected number l_x of survivors at age x, from the initial group of 1000 individuals are shown in Table 8.1. The column of l_x is useful to find l_{x+1} and d_x for $x = 30$ to $x = 34$ and we need l_{x+1} for $x = 30$ to $x = 34$ to decide the fund per surviving policy, as reported in the last column of Table 8.2. Hence, the columns for both l_x and l_{x+1} are included in the table.

We need these values to decide the expected claim payments at the end of each policy year for 5 years, and to decide the number of policies in force at the beginning and at the end of each year, for 5 years. We now study the cash flow corresponding to the portfolio of these 1000 policies.

Age x	p_x	q_x	l_x	d_x	l_{x+1}
30	0.998116	0.001884	1000.0000	1.8838	998.1162
31	0.997936	0.002064	998.1162	2.0606	996.0556
32	0.997738	0.002262	996.0556	2.2535	993.8021
33	0.997521	0.002479	993.8021	2.4640	991.3381
34	0.997283	0.002717	991.3381	2.6935	988.6445

Table 8.1 Values of p_x, q_x, l_x, d_x and l_{x+1}

The total premiums collected at the commencement of the policy will be $1000 \times 172.58 = 172580.00$. At the end of the first year, the accumulated value of this, with interest at a rate $i = \exp(\delta) - 1 = 0.0512711$, is $(1 + i) \times 172580$ which is 181428.40. There will be $d_{30} = 1.8838$ expected number of deaths in the first year which result in a claim amount of 1883.82 units. Hence, the fund at the end of the year is given by, $181428.40 - 1883.82 = 179544.5$ units. At the beginning of the second year, the premiums paid by the 998.1162 expected number of survivors will be added to this amount giving a total fund of 351799.4 units. The amount during the second year at the annual rate of interest i, corresponding to this fund will be 369836.6. There will be deductions of an amount of 2056.71 units corresponding to the death claims for 2.0606 expected number of deaths. Thus, the fund at the end of the second year is $369836.6 - 2056.71 = 367776.9$ units. Similar calculations can be done for the remaining three years. We use the second part of Code 8.4.1 to find these fund values, which are presented in Table 8.2.

n	Premiums received	Total fund at the beginning of the year	Amount at the end of the year	Death claims	Fund at the end of the year	Fund per survivor
1	172580.0	172580.0	181428.4	1883.82	179544.5	179.88
2	172254.9	351799.4	369836.6	2056.71	367779.9	369.24
3	171899.3	539679.1	567349.1	2244.64	565104.4	568.63
4	171510.4	736614.8	774381.8	2448.73	771933.1	778.68
5	171085.1	943018.2	991367.8	2670.20	988697.6	1000.05

Table 8.2 Fund values

The figures in the last column give the value of the fund for each surviving policy. These are obtained by dividing the fund at the end of the year, by the number of survivors at the end of the year as given in the last column of Table 8.1. The values in the last column of Table 8.2 indicate the expected future obligations of the insurance company towards each active policy. The company must keep this much fund at hand to pay the claim if it arises. It is thus the policy value or the reserve for each surviving policy. It is to be noted that the number in the 5th row of the last column is 1000.05. It should be 1000, due to the accumulation of computational errors, we get it as 1000.05. The insurer has to pay 1000 units to each survivor at the end of the five years, as the underlying policy is the five year endowment policy.   ■

Thus, Table 8.2 shows the expected cash flow corresponding to a group of 1000 policies, all being 5-year endowment insurances. Further it depicts that the fund generated via the premiums and the interest earned on it, is sufficient to pay the benefit whenever the claim arises, thus confirming that the calculation of net premium 172.58 for the benefit of 1000 units is correct. In the next section we will show that these figures can be found as,

$$\text{Accumulated value at time } t \text{ of past premiums}$$
$$- \text{ Accumulated value at } t \text{ of past benefits}$$

that is, using the retrospective approach. We will discuss in the next section how these figures can also be found using the prospective approach. It is important to note that the retrospective reserve is the amount of money the company has accumulated, whereas the prospective reserve is the money needed to meet the future liabilities.

Computation of reserves is required for various purposes, for example,
(i) to pay surrender values or transfer values in a pension fund,
(ii) to work out the revised premium or sum assured if a policy is altered or converted to another type,
(iii) for inclusion in statutory returns to supervisory bodies for the purpose of demonstrating the solvency of the office,
(iv) for internal office calculations to decide the bonus rates of with-profit policies, the distribution of profits to share holders, and so on.

Insurance laws impose a variety of reserve requirements on insurance companies. There are two types of reserves—policy reserves and contingency reserves. Policy reserves are the reserves reported in Table 8.2. Contingency reserves are the funds to be used in adverse and unexpected conditions. An insurance company has to pay death claims even if conditions occur that are less favorable than those expected when it calculated the premium rates. Reserves displayed in the last column of Table 8.2 may change if either the rate of interest or the survival probabilities are different from those assumed at the time of determination of the premium. Mortality statistics as exhibited in life tables show the overall rates of mortality that can be expected, but fluctuations may occur in some instances, for example, in the case of an epidemic. Such unexpected occurrences could have an adverse effect on the mortality rates. Moreover, insurers have limited control over the rates of return they earn on their investments and consequently a company may not be able to earn the rate of investment return it anticipated when it calculated the premium rates. Additionally, operating expenses may rise faster than the insurer expected when premium rates are set. A loading is added to the net premium to cover such unusual occurrences. These extra amounts are useful to set up contingency reserves. We will not study this type of reserve in this book.

In the following sections, we discuss in detail reserve calculations, using both the prospective and retrospective approaches, for fully continuous policy, fully discrete policy, semi-continuous policy and reserves corresponding to true m-thly benefit premiums. It is assumed that the mortality and interest rates adopted at policy issue for the determination of benefit premiums continue to be appropriate and are used in the determination of reserves. Reserve based on benefit premium, that is, premium obtained using equivalence principle, is known as benefit reserve. In this book, it is referred to simply as reserve.

8.2   Fully Continuous Reserves

In this section, we discuss the determination of reserves for fully continuous policies. We first study the prospective approach.

Prospective Reserve: Suppose (x) signs a fully continuous policy. We want to find the future obligations to the insurance company corresponding to this policy after t units of time from the time of signing the contract. As we have defined the loss at issue random variable in the determination of premiums, we define a random variable specifying the present value at t of the prospective loss to the insurance company corresponding to this policy. It is denoted by ${}_tL(T(x)) \equiv {}_tL(T)$ and is defined as,

$$
\begin{aligned}
{}_tL(T) \;=\; & \text{Present value random variable at } t \text{ of future benefits and of} \\
& \text{future expenses} \\
- \;& \text{Present value random variable at } t \text{ of future premiums}
\end{aligned}
\tag{8.2.1}
$$

Premiums in this formulation are assumed to be benefit premiums. Note that ${}_0L(T(x))$ is the same as the loss at issue random variable defined in Chapter 7. The benefit reserve at time t of the contract, calculated prospectively, that is, with reference to future cash flows, is denoted by ${}_tV$ and is defined as ${}_tV = E({}_tL(T))$. The letter V comes from the word 'policy value' used for reserve. Thus,

$$
\begin{aligned}
{}_tV \;=\; & \text{Actuarial present value at } t \text{ of future benefits and expenses} \\
- \;& \text{Actuarial present value at } t \text{ of future premiums}
\end{aligned}
\tag{8.2.2}
$$

Here E denotes the conditional expectation of ${}_tL(T)$ given that $T(x) > t$.

If expenses are ignored, the reserve is the difference between the actuarial present values of future benefits and future premiums. The mortality, interest rate and expense assumptions used to evaluate ${}_tV$ are known as the reserving basis. This may or may not agree with the premium basis, that is, assumptions regarding mortality, interest rates and expenses while determining the premium. If these bases agree and if there are no expenses, the reserves obtained are known as net premium reserves. By convention, the reserve ${}_tV$ is calculated just before receipt of any premium then due. The reserve just after payment of this premium is, ${}_tV + P - e$, where P is the premium paid at t and e is the expense incurred for that transaction. When $t = 0$, ${}_0V = 0$, as the premium is determined using the equivalence principle.

For various types of insurance models, appropriate suffixes are attached to the symbol ${}_tV$. As an illustration, we discuss how to find the reserve for a whole life insurance of unit benefit issued to (x) on a fully continuous basis with an annual benefit premium of $\overline{P}(\overline{A}_x)$. Using the definition of ${}_tL$ as given in Eq. 8.2.1, the prospective loss random variable for the whole life insurance is defined as follows.

$$
{}_tL(T) \;=\; v^{T(x)-t} \;-\; \overline{P}(\overline{A}_x)\, \overline{a}_{\overline{T(x)-t|}}
$$

Thus, the first term corresponds to the present value at t of the benefit payments payable at the moment of death while the second term corresponds to the present value at t of the future premium payments at the rate of $\overline{P}(\overline{A}_x)$ per annum, the period of such payments being the random variable $(T(x) - t)$. In international actuarial notation, the reserve is denoted by ${}_t\overline{V}(\overline{A}_x)$ and is obtained as a conditional expectation of ${}_tL$ given that $T(x) > t$. Such a conditional event is in view of the fact that the valuation of the policy will be done only if it is in force. Thus,

$$
\begin{aligned}
{}_t\overline{V}(\overline{A}_x) &= E({}_tL(T)|T(x) > t) \\
&= E(V^{T(x)-t}|T(x) > t) - \overline{P}(\overline{A}_x)E(\overline{a}_{\overline{T(x)-t|}}|T(x) > t) \\
&= \overline{A}_{x+t} - \overline{P}(\overline{A}_x)\overline{a}_{x+t}
\end{aligned}
\tag{8.2.3}
$$

In the last expression, we use the result that in aggregate mortality setup the conditional distribution of $T(x) - t$ given $T(x) > t$ is the same as the distribution of $T(x + t)$, the future life time of $(x + t)$. For example,

$$
P[T(50) > 15|T(50) > 10] = \frac{{}_{15}p_{50}}{{}_{10}p_{50}} = \frac{{}_{10}p_{50}\, {}_5p_{60}}{{}_{10}p_{50}} = {}_5p_{60} = P[T(60) > 5]
$$

Thus, the probability that a fifty year old survives to age 65 given that he has survived to age 60 is the same as the probability that a 60 year old survives to age 65. Similarly,

$$
\begin{aligned}
P[K(30) = 12|K(30) \geq 5] &= \frac{{}_{12|}q_{30}}{{}_5p_{30}} = \frac{{}_{12}p_{30}q_{42}}{{}_5p_{30}} = \frac{{}_5p_{30}\, {}_7p_{35}\, q_{42}}{{}_5p_{30}} \\
&= {}_7p_{35}q_{42} = {}_{7|}q_{35} = P[K(35) = 7]
\end{aligned}
$$

Such results are not true in the select survival setup, as defined in Section 4.6. In the entire discussion in this chapter, we assume an aggregate mortality pattern. By following the steps similar to those adopted to obtain the variance of the loss at issue random variable in Chapter 7, the variance of ${}_tL$ is obtained as follows. The expression for ${}_tL$ can be rewritten as

$$
\begin{aligned}
{}_tL(T) &= v^{T(x)-t}\left[1 + (\overline{P}(\overline{A}_x)/\delta)\right] - \overline{P}(\overline{A}_x)/\delta \\
\Rightarrow\ Var({}_tL|T(x) > t) &= \left(1 + \overline{P}(\overline{A}_x)/\delta\right)^2 Var(v^{T(x)-t}|T(x) > t) \\
&= \left(1 + \overline{P}(\overline{A}_x)/\delta\right)^2 \left({}^2\overline{A}_{x+t} - \overline{A}^2_{x+t}\right)
\end{aligned}
\tag{8.2.4}
$$

The formula ${}_t\overline{V}(\overline{A}_x) = \overline{A}_{x+t} - \overline{P}(\overline{A}_x)\overline{a}_{x+t}$, as derived in Eq. 8.2.3, can be expressed in different forms as follows.

$$
{}_t\overline{V}(\overline{A}_x) = \left[\frac{\overline{A}_{x+t}}{\overline{a}_{x+t}} - \overline{P}(\overline{A}_x)\right]\overline{a}_{x+t} = [\overline{P}(\overline{A}_{x+t}) - \overline{P}(\overline{A}_x)]\overline{a}_{x+t}
\tag{8.2.5}
$$

This formula for reserve is known as the premium difference formula. A second formula is obtained by factoring the actuarial present value of future benefits out of the prospective formula. Thus we have,

$$
{}_t\overline{V}(\overline{A}_x) = \left[1 - \overline{P}(\overline{A}_x)\frac{\overline{a}_{x+t}}{\overline{A}_{x+t}}\right]\overline{A}_{x+t} = \left[1 - \frac{\overline{P}(\overline{A}_x)}{\overline{P}(\overline{A}_{x+t})}\right]\overline{A}_{x+t}
\tag{8.2.6}
$$

This is called a paid up insurance formula. The formula for reserve can be expressed in terms of a single actuarial function as follows. Expressing $\overline{A}_x$ and $\overline{P}(\overline{A}_x)$ in terms of annuity function we obtain,

$$_t\overline{V}(\overline{A}_x) \;\; = \;\; 1 - \delta\overline{a}_{x+t} - (1/\overline{a}_x - \delta)\,\overline{a}_{x+t} = 1 - \overline{a}_{x+t}/\overline{a}_x \tag{8.2.7}$$

This is known as *annuity function* formula. In Eq. 8.2.7, using $\overline{a}_{x+t} = (1 - \overline{A}_{x+t})/\delta$, we obtain

$$_t\overline{V}(\overline{A}_x) \;\; = \;\; (\overline{A}_{x+t} - \overline{A}_x)/(1 - \overline{A}_x) \tag{8.2.8}$$

This expression of $_t\overline{V}(\overline{A}_x)$ is useful to compute the reserve when $\overline{A}_{x+t}$ is known for various values of t. In Eq. 8.2.5, if we substitute $\overline{a}_{x+t} = (\overline{P}(\overline{A}_{x+t}) + \delta)^{-1}$, we get,

$$
\begin{aligned}
_t\overline{V}(\overline{A}_x) \;\; &= \;\; [\overline{P}(\overline{A}_{x+t}) - \overline{P}(\overline{A}_x)]\,\overline{a}_{x+t} \\
&= \;\; (\overline{P}(\overline{A}_{x+t}) - \overline{P}(\overline{A}_x))/(\overline{P}(\overline{A}_{x+t}) + \delta)
\end{aligned} \tag{8.2.9}
$$

The following example illustrates the computation of prospective reserve for a fully continuous whole life insurance.

Example 8.2.1. For a fully continuous whole life insurance of 1000 issued to (25), with premiums payable as whole life annuity, it is given that $\delta = 0.05$ and mortality follows Gompertz' law with $B = 0.0001151$ and $C = 1.096$. Calculate prospective reserve for $t = 5, 10$ and 15.

Solution: According to the international actuarial notation, we have to find $1000\,_t\overline{V}(\overline{A}_{25})$ for $t = 5, 10$ and 15. We can use any one of the formulae listed above. However, it is simple to use Eq. 8.2.8, as we need only the values of $\overline{A}_x$ for various x. From Eq. 8.2.8, the formula for prospective reserve is

$$_t\overline{V}(\overline{A}_x) \;\; = \;\; (\overline{A}_{x+t} - \overline{A}_x)/(1 - \overline{A}_x)$$

In Example 5.3.7, we have computed $\overline{A}_x$ for $x = 25, 30, 35$ and 40 for Gompertz' law with given values of parameters. These are shown in the left panel of Table 8.3. Using it we find $_t\overline{V}(\overline{A}_{25})$ for $t = 5, 10$ and 15. The results are shown in the right panel of Table 8.3. These

Age	$1000\overline{A}_x$	t	$1000\,_t\overline{V}(\overline{A}_{25})$
25	152.54	0	0.00
30	189.12	5	43.17
35	232.70	10	94.59
40	283.72	15	154.80

Table 8.3 Gompertz' law: Prospective reserve

values are obtained using Code 8.4.2. Values of $1000\,_t\overline{V}(\overline{A}_{25})$ indicate the expected financial liability of the insurance company at different time points in the period when policy is in force. Thus, at $t = 5$, the expected fund per surviving policy that insurer must have is 43.17 to pay the claim of 1000 units if it is made at that time. It is to be noted that at $t = 0$, the reserve is 0, as the premiums are determined using the equivalence principle and the prospective loss random variable ${}_tL(T)$ at $t = 0$ is the loss at issue random variable L. Further, as t increases, the reserve increases, because the probability of claim increases in whole life insurance as t increases. ∎

The prospective loss random variables and the corresponding reserves for an n-year term insurance and n-year endowment insurance are defined on similar lines as those for the whole life insurance. To write these expressions, recall the following expressions for present value random variable Z of the unit benefit as discussed in Chapter 5. For the n-year term insurance, Z is given by,

$$Z = \begin{cases} v^T, & \text{if } T \le n \\ 0, & \text{if } T > n \end{cases}$$

For an n-year endowment insurance, Z is given by,

$$Z_T = \begin{cases} v^T, & \text{if } \quad T \le n \\ v^n, & \text{if } \quad T > n \end{cases}$$

For all these insurance products, suppose premiums are payable as n-year temporary continuous life annuity. Then from Chapter 6, the present value random variable Y corresponding to this annuity is given by,

$$Y = \begin{cases} \overline{a}_{\overline{T}|}, & \text{if } \quad 0 \le T < n \\ \overline{a}_{\overline{n}|}, & \text{if } \quad T \ge n \end{cases}$$

As defined at the beginning of this section, the prospective loss random variable ${}_tL$ for $t \le n$ for n-year term insurance is given by,

$${}_tL(T) = \begin{cases} v^{T(x)-t} - \overline{P}(\overline{A}^1_{x:\overline{n}|})\, \overline{a}_{\overline{T(x)-t}|}, & \text{if } \quad t < T(x) < n \\ 0 - \overline{P}(\overline{A}^1_{x:\overline{n}|})\, \overline{a}_{\overline{n-t}|}, & \text{if } \quad T(x) \ge n \end{cases}$$

At $t = n$, ${}_tL = 0$, since by convention, $\overline{a}_{\overline{0}|} = 0$. Thus, the prospective loss random variable for n-year term insurance is degenerate at 0 if $t = n$. It is logically appealing since in an n-year term insurance, the insurer has no liability of paying the claims when the term of n years is over.

For n-year endowment insurance, ${}_tL$ for $t \le n$ is given by,

$${}_tL = \begin{cases} v^{T(x)-t} - \overline{P}(\overline{A}_{x:\overline{n}|})\, \overline{a}_{\overline{T(x)-t}|}, & \text{if } \quad t < T(x) < n \\ v^{n-t} - \overline{P}(\overline{A}_{x:\overline{n}|})\, \overline{a}_{\overline{n-t}|}, & \text{if } \quad T(x) \ge n \end{cases}$$

Note that the prospective loss random variable for an n-year endowment insurance is degenerate at 1 if $t = n$. It is also logically appealing, since in this insurance product, the insurer has a liability of paying 1 unit at the end of the term, corresponding to each surviving policy.

The complexity in the definition of ${}_tL(T)$ increases, when the number of premium paying years is restricted to h years in the whole life insurance or in an n-year term or endowment insurance, with $h < n$. For the whole life insurance, when the premiums are payable as h-year temporary continuous life annuity, ${}_tL(T)$ is defined as follows. For $t < h$,

$$
{}_tL(T) = \begin{cases} v^{T(x)-t} - {}_h\overline{P}(\overline{A}_x)\,\overline{a}_{\overline{T(x)-t|}}, & \text{if } \quad t < T(x) < h \\ v^{T(x)-t} - {}_h\overline{P}(\overline{A}_x)\,\overline{a}_{\overline{h-t|}}, & \text{if } \quad t \leq h < T(x) \end{cases}
$$

Suppose $t = h$, thus, it is implicitly assumed that $T(x) > h$. In this case, ${}_tL(T)$ is given by,

$$
{}_tL = v^{T(x)-h} - {}_h\overline{P}(\overline{A}_x)\,\overline{a}_{\overline{h-h|}} = v^{T(x)-h}
$$

Note that it is combined with the second equation of the case of $t \leq h$. For $t > h$, the present value random variable of annuity payments is 0 as the premium paying term is over. Hence, ${}_tL$ is given by,

$$
{}_tL = v^{T(x)-t}
$$

On similar lines for an n-year endowment insurance, when premiums are payable as h-year temporary continuous life annuity, ${}_tL$ for the case $t \leq h$ is given by,

$$
{}_tL(T) = \begin{cases} v^{T(x)-t} - {}_h\overline{P}(\overline{A}_{x:\overline{n|}})\,\overline{a}_{\overline{T(x)-t|}}, & \text{if} \quad t < T(x) < h \\ v^{T(x)-t} - {}_h\overline{P}(\overline{A}_{x:\overline{n|}})\,\overline{a}_{\overline{h-t|}}, & \text{if } \quad t \leq h < T(x) < n \\ v^{n-t} - {}_h\overline{P}(\overline{A}_{x:\overline{n|}})\,\overline{a}_{\overline{h-t|}}, & \text{if } \quad t \leq h \leq n < T(x) \end{cases}
$$

For $h < t \leq n$, it is given by,

$$
{}_tL(T) = \begin{cases} v^{T(x)-t}, & \text{if } \quad T(x) < n \\ v^{n-t}, & \text{if } \quad T(x) \geq n \end{cases}
$$

For an n-year term insurance, when the premiums are payable as h-year temporary continuous life annuity, ${}_tL(T)$ is similar to that for an n-year endowment with the change that v^{n-t} is replaced by 0, as there is no benefit payment if death occurs after n years.

In Table 8.4, we summarize these expressions for the prospective loss random variable ${}_tL$ for some products.

The conditional expectation of ${}_tL(T)$, conditional on the event that $T(x) > t$, gives the formula for the prospective reserve. We again use the fact that in aggregate mortality pattern, the conditional distribution of $T(x) - t$ is the same as the distribution of $T(x+t)$. Thus, the expectation of ${}_tL(T)$ with respect to the distribution of $T(x+t)$, gives the formulae for the prospective reserve. Table 8.5 displays the international actuarial notation and the formulae for reserve using the prospective approach for a variety of insurance models.

Product	$_tL(T)$
Whole Life	$v^{T(x)-t} - \overline{P}(\overline{A}_x)\,\overline{a}_{\overline{T(x)-t\|}}$
n-Year term	$v^{T(x)-t} - \overline{P}(\overline{A}^1_{x:\overline{n}\|})\,\overline{a}_{\overline{T(x)-t\|}}, \qquad T(x) < n$ $0 - \overline{P}(\overline{A}^1_{x:\overline{n}\|})\,\overline{a}_{\overline{n-t\|}}, \qquad T(x) \geq n$
n-Year endowment	$v^{T(x)-t} - \overline{P}(\overline{A}_{x:\overline{n}\|})\,\overline{a}_{\overline{T(x)-t\|}}, \qquad T(x) < n$ $v^{n-t} - \overline{P}(\overline{A}_{x:\overline{n}\|})\,\overline{a}_{\overline{n-t\|}}, \qquad T(x) \geq n$
h-Payment years, whole life	$v^{T(x)-t} - {_h}\overline{P}(\overline{A}_x)\,\overline{a}_{\overline{T(x)-t\|}}, \qquad T(x) < h$ $v^{T(x)-t} - {_h}\overline{P}(\overline{A}_x)\,\overline{a}_{\overline{h-t\|}}, \qquad t \leq h < T(x)$ $v^{T(x)-t}, \qquad h < t < T(x)$
h-Payment years, n-year endowment	$v^{T(x)-t} - {_h}\overline{P}(\overline{A}_{x:\overline{n}\|})\,\overline{a}_{\overline{T(x)-t\|}}, \quad t < T(x) < h$ $v^{T(x)-t} - {_h}\overline{P}(\overline{A}_{x:\overline{n}\|})\,\overline{a}_{\overline{h-t\|}}, \quad t \leq h < T(x) < n$ $v^{n-t} - {_h}\overline{P}(\overline{A}_{x:\overline{n}\|})\,\overline{a}_{\overline{h-t\|}}, \quad t \leq h \leq n < T(x)$ $v^{T(x)-t}, \qquad h < t < T(x) < n$ $v^{n-t}, \qquad h < t \leq n < T(x)$

Table 8.4 Prospective loss random variable: Fully continuous policy

From the formulae as derived in Section 5.3, it is clear that, at $t = 0$, $\overline{A}^1_{x:\overline{t}\|} = 0$ and $\overline{A}_{x:\overline{t}\|} = 1$. Similarly, $\overline{a}_{x:\overline{0}\|} = 0$. With these values, note that the reserve at $t = n$ for an n-year term insurance is 0 while for an n-year endowment, it is 1.

We now proceed to derive the formulae for the reserve by a retrospective approach.

Retrospective Reserve: To derive the expressions for retrospective reserves, we adopt a procedure similar to that in Example 8.1.1. This procedure consists of finding the income and expenditure for each year and carrying over the balance of the last year to the next year. Symbolically, the procedure can be described as follows. For illustration we take $n = 3$, that is, 3-year endowment insurance and denote the premium simply by P. Suppose a portfolio consists of l_x policies. At the beginning of the year the collection via premiums is $l_x P$. Its accumulation at the end of the year is $l_x P(1 + i)$. The expected number of deaths in the first year is d_x, hence if the benefit amount is 1 unit, the expected claim amount is d_x. Thus, at the end of the first year, the fund or the balance is

$$l_x P(1+i) - d_x = l_x(1+i)\{P - vq_x\}$$
$$= l_x(1+i)\{P\ddot{a}_{x:\overline{1}\|} - A^1_{x:\overline{1}\|}\} \qquad (8.2.10)$$

Product	Notation	Prospective reserve
Whole life	${}_t\overline{V}(\overline{A}_x)$	$\overline{A}_{x+t} - \overline{P}(\overline{A}_x)\,\overline{a}_{x+t}$
n-Year term	${}_t\overline{V}(\overline{A}^{1}_{x:\overline{n}\rvert})$	$\overline{A}^{1}_{x+t:\overline{n-t}\rvert} - \overline{P}(\overline{A}^{1}_{x:\overline{n}\rvert})\,\overline{a}_{x+t:\overline{n-t}\rvert}, \qquad t \le n$
n-Year endowment	${}_t\overline{V}(\overline{A}_{x:\overline{n}\rvert})$	$\overline{A}_{x+t:\overline{n-t}\rvert} - \overline{P}(\overline{A}_{x:\overline{n}\rvert})\,\overline{a}_{x+t:\overline{n-t}\rvert}, \qquad t \le n$
h-Payment years, whole life	${}^{h}_{t}\overline{V}(\overline{A}_x)$	$\overline{A}_{x+t} - {}_h\overline{P}(\overline{A}_x)\,\overline{a}_{x+t:\overline{h-t}\rvert}, \qquad t \le h$ $\overline{A}_{x+t}, \qquad t > h$
h-Payment years, n-year endowment	${}^{h}_{t}\overline{V}(\overline{A}_{x:\overline{n}\rvert})$	$\overline{A}_{x+t:\overline{n-t}\rvert} - {}_h\overline{P}(\overline{A}_{x:\overline{n}\rvert})\,\overline{a}_{x+t:\overline{h-t}\rvert}, \quad t \le h < n$ $\overline{A}_{x+t:\overline{n-t}\rvert}, \qquad h < t \le n$

Table 8.5 Notation and formulae for fully continuous prospective benefit reserves

At the beginning of the second year, the collection via premiums is $l_{x+1}P$. Carrying forward the last year's balance, the fund at the beginning of the second year is $l_{x+1}P + l_x P(1+i) - d_x$. Its accumulation at the end of the year is $\{l_{x+1}P + l_x P(1+i) - d_x\}(1+i)$. The expected number of deaths in the second year is d_{x+1}, hence the expected claim amount is d_{x+1}. Thus, at the end of the second year, the balance is

$$
\begin{aligned}
& \{l_{x+1}P + l_x P(1+i) - d_x\}(1+i) - d_{x+1} \\
=\ & Pl_x(1+i)^2\{1 + vp_x\} - l_x(1+i)^2\{vq_x + v^2 p_x q_{x+1}\} \\
=\ & l_x(1+i)^2\{P\ddot{a}_{x:\overline{2}\rvert} - A^{1}_{x:\overline{2}\rvert}\}
\end{aligned}
\tag{8.2.11}
$$

Continuing in this manner, at the end of the third year, the fund F_3 is

$$
F_3 = \{l_{x+2}P + [l_{x+1}P + l_x P(1+i) - d_x](1+i) - d_{x+1}\}(1+i) - d_{x+2}
$$

The terms in F_3 can be rearranged as follows.

$$
\begin{aligned}
F_3 =\ & \{Pl_x(1+i)^3 + Pl_{x+1}(1+i)^2 + Pl_{x+2}(1+i)\} \\
& - \{d_x(1+i)^2 + d_{x+1}(1+i) + d_{x+2}\} \\
=\ & Pl_x(1+i)^3\{1 + vp_x + v^2\,{}_2p_x\} \\
& - l_x(1+i)^3\{vq_x + v^2 p_x q_{x+1} + v^3\,{}_2p_x q_{x+2}\} \\
=\ & l_x(1+i)^3\Big\{P\sum_{j=0}^{2} v^j\,{}_jp_x - \sum_{j=0}^{2} v^{j+1}\,{}_jp_x q_{x+j}\Big\} \\
=\ & l_x(1+i)^3\{P\ddot{a}_{x:\overline{3}\rvert} - A^{1}_{x:\overline{3}\rvert}\}
\end{aligned}
\tag{8.2.12}
$$

The expressions in Eqs 8.2.10, 8.2.11 and 8.2.12 can be interpreted as,

$$\text{Accumulation of inflow via premiums}$$
$$- \text{ Accumulation of outflow via claim payments}$$

Funds at the end of the first, second and third year are to be shared by l_{x+1}, l_{x+2} and l_{x+3} surviving policies respectively. Hence, dividing Eqs 8.2.10, 8.2.11 and 8.2.12 by l_{x+1}, l_{x+2} and l_{x+3} respectively, and after some algebra, we get the expressions for the fund per surviving policy at the end of the first, second and third year. These are respectively given by,

$$\frac{1}{{}_1E_x}[P\ddot{a}_{x:\overline{1|}} - A^{\;1}_{x:\overline{1|}}], \qquad \frac{1}{{}_2E_x}[P\ddot{a}_{x:\overline{2|}} - A^{\;1}_{x:\overline{2|}}] \quad \& \quad \frac{1}{{}_3E_x}[P\ddot{a}_{x:\overline{3|}} - A^{\;1}_{x:\overline{3|}}]$$

These are the expressions of reserve using a retrospective approach for $t = 1, 2$ and 3. It is easy to check that, with formula $P = A_{x:\overline{3|}}/\ddot{a}_{x:\overline{3|}}$ for premium P in 3-year endowment insurance,

$$\frac{1}{{}_3E_x}[P\ddot{a}_{x:\overline{3|}} - A^{\;1}_{x:\overline{3|}}] = \frac{1}{{}_3E_x}[A_{x:\overline{3|}} - A^{\;1}_{x:\overline{3|}}] = 1$$

We adopt similar arguments to derive the expression for the retrospective reserve for various fully continuous policies with some modifications to suit the continuous setup.

Suppose a portfolio consists of l_x policies, each being a whole life insurance contract with unit benefit, payable at the moment of death, where l_x denotes the expected number of survivors to age x. Each of these policies is purchased by paying the annual premium, payable as continuous whole life annuity. Thus, the annual premium is given by $\overline{P}(\overline{A}_x)$. The reserve using a retrospective approach is defined as,

$$\text{RV}(t) = \text{Accumulated value of premiums at } t$$
$$- \text{ Accumulated value of death benefits paid by time } t$$

To find the accumulated value of premiums at t we proceed as follows. For $u \in (0, t)$, l_{x+u} denotes the number of survivors at u, each one of them will be paying the premium continuously at the rate of $\overline{P}(\overline{A}_x)$, per annum. The fund collected via premiums at u will accumulate for a time period $t - u$. Hence, the accumulated value at t of premiums is given by,

$$\int_0^t \overline{P}(\overline{A}_x)l_{x+u}(1+i)^{t-u}du$$

To find the accumulated value of death benefits paid by time t, we adopt a similar approach. The density of deaths at u is given by $l_{x+u}\mu_{x+u}$. If the death benefit is 1 unit, the accumulated value at t of the death benefits paid by time t is given by,

$$\int_0^t l_{x+u}\mu_{x+u}(1+i)^{t-u}du$$

Hence, the reserve corresponding to each of l_{x+t} surviving policies is given by,

$$_t\overline{V}(\overline{A}_x) = \frac{\int_0^t \overline{P}(\overline{A}_x)l_{x+u}(1+i)^{t-u}du \; - \; \int_0^t l_{x+u}\mu_{x+u}(1+i)^{t-u}du}{l_{x+t}}$$

$$= \frac{l_x\,(1+i)^t\overline{P}(\overline{A}_x)\{\int_0^t v^u \; _up_x\,du\} - l_x(1+i)^t\{\int_0^t v^u \; _up_x\,\mu_{x+u}\,du\}}{l_{x+t}}$$

$$= \frac{\overline{P}(\overline{A}_x)\,\overline{a}_{x:\overline{t}|} \; - \; \overline{A}^1_{x:\overline{t}|}}{(1+i)^{-t}\,_tp_x} = \frac{\overline{P}(\overline{A}_x)\,\overline{a}_{x:\overline{t}|} \; - \; \overline{A}^1_{x:\overline{t}|}}{_tE_x} \tag{8.2.13}$$

$$= \overline{P}(\overline{A}_x)\,\overline{S}_{x:\overline{t}|} \; - \; _t\overline{K}_x$$

where $\overline{S}_{x:\overline{t}|} = \overline{a}_{x:\overline{t}|}/_tE_x$ and $_t\overline{K}_x = \overline{A}^1_{x:\overline{t}|}/\,_tE_x$. Note that $\overline{S}_{x:\overline{t}|}$ can be expressed as $\overline{S}_{x:\overline{t}|} = (1+i)^t\overline{a}_{x:\overline{t}|}/\,_tp_x$. It denotes the accumulated value of premiums for 1 unit benefit per annum, paid continuously for at most t years, provided the policy is active for t years after signing the contract. Recall that $\overline{S}_{\overline{n}|} = (1+i)^n\overline{a}_{\overline{n}|}$ denotes the accumulated value of the n-year continuous annuity certain, while $\overline{S}_{x:\overline{t}|} = \overline{a}_{x:\overline{t}|}/\,_tE_x$ denotes the accumulated value of t-year continuous temporary life annuity. On similar lines, $_t\overline{K}_x = (1+i)^t\overline{A}^1_{x:\overline{t}|}/_tp_x$ denotes the accumulated value of the benefit payments by time t, provided the policy is active for t years after signing the contract. Hence, $_t\overline{K}_x$ is known as the accumulated cost of insurance.

In the next example we compute the retrospective reserve when the reserving basis is as in Example 8.2.1.

Example 8.2.2. Find the retrospective reserve for the whole life insurance with a benefit of 1000 units when the reserving basis is the same as in Example 8.2.1. Verify that both the prospective and retrospective reserves are the same.

Solution: The reserve using retrospective approach is given by,

$$_t\overline{V}(\overline{A}_x) = (\overline{P}(\overline{A}_x)\,\overline{a}_{x:\overline{t}|} \; - \; \overline{A}^1_{x:\overline{t}|})/_tE_x$$

We have the values of $\overline{A}_x$ for $x = 25, 30, 35$ and 40. To use these, we express the terms in the formula for reserve using the retrospective approach in terms of $\overline{A}_x$. Note that $\overline{P}(\overline{A}_{25}) = \delta\,\overline{A}_{25}/(1-\overline{A}_{25}) = 0.00899$. Recall that $\overline{A}^1_{x:\overline{t}|}$ and $\overline{A}_{x:\overline{t}|}$ can be expressed as

$$\overline{A}^1_{x:\overline{t}|} = \overline{A}_x - {_tE_x}\,\overline{A}_{x+t} \quad \text{and} \quad \overline{A}_{x:\overline{t}|} = \overline{A}_x - {_tE_x}\,\overline{A}_{x+t} + {_tE_x}$$

From $\overline{A}_{x:\overline{t}|}$, we find $\overline{a}_{x:\overline{t}|} = (1-\overline{A}_{x:\overline{t}|})/\delta$. Thus, it is enough to compute $_tE_x = v^t\,_tp_x = e^{-\delta t}\,_tp_x$ for Gompertz' law for $x = 25$ and for $t = 5, 10$ and 15. Using the survival function we get $_tE_x$ for theses t values. Using these values and the values of $\overline{A}_x$ for $x = 25, 30, 35$ and 40, we find $_t\overline{V}(\overline{A}_{25})$. Code 8.4.3 is used for computing these reserves, which are reported in Table 8.6.

From Tables 8.3 and 8.6, we see that the values of the reserve using both the approaches are the same. It is to be noted that the reserving basis is the same as in Example 8.2.1 and it is also the same as in premium determination. ∎

t	$_tE_{25}$	$1000\ _tV(\overline{A}_{25})$
5	0.7732	43.16
10	0.5953	94.59
15	0.4553	154.80

Table 8.6 Gompertz' law: Retrospective reserve

The retrospective reserves for other insurance products are obtained on similar lines as those for the whole life insurance by considering the accumulated value at t of the premiums received and the accumulated value at t of the benefits paid. The retrospective reserve for the n-year term insurance is derived exactly along the same lines as the reserve for the whole life insurance in Eq. 8.2.13, with the only change that $\overline{P}(\overline{A}_x)$ is replaced by $\overline{P}(\overline{A}^1_{x:\overline{n}|})$. Here $t \leq n$, as the policy is in force only for n years. Thus, for $t \leq n$ the accumulated value at t of the premiums received is given by,

$$\int_0^t \overline{P}(\overline{A}^1_{x:\overline{n}|})l_{x+u}(1+i)^{t-u}du$$

and the accumulated value at t of the benefits paid is given by,

$$\int_0^t l_{x+u}\mu_{x+u}(1+i)^{t-u}du$$

Thus, the retrospective reserve for n-year term insurance is given by,

$$_t\overline{V}(\overline{A}^1_{x:\overline{n}|}) \;=\; \left(\overline{P}(\overline{A}^1_{x:\overline{n}|})\overline{a}_{x:\overline{t}|} \;-\; \overline{A}^1_{x:\overline{t}|}\right)\Big/ \,_tE_x \ , \qquad t \leq n$$

At $t = n$, $\quad _n\overline{V}(\overline{A}^1_{x:\overline{n}|}) \;=\; \left(\dfrac{\overline{A}^1_{x:\overline{n}|}}{\overline{a}_{x:\overline{n}|}}\overline{a}_{x:\overline{n}|} \;-\; \overline{A}^1_{x:\overline{n}|}\right)\Big/ \,_tE_x \;=\; 0$

as expected. Similarly, the retrospective reserve for the n-year endowment insurance is given by,

$$_t\overline{V}(\overline{A}_{x:\overline{n}|}) \;=\; \left(\overline{P}(\overline{A}_{x:\overline{n}|})\overline{a}_{x:\overline{t}|} \;-\; \overline{A}^1_{x:\overline{t}|}\right)\Big/ \,_tE_x \ , \qquad t \leq n$$

It is easy to check that at $t = n$, it is 1 as expected.

In whole life insurance, if the number of premium paying years is at most h, then the derivation is similar to the case when premiums are payable as whole life annuity, with the change that the upper limit in the integral $\int_0^t \overline{P}(\overline{A}_x)l_{x+u}(1+i)^{t-u}du$ is replaced by h, as the collection via premiums is for h years only. Similar changes are to be made if the premiums are payable as h-year temporary annuity in n-year term insurance or n-year endowment insurance. Table 8.7 displays the formulae for reserve using retrospective approach for a variety of insurance models discussed in Chapter 5.

Note that all the formulae derived above are for the reserve when the benefit is 1 unit. The reserve for a benefit of b units is obtained simply by multiplying the formulae by b.

Product	Notation	Retrospective reserve
Whole life	$_t\overline{V}(\overline{A}_x)$	$\overline{P}(\overline{A}_x)\overline{S}_{x:\overline{t}\rvert} - {}_t\overline{K}_x$
n-Year term	$_t\overline{V}(\overline{A}^1_{x:\overline{n}\rvert})$	$\overline{P}(\overline{A}^1_{x:\overline{n}\rvert})\overline{S}_{x:\overline{t}\rvert} - {}_t\overline{K}_x,\ t \le n$
n-Year endowment	$_t\overline{V}(\overline{A}_{x:\overline{n}\rvert})$	$\overline{P}(\overline{A}_{x:\overline{n}\rvert})\overline{S}_{x:\overline{t}\rvert} - {}_t\overline{K}_x,\ t \le n$
h-Payment years, whole life	$^h_t\overline{V}(\overline{A}_x)$	$_h\overline{P}(\overline{A}_x)\overline{S}_{x:\overline{t}\rvert} - {}_t\overline{K}_x,\ \ t \le h$ $\{_h\overline{P}(\overline{A}_x)\overline{a}_{x:\overline{h}\rvert} - \overline{A}^1_{x:\overline{t}\rvert}\}/{}_tE_x,\ \ t > h$
h-Payment years, n-year endowment	$^h_t\overline{V}(\overline{A}_{x:\overline{n}\rvert})$	$_h\overline{P}(\overline{A}_{x:\overline{n}\rvert})\overline{S}_{x:\overline{t}\rvert} - {}_t\overline{K}_x,\ \ t \le h$ $\{_h\overline{P}(\overline{A}_{x:\overline{n}\rvert})\overline{a}_{x:\overline{h}\rvert} - \overline{A}^1_{x:\overline{t}\rvert}\}/{}_tE_x,$ $h < t \le n$

Table 8.7 Notation and formulae for fully continuous retrospective reserve

Remark 8.2.1. The prospective reserve of a policy refers to future cash flows, assuming that the contract will not be surrendered. In practice, the policy holder may wish to surrender the contract and in that event, at least in the early years of the policy, will probably expect a surrender value related to the accumulation of his premiums less expenses and the cost of the life insurance cover. Such a surrender value is related to the retrospective reserve of the contract.

Remark 8.2.2. The equality of prospective and retrospective policy values will hold only if both reserves and the premiums are based on the same rates of interest and the same mortality law. When these requirements are satisfied, either method, whichever is simpler, can be used to find the reserve.

The next example illustrates the computation of prospective as well as retrospective reserve for some fully continuous policies.

Example 8.2.3. Suppose the mortality pattern is according to Gompertz' law given by $\mu_x = BC^x$ with $B = 0.0001151$ and $C = 1.096$. Suppose the force of interest is $\delta = 0.05$. Compute the reserve for $t = 1$ to 5, using both the approaches, corresponding to the following fully continuous policies issued to (30), with a benefit of 1000 units. (i) 5-year term insurance, (ii) 5-year endowment insurance. For (i) and (ii) assume that the premiums are payable as 5-year temporary continuous life annuity. (iii) 5-year endowment insurance with a premium paying term of 2 years. (iv) Whole life insurance with a premium paying term of 3 years.

Solution: From Table 8.5, the formulae of prospective reserves for (i) and (ii) respectively are as follows.

$$_tV(\overline{A}^{\,1}_{x:\overline{n}|}) = \overline{A}^{\,1}_{x+t:\overline{n-t}|} - \overline{P}(\overline{A}^{\,1}_{x:\overline{n}|})\,\overline{a}_{x+t:\overline{n-t}|}, \quad t \le n$$

$$\text{and} \quad _tV(\overline{A}_{x:\overline{n}|}) = \overline{A}_{x+t:\overline{n-t}|} - \overline{P}(\overline{A}_{x:\overline{n}|})\,\overline{a}_{x+t:\overline{n-t}|}, \quad t \le n$$

From these formulae, we note that we require the following actuarial present values. $\overline{A}^{\,1}_{30+t:\overline{5-t}|}$, $\overline{A}_{30+t:\overline{5-t}|}$ and $\overline{a}_{30+t:\overline{5-t}|}$ for $t = 0$ to 4. For $t = 5$, the values are $0, 1$ and 0 respectively. Values with $t = 0$ are useful to find the premiums for these insurance products. These actuarial present values are computed as in Example 5.3.7 and Example 6.3.2. We use Part II of Code 8.4.4 to find these values. These are reported in Table 8.8.

t	$\overline{A}^{\,1}_{30+t:\overline{5-t}|}$	$\overline{A}_{30+t:\overline{5-t}|}$	$\overline{a}_{30+t:\overline{5-t}|}$
0	0.0100	0.7799	4.4018
1	0.0085	0.8195	3.6098
2	0.0069	0.8612	2.7762
3	0.0049	0.9051	1.8985
4	0.0026	0.9513	0.9741
5	0.0000	1.0000	0.0000

Table 8.8 Gompertz' law: Actuarial present values

Dividing the second and third entry in the first row by the last entry in the first row, we get the following values of premium for unit benefit.

$$\overline{P}(\overline{A}^{\,1}_{30:\overline{5}|}) = 0.002261 \quad \text{and} \quad \overline{P}(\overline{A}_{30:\overline{5}|}) = 0.1772$$

Multiplying the last column by $\overline{P}(\overline{A}_{x:\overline{n}|})$ and then subtracting from the second column, we get the reserve for a 5-year term insurance. Similarly, we find the reserve for the endowment insurance. Since the benefit amount is 1000 units, we multiply these values by 1000. These reserves for 1000 unit benefit are obtained using Part II of Code 8.4.4 and are reported in the second and third columns of Table 8.9.

t	$1000\ _tV(\overline{A}^{\,1}_{30:\overline{n}|})$	$1000\ _tV(\overline{A}_{30:\overline{n}|})$	$1000\ _t^2V(\overline{A}_{30:\overline{n}|})$	$1000\ _t^3V(\overline{A}_{30})$
1	0.39	179.93	419.46	67.95
2	0.61	369.30	861.19	139.35
3	0.64	568.69	905.07	214.39
4	0.45	778.70	951.29	223.40
5	0.00	1000.00	1000.00	232.70

Table 8.9 Gompertz' law: Prospective reserve

Note that as t increases, the reserves for 5-year endowment insurance increase and for $t = 5$ it is 1000 units as in this product, the insurer has the liability to pay the benefit of 1000 to each surviving policy. On the other hand, the reserve for a 5-year term does not have a monotone behaviour, it increases for $t = 1, 2, 3$ and then decreases at $t = 4$ and is 0 at $t = 5$, appealing logically as at the end of the term, insurer has no liability to pay the benefit.

(iii) When the premium paying term is 2 years, we first find the premium payable in this setup. The formula for premium is given by,

$$_2\overline{P}(\overline{A}_{30:\overline{5}|}) \;=\; \overline{A}_{30:\overline{5}|}/\overline{a}_{30:\overline{2}|} \;=\; \delta\overline{A}_{30:\overline{5}|}/(1 - \overline{A}_{30:\overline{2}|}) \;=\; 0.4106$$

It is quite a bit higher than the premium payable as 5-year temporary annuity, as expected. The formula for reserve in Table 8.5 is given by,

$$_t^h\overline{V}(\overline{A}_{x:\overline{n}|}) \;=\; \begin{cases} \overline{A}_{x+t:\overline{n-t}|} - {}_h\overline{P}(\overline{A}_{x:\overline{n}|})\,\overline{a}_{x+t:\overline{h-t}|}, & \text{if } \;\; t \leq h < n \\ \overline{A}_{x+t:\overline{n-t}|}, & \text{if } \;\; h < t \leq n \end{cases}$$

We thus require $\overline{a}_{30+t:\overline{2-t}|}$ for $t = 1$ and 2. These values are 0.974428 and 0 respectively. We compute the values of reserve, using Part III of Code 8.4.4. These are then multiplied by 1000 and reported in the fourth column of Table 8.9. Note that with a limited number of years for premium payments, the reserve is high for all values of t.

(iv) To find the prospective reserve for whole life insurance with a premium paying period of $h = 3$ years, the formula for reserve in Table 8.5 is given by,

$$_t^h\overline{V}(\overline{A}_x) \;=\; \begin{cases} \overline{A}_{x+t} - {}_h\overline{P}(\overline{A}_x)\,\overline{a}_{x+t:\overline{h-t}|}, & \text{if } \;\; t \leq h \\ \overline{A}_{x+t}, & \text{if } \;\; t > h \end{cases}$$

We find the required actuarial present values and the premium. For the given mortality law and $\delta = 0.05$, we compute $\overline{A}_x$ for $x = 30$ to 35 and $\overline{a}_{30+t:\overline{3-t}|}$ for $t = 0, 1, 2, 3$. Then the premium is given by,

$$_3\overline{P}(\overline{A}_{30}) \;=\; \overline{A}_{30}/\overline{a}_{30:\overline{3}|} \;=\; 0.1891/2.7778 \;=\; 0.06808$$

The values of $\overline{A}_x$ are $0.1973, 0.2057, 0.2144, 0.2234, 0.2327$ for $x = 31$ to 35 respectively. The values of $\overline{a}_{30+t:\overline{3-t}|}$ for $t = 1, 2, 3$ are $1.8993, 0.9743$ and 0 respectively. With these we get the values of prospective reserve using the formula

$$_t^3\overline{V}(\overline{A}_{30}) \;=\; \overline{A}_{30+t} - {}_h\overline{P}(\overline{A}_{30})\,\overline{a}_{30+t:\overline{3-t}|}, \quad \text{for } \;\; t = 1, 2, 3$$

For $t = 4$ and 5, it is A_{34} and A_{35} respectively. These reserve values are then multiplied by 1000. We use Part IV of Code 8.4.4 to compute these values which are reported in the last column of Table 8.9.

We now proceed to the calculation of reserve with the retrospective approach. The formulae of retrospective reserves for (i) and (ii) from Table 8.7 are given by,

$$_t\overline{V}(\overline{A}{}^{\,1}_{x:\overline{n}|}) = \overline{P}(\overline{A}{}^{\,1}_{x:\overline{n}|})\overline{S}_{x:\overline{t}|} - {}_t\overline{K}_x, \; t \leq n$$

$$\text{and } \;\; _t\overline{V}(\overline{A}_{x:\overline{n}|}) = \overline{P}(\overline{A}_{x:\overline{n}|})\overline{S}_{x:\overline{t}|} - {}_t\overline{K}_x, \; t \leq n$$

We require the following actuarial present values. $\overline{A}{}^{\,1}_{30:\overline{t}|}$, $_t E_{30}$ and $\overline{a}_{30:\overline{t}|}$ for $t = 1$ to 5. These actuarial present values are obtained using the approach adopted in Example 5.3.7 and Example 6.3.2. We use Part V of Code 8.4.4 to compute these values, which are reported in Table 8.10.

t	$\overline{A}^{\,1}_{30:\overline{t}|}$	$_tE_{30}$	$\overline{a}_{30:\overline{t}|}$
1	0.0018	0.9494	0.9745
2	0.0037	0.9013	1.8997
3	0.0057	0.8554	2.7778
4	0.0078	0.8116	3.6111
5	0.0100	0.7700	4.4018

Table 8.10 Gompertz' law: Actuarial present values for retrospective reserve

The values of the premium are already computed. From the formula for the reserve with a retrospective approach for a 5-year term, we note that to find the values, we have to multiply each element in the last column of the Table 8.10 by the premium $\overline{P}(\overline{A}^{\,1}_{30:\overline{5}|}) = 0.002261$ and from that subtract the corresponding elements of the second column of Table 8.10. The resulting difference is then divided by the corresponding elements of the third column of Table 8.10. For the 5-year endowment, the procedure is the same as that for the 5-year term insurance, with the change that the value of premium is to be taken as $\overline{P}(\overline{A}_{30:\overline{5}|}) = 0.1772$. Since the benefit amount is 1000 units, we multiply these reserve values by 1000. The reserves for a 1000 unit benefit are reported in the second and the third columns of Table 8.11. We use Part V of Code 8.4.4 to compute these values.

(iii) When the premium paying term is 2 years, the premium is given by, $_2\overline{P}(\overline{A}_{30:\overline{5}|}) = 0.4106$. The formula for the retrospective reserve is given by,

$$_t^2\overline{V}(\overline{A}_{30:\overline{5}|}) = \frac{_2\overline{P}(\overline{A}_{30:\overline{5}|})\overline{a}_{x:\overline{t}|} - \overline{A}^{\,1}_{30:\overline{t}}}{_tE_{30}}, \quad t = 1,2$$

$$\text{and } _t^2\overline{V}(\overline{A}_{30:\overline{5}|}) = \frac{_2\overline{P}(\overline{A}_{30:\overline{5}|})\overline{a}_{x:\overline{2}|} - \overline{A}^{\,1}_{30:\overline{t}}}{_tE_{30}}, \quad t = 3,4,5$$

We have calculated all these quantities above. The reserve values after multiplying by 1000, are reported in the fourth column of Table 8.11. We use Part VI of Code 8.4.4 to compute these values.

(iv) To find the retrospective reserve for whole life insurance with 3-year premium payments, we have the following formulae.

$$\frac{_3\overline{P}(\overline{A}_{30})\overline{a}_{30:\overline{t}|} - \overline{A}^{\,1}_{30:\overline{t}}}{_tE_{30}}, \quad t = 1,2,3 \quad \text{and} \quad \frac{_3\overline{P}(\overline{A}_{30})\overline{a}_{30:\overline{3}|} - \overline{A}^{\,1}_{30:\overline{t}}}{_tE_{30}}, \quad t = 4,5$$

We have calculated all these quantities above. The reserve values after multiplying by 1000, are reported in the last column of Table 8.11. We use Part VII of Code 8.4.4 to compute these values.

Again note that the values of prospective and retrospective reserves match perfectly.   ∎

The following are some more examples to clarify the concept of reserve and related formulae.

t	$1000\ {}_t\overline{V}(\overline{A}\,{}^{1}_{x:\overline{n}|})$	$1000\ {}_t\overline{V}(\overline{A}_{x:\overline{n}|})$	$1000\ {}_t^{2}\overline{V}(\overline{A}_{x:\overline{n}|})$	$1000\ {}_t^{3}\overline{V}(\overline{A}_{30})$
1	0.39	179.93	419.46	67.95
2	0.61	369.30	861.19	139.35
3	0.64	568.69	905.07	214.39
4	0.45	778.70	951.29	223.40
5	0.00	1000.00	1000.00	232.70

Table 8.11 Gompertz' law: Retrospective reserve

Example 8.2.4. Three expressions are given below for ${}_t\overline{V}(\overline{A}_x)$. Which is the correct expression?

(I) $(\overline{A}_{x+t} - \overline{A}_x)/(1 - \overline{A}_{x+t})$ (II) $\overline{P}(\overline{A}_x)\overline{S}_{x:\overline{t}|} - \overline{P}(\overline{A}\,{}^{1}_{x:\overline{t}|})\overline{a}_{x:\overline{t}|}/\ {}_tE_x$ (III) $(\overline{P}(\overline{A}_{x+t}) - \overline{P}(\overline{A}_x))\ \overline{a}_x$

Solution: By Eq. 8.2.8,

$$_t\overline{V}(\overline{A}_x) \ = \ (\overline{A}_{x+t} - \overline{A}_x)/(1 - \overline{A}_x)$$

Thus (I) is false. From Table 8.7,

$$_t\overline{V}(\overline{A}_x) \ = \ \overline{P}(\overline{A}_x)\overline{S}_{x:\overline{t}|} - \frac{\overline{A}\,{}^{1}_{x:\overline{t}|}}{_tE_x} \ = \ \overline{P}(\overline{A}_x)\overline{S}_{x:\overline{t}|} - \frac{\overline{P}(\overline{A}\,{}^{1}_{x:\overline{t}|})\overline{a}_{x:\overline{t}|}}{_tE_x}$$

Thus (II) is true. By Eq. 8.2.5,

$$_t\overline{V}(\overline{A}_x) \ = \ (\overline{P}(\overline{A}_{x+t}) - \overline{P}(\overline{A}_x))\ \overline{a}_{x+t}$$

Thus (III) is false. ∎

Example 8.2.5. L is the loss at issue random variable for a fully continuous whole life insurance of 1 issued to (30) with the net premium determined by the equivalence principle. It is given that $\overline{A}_{50} = 0.7$, ${}^{2}\overline{A}_{30} = 0.3$ and $Var(L) = 0.2$. Calculate ${}_{20}\overline{V}(\overline{A}_{30})$.

Solution: By definition,

$$Var(L) \ = \ \frac{{}^{2}\overline{A}_{30} - \overline{A}_{30}^{2}}{(\delta\overline{a}_{30})^2} \ = \ \frac{{}^{2}\overline{A}_{30} - \overline{A}_{30}^{2}}{(1 - \overline{A}_{30})^2}$$

It is given that $Var(L) = 0.2$ and ${}^{2}\overline{A}_{30} = 0.3$. Hence, we get $\overline{A}_{30} = 0.5$. Consequently by Eq. 8.2.8,

$$_{20}\overline{V}(\overline{A}_{30}) = (\overline{A}_{50} - \overline{A}_{30})/(1 - \overline{A}_{30}) = (0.7 - 0.5)/(1 - 0.5) = 0.4$$

∎

Example 8.2.6. Calculate $\overline{a}_{50:\overline{20}|}$, given the following data.

(i) $\overline{P}(\overline{A}_{50}) \ = \ 0.02781$,   (ii) ${}_{20}\overline{V}(\overline{A}_{50}) \ = \ 0.21512$,   (iii) ${}_{20}E_{50} \ = \ 0.18072$ and (iv) ${}_{20}\overline{K}_{50} = 1.28895$.

Solution: By Eq. 8.2.5,

$$_{20}\overline{V}(\overline{A}_{50}) = \overline{P}(\overline{A}_{50})\,\overline{S}_{50:\overline{20|}} - {_{20}}\overline{K}_{50} = \overline{P}(\overline{A}_{50})\,\frac{\overline{a}_{50:\overline{20|}}}{_{20}E_{50}} - {_{20}}\overline{K}_{50}$$

Thus, from the given data we get, $\overline{a}_{50:\overline{20|}} = 9.77$.   ∎

Example 8.2.7. Calculate $_t\overline{V}(\overline{A}_x)$ given the following information.
(i) $_t\overline{K}_x = 0.30$, (ii) $_tE_x = 0.45$, (iii) $\overline{A}_{x+t} = 0.52$.

Solution: Being given the factor $_t\overline{K}_x$, it is natural to consider the retrospective form of the reserve as a possible approach to the solution. From Table 8.7, $_t\overline{V}(\overline{A}_x) = \overline{P}(\overline{A}_x)\,\overline{S}_{x:\overline{t|}} - {_t}\overline{K}_x$. But from the given information we cannot get $\overline{P}(\overline{A}_x)$ or $\overline{S}_{x:\overline{t|}}$ and this approach is not useful. Since this is a fully continuous whole life policy, there are various ways of representing the reserve. The insurance form is $_t\overline{V}(\overline{A}_x) = (\overline{A}_{x+t} - \overline{A}_x)/(1 - \overline{A}_x)$. We are given $\overline{A}_{x+t}$, hence from the given information we find $\overline{A}_x$. Now,

$$_t\overline{K}_x = \overline{A}\,^1_{x:\overline{t|}}/\,{_tE_x} \quad \Rightarrow \quad \overline{A}\,^1_{x:\overline{t|}} = {_t}\overline{K}_x\,{_tE_x} = 0.135$$

$$\Rightarrow \quad \overline{A}_x = \overline{A}\,^1_{x:\overline{t|}} + {_tE_x}\,\overline{A}_{x+t} = 0.369$$

$$\Rightarrow \quad _t\overline{V}(\overline{A}_x) = (\overline{A}_{x+t} - \overline{A}_x)/(1 - \overline{A}_x) = 0.24 \qquad ∎$$

In the entire section we discussed the computation of reserve for insurance products. Reserve for annuity products can be defined and computed along similar lines. For example, for n-year deferred whole life annuity, when premiums are paid as n-year temporary life annuity, the reserve denoted by, $_t\overline{V}(_{n|}\overline{a}_x)$ is given by,

$$_t\overline{V}(_{n|}\overline{a}_x) = \begin{cases} _{n-t|}\overline{a}_{x+t} - \overline{P}(_{n|}\overline{a}_x)\,\overline{a}_{x+t:\overline{n-t|}} & t \le n \\ \overline{a}_{x+t} & t > n \end{cases}$$

In the next section, we proceed to study the reserve for the fully discrete policy.

8.3   Fully Discrete Reserves

Computation of reserve for fully discrete policies is already illustrated in Example 8.1.1. In this section, we proceed to derive the formulae for reserve for various insurance products. The entire development is similar to that in Section 2, with the modifications to suit the discrete setup. The main change is that the underlying random variable is $K(x)$ instead of $T(x)$; thus, the expectations are with respect to the distribution of $K(x)$.

We first study the prospective approach. Suppose (x) signs a fully discrete policy. We want to find the future obligations to the insurance company corresponding to this policy after k units of time from the time of signing the contract, k being a positive integer. As we have defined $_tL$, we define a random variable $_kL$ specifying the present value at k of the prospective loss to the insurance company corresponding to this policy. Prospective loss random variable $_kL(K(x)) \equiv {_kL}(K)$ for whole life insurance is defined as,

$$_kL(K) = v^{(K(x)-k)+1} - P_x\,\ddot{a}_{\overline{(K(x)-k)+1|}}$$

The benefit reserve, denoted by $_kV_x$, is the conditional expectation of the prospective loss $_kL$, given that $K(x) = k, k+1, \cdots$. Thus,

$$_kV_x \;=\; E(_kL(K)|K(x) = k, k+1, \cdots) \tag{8.3.1}$$

As in the previous section, we use the result that in aggregate mortality setup, the conditional distribution of $K(x) - k$ given $K(x) \geq k$ is the same as the distribution of $K(x + k)$. Thus, for the whole life insurance policy with unit benefit, payable at the end of the year of death, premiums payable as whole life annuity due, the formula for the prospective benefit reserve is,

$$_kV_x \;=\; A_{x+k} - P_x \, \ddot{a}_{x+k} \tag{8.3.2}$$

As in Section 2, this formula is the actuarial present value of the future benefits minus the actuarial present value of the future benefit premiums. Analogous to Eq. 8.2.4, we have

$$
\begin{aligned}
Var[\,_kL|K(x) \geq k] \;&=\; Var\left[v^{[K(x)-k]+1} \left(1 + (P_x/d)\right) |K(x) \geq k \right] \\
&=\; (1 + (P_x/d))^2 \, Var[v^{[K(x)-k]+1}|K(x) \geq k] \\
&=\; (1 + (P_x/d))^2 \left[\, ^2A_{x+k} - (A_{x+k})^2 \right] \tag{8.3.3}
\end{aligned}
$$

From the formula for the reserve as given in Eq. 8.3.2, we obtain the premium difference formula and paid up insurance formula. These are as follows.

$$
\begin{aligned}
\text{Premium difference formula:} \quad _kV_x \;&=\; (P_{x+k} - P_x)\, \ddot{a}_{x+k} \tag{8.3.4} \\
\text{Paid up insurance formula:} \quad _kV_x \;&=\; (1 - P_x/P_{x+k})\, A_{x+k} \tag{8.3.5}
\end{aligned}
$$

As in the fully continuous case, formula for the reserve as given in Eq. 8.3.2, can be expressed either in terms of the annuity symbol or the premium symbol or the symbol of actuarial present value of benefit. These are given below and are obtained by using the relations $A_x = 1 - d\ddot{a}_x$ and $1/\ddot{a}_x = P_x + d$.

$$
\begin{aligned}
_kV_x \;&=\; 1 - d\ddot{a}_{x+k} - ((1/\ddot{a}_x) - d)\, \ddot{a}_{x+k} = 1 - \ddot{a}_{x+k}/\ddot{a}_x \tag{8.3.6} \\
_kV_x \;&=\; 1 - (P_x + d)/(P_{x+k} + d) = (P_{x+k} - P_x)/(P_{x+k} + d) \tag{8.3.7} \\
\text{and} \quad _kV_x \;&=\; 1 - (1 - A_{x+k})/(1 - A_x) = (A_{x+k} - A_x)/(1 - A_x) \tag{8.3.8}
\end{aligned}
$$

These expressions are useful in the computation of reserves depending on the available data. For a fully discrete policy, the inflow and the outflow of money to the insurance company is at discrete time points, hence the policy is also valued at discrete time points. As a consequence, we find the expression of reserve at integer values of k.

The reserves for other types of insurance models, such as n-year term insurance, n-year endowment insurance, can be defined on similar lines as those for the whole life insurance. In Table 8.12, we summarize the expression for the prospective loss random variable $_kL(K)$ for some insurance products. These expressions are derived on similar lines as those for the fully continuous policy, with the major change being, replacement of $T(x)$ by $K(x) + 1$.

The conditional expectation of $_kL(K)$, conditional on $K(x) \geq k$, gives the formula for the prospective reserve. Again in the aggregate mortality pattern, the conditional distribution of

Product	$_kL(K)$
Whole life	$v^{K(x)+1-k} \;-\; P_x\,\ddot{a}_{\overline{K(x)+1-k|}}$
n-Year term	$v^{K(x)+1-k} - P^1_{x:\overline{n|}}\,\ddot{a}_{\overline{K(x)+1-k|}},\;\; K(x)=k,\cdots,n-1$ $0 - P^1_{x:\overline{n|}}\,\ddot{a}_{\overline{n-k|}},\;\; K(x)=n,\cdots,$
n-Year endowment	$v^{K(x)+1-k} - P_{x:\overline{n|}}\,\ddot{a}_{\overline{K(x)+1-k|}},\;\; K(x)=k,\cdots,n-1$ $v^{n-k} - P_{x:\overline{n|}}\,\ddot{a}_{\overline{n-k|}}\;,\;\; K(x)=n,n+1,\cdots,$
h-Payments whole life	$v^{K(x)+1-k} -_h P_x\,\ddot{a}_{\overline{K(x)+1-k|}}\;,\;\; k \le K(x)+1 < h$ $v^{K(x)+1-k} -_h P_x\,\ddot{a}_{\overline{h-k|}}\;,\;\; k \le h < K(x)+1$ $v^{K(x)+1-k},\;\;\;\;\;\;\;\;\;\;\;\;\;\; k > h$
h-Payments n-Year endowment	$v^{K(x)+1-k} -_h P_{x:\overline{n|}}\,\ddot{a}_{\overline{K(x)+1-k|}}\;,\; k \le K(x)+1 < h$ $v^{K(x)+1-k} -_h P_{x:\overline{n|}}\,\ddot{a}_{\overline{h-k|}}\;,\;\;\; k \le h < K(x)+1 \le n$ $v^{n-k} -_h P_{x:\overline{n|}}\,\ddot{a}_{\overline{h-k|}}\;,\;\;\;\;\; k \le h < n < K(x)+1$ $v^{K(x)+1-k}\;,\;\;\;\;\;\;\;\;\; h < k \le K(x)+1 \le n$ $v^{n-k}\;,\;\;\;\;\;\;\;\;\;\;\;\;\; h \le k \le n < K(x)+1$

Table 8.12 Prospective loss random variable: Fully discrete policy

$K(x)-k$ is the same as the distribution of $K(x+k)$. Thus, the expectation of $_kL$ with respect to the distribution of $K(x+k)$, gives the formulae for the prospective reserve. Table 8.13 displays the international actuarial notation and the formulae for reserve using prospective approach for a variety of insurance models. The values of the reserves at the end of the term, that is, at $k = n$, are 0 and 1 for n-year term insurance and n-year endowment insurance respectively.

The following example illustrates the computation of prospective reserve for fully discrete policies.

Example 8.3.1. Suppose the force of mortality follows Makeham's law given by $\mu_x = A + BC^x$ with $A = 0.0007, B = 0.0001151$ and $C = 1.096$. Suppose $i = 0.05$ and for the same law and i, $A_{40} = 0.27975$. Calculate prospective reserve for the whole life insurance issued to (30), with a benefit of 1000 units, for $k = 1$ to 10, when premiums are payable as whole life annuity due.

Solution: It is given that $A_{40} = 0.27975$. We then use the backward recurrence relation $A_x = vq_x + vp_x A_{x+1}$ derived in Eq. 5.4.3 to find A_x for $x = 30$ to 39. Values for A_x are obtained using Part II of Code 8.4.5. Once we have these values, we use the formula $_kV_x = (A_{x+k} - A_x)/(1 - A_x)$ for the prospective reserve as given in Eq. 8.3.8. We use Part II of Code 8.4.5 to obtain values of $_kV_x$. Table 8.14 displays values of the reserve for $k = 1$ to

Product	Notation	Prospective reserve
Whole life	$_kV_x$	$A_{x+k} - P_x\,\ddot{a}_{x+k}$
n-Year term	$_kV^1_{x:\overline{n}\mid}$	$A^1_{x+k:\overline{n-k}\mid} - P^1_{x:\overline{n}\mid}\,\ddot{a}_{x+k:\overline{n-k}\mid},\qquad k \le n$
n-Year endowment	$_kV_{x:\overline{n}\mid}$	$A_{x+k:\overline{n-k}\mid} - P_{x:\overline{n}\mid}\,\ddot{a}_{x+k:\overline{n-k}\mid},\qquad k \le n$
h-Payments, whole life	$^h_kV_x$	$A_{x+k} - {_hP_x}\,\ddot{a}_{x+k:\overline{h-k}\mid},\qquad k \le h$ $A_{x+k},\qquad k > h$
h-Payments, n-year endowment	$^h_kV_{x:\overline{n}\mid}$	$A_{x+k:\overline{n-k}\mid} - {_hP_{x:\overline{n}\mid}}\,\ddot{a}_{x+k:\overline{h-k}\mid},\qquad k \le h < n$ $A_{x+k:\overline{n-k}\mid},\qquad h < k \le n$

Table 8.13 Notation and formulae for fully discrete prospective benefit reserves

10. For $k = 0$, it is 0 as we calculate the premium by the equivalence principle. It is clear from Eq. 8.3.8 also. Further, note that the reserve increases as k increases. ∎

k	1	2	3	4	5
$1000\,_kV_{30}$	10.12	20.59	31.40	42.56	54.08
k	6	7	8	9	10
$1000\,_kV_{30}$	65.95	78.18	90.77	103.72	117.02

Table 8.14 Makeham's law: $1000A_x$ and prospective reserve $1000\,_kV_{30}$

We now derive the formulae for the reserve using the retrospective approach.

Retrospective Reserve: Computation of retrospective reserve for 5-year endowment insurance is already illustrated in Example 8.1.1. In Section 2, the derivation of the formula for retrospective reserve is illustrated for 3-year endowment insurance. We follow the same procedure to derive the formulae for the retrospective reserve for various discrete policies. We begin with the whole life insurance with a benefit of 1 unit to be paid at the end of the year of death and premiums payable as whole life annuity due. The formula is derived by accumulating the funds of l_x identical policies until time k and sharing the money among the survivors. The interest is assumed to be earned at the rate i per annum. The accumulated fund F at time k

is derived below.

$$
\begin{aligned}
F &= \{l_x P_x (1+i)^k + l_{x+1} P_x (1+i)^{k-1} + \cdots + l_{x+k-1} P_x (1+i)\} \\
&\quad - \{d_x (1+i)^{k-1} + d_{x+1}(1+i)^{k-2} + \cdots + d_{x+k-1}\} \\
&= l_x \, P_x (1+i)^k \{1 + v p_x + v^2 p_x p_{x+1} + \cdots + v^{k-1} \,_{k-1}p_x\} \\
&\quad - l_x (1+i)^k \{v q_x + v^2 p_x q_{x+1} + \cdots + v^k \,_{k-1}p_x q_{x+k-1}\} \\
&= l_x \, P_x (1+i)^k \left\{ \sum_{j=0}^{k-1} v^j \,_{j}p_x \right\} - l_x (1+i)^k \left\{ \sum_{j=0}^{k-1} v^{j+1} \,_{j}p_x q_{x+j} \right\}
\end{aligned}
$$

Thus, $F = l_x (1+i)^k \left[P_x \ddot{a}_{x:\overline{k}|} - A^1_{x:\overline{k}|} \right]$. Dividing F by l_{x+k} we get the retrospective reserve as follows.

$$
\begin{aligned}
_kV_x &= \frac{l_x (1+i)^k}{l_{x+k}} [P_x \ddot{a}_{x:\overline{k}|} - A^1_{x:\overline{k}|}] \\
&= \frac{1}{_kE_x}[P_x \ddot{a}_{x:\overline{k}|} - A^1_{x:\overline{k}|}] \;=\; P_x \ddot{S}_{x:\overline{k}|} \;-\; _kK_x
\end{aligned} \tag{8.3.9}
$$

Note that this expression is similar to that in Eq. 8.2.13. In Eq. 8.3.9, $_kK_x = A^{\,1}_{x:\overline{k}|}/{}_kE_x = (1+i)^k \ddot{a}_{x:\overline{k}|}/{}_kp_x$ denotes the accumulated cost of insurance.. Similarly, $\ddot{S}_{x:\overline{k}|} = \ddot{a}_{x:\overline{k}|}/{}_kE_x = (1+i)^k \ddot{a}_{x:\overline{k}|}/{}_kp_x$ denotes the accumulated value of 1 unit benefit per annum, paid at the beginning of the year for at most k years, provided the policy is active for k years. Note that,

$$
\ddot{S}_{x:\overline{k}|} \;=\; 1/(A^{\,1}_{x:\overline{k}|}/\ddot{a}_{x:\overline{k}|}) \;=\; 1/P^{\,1}_{x:\overline{k}|}
$$

where $P_{\,x:\overline{k}|}^{1}$ is the premium for 1 unit benefit in k-year pure endowment insurance. Further from Eq. 8.3.9 we have,

$$
\begin{aligned}
_kV_x &= \frac{P_x \ddot{a}_{x:\overline{k}|} - A^1_{x:\overline{k}|}}{_kE_x} = \frac{\ddot{a}_{x:\overline{k}|}}{_kE_x}\left\{ P_x - \frac{A^1_{x:\overline{k}|}}{\ddot{a}_{x:\overline{k}|}} \right\} \\
&= (P_x - P^1_{x:\overline{k}|})/P^{\,1}_{x:\overline{k}|}
\end{aligned} \tag{8.3.10}
$$

The formula in Eq. 8.3.10 expresses $_kV_x$ in terms of premium symbols.

The retrospective reserves for other insurance products are obtained along similar lines as those for whole life insurance by considering the accumulated value at k of the premiums received and the accumulated value at k of the benefits paid. The expression for premium changes from product to product but the rest of the arguments remain the same. For h-payment whole life insurance and for h-payment n-year endowment insurance, the derivation is similar to that for the fully continuous policies. The accumulation of the fund is for h years only, with the payment of premiums at the beginning of the year. Table 8.15 displays the formulae for reserve using retrospective approach for a variety of insurance models discussed in Chapter 5.

Example 8.3.2. Suppose the force of mortality follows Makeham's law given by $\mu_x = A + BC^x$ with $A = 0.0007, B = 0.0001151$ and $C = 1.096$. Suppose $i = 0.05$ and for the same mortality

Product	Notation	Retrospective reserve
Whole life	$_kV_x$	$P_x \ddot{S}_{x:\overline{k}|} - {}_kK_x$
n-Year term	$_kV^1_{x:\overline{n}|}$	$P^1_{x:\overline{n}|} \ddot{S}_{x:\overline{k}|} - {}_kK_x, \; k \le n$
n-Year endowment	$_kV_{x:\overline{n}|}$	$P_{x:\overline{n}|} \ddot{S}_{x:\overline{k}|} - {}_kK_x, \qquad k \le n$
h-Payments, whole life	$_k^hV_x$	$_hP_x \ddot{S}_{x:\overline{k}|} - {}_kK_x, \qquad k \le h$ $\{_hP_x \ddot{a}_{x:\overline{h}|} - A^1_{x:\overline{k}|}\}/{}_kE_x \;, \qquad k > h$
h-Payments, n-year endowment	$_k^hV_{x:\overline{n}|}$	$_hP_{x:\overline{n}|} \ddot{S}_{x:\overline{k}|} - {}_kK_x, \qquad k < h$ $\{_hP_{x:\overline{n}|} \ddot{a}_{x:\overline{h}|} - A^1_{x:\overline{k}|}\}/{}_kE_x \;, \qquad h < k \le n$

Table 8.15 Notation and formulae for fully discrete retrospective reserves

law and i, $A_{40} = 0.27975$. Calculate the retrospective reserve for the whole life insurance issued to (30) with a benefit of 1000, for $k = 1$ to 10, when premiums are payable as whole life annuity due.

Solution: To find the reserve using the retrospective approach, we have from Eq. 8.3.9,

$$_kV_x = (P_x \, \ddot{a}_{x:\overline{k}|} - A^1_{x:\overline{k}|})/{}_kE_x$$

We have the values of A_x for $x = 30$ to 40. As in Example 8.2.2, to use these values we express the functions in this formula in terms of A_x. Note that $P_{30} = d \, A_{30}/(1 - A_{30}) = 0.01075838$, with $d = 0.04761905$ corresponding to $i = 0.05$. $A^1_{30:\overline{k}|}$ and $A_{30:\overline{k}|}$ can be expressed as

$$A^1_{30:\overline{k}|} = A_{30} - {}_kE_{30} \, A_{30+k} \quad \& \quad A_{30:\overline{k}|} = A_{30} - {}_kE_{30} \, A_{x+30} + {}_kE_{30}$$

From $A_{30:\overline{k}|}$, we find $\ddot{a}_{30:\overline{k}|} = (1 - A_{30:\overline{k}|})/d$. Thus, it is enough to compute $_kE_{30} = v^k \, {}_kp_{30}$ for Makeham's law for $x = 30$ and for $k = 1$ to 10. Using the survival function of Makeham's law, we get $_kE_{30}$ for these k values. Using these values and values of A_x for $x = 30$ to 40, we find $_kV_{30}$. Table 8.16 reports these values. We use Code 8.4.6 to obtain these values.   ∎

Comparing these reserve values with those in Table 8.14, we note that the values of reserve using both the approaches are the same.   ∎

k	1	2	3	4	5
$1000\ _kV_{30}$	10.12	20.59	31.40	42.56	54.08
k	6	7	8	9	10
$1000\ _kV_{30}$	65.95	78.18	90.77	103.72	117.02

Table 8.16 Makeham's law: Retrospective reserve $1000\ _kV_{30}$

Example 8.3.3. Compute the reserve corresponding to the following fully discrete policies issued to (30) for $k = 1$ to 5. (i) 5-year term insurance. (ii) 5-year endowment insurance. For (i) and (ii) assume that the premiums are payable as 5-year temporary life annuity due. (iii) 5-year endowment insurance with a premium paying term of 2 years. (iv) Whole life insurance with a premium paying term of 3 years. Suppose the benefit amount is 1000 units and the mortality pattern is according to Gompertz' law given by $\mu_x = BC^x$ with $B = 0.0001151$ and $C = 1.096$. Suppose the force of interest is $\delta = 0.05$. Use both the approaches.

Solution: From Table 8.13, the formulae for the prospective reserve for a 5-year term insurance and a 5-year endowment insurance respectively are as follows.

$$_kV^1_{30:\overline{5}|} = A^{\ 1}_{30+k:\overline{5-k}|} - P^1_{30:\overline{5}|}\ \ddot{a}_{30+k:\overline{5-k}|}, \quad k \le 5$$

$$_kV_{30:\overline{5}|} = A_{30+k:\overline{5-k}|} - P_{30:\overline{5}|}\ \ddot{a}_{30+k:\overline{5-k}|}, \quad k \le 5$$

From the formulae for prospective reserves for (i) and (ii), we require the following actuarial present values. $A^{\ 1}_{30+k:\overline{5-k}|}$, $A_{30+k:\overline{5-k}|}$ and $\ddot{a}_{30+k:\overline{5-k}|}$ for $k = 0$ to 4. For $k = 5$, the values are 0, 1 and 0 respectively. Values with $k = 0$ are useful to find the premiums for these insurance products. These actuarial present values are obtained using the same procedures as in Example 5.4.4. We use Part III of Code 8.4.7 to obtain these values, which are reported in Table 8.17.

k	$A^{\ 1}_{30+k:\overline{5-k}|}$	$A_{30+k:\overline{5-k}|}$	$\ddot{a}_{30+k:\overline{5-k}|}$
0	0.0097	0.7797	4.5177
1	0.0083	0.8193	3.7051
2	0.0067	0.8610	2.8496
3	0.0048	0.9050	1.9489
4	0.0026	0.9512	1.0000
5	0.0000	1.0000	0.0000

Table 8.17 Gompertz' law: Actuarial present values for discrete policies

Dividing the second and the third entry in the first row by the last entry in the first row, we get the following values of premium for unit benefit.

$$P^1_{30:\overline{5}|} = 0.002149 \quad \text{and} \quad P_{30:\overline{5}|} = 0.172580$$

To get the reserve for a 5-year term insurance, we proceed along the following two steps. (a) Multiply the last column by $P^1_{30:\overline{5}|}$. (b) Subtract the resulting column from the second

column. Similarly we find the reserve for the other product. Since the benefit amount is 1000 units, we multiply these values by 1000. The reserves for 1000 unit benefit are obtained using Part IV of Code 8.4.7 and are reported in the second and the third columns of Table 8.18.
(iii) When the premium paying term is 2 years, we first find the premium payable in this setup. The formula for premium is given by,

$$_2P_{30:\overline{5}|} = A_{30:\overline{5}|}/\ddot{a}_{30:\overline{2}|} = (d\,A_{30:\overline{5}|})/(1 - A_{30:\overline{2}|}) = 0.399945$$

quite higher than the premium payable as 5-year temporary annuity due. We also require $\ddot{a}_{30+k:\overline{2-k}|}$ for $k = 1$ and 2. By definition these values are 1 and 0 respectively. Using the formula given in Table 8.13, we compute the values of reserve. For $k = 1$, it is $A_{31:\overline{4}|} - {_2P_{30:\overline{5}|}}$ and for $k = 2$ to 5, it is $A_{30+k:\overline{5-k}|}$. These are then multiplied by 1000 and reported in the fourth column of Table 8.18. We use Part V of Code 8.4.7 to obtain these values. Note that with limited number of premium payments, the reserve is high for all values of k.
(iv) To find the prospective reserve for whole life insurance with a premium paying period of $h = 3$ years, we find the required actuarial present values and the premium. For the given mortality law and $\delta = 0.05$, we compute A_x for $x = 30$ to 35 and $\ddot{a}_{30+k:\overline{3-k}|}$ for $k = 0, 1, 2, 3$. The values of A_x are $0.1845, 0.1924, 0.2007, 0.2091, 0.2179, 0.2270$ for $x = 30$ to 35 respectively. The values of $\ddot{a}_{30+k:\overline{3-k}|}$ for $k = 0, 1, 2, 3$ are $2.8507, 1.9493, 1$ and 0 respectively. With these values, the premium is given by,

$$_3P_{30} = A_{30}/\ddot{a}_{30:\overline{3}|} = 0.184465/2.8507 = 0.0647$$

We then use the formula $_k^3V_{30} = A_{30+k} - {_3P_{30}}\,\ddot{a}_{30+k:\overline{3-k}|}$ to find the values of prospective reserve for $k = 1, 2, 3$. For $k = 4$ and 5, it is A_{34} and A_{35}. These reserve values are multiplied by 1000 and are reported in the last column of Table 8.18. We use Part VI of Code 8.4.7 to obtain these values.

k	Term	Endowment	2-Payments endowment	3-Payments whole life
1	0.38	179.88	419.36	66.27
2	0.59	369.23	861.02	135.91
3	0.62	568.62	904.95	209.11
4	0.43	778.65	951.23	217.89
5	0.00	1000.00	1000.00	226.97

Table 8.18 Gompertz' law: Prospective reserve for discrete policies

We now proceed to obtain the reserves with a retrospective approach for the same insurance products. From Table 8.15, for (i) and (ii) the formulae for the reserve with a retrospective approach are as follows.

$$_kV^1_{30:\overline{5}|} = (P^1_{30:\overline{5}|}\ddot{a}_{30:\overline{k}|} - A^1_{30:\overline{k}|})/{_kE_{30}}, \quad k = 1, 2, 3, 4, 5$$

$$_kV_{30:\overline{5}|} = (P_{30:\overline{5}|}\ddot{a}_{30:\overline{k}|} - A^1_{30:\overline{k}|})/{_kE_{30}}, \quad k = 1, 2, 3, 4, 5$$

In Example 4.3.4 we have derived the probability mass function of $K(30)$ for Gompertz' law. It is given by,

$$P[K(x) = k] = \exp[-mC^x(C^k - 1)] - \exp[-mC^x(C^{k+1} - 1)]$$

Using it we find the values of $A^1_{30:\overline{k}|}$ and $_kE_{30}$ for $k = 1$ to 5. These two together give the values of $A_{30:\overline{k}|}$ for $k = 1$ to 5 and hence the values of $\ddot{a}_{30:\overline{k}|}$. These actuarial present values are obtained using the approach adopted in Example 5.4.4 and Part VII of Code 8.4.7. These are reported in Table 8.19.

k	$A^1_{30:\overline{k}|}$	$_kE_{30}$	$\ddot{a}_{30:\overline{k}|}$
1	0.0018	0.9494	1.0000
2	0.0037	0.9013	1.9494
3	0.0056	0.8554	2.8507
4	0.0076	0.8116	3.7061
5	0.0097	0.7700	4.5177

Table 8.19 Gompertz' law: Actuarial present values for retrospective reserve

We have already computed the values of premiums. Using these values and the formulae in Table 8.15, we get the reserve with the retrospective approach for all k and for the 5-year term and the 5-year endowment insurance. For 2-payments, 5-year endowment insurance, the retrospective reserve is given by,

$$
{}^2_kV_{30:\overline{5}|} = \begin{cases} ({}_2P_{30:\overline{5}|}\ddot{a}_{30:\overline{k}|} - A^1_{30:\overline{k}|})/{}_kE_{30}, & \text{if} \quad k = 1,2 \\ ({}_2P_{30:\overline{5}|}\ddot{a}_{30:\overline{2}|} - A^1_{30:\overline{k}|})/{}_kE_{30}, & \text{if} \quad k = 3,4,5 \end{cases}
$$

For 3-payments, whole life insurance, the retrospective reserve is given by,

$$
{}^3_kV_{30} = \begin{cases} ({}_3P_{30}\ddot{a}_{30:\overline{k}|} - A^1_{30:\overline{k}|})/{}_kE_{30}, & \text{if} \quad k = 1,2,3 \\ ({}_3P_{30}\ddot{a}_{30:\overline{3}|} - A^1_{30:\overline{k}|})/{}_kE_{30}, & \text{if} \quad k = 4,5 \end{cases}
$$

These are then multiplied by 1000 as the benefit amount is 1000 units and are reported in Table 8.20. We use Part VIII, IX and X of Code 8.4.7 to obtain these values.

Note that values of the reserve for all the insurance products are the same for both the prospective and retrospective approaches.   ∎

The following are some more examples illustrating the concept of fully discrete reserve.

Example 8.3.4. A fully discrete whole life insurance of 1 is issued to (25). The premiums are paid annually up to age 65. The net premium during the first 10 years is P_{25}, followed by a different level annual premium payable during the next 30 years. It is given that $A_{35} = 0.3, P_{25} = 0.01$ and $d = 0.06$. Calculate the reserve at the end of the tenth year.

k	Term	Endowment	2-Payments endowment	3-Payments whole life
1	0.38	179.88	419.36	66.27
2	0.59	369.23	861.02	135.91
3	0.62	568.62	904.95	209.11
4	0.43	778.65	951.23	217.89
5	0.00	1000.00	1000.00	226.97

Table 8.20 Gompertz' law: Retrospective reserve for discrete policies

Solution: Using the prospective approach to compute the reserve, we have,

$$_{10}V = A_{35} - P_{25}\ddot{a}_{35} = 0.3 - (0.01)\left((1 - 0.3)/0.06\right) = 0.183$$

In the calculation of reserve for $t = 10$, the different level premium does not matter. Note that the reserve is simply denoted by $_{10}V$ as the underlying insurance product is not among the standard ones. The prefix 10 indicates the valuation of the policy at $k = 10$. ∎

Example 8.3.5. Suppose $_kL$ is the prospective loss random variable for a fully discrete whole life insurance of 1000 issued to (x). It is given that $A_x = 0.125$, $A_{x+k} = 0.4$, $^2A_{x+k} = 0.2$, $d = 0.05$. Assume that the premiums are calculated on the basis of the equivalence principle. (i) Calculate $E(_kL)$, $Var(_kL)$ and the aggregate reserve at a time k for 100 policies of this type. (ii) An insurer has 100 policies of this type at a time k. Assume the losses on the 100 policies are independent. (a) Calculate the aggregate amount the insurer must have at time k so that the probability of meeting its obligations for these policies is 0.95.
(b) The premium is changed at a time k. Calculate the level annual premium the insurer must charge beginning at a time k on each policy so that together with the reserve calculated in (i), the probability of meeting its obligations for these policies is 0.95. Use the approximation that the aggregate loss is normally distributed in (a) and (b).

Solution: (i) From the given information

$$\begin{aligned}
\ddot{a}_x &= (1 - A_x)/d = 17.5, \quad \Rightarrow \quad 1000P_x = 1000A_x/\ddot{a}_x = 7.142860 \\
\ddot{a}_{x+k} &= (1 - A_{x+k})/d = 12 \\
E(_kL) &= 1000A_{x+k} - 1000P_x\ddot{a}_{x+k} = 314.2857, \quad \text{by definition} \\
Var(_kL) &= 1000^2(1 + P_x/d)^2(\,^2A_{x+k} - A_{x+k}^2) = 52,244.898 \\
\text{and } 100E(_kL) &= 31428.57, \quad \text{reserve for 100 policies.}
\end{aligned}$$

(ii) (a) Suppose, $S = \sum_{j=1}^{100} {}_kL^{(j)}$. Then,

$$\mu = E(S) = 100E(_kL) = 31428.57, \quad \sigma^2 = Var(S) = 100Var(_kL) = 5224,489.8$$

By normal approximation,

$$P[S > c] = 0.05 \quad \Rightarrow \quad (c - \mu)/\sigma = 1.645 \quad \Rightarrow \quad c = 35,188.57$$

is the amount that the insurer must have at a time k so that the probability of meeting its obligations for these policies is 0.95.

(b) Suppose $\hat{P}$ is the premium to be determined for the insurance of 1000 units. Then,

$$E(\ _k\hat{L}) = 1000A_{x+k} - \hat{P}\,\ddot{a}_{x+k} = 400 - 12\hat{P}$$

$$Var(\ _k\hat{L}) = \left(1000 + (\hat{P}/d)\right)^2 (\ ^2A_{x+k} - A_{x+k}^2) = (200 + 4\hat{P})^2$$

Suppose $\hat{S} = \sum_{j=1}^{100} \ _k\hat{L}^{(j)}$, then $\hat{\mu} = E(\hat{S}) = 100E(\ _k\hat{L}) = 40{,}000 - 1200\hat{P}$ and $\hat{\sigma}^2 = Var(\hat{S}) = 100(200 + 4\hat{P})^2$. Again by normal approximation,

$$P[\hat{S} > c_1] = 0.05 \;\Rightarrow\; (c_1 - \hat{\mu})/\hat{\sigma} = 1.645 \;\Rightarrow\; c_1 = \hat{\mu} + 1.645\,\hat{\sigma}$$
$$\Rightarrow\; c_1 = 100E(_k L) = 31428.57 \quad \text{by (i)}$$
$$\Rightarrow\; \hat{P} = 10.46$$

$\blacksquare$

Example 8.3.6. Determine an expression in actuarial present values and benefit premiums for the $Var(_k L | K(x) \geq k)$ for a fully discrete n-year endowment insurance with a unit benefit.

Solution: For a fully discrete n-year endowment insurance with a unit benefit, the prospective loss random variable is given by,

$$_k L = \begin{cases} v^{K(x)-k+1}\left(1 + \frac{P_{x:\overline{n|}}}{d}\right) - \frac{P_{x:\overline{n|}}}{d}, & \text{if } K(x) = k, k+1, \cdots, n-1 \\[2mm] v^{n-k}\left(1 + \frac{P_{x:\overline{n|}}}{d}\right) - \frac{P_{x:\overline{n|}}}{d}, & \text{if } K(x) = n, n+1, \cdots \end{cases}$$

Hence,

$$Var(_k L | K(x) \geq k) = \left(1 + (P_{x:\overline{n|}}/d)\right)^2 [\ ^2A_{x+k:\overline{n-k|}} - (A_{x+k:\overline{n-k|}})^2]$$

$\blacksquare$

Example 8.3.7. It is given that $A_{x:\overline{n|}} = 0.20$ and $d = 0.08$. Calculate $_{n-1}V_{x:\overline{n|}}$.

Solution: By definition,

$$_{n-1}V_{x:\overline{n|}} = A_{x+n-1:\overline{1|}} - P_{x:\overline{n|}}\ddot{a}_{x+n-1:\overline{1|}} = (\ vq_{x+n-1} + vp_{x+n-1}) - P_{x:n}$$
$$= v - (dA_{x:\overline{n|}})/(1 - A_{x:\overline{n|}}) = 0.92 - (0.08)(0.20)/0.80 = 0.90$$

$\blacksquare$

Example 8.3.8. It is given that $\ddot{a}_{x+t} = 10$, $_t V_x = 0.100$, $_{t+1}V_x = 0.127$ and $P_{x+t+1} = 0.043$. Find d.

Solution : Observe that

$$0.100 = \ _t V_x = 1 - \ddot{a}_{x+t}/\ddot{a}_x \;\Rightarrow\; \ddot{a}_x = 11.11$$
$$0.127 = \ _{t+1}V_x = 1 - \ddot{a}_{x+t+1}/\ddot{a}_x \;\Rightarrow\; \ddot{a}_{x+t+1} = 9.70$$
$$\text{and } 0.043 = P_{x+t+1} = (1/\ddot{a}_{x+t+1}) - d \;\Rightarrow\; d = 0.06$$

$\blacksquare$

Example 8.3.9. For a special fully discrete modified premium whole life insurance of 1 issued to (45), the following information is given. (i) The benefit premium is P_1 for the first 20 years and P_2 thereafter. (ii) Benefit premiums P_1 and P_2 are determined according to the equivalence principle. (iii) $A_{45} = 0.25$, $A_{65} = 0.50$, $\ddot{a}_{45} = 15.71$, $\ddot{a}_{65} = 10.60$. (iv) $\ddot{a}_{45:\overline{20|}} = 12.43$. (v) If the benefit premium after year 20 is still P_1 and the benefit after year 20 is reduced to 0.75, then $_{20}V$ is unchanged. Calculate $_{20}V$.

Solution: By the equivalence principle P_1 and P_2 are determined by the equation,

$$0.25 = A_{45} = P_1\ddot{a}_{45:\overline{20|}} + P_2[\ddot{a}_{45} - \ddot{a}_{45:\overline{20|}}] = 12.43P_1 + 3.28P_2$$

From the given information and prospective reserve approach, we get

$$
\begin{aligned}
_{20}V \;&=\; A_{65} - P_2\ddot{a}_{65} = 0.75A_{65} - P_1\ddot{a}_{65} \\
&\Rightarrow\; 0.5 - 10.6P_2 = 0.375 - 10.6P_1 \;\Rightarrow\; P_1 = P_2 - 0.0118
\end{aligned}
$$

Solving the two equations in P_1 and P_2, we get,

$$P_2 = 0.0252 \quad\Rightarrow\quad {}_{20}V = A_{65} - P_2\ddot{a}_{65} = 0.23$$

∎

Example 8.3.10. Calculate $_{20}V_{25}$ given that (i) $_{10}V_{25} = 0.1$ and (ii) $_{10}V_{35} = 0.2$.

Solution: Using the annuity form of the whole life reserve, note that

$$
\begin{aligned}
_{10}V_{25} \;&=\; 1 - \ddot{a}_{35}/\ddot{a}_{25} \quad\text{and}\quad {}_{10}V_{35} = 1 - \ddot{a}_{45}/\ddot{a}_{35} \\
&\Rightarrow\; \ddot{a}_{35}/\ddot{a}_{25} = 0.9 \quad\text{and}\quad \ddot{a}_{45}/\ddot{a}_{35} = 0.8 \\
&\Rightarrow\; {}_{20}V_{25} = 1 - \frac{\ddot{a}_{45}}{\ddot{a}_{25}} = 1 - \left(\frac{\ddot{a}_{35}}{\ddot{a}_{25}}\right)\left(\frac{\ddot{a}_{45}}{\ddot{a}_{35}}\right) = 0.28
\end{aligned}
$$

∎

Example 8.3.11. It is given that (i) $P_x = 0.00646$, (ii) $P_{x:\overline{n|}} = 0.04166$, (iii) $P^1_{x:\overline{n|}} = 0.00211$ and (iv) $P_{x+n} = 0.01426$. Calculate i.

Solution: i is involved in a number of formulae for reserve. The one useful here is $_nV_x = (P_{x+n} - P_x)/(P_{x+n} + d)$. Using Eq. 8.3.10 we can also obtain $_nV_x$ as

$$_nV_x = \frac{P_x - P^1_{x:\overline{n|}}}{_nE_x} = \frac{P_x - P^1_{x:\overline{n|}}}{P_{x:\overline{n|}} - P^1_{x:\overline{n|}}} = \frac{0.00646 - 0.00211}{0.04166 - 0.00211} = 0.110$$

Then

$$0.11 = {}_nV_x = \frac{P_{x+n} - P_x}{P_{x+n} + d} = \frac{0.01426 - 0.00646}{0.01426 + d}$$

$$\Rightarrow\; d = 0.0566 \;\Rightarrow\; i = d/(1-d) = 0.06$$

∎

Example 8.3.12. Calculate $_{10}^{20}V_x$ when (i) $P_x = 0.01212$, (ii) $_{20}P_x = 0.01508$, (iii) $P_{x:\overline{10}|}^{\ 1} = 0.06942$ and (iv) $_{10}V_x = 0.11430$.

Solution: Using the formula for retrospective reserve

$$
\begin{aligned}
{}_{10}^{20}V_x &= {}_{20}P_x\,\ddot{S}_{x:\overline{10}|} - {}_{10}K_x \quad \& \quad {}_{10}V_x = P_x\,\ddot{S}_{x:\overline{10}|} - {}_{10}K_x \\
&\Rightarrow {}_{10}^{20}V_x - {}_{10}V_x = [\,{}_{20}P_x - P_x\,]\,\ddot{S}_{x:\overline{10}|} = [\,{}_{20}P_x - P_x\,]/P_{x:\overline{10}|}^{\ 1} \\
&\Rightarrow {}_{10}^{20}V_x = {}_{10}V_x + [\,{}_{20}P_x - P_x\,]/P_{x:\overline{10}|}^{\ 1} = 0.157
\end{aligned}
$$

∎

The reserve for discrete annuity products are defined analogously. For example, the reserve for an n-year deferred whole life annuity due, purchased by paying premiums as an n-year temporary annuity due is denoted by $_kV({}_{n|}\ddot{a}_x)$ and is given by,

$$
{}_kV({}_{n|}\ddot{a}_x) = \begin{cases} {}_{n-k|}\ddot{a}_{x+k} - P({}_{n|}\ddot{a}_x)\,\ddot{a}_{x+k:\overline{n-k|}}, & k < n \\ \ddot{a}_{x+k}, & k \geq n \end{cases}
$$

We now discuss the variations needed in the following two cases. (i) Semi-continuous policies and (ii) when the premiums are paid as m-thly annuity due.

Reserve for a Semi-continuous Policy: Reserves for semi-continuous policies are obtained by combining the appropriate parts from the two types of policies discussed above. For example, for an n-year endowment insurance, when the benefit is to be paid at the moment of death and the premiums are payable as n-year temporary life annuity due, the prospective reserve is given by,

$$
{}_kV(\overline{A}_{x:\overline{n}|}) = \overline{A}_{x+k:\overline{n-k|}} - P(\overline{A}_{x:\overline{n}|})\,\ddot{a}_{x+k:\overline{n-k|}}, \qquad \text{for } k \leq n
$$

The expression for retrospective reserve is given by,

$$
{}_kV(\overline{A}_{x:\overline{n}|}) = \{P(\overline{A}_{x:\overline{n}|})\ddot{a}_{x:\overline{k}|} - \overline{A}_{x:\overline{k}|}^{1}\}/{}_kE_x, \qquad \text{for } k \leq n
$$

Reserve Corresponding to Fractional Premiums: These reserves are derived on similar lines as those for annual premiums. In a prospective reserve, the second term taking into account the present value at t of the future premiums gets changed to incorporate the inflow to the company by m-thly annuity due. In retrospective reserve, the first term taking into account the accumulated value of the premiums at t needs to be changed to incorporate the accumulation of premiums by m-thly annuity due. For example, for whole life insurance, when the benefit is to be paid at the end of year of death and the premiums are payable as m-thly whole life annuity due, the prospective reserve is given by,

$$
{}_kV_x^{(m)} = A_{x+k} - P_x^{(m)}\,\ddot{a}_{x+k}^{(m)}
$$

The expression for retrospective reserve is given by,

$$
{}_kV_x^{(m)} = \{P_x^{(m)}\ddot{a}_{x:\overline{k}|}^{(m)} - A_{x:\overline{k}|}^{1}\}/{}_kE_x
$$

For whole life insurance, using the formulae for prospective reserves $_kV_x^{(m)}$ and $_kV_x$, we get a relation between the two as follows. It is useful to compute the reserve corresponding to fractional premiums, once we know the reserve corresponding to the annual premium.

$$
\begin{aligned}
_kV_x^{(m)} - {}_kV_x &= \{A_{x+k} - P_x^{(m)}\, \ddot{a}_{x+k}^{(m)}\} - \{A_{x+k} - P_x\, \ddot{a}_{x+k}\} \\
&= P_x\, \ddot{a}_{x+k} - P_x^{(m)}\, \ddot{a}_{x+k}^{(m)} \\
&= \frac{A_x}{\ddot{a}_x^{(m)}} \frac{\ddot{a}_x^{(m)}}{\ddot{a}_x}\ddot{a}_{x+k} - P_x^{(m)}\, \ddot{a}_{x+k}^{(m)} \\
&= P_x^{(m)}\left\{ \frac{\alpha(m)\ddot{a}_x - \beta(m)}{\ddot{a}_x}\ddot{a}_{x+k} - \alpha(m)\ddot{a}_{x+k} + \beta(m)\right\} \\
&= P_x^{(m)}\beta(m)\{1 - (\ddot{a}_{x+k}/\ddot{a}_x)\} = P_x^{(m)}\beta(m)\, {}_kV_x \\
\Rightarrow\quad _kV_x^{(m)} &= {}_kV_x(1 + P_x^{(m)}\beta(m)) \tag{8.3.11}
\end{aligned}
$$

In this derivation, it is assumed that the mortality pattern in each unit age interval is according to the uniform distribution.

Example 8.3.13. Suppose the force of mortality follows Makeham's law given by $\mu_x = A + BC^x$ with $A = 0.0007, B = 0.0001151$ and $C = 1.096$. Suppose $i = 0.05$ and for the same mortality law and i, $A_{40} = 0.27975$. Calculate the prospective reserve for the whole life insurance issued to (30) with a benefit of 1000, for $k = 1$ to 10, when premiums are payable as m-thly whole life annuity due with $m = 2, 4, 12$. Assume that in each unit interval, the mortality pattern satisfies the uniformity assumption.

Solution: For the specified mortality pattern and interest rate we have computed values of A_x for $x = 30$ to 40 and $_kV_{30}$ for $k = 1$ to 10 in Example 8.3.1. We use Parts II and III of Code 8.4.8 to find these values. Then we use the formula as derived in Eq. 8.3.11 to compute m-thly reserves. The m-thly premium is given by the formula,

$$
P_{30}^{(m)} = A_{30}/\ddot{a}_{30}^{(m)} = A_{30}/(\alpha(m)\ddot{a}_{30} - \beta(m)), \quad \text{where } \ddot{a}_{30} = (1 - A_{30})/d
$$

Table 8.21 reports these values for $m = 2, 4$, and 12. These are obtained using Part IV of Code 8.4.8.

m	$i^{(m)}$	$d^{(m)}$	$\alpha(m)$	$\beta(m)$	$\ddot{a}_{30}^{(m)}$	$P_{30}^{(m)}$
2	0.04939	0.04820	1.00015	0.25617	16.87628	0.01092
4	0.04909	0.04849	1.00019	0.38272	16.75038	0.01100
12	0.04889	0.04869	1.00020	0.46651	16.66678	0.01106

Table 8.21 Makeham's law: $P_{30}^{(m)}$

Using the values of $_kV_{30}$ for $k = 1$ to 10 as obtained in Example 8.3.1 and the formula derived in Eq. 8.3.11, we find the values of six monthly, quarterly and monthly reserves, using Part V of Code 8.4.8. Table 8.22 reports the values, along with $_kV_{30}$.

k	Annual	Six monthly	Quarterly	Monthly
1	10.12	10.15	10.17	10.18
2	20.59	20.65	20.68	20.70
3	31.40	31.49	31.53	31.56
4	42.56	42.68	42.74	42.78
5	54.08	54.23	54.30	54.36
6	65.95	66.13	66.23	66.29
7	78.18	78.40	78.51	78.58
8	90.77	91.02	91.15	91.24
9	103.72	104.01	104.15	104.25
10	117.02	117.35	117.52	117.63

Table 8.22 Makeham's law: $1000\ {}_{k}V_{30}^{(m)}$

From the table we note that the reserves differ marginally for various values of m.   ∎

Table 8.23 displays the international actuarial notation and the formulae for reserve using the prospective approach for fully discrete policies when the premiums are payable m-thly.

Product	Notation	Prospective reserve
Whole life	${}_{k}V_{x}^{(m)}$	$A_{x+k} - P_{x}^{(m)}\, \ddot{a}_{x+k}^{(m)}$
n-Year term	${}_{k}V_{x:\overline{n}|}^{1(m)}$	$A\,{}^{\ 1}_{x+k:\overline{n-k}|} - P_{x:\overline{n}|}^{1(m)}\, \ddot{a}_{x+k:\overline{n-k}|}^{(m)},\ \ k \le n$
n-Year endowment	${}_{k}V_{x:\overline{n}|}^{(m)}$	$A_{x+k:\overline{n-k}|} - P_{x:\overline{n}|}^{(m)}\, \ddot{a}_{x+k:\overline{n-k}|}^{(m)},\ \ \ k \le n$
h-Payments, whole life	${}_{k}^{h}V_{x}^{(m)}$	$A_{x+k} - {}_{h}P_{x}^{(m)}\, \ddot{a}_{x+k:\overline{h-k}|}^{(m)},\ \ \ \ k \le h$ $A_{x+k},\ \ \ \ \ \ \ \ \ \ \ \ \ \ k > h,$
h-Payments, n-year endowment	${}_{k}^{h}V_{x:\overline{n}|}^{(m)}$	$A_{x+k:\overline{n-k}|} - {}_{h}P_{x:\overline{n}|}^{(m)}\, \ddot{a}_{x+k:\overline{h-k}|}^{(m)},$ $k \le h < n$ $A_{x+k:\overline{n-k}|},\ \ \ \ \ \ \ h < k \le n$

Table 8.23 Notation and formulae: Fully discrete prospective benefit reserves with m-thly premiums

Table 8.24 displays the international actuarial notation and the formulae for reserve using the retrospective approach for fully discrete policies when the premiums are payable m-thly.

Example 8.3.14. It is given that deaths are uniformly distributed over each year of age,

Product	Notation	Retrospective reserve
Whole life	$_kV_x^{(m)}$	$P_x^{(m)}\ddot{S}_{x:\overline{k}|}^{(m)} - {}_kK_x$
n-Year term	$_kV_{x:\overline{n}|}^{1(m)}$	$P_{x:\overline{n}|}^{1(m)}\ddot{S}_{x:\overline{k}|}^{(m)} - {}_kK_x,\ k \le n$
n-Year endowment	$_kV_{x:\overline{n}|}^{(m)}$	$P_{x:\overline{n}|}^{(m)}\ddot{S}_{x:\overline{k}|}^{(m)} - {}_kK_x, \qquad k \le n$
h-Payments, whole life	$_k^hV_x^{(m)}$	$_hP_x^{(m)}\ddot{S}_{x:\overline{k}|}^{(m)} - {}_kK_x, \qquad k \le h$ $\{_hP_x^{(m)}\ddot{a}_{x:\overline{h}|}^{(m)} - A_{x:\overline{k}|}^1\}/{}_kE_x,$ $\qquad\qquad k > h$
h-Payments, n-year endowment	$_k^hV_{x:\overline{n}|}^{(m)}$	$_hP_{x:\overline{n}|}^{(m)}\ddot{S}_{x:\overline{k}|}^{(m)} - {}_kK_x, \qquad k < h$ $\{_hP_{x:\overline{n}|}^{(m)}\ddot{a}_{x:\overline{h}|}^{(m)} - A_{x:\overline{k}|}^1\}/{}_kE_x,$ $\qquad\qquad h < k \le n$

Table 8.24 Notation and formulae: Fully discrete retrospective benefit reserves with m-thly premiums

$i = 0.5, 1000\ _5V_{25} = 44.71$, $\ddot{a}_{25}^{(4)} = 17.031204$ and $A_{25} = 0.17092$. Calculate $1000(\ _5V_{25}^{(4)} - {}_5V_{25})$.

Solution: For the whole life insurance, under the assumption of uniformity for fractional ages we have a relation between $_kV_x^{(m)}$ and $_kV_x$ as,

$$
\begin{aligned}
_kV_x^{(m)} - {}_kV_x &= P_x^{(m)}\beta(m)V_x \\
\Rightarrow\ _5V_{25}^{(4)} - {}_5V_{25} &= \beta(4)P_{25}^{(4)}\ {}_5V_{25} \\
&= \beta(4)(A_{25}\ {}_5V_{25})/\ddot{a}_{25}^{(4)} = 0.00017 \\
\Rightarrow\ 1000(_5V_{25}^{(4)} - {}_5V_{25}) &= 0.17
\end{aligned}
$$

Thus, in this example also the difference between the valuation of the policy at 5 when premiums are paid annually or quarterly, is marginal. ∎

In the entire discussion so far, we focused on the computation of reserves involving net premiums. Reserves corresponding to gross premiums are computed on similar lines. In this setup, as stated in Eq. 8.2.1, a random variable $_tL$ is defined as,

$$
\begin{aligned}
_tL\ =\ &\text{Present value random variable at } t \text{ of future benefits} \\
&\text{and of future expenses} \\
-\ &\text{Present value random variable at } t \text{ of future gross premiums}
\end{aligned}
$$

The reserve $_tV$ is defined as,

$$_tV \;=\; E(_tL) = \text{Actuarial present value at } t \text{ of future benefits and expenses}$$

$$- \text{ Actuarial present value at } t \text{ of future gross premiums,}$$

where E denotes the conditional expectation of $_tL$ given that $T(x) > t$ or $K(x) \geq t$ depending on the nature of the policy. We use setup of Example 7.6.1 to illustrate the computation of reserve corresponding to gross premiums.

Example 8.3.15. It is given that $q_{28} = 0.135$, $q_{29} = 0.146$, $q_{30} = 0.159$, $q_{31} = 0.173$, $q_{32} = 0.188$ and $i = 0.05$. The annual premiums are payable as 3-year temporary discrete annuity due for benefit of 50,000, payable at the end of year of death, in a 5-year term insurance to (28). Expenses, paid at the beginning of the policy year, are as shown in Table 8.25. The settlement expenses, paid at the end of the year of death, are 50 per policy plus 10 per 1000 of insurance. Find the prospective reserve for this policy at $k = 1$ to 5.

Year	Per policy	Per 1000 of insurance	Fraction of premium
First	50	6	0.25
Second	25	3	0.15
Third	25	3	0.15

Table 8.25 Annual expenses

Solution: The policy in this illustration is a 5-year term insurance with a benefit of $50,000$ and gross premiums are payable for at most 3 years. Using the general formulation of the reserve as stated above and the solution of Example 7.6.1, the reserve at $k = 1, 2$ and 3 is given by,

$$50000 A^{1}_{28+k:\overline{5-k}|} + 550 A^{1}_{28+k:\overline{5-k}|} + (175 + 0.15G)\ddot{a}_{28+k:\overline{3-k}|} - G\ddot{a}_{28+k:\overline{3-k}|}$$

For $k = 4$ and 5, the actuarial present value of inflow via premiums is 0 and also the actuarial present value of renewal expenses is 0. Hence for $k = 4$ and 5, the reserve is given by,

$$50000 A^{1}_{28+k:\overline{5-k}|} + 550 A^{1}_{28+k:\overline{5-k}|}$$

For the given data we have computed the value of G as 13033.91 in Example 7.6.1. For the given q_x values and $i = 0.05$, values of $A^{1}_{28+k:\overline{5-k}|}$ for $k = 1$ to 5 are $0.46141, 0.39635, 0.30578, 0.17905$ and 0 respectively. Similarly, the values of $\ddot{a}_{28+k:\overline{3-k}|}$ for $k = 1$ to 3 are $1.81333, 1$ and 0 respectively. With these values we obtain the values of the reserve. These are as shown in Table 8.26.

k	1	2	3	4	5
Reserve	3552.01	9131.49	15457.34	9050.86	0

Table 8.26 Reserve with gross premiums

Note that as in the case of reserves for other term insurance products, the reserve increases for some period and then decreases and is 0 at the end of the term. ∎

The next section presents R codes used to compute the reserves and the related monetary functions in Sections 1, 2 and 3.

8.4   R Codes

The valuation of policy, that is, computation of reserves is an important actuarial calculation studied in this chapter. From the formulae of reserve derived in Sections 2 and 3, we notice that to compute the reserve we need some 'A' functions and 'a' functions. In Chapters 5 and 6, we have discussed how to compute these for various mortality patterns. Some additional functions to those codes help us to write codes to compute the reserve.

Code 8.4.1. In Section 1 we have introduced the concept of reserve with Example 8.1.1. We use this code to compute the required values in this example. It consists of two parts. We use the first part of this code to compute the values reported in Table 8.1 corresponding to the given mortality law. These values along with the value of δ and the second part of the code are used to compute the fund values in Table 8.2.

```
# Part I
B=0.0001151; C=1.096; m=B/log(C); del=0.05; int=exp(del)-1
x=30:35; u=length(x); y=exp(-m*(C^x-1)); p=c()
for (i in 1:(u-1))
{
p[i]=y[i+1]/y[i]
}
q=1-p; x1=30:34; l1=1000;l=c(l1,2:u)
for (i in 2:u)
{
l[i]=l[i-1]*p[i-1]
}
l2=l[-length(l)]; l3=l[-1]; de=l2*q
d=round(data.frame(p,q),6)
d1=round(data.frame(l2,de,l3),4)
d2=data.frame(x1,d,d1);d2 # Table 8.1
# Part II
Pr=172.58; f1=Pr*l2; dec=de*l2; f2=f3=f4=c()
f2[1]=f1[1]; f3[1]=f2[1]*(1+int); f4[1]=f3[1]-dec[1]
for(i in 2:5)
{
f2[i]=f1[i]+f4[i-1]
```

```
f3[i]=f2[i]*(1+int)
f4[i]=f3[i]-dec[i]
}
f5=f4/13; year=1:5
d3=round(data.frame(year,f1,f2,f3,dec,f4,f5),2)
d3 # Table 8.2
```

∎

Code 8.4.2. In Example 8.2.1, we have computed the prospective reserve for a fully continuous whole life insurance of 1000 issued to (25), with premiums payable as whole life annuity. It is given that $\delta = 0.05$ and mortality follows Gompertz' law with $B = 0.0001151$ and $C = 1.096$. We use this code to compute the values of prospective reserve for $t = 5, 10$ and 15, which are reported in Table 8.3.

```
# Part I: Mortality and interest patterns
B=0.0001151; C=1.096; m=B/log(C); del=0.05
la=(-del/log(C))+1; x=c(25,30,35,40); alx=m*C^x
# Part II: Prospective reserve for whole life insurance
w=PV=c()
for(j in 1:length(x))
{
w[j]=exp(alx[j])*gamma(la)*alx[j]^(1-la)*(1-pgamma(1,la,alx[j]))
}
w1=round(1000*w,2);w1 # APV in whole life insurance
# Left panel of Table 8.3
for(j in 1:length(x))
{
PV[j]=1000*(w[j]-w[1])/(1-w[1])
}
PV1=round(PV,2); PV1 # Right panel of Table 8.3
```

∎

Code 8.4.3. In Example 8.2.2, we have obtained the retrospective reserve for the whole life insurance with a benefit of 1000 units when the reserving basis is the same as in Example 8.2.1. We use this code to compute the retrospective reserves which are reported in Table 8.6.

```
# Part I: Mortality and interest parameters
B=0.0001151; C=1.096; m=B/log(C); del=0.05; v=exp(-del)
la=(-del/log(C))+1; x=25; t=c(0,5,10,15); alx=m*C^(x+t)
# Part II: Retrospective reserve for whole life insurance
w=c(); # vector to store APV of whole life insurance
for(j in 1:length(t))
```

```
{
w[j]=exp(alx[j])*gamma(la)*alx[j]^(1-la)*(1-pgamma(1,la,alx[j]))
}
Pr=del*w[1]/(1-w[1]);Pr # Premium
tEx=exp(-del*t-alx[1]*(C^t-1))
tA=w[1]-tEx*w # APV of term insurance
tE=tA+tEx # APV of endowment insurance
abarxt=(1-tE)/del # APV of temporary life annuity
tV=round(1000*(Pr*abarxt-tA)/tEx,2) # Retrospective reserve
tEx1=round(tEx[-1],4)
d=data.frame(t[-1],tEx1,tV[-1]); d # Table 8.6
```

■

Code 8.4.4. In Example 8.2.3, we have obtained the prospective and the retrospective reserve for the term, endowment and whole life insurance with limited term of premium payments, for the benefit of 1000 units. It is given that the mortality pattern is according to Gompertz' law and the force of interest is $\delta = 0.05$. We use this code to compute these reserves. The code is divided into parts. Parts II to IV are concerned with the prospective reserves while the remaining are for the retrospective reserves. The monetary functions computed using this code are reported in Table 8.8, Table 8.9, Table 8.10 and Table 8.11.

```
# Part I: Mortality and interest parameters
B=0.0001151; C=1.096; m=B/log(C); del=0.05; v=exp(-del)
la=(-del/log(C))+1; x=30; t=0:5; alx=m*C^(x+t)
# Part II: Prospective reserve: Term and endowment insurance
T=P=U=c()
for(i in 1:length(t))
{
U[i]=pgamma(C^(5-t[i]),la,alx[i])-pgamma(1,la,alx[i])
T[i]=exp(alx[i])*gamma(la)*alx[i]^(1-la)*U[i]
P[i]=exp(-del*(5-t[i])-alx[i]*(C^(5-t[i])-1))
}
E=T+P # Endowment insurance
A=(1-E)/del # temporary annuity
d=round(data.frame(t,T,E,A),4);d # Table 8.8
PrT=T[1]/A[1]; PrT; PrE=E[1]/A[1]; PrE
tVT=1000*(T-PrT*A); tVT # Column 2 of Table 8.9
tVE=1000*(E-PrE*A); tVE # Column 3 of Table 8.9
# Part III
U1=pgamma(C^2,la,alx[1])-pgamma(1,la,alx[1])
T1=exp(alx[1])*gamma(la)*alx[1]^(1-la)*U1
```

```
P1=exp(-del*2-alx[1]*(C^2-1))
E1=T1+P1; A1=(1-E1)/del; A1
PrE2=E[1]/A1; PrE2 # 2-year premium
U2=pgamma(C,la,alx[2])-pgamma(1,la,alx[2])
T2=exp(alx[2])*gamma(la)*alx[2]^(1-la)*U2
P2=exp(-del-alx[2]*(C-1))
E2=T2+P2; A2=(1-E2)/del; A2
tVE2=1000*(E[2]-PrE2*A2); tVE2 # Column 4 of Table 8.9
tVE2vec=c(tVE2,1000*E[3:6]); tVE2vec
# Part IV
w=c() # vector to store APV of whole life insurance
for(j in 1:length(t))
{
w[j]=exp(alx[j])*gamma(la)*alx[j]^(1-la)*(1-pgamma(1,la,alx[j]))
}
w
U3=pgamma(C^3,la,alx[1])-pgamma(1,la,alx[1])
T3=exp(alx[1])*gamma(la)*alx[1]^(1-la)*U3
P3=exp(-del*3-alx[1]*(C^3-1))
E3=T3+P3; A3=(1-E3)/del; A3
PrW3=w[1]/A3; PrW3
U4=pgamma(C^2,la,alx[2])-pgamma(1,la,alx[2])
T4=exp(alx[2])*gamma(la)*alx[2]^(1-la)*U4
P4=exp(-del*2-alx[2]*(C^2-1))
E4=T4+P4; A4=(1-E4)/del;A4
U5=pgamma(C,la,alx[3])-pgamma(1,la,alx[3])
T5=exp(alx[3])*gamma(la)*alx[3]^(1-la)*U5
P5=exp(-del-alx[3]*(C-1))
E5=T5+P5; A5=(1-E5)/del; A5
abarxt=c(A4,A5,0)
tVW3=1000*(w[2:4]-PrW3*abarxt); tVW3
tVW3vec=c(tVW3,1000*c(w[5],w[6]))
tVW3vec # Column 5 of Table 8.9
d1=round(data.frame(t[2:6],tVT[2:6],tVE[2:6],tVE2vec,tVW3vec),2)
d1 # Table 8.9
# Part V: Retrospective reserves for term and endowment
T6=P6=U6=c(); t1=t[-1]
for(i in 1:length(t1))
{
```

```
U6[i]=pgamma(C^t1[i],la,alx[1])-pgamma(1,la,alx[1])
T6[i]=exp(alx[1])*gamma(la)*alx[1]^(1-la)*U6[i]
P6[i]=exp(-del*t1[i]-alx[1]*(C^t1[i]-1))
}
E6=T6+P6 # Endowment insurance
A6=(1-E6)/del # temporary annuity
d2=round(data.frame(t1,T6,P6,A6),4);d2 # Table 8.10
tRVT=1000*(PrT*A6-T6)/P6;tRVT # Column 2 of Table 8.11
tRVE=1000*(PrE*A6-T6)/P6;tRVE # Column 3 of Table 8.11
# Part VI: Endowment insuramce, 2 premiums
a1=1000*(PrE2*A6[1:2]-T6[1:2])/P6[1:2]; a1
a2=1000*(PrE2*A6[2]-T6[3:5])/P6[3:5]; a2
tRVE2=c(a1,a2); tRVE2 # Column 4 of Table 8.11
# Part VII: whole life insuramce, 3 premiums
a3=1000*(PrW3*A6[1:3]-T6[1:3])/P6[1:3]; a3
a4=1000*(PrW3*A6[3]-T6[4:5])/P6[4:5]; a4
tRVW3=c(a3,a4); tRVW3 # Column 5 of Table 8.11
d3=round(data.frame(t1,tRVT,tRVE,tRVE2,tRVW3),2);d3 # Table 8.11
```

∎

Code 8.4.5. In Example 8.3.1, we have obtained values of prospective reserve for whole life insurance issued to (30), with the benefit of 1000 units, for $k = 1$ to 10, when premiums are payable as whole life annuity due. The force of mortality is Makeham's law with $A_{40} = 0.27975$ and $i = 0.05$. We use this code to compute the prospective reserve values. These are reported in Table 8.14.

```
# Part I: Parameters of Mortality law and interest rate
A=0.0007; B=0.0001151; C=1.096; m=B/log(C); int=0.05; v=1/(1+int)
# Part II: Values of Ax
x=30:40; y=exp(A*x-m*(C^x-1)); l=length(x); p=1:(l-1); A40=0.27975
for (i in 1:(l-1))
{
p[i]=y[i+1]/y[i]
}
p1=1-p; A1=c(1:(l-1),A40)
for(i in 1:l)
{
A1[l-i]=v*p1[l-i] + v*p[l-i]*A1[l-i+1]
}
# Part III: Values of kVx
k=0:10; kPV=1000*(A1-A1[1])/(1-A1[1])
```

```
d=round(data.frame(k,kPV),2); d # Table 8.14
```

■

Code 8.4.6. In Example 8.3.2, it is given that the force of mortality follows Makeham's law, $A_{40} = 0.27975$ and $i = 0.05$. We use this code to calculate the retrospective reserve for the whole life insurance issued to (30) with a benefit of 1000, for $k = 1$ to 10, when premiums are payable as whole life annuity due. Table 8.16 reports these values.

```
# Part I: Parameters of mortality law and interest rate
A=0.0007; B=0.0001151; C=1.096; m=B/log(C); int=0.05; v=1/(1+int)
d=1-v
# Part II: Values of Ax
x=30:40; y=exp(A*x-m*(C^x-1)); l=length(x); p=1:(l-1); A40=0.27975
for (i in 1:(l-1))
{
p[i]=y[i+1]/y[i]
}
p1=1-p; A1=c(1:(l-1),A40)
for(i in 1:l)
{
A1[l-i]=v*p1[l-i] + v*p[l-i]*A1[l-i+1]
}
# Part III: Values of retrospective reserve
k=1:10; P30=d*A1[1]/(1-A1[1]); P30
kpx=exp(-A*k-m*C^(x[1])*(C^k-1)); kEx= v^k*kpx
Tk=A1[1]-kEx*A1[-1]; Ek=Tk+kEx; abark=(1-Ek)/d
kRVx=1000*(P30*abark-Tk)/kEx
d1=round(data.frame(k,kRVx),2); d1 # Table 8.16
```

■

Code 8.4.7. In Example 8.3.3, we have computed the prospective and retrospective reserves corresponding to the following fully discrete policies issued to (30) for $k = 1$ to 5. (i) 5-year term insurance and (ii) 5-year endowment insurance, when premiums are payable as 5-year temporary life annuity due in both policies. (iii) 5-year endowment insurance with a premium paying term of 2 years. (iv) Whole life insurance with a premium paying term of 3 years. The underlying mortality pattern is Gompertz' law and the force of interest is $\delta = 0.05$. We use this code to compute the reserve values. The values of all the monetary functions are reported in Table 8.17, Table 8.18, Table 8.19 and Table 8.20.

```
# Part I: Parameters of mortality law and interest rate
B=0.0001151; C=1.096; m=B/log(C); del=0.05; v=exp(-del)
d=1-v; x=30; k=0:5; l=0:(100-min(x+k))
```

```
# Part II: Distribution of K(x)
y=u=matrix(nrow=length(l),ncol=length(k))# pmf of K(x)
for(i in 1:length(l))
{
for(j in 1:length(k))
{
u[i,j]=exp(-m*C^(x+k[j])*(C^(l[i]+1)-1))
y[i,j]=exp(-m*C^(x+k[j])*(C^l[i]-1))-u[i,j]
}
}
apply(y,2,sum)
# Part III: APVs of benefits and annuity
b=v^(l+1);w=pu=c(); t=matrix(nrow=length(l),ncol=length(k))
for(j in 1:length(k))
{
w[j]=sum(y[,j]*b)
t[,j]=cumsum(y[,j]*b)
pu[j]=v^(5-k[j])*exp(-m*C^(x+k[j])*(C^(5-k[j])-1))
}
w # Whole life insurance
T=c(t[5,1],t[4,2],t[3,3],t[2,4],t[1,5],0)# Term insurance
T # column 2 of Table 8.17
E=T+pu; E # Endowment insurance # column 3 of Table 8.17
adot=(1-E)/d; adot
d1=round(data.frame(k,T,E,adot),4); d1 # Table 8.17
# Part IV: Premiums and prospective reserves, term and endowment
PrT=T[1]/adot[1]; PrT; PrE=E[1]/adot[1]; PrE
PRT=1000*(T[-1]-PrT*adot[-1]); PRT # Column 2 of Table 8.18
PRE=1000*(E[-1]-PrE*adot[-1]); PRE # Column 3 of Table 8.18
# Part V:Premium and prospective reserves, endowment,2 premiums
E302=t[2,1]+v^2*exp(-m*C^(30)*(C^2-1)); E302
Pr2E30=d*E[1]/(1-E302); Pr2E30
PRE2=E[2]-Pr2E30
PRE2vec=1000*c(PRE2,E[3:6]); PRE2vec # Column 4 of Table 8.18
# Part VI:Premiums and prospective reserves,whole life,3 premiums
w1=w[1:6]; w1
E312=t[2,2]+v^2*exp(-m*C^(31)*(C^2-1)); E312
adot312=(1-E312)/d; adot312
E303=t[3,1]+v^3*exp(-m*C^(30)*(C^3-1)); E303
```

```
adot303 =(1-E303)/d; adot303; PrW3=w1[1]/adot303; PrW3
adot3=c(adot312,1,0); PRW3=w1[2:4]-PrW3*adot3
PRW3vec=1000*c(PRW3,w1[5:6]); PRW3vec # Column 5 of Table 8.18
d2=round(data.frame(k[-1],PRT,PRE,PRE2vec,PRW3vec),2)
d2 # Table 8.18
# Part VII: Retrospective reserves, APVs of benefits and annuity
Te=t[1:5,1]; E30k=v^k[-1]*exp(-m*C^(30)*(C^k[-1]-1))
En=Te+E30k; adot30k=(1-En)/d
d3=round(data.frame(k[-1],Te,E30k,adot30k),4);d3 # Table 8.19
# Part VIII: Retrospective reserves, term and endowment
RRT=1000*(PrT*adot30k-Te)/E30k; RRT # Column 2 of Table 8.20
RRE=1000*(PrE*adot30k-Te)/E30k; RRE # Column 3 of Table 8.20
# Part IX: Retrospective reserve, endowment, 2 premiums
RRE21=1000*(Pr2E30*adot30k[1:2]-Te[1:2])/E30k[1:2]; RRE21
RRE22=1000*(Pr2E30*adot30k[2]-Te[3:5])/E30k[3:5]; RRE22
RRE2=c(RRE21,RRE22); RRE2 # Column 4 of Table 8.20
# Part X: Retrospective reserve, whole life, 3 premiums
RRW31=1000*(PrW3*adot30k[1:3]-Te[1:3])/E30k[1:3]; RRW31
RRW32=1000*(PrW3*adot30k[3]-Te[4:5])/E30k[4:5]; RRW32
RRW3=c(RRW31,RRW32); RRW3 # Column 5 of Table 8.20
d4=round(data.frame(k[-1],RRT,RRE,RRE2,RRW3),2); d4 # Table 8.20
```

∎

Code 8.4.8. In Example 8.3.13, the force of mortality follows Makeham's law, $A_{40} = 0.27975$ and $i = 0.05$. We use this code to compute the prospective reserve for whole life insurance issued to (30) with benefit of 1000, for $k = 1$ to 10, when premiums are payable as m-thly whole life annuity due with $m = 2, 4, 12$. Table 8.21 and Table 8.22 report the values obtained using this code.

```
# Part I: Parameters of mortality law and interest rate
A=0.0007; B=0.0001151; C=1.096; m1=B/log(C); int=0.05; v=1/(1+int)
d=int/(1+int)
# Part II: Values of Ax
x=30:40; y=exp(A*x-m1*(C^x-1)); l=length(x); p=1:(l-1); A40=0.27975
for (i in 1:(l-1))
{
p[i]=y[i+1]/y[i]
}
p1=1-p; A1=c(1:(l-1),A40)
for(i in 1:l)
```

```
{
A1[l-i]=v*p1[l-i] + v*p[l-i]*A1[l-i+1]
}
# Part III: Values of kVx
kPV=(A1-A1[1])/(1-A1[1])
# Part IV: Values of m-thly premiums
m=c(2,4,12); im=m*((1+int)^(1/m)-1); dm=m*(1-v^(1/m))
alm=(int*d)/(im*dm); bm=(int-im)/(im*dm)
adot30=(1-A1[1])/d; adot30m=alm*adot30-bm
Pr30m=A1[1]/adot30m; Pr30m # m-thly premiums
d1=round(data.frame(m,im,dm,alm,bm,Pr30m),5); d1 # Table 8.21
# Part V: Values of m-thly prospective reserves
k=1:10; kPVm=matrix(nrow=length(k),ncol=length(m))
for(i in 1:length(k))
{
for(j in 1:length(m))
{
kPVm[i,j]=kPV[-1][i]*(1+Pr30m[j]*bm[j])
}
}
PV=round(1000*cbind(kPV[-1],kPVm),2); PV # Table 8.22
```

■

8.5   Conceptual Exercises

8.5.1 It is given that $q_{25} = 0.145$, $q_{26} = 0.141$, $q_{27} = 0.156$, $q_{28} = 0.174$, $q_{29} = 0.185$ and
$i = 0.06$. Compute the reserve for k = 1 to 5, using both the approaches, corresponding
to the following fully discrete policies with a benefit of 1000 units issued to (25).
(a) 5-year term insurance, (b) 5-year pure endowment insurance and (c) 5-year endowment
insurance. For (a), (b) and (c) assume that the premiums are payable as 5-year temporary
annuity due. (d) 5-year endowment insurance with a premium paying term of 2 years.
(e) 5-year term insurance with a premium paying term of 2 years.

8.5.2 Compute the reserves for all the policies in Exercise 8.5.1, if the benefit is payable at the
moment of death. State the assumptions you make. Compare the reserve values with
those obtained in Exercise 8.5.1.

8.5.3 Compute the reserves for all the policies in Exercise 8.5.1, if the premiums are payable
as quarterly annuity due. State the assumptions you make. Compare the reserve values
with those obtained in Exercise 8.5.1.

8.5.4 Write the expressions for variance of the prospective random variable in an n-year term insurance and n-year endowment in fully continuous policies.

8.5.5 Find the values of $_nV_x$ and $_nV_{x+n}$, given that,

$$1 - A_{x+2n} = A_{x+2n} - A_{x+n} = A_{x+n} - A_x.$$

8.5.6 Calculate $_1V_{40}$ given that $P_{40} = 0.01536$, $p_{40} = 0.99647$ and $i = 0.05$.

8.5.7 Given that $P_x = 0.02$, $_nV_x = 0.06$ and $P_{x:\overline{n}|}^{1} = 0.25$, find $P_{\ x:\overline{n}|}^{1}$.

8.5.8 If $_{10}V_{35} = 0.150$ and $_{20}V_{35} = 0.354$, calculate $_{10}V_{45}$.

8.5.9 Under the assumption of a uniform distribution of deaths in each year of age, which of the following are correct?

(a) $_kV(\overline{A}_{x:\overline{n}|}) = (i/\delta) \, _kV_{x:\overline{n}|}$

(b) $_kV(\overline{A}_x) = (i/\delta) \, _kV_x$

(c) $_kV(\overline{A}^{\ 1}_{x:\overline{n}|}) = (i/\delta) \, _kV^{\ 1}_{x:\overline{n}|}$

8.5.10 For a fully discrete 10-payment whole life insurance of Rs 1 lakh issued to (30), it is given that $i = 0.06$, $q_{39} = 0.012, q_{40} = 0.014$ and $q_{41} = 0.016$. The level annual benefit premium is Rs 2000 and the benefit reserve at the end of year 9 is $32,000$. Compute $100000 A_{41}$.

8.5.11 For a fully discrete whole life insurance of 1000 issued to (30), it is given that $1000 A_{30} = 240$, $1000 \, _5V_{30} = 55, 1000 \, _6V_{30} = 66$, $1000 \, _7V_{30} = 78$ and $q_{35} = 0.01$. Find q_{36}.

8.5.12 The benefit reserve at the end of year n for a fully discrete whole life insurance of 1 unit benefit issued to (x) is 0.563. If $P_x = 0.09$ and $P_{x:\overline{n}|}^{1} = 0.00864$, then find $P_{\ x:\overline{n}|}^{1}$.

8.5.13 For $0 < t \le m$, show that,

(a) $\overline{P}(\overline{A}_{x:\overline{m+n}|}) = \overline{P}(\overline{A}^{\ 1}_{x:\overline{m}|}) + \overline{P}_{x:\overline{m}|}^{1} \, _m\overline{V}(\overline{A}_{x:\overline{m+n}|})$

(b) $_t\overline{V}(\overline{A}_{x:\overline{m+n}|}) = \, _t\overline{V}(\overline{A}^{\ 1}_{x:\overline{m}|}) + \, _t\overline{V}_{x:\overline{m}|}^{1} \, _m\overline{V}(\overline{A}_{x:\overline{m+n}|})$

8.5.14 Show that $_kV_x^{(m)} = \dfrac{\{\alpha(m) - d\beta(m)\} \, _kV_x \, \ddot{a}_x}{\ddot{a}_x^{(m)}}$.

8.6   Computational Exercises

For the following questions, use the mortality pattern and interest pattern that you have adopted in the computational exercise 5.8.2. State the assumptions, if needed.

8.6.1 Find the prospective and retrospective reserves at $t = 1$ to 5 for the following fully continuous policies, issued to (30), with a benefit of 1000. (i) 5-year term insurance and (ii) 5-year endowment insurance. For (i) and (ii) assume that the premiums are payable as 5-year temporary continuous life annuity. (iii) Find prospective and retrospective reserves at $t = 1$ to 5 for 5-year endowment insurance with a premium paying term of 2 years. (iv) Find prospective and retrospective reserves at $t = 1$ to 5 for whole life insurance, with the premium payable as whole life annuity and when the premium paying term is 3 years.

8.6.2 Find the prospective and retrospective reserves at $t = 1$ to 5 for the following fully discrete policies, issued to (30), with a benefit of 1000. (i) 5-year term insurance and (ii) 5-year endowment insurance. For (i) and (ii) assume that the premiums are payable as 5-year temporary life annuity due. (iii) Find the prospective and retrospective reserves at $t = 1$ to 5 for the 5-year endowment insurance with a premium paying term of 2 years. (iv) Find the prospective and retrospective reserves at $t = 1$ to 5 for whole life insurance, with the premium payable as whole life annuity due and when the premium paying term is 3 years.

8.6.3 In both the exercises, verify that (i) both the prospective and retrospective approaches give the same results, (ii) for a 5-year term insurance, the reserve at $t = 5$ is 0 and (iii) for a 5-year endowment insurance, the reserve at $t = 5$ is 1.

8.6.4 Calculate the prospective reserve for the whole life insurance issued to (30) with a benefit of 1000, for $k = 1$ to 10, when premiums are payable as m-thly whole life annuity due with $m = 2, 4, 12$. Assume that in each unit interval the mortality pattern satisfies the uniformity assumption.

8.7   Multiple Choice Questions

Note: Unless specified otherwise, you have to identify which of the options is correct. Answers are given in the solutions of conceptual exercises.

8.7.1 The prospective reserve $_5\overline{V}(\overline{A}_{30})$ after 5 years from signing the contract, corresponding to the whole life insurance of unit benefit issued to (30) on a fully continuous basis, with an annual benefit premium of $\overline{P}(\overline{A}_x)$ is

(a) $\overline{A}_{35} - \overline{P}(\overline{A}_{30})\overline{a}_{35}$
(b) $\overline{A}_{35} - \overline{P}(\overline{A}_{35})\overline{a}_{35}$
(c) $\overline{A}_{30} - \overline{P}(\overline{A}_{30})\overline{a}_{35}$
(d) $\overline{A}_{35} - \overline{P}(\overline{A}_{30})\overline{a}_{30}$

8.7.2 The prospective reserve $_5\overline{V}(\overline{A}_{30})$ after 5 years from signing the contract, corresponding to the whole life insurance of unit benefit issued to (30) on a fully continuous basis, with an annual benefit premium of $\overline{P}(\overline{A}_{30})$ is given by the following four formulae.
(I) $[\overline{P}(\overline{A}_{35}) - \overline{P}(\overline{A}_{30})]\overline{a}_{35}$
(II) $[1 - \overline{P}(\overline{A}_{30})/\overline{P}(\overline{A}_{35})]\,\overline{A}_{35}$ (III) $1 - \overline{a}_{35}/\overline{a}_{30}$
(IV) $(\overline{P}(\overline{A}_{35}) - \overline{P}(\overline{A}_{30}))/(\overline{P}(\overline{A}_{35}) + \delta)$

(a) Only (I) is true
(b) Only (I) and (II) are true
(c) Only (I), (II) and (III) are true
(d) All four are true

8.7.3 The prospective reserve $_5\overline{V}(\overline{A}_{30})$ after 5 years from signing the contract, corresponding to the whole life insurance of the unit benefit issued to (30) on a fully continuous basis, with an annual benefit premium of $\overline{P}(\overline{A}_{30})$ is given by

(a) $(\overline{A}_{30} - \overline{A}_{35})/(1 - \overline{A}_{35})$
(b) $(\overline{A}_{35} - \overline{A}_{30})/(1 - \overline{A}_{30})$
(c) $(\overline{A}_{35} - \overline{A}_{30})/(1 - \bar{a}_{30})$
(d) $(\bar{a}_{35} - \bar{a}_{30})/(1 - \bar{a}_{30})$

8.7.4 The following are two statements. (I) Reserve at $t = 10$ for 10-year, 1 unit benefit term insurance is 0. (II) Reserve at $t = 10$ for 10-year, 1 unit benefit endowment insurance is 1.

(a) (I) is true but (II) is false
(b) (I) is false but (II) is true
(c) Both (I) and (II) are false
(d) Both (I) and (II) are true

8.7.5 It is given that $\overline{A}_{40} = 0.5$, $_{10}E_{30} = 0.4$ and $\overline{A}^{\,1}_{30:\overline{10}|} = 0.13$, then $_{10}\overline{V}(\overline{A}_{30})$

(a) cannot be computed in view of insufficient information
(b) is 0.2537
(c) is 0.2018
(d) is 0.3045

8.7.6 It is given that $\overline{A}_{50} = 0.7$, $^2\overline{A}_{30} = 0.3$ and $Var(L) = 0.2$. Calculate $_{20}\overline{V}(\overline{A}_{30})$.

(a) cannot be computed in view of insufficient information
(b) is 0.3
(c) is 0.2
(d) is 0.4

8.7.7 It is given that $A_{32} = 0.1972, p_{31} = 0.9986$ and $i = 0.05$. Then (I) $1000\,_1V_{31} = 10.26$. (II) $1000\,_0V_{31} = 9.07$.

(a) (I) is true but (II) is false
(b) (I) is false but (II) is true
(c) Both (I) and (II) are false
(d) Both (I) and (II) are true

8.7.8 The following are two statements. (I) $_1V_{35} = A_{36} - P_{36}\ddot{a}_{36}$
(II) $_1V_{35} = A_{36} - P_{35}\ddot{a}_{36}$

(a) (I) is true but (II) is false
(b) (I) is false but (II) is true
(c) Both (I) and (II) are false

(d) Both (I) and (II) are true

8.7.9 The following are three formulae for $_4V_{44}$. (I) $1 - \ddot{a}_{48}/\ddot{a}_{44}$
(II) $(P_{44} - P_{40})/(P_{44} + d)$ (III) $(P_{44} - P_{40})/(P_{44} + v)$

(a) Only (I) is correct
(b) Only (I) and (II) are correct
(c) Only (I) and (III) are correct
(d) All three are incorrect

8.7.10 The following are four formulae for $_5V_{45}$. (I) $(P_{50} - P_{45})\,\ddot{a}_{50}$
(II) $(P_{50} - P_{45})\,\ddot{a}_{45}$ (III) $(1 - P_{45}/P_{50})\,A_{50}$ (IV) $(1 - P_{45}/P_{50})\,\ddot{a}_{50}$

(a) Only (I) is correct
(b) Only (III) is correct
(c) Only (II) and (IV) are correct
(d) Only (I) and (III) are correct

8.7.11 A fully discrete whole life insurance of Rs 1 lakh is issued to (30). The premiums are
paid annually up to age 70. The net premium during the first 10 years is P_{30}, followed
by a different level annual premium payable during the next 40 years. It is given that
$A_{40} = 0.4, P_{30} = 0.02$ and $d = 0.06$. The reserve at the end of the tenth year

(a) cannot be computed in view of insufficient information
(b) is 20000
(c) is 0.2
(d) is 26667.67

8.7.12 If $A_{25:\overline{10|}} = 0.40$ and $d = 0.06$, then $_9V_{25:\overline{10|}}$ and $_{10}V_{25:\overline{10|}}$

(a) cannot be computed in view of insufficient information
(b) are respectively given by 0.9 and 1
(c) are respectively given by 0.9 and 0
(d) are respectively given by 0.02 and 1

8.7.13 If $\ddot{a}_{50} = 8$ and $_{10}V_{40} = 0.2$, then $\ddot{a}_{40}$

(a) is 9
(b) is 10
(c) is 12
(d) cannot be computed in view of insufficient information

8.7.14 If $\ddot{a}_{50} = 8, A_{40} = 0.4$ and $v = 0.94$, then $_{10}V_{40}$

(a) is 1
(b) is 0.8
(c) is 0.2
(d) cannot be computed in view of insufficient information

8.7.15 If $_5V_{30} = 0.3$ and $_5V_{35} = 0.2$, then $_{10}V_{30}$

(a) is 0.40
(b) is 0.44

(c) is 0.56
(d) cannot be computed in view of insufficient information

8.7.16 If $P_{35} = 0.006$, $P_{35:\overline{5}|} = 0.042$ and $P^{\,1}_{35:\overline{5}|} = 0.002$, then $_5V_{35}$

  (a) is 0.1
  (b) is 0.4
  (c) is 0.5
  (d) cannot be computed in view of insufficient information

8.7.17 If $P_{35} = 0.006$ and $P_{45} = 0.008$ then $_{10}V_{35}$

  (a) is 0.1
  (b) is 0.4
  (c) is 0.5
  (d) cannot be computed in view of insufficient information

8.7.18 If $P_{35} = 0.02$, $P_{30} = 0.01$ and $_5V_{30} = 0.1$, then v

  (a) is 0.80
  (b) is 0.92
  (c) is 0.08
  (d) cannot be computed in view of insufficient information

8.7.19 If $_{10}V_{25} = 0.1$, $P_{25} = 0.02$, $_5P_{25} = 0.025$ and $\ddot{S}_{25:\overline{10}|} = 20$, then $_{10}^{5}V_{25}$

  (a) is 0.89
  (b) is 0.11
  (c) is 0.20
  (d) cannot be computed in view of insufficient information

8.7.20 The reserve $_kV^{(m)}_{x:\overline{n}|}$ for an n-Year endowment insurance of one unit benefit, when premiums are payable as m-thly n-year temporary annuity due, when $k \le n$, is given by

  (a) $A^{(m)}_{x+k:\overline{n-k}|} - P^{(m)}_{x:\overline{n}|}\, \ddot{a}_{x+k:\overline{n-k}|}$

  (b) $A_{x+k:\overline{n-k}|} - P^{(m)}_{x:\overline{n}|}\, \ddot{a}^{(m)}_{x+k:\overline{n-k}|}$

  (c) $A_{x+k:\overline{n-k}|} - P_{x:\overline{n}|}\, \ddot{a}^{(m)}_{x+k:\overline{n-k}|}$

  (d) $A_{x+k:\overline{n-k}|} - P^{(m)}_{x:\overline{n}|}\, \ddot{a}_{x+k:\overline{n-k}|}$

Chapter 9

Multiple Life Contracts

Key Terms: Gompertz' mortality law, Joint life status, Last survivor status, Makeham's mortality law, Monetary functions for multiple lives, Time until failure of the status.

9.1   Introduction

In Chapter 7 we have studied how to find the premiums for a variety of insurance products and some annuity products. In all these insurance and annuity products, the benefit is payable to a single individual. However, group insurance is a popular product wherein more than one life is involved in a policy contract. For instance, both husband and wife are insured through a policy. Similarly, the full family could be insured. In an organization, all the employees can be insured together. In these cases we come across with what is known as a 'Group insurance' or 'Multiple Life Contracts'. Thus, a group insurance is a type of insurance plan which covers a group of people under the same plan. It provides the same coverage to all the members of the group irrespective of their age, sex, occupation, economic status, and so on. Beneficiaries of a member in the group receive the same benefits in the case of death of the policy holder.

In group insurance, the important issue is to decide when and how the benefit is to be paid. When a single life is involved in the insurance product, then the benefit is payable at the moment of death or at the end of year of death of the individual, that is, when the status of the individual as 'being alive' terminates. In an annuity product, annuity payments cease when the individual dies, that is, when the status of the individual as 'being alive' terminates.

While dealing with group insurance or annuity products, we have to define a so called status of the group, termination of which determines the time of death benefit payment in insurance products and time to stop survival benefits in annuity products. Thus, the first step while dealing with multiple lives is to define the status of a group and an extension of the definition of $T(x)$ or $K(x)$ to a group consisting of more than one individual. The following two are the commonly used approaches. Suppose $(x_1, \cdots, x_m)$ denotes a group of m individuals, where x_i represents the age of the ith member at the time of policy issue. Suppose $T(x_i)$ denotes the future life time of individual i.

Joint Life Status: The joint life status exists as long as all members of the group are alive and ends when the first death takes place. It is denoted by $(x_1 x_2 \cdots x_m)$. Note that the x_i's are not separated by commas but written like a product. The random variable $T(x_1 \cdots x_m)$ denotes the time until failure of the status. It is defined as follows.

$$T(x_1 \cdots x_m) \;=\; \min\{T(x_1), T(x_2), \cdots, T(x_m)\}$$

Last Survivor Status: As the term indicates, this status exists as long as at least one member is alive and fails upon the last death. It is denoted by $(\overline{x_1 \cdots x_m})$. In the last survivor status, the random variable $T(\overline{x_1 \cdots x_m})$ describing the time-until failure of the status is defined as follows.

$$T(\overline{x_1 \cdots x_m}) \;=\; \max\{T(x_1), T(x_2), \cdots, T(x_m)\}$$

Many pension plans have joint and last survivor annuity options. In government pension plans, after the death of the husband, his wife gets a part of the pension.

After defining the status of the group, the next step is to obtain the distribution of time until failure of the status. Once it is defined, the entire theory developed in Chapters 5, 6, 7 and 8 can be easily generalized to define the actuarial present value of the benefit to be paid at the moment of termination of a status of the group and to find premiums and reserves. The main aim of this chapter is to express the distribution of time to termination of the status of the group, in terms of the distribution of time to failure of the status as 'being alive' for each individual of the group. With such a relation, the actuarial present values of benefit in group insurance or group annuity can be obtained using the knowledge of distribution of $T(x)$ corresponding to each individual in the group.

In Section 2, we derive the distribution of the future lifetime of the joint life status and in Section 3, that of the last survivor status. In both the cases, we express the probability of survival or failure of the status in terms of the probabilities for the individual lives in a group.

9.2   Joint Life Status

We have noted that for the joint life status

$$T(x_1 \cdots x_m) = \min\{T(x_i), i = 1, \cdots, m\}$$

and for the last survivor status,

$$T(\overline{x_1 \cdots x_m}) = \max\{T(x_i), i = 1, \cdots, m\}$$

If $Y_1, \cdots, Y_n$ are independent and identically distributed random variables, then the distribution function or the survival function of

$$Y_{(1)} \;=\; \min\{Y_i, i = 1, \cdots, n\} \quad \text{and} \quad Y_{(n)} \;=\; \max\{Y_i, i = 1, \cdots, n\}$$

can be obtained easily. However, in the group of individuals involved in the insurance contract, there is some association among the members of the group, for example, they may be members of a family, husband and wife or workers working in the same company in the same environment. As a result $T(x_1), \cdots, T(x_m)$ cannot be considered as independent random variables. In practice, however, it is very difficult to model the dependence of these time-until-death random variables. Hence, in obtaining multiple life functions, an assumption is made that $T(x_1), T(x_2), \cdots, T(x_m)$ are independent random variables. Under this assumption, we express the distribution of $T(x_1 \cdots x_m)$ and $T(\overline{x_1 \cdots x_m})$ in terms of the distribution of $T(x_1), \cdots, T(x_m)$. To avoid the complexity of notation, we take $m = 2$. The extension to $m > 2$ is straightforward.

In international actuarial notation, the distribution function of time $T(xy)$ to failure of a joint life status of a group of two individuals of ages x and y, is given by

$$
\begin{aligned}
{}_t q_{xy} \;=\; F_{T(xy)}(t) \;&=\; P[\min\{T(x), T(y)\} \leq t] \\
&=\; 1 - P[\min\{T(x), T(y)\} \geq t] \\
&=\; 1 - P[T(x) > t]P[T(y) > t], \quad \text{(in view of independence assumption)} \\
&=\; 1 - {}_t p_x \, {}_t p_y
\end{aligned}
$$

Thus, under the assumption of independence, the survival function ${}_t p_{xy}$ and the probability density function $f_{T(xy)}(t)$ of the time to failure of the joint life status, are given by,

$$
{}_t p_{xy} \;=\; {}_t p_x \, {}_t p_y \tag{9.2.1}
$$

$$
\begin{aligned}
f_{T(xy)}(t) \;&=\; \frac{d}{dt}(1 - {}_t p_x \, {}_t p_y) = -\, {}_t p_x (-\, {}_t p_y \mu_{y+t}) - {}_t p_y (-\, {}_t p_x \mu_{x+t}) \\
&=\; {}_t p_x \, {}_t p_y (\mu_{x+t} + \mu_{y+t}) \tag{9.2.2}
\end{aligned}
$$

From the expression of the survival function and the density function, it is easy to see that the force of failure of the joint life status at time t, denoted by $\mu_{xy}(t)$ is,

$$
\mu_{xy}(t) = f_{T(xy)}(t)/(1 - F_{T(xy)}(t)) = \mu_{x+t} + \mu_{y+t} \tag{9.2.3}
$$

Thus, the force of failure for the joint life status is the sum of the forces of mortality for the individuals if their future lifetimes are independent.

The curtate future lifetime of the joint life status, denoted by $K(xy)$, is defined as follows. Note that $K(xy) = k$, if the joint life status terminates in year $k + 1$, that is, if either $K(x) = k$ and $K(y) > k$ or $K(y) = k$ and $K(x) > k$ or $K(x) = k$ and $K(y) = k$. Thus, $K(xy) = \min\{K(x), K(y)\}$. We have noted in Chapter 4 that $K(x)$ is the largest integer strictly smaller than $T(x)$. We now investigate if we have a similar relation between $T(xy)$ and $K(xy)$. Observe that if $T(xy) = 7.8$, then one of (x) or (y) survive 7 complete years and hence $K(xy) = 7$. If $T(xy) = 0.8$, then one of (x) or (y) survive 0 complete years and hence $K(xy) = 0$. Similarly if $T(xy) = 10$, then one of (x) or (y) survive 9 complete years and hence $K(xy) = 9$. Thus, $K(xy)$ is the largest integer strictly smaller than $T(xy)$. As a consequence, $P[T(xy) \leq t] = P[K(xy) \leq t^*]$, where t^* is the largest integer strictly smaller than t.

In other words, $K(xy) = k$, if both lives survive k years, but both do not survive $k+1$ years. Such interpretation is helpful to find the probability mass function of $K(xy)$. It is obtained below.

$$
\begin{aligned}
P[K(xy) = k] \;&=\; {}_k p_{xy} - {}_{(k+1)} p_{xy} \;=\; {}_k p_{xy} - {}_k p_{xy}\, p_{x+k\;y+k} \\
&=\; {}_k p_{xy}(1 - p_{x+k\;y+k}) \;=\; {}_k p_{xy}\, q_{x+k\;y+k}\;, \quad k = 0, 1, \cdots
\end{aligned}
$$

To express this probability in terms of single life functions, we note that,

$$\begin{aligned}
q_{x+k\ y+k} &= (1 - p_{x+k\ y+k}) = 1 - p_{x+k}p_{y+k} \\
&= 1 - (1 - q_{x+k})(1 - q_{y+k}) = q_{x+k} + q_{y+k} - q_{x+k}q_{y+k} \\
\Rightarrow\ P[K(xy) = k] &= {}_kp_x\ {}_kp_y(q_{x+k} + q_{y+k} - q_{x+k}q_{y+k}),\ k = 0, 1, \cdots
\end{aligned}$$
(9.2.4)

Alternatively, the probability mass function of $K(xy)$ can also be derived as follows.

$$\begin{aligned}
P[K(xy) = k] &= P[K(x) = k, K(y) > k]\ +\ P[K(y) = k, K(x) > k] \\
&+\ P[K(x) = k, K(y) = k] \\
&= P[K(x) = k, K(y) \geq k + 1]\ +\ P[K(y) = k, K(x) \geq k + 1] \\
&+\ P[K(x) = k, K(y) = k] \\
&= {}_{k+1}p_y\ {}_kp_x q_{x+k}\ +\ {}_{k+1}p_x\ {}_kp_y q_{y+k}\ +\ {}_kp_x\ {}_kp_y\ q_{x+k}q_{y+k} \\
&= {}_kp_y p_{y+k}\ {}_kp_x q_{x+k}\ +\ {}_kp_x p_{x+k}\ {}_kp_y q_{y+k}\ +\ {}_kp_x\ {}_kp_y q_{x+k}q_{y+k} \\
&= {}_kp_x\ {}_kp_y[(1 - q_{y+k})q_{x+k} + (1 - q_{x+k})q_{y+k} + q_{x+k}q_{y+k}] \\
&= {}_kp_x\ {}_kp_y(q_{x+k} + q_{y+k} - q_{x+k}q_{y+k}),\ k = 0, 1, \cdots
\end{aligned}$$

Once we have derived the formulae for the probability density function of $T(xy)$ and probability mass function of $K(xy)$, that is, extensions of the results in Sections 4.2 and 4.3 to multiple lives, insurances and annuities as discussed in Chapters 5 and 6 respectively for an individual life, can be defined easily for a group status. For example, with the definition of the joint life status for multiple lives, an annuity is payable as long as the status survives, while an insurance is payable upon the failure of the status.

The actuarial present value of unit amount payable immediately on the failure of the joint life status in whole life insurance, is denoted by $\overline{A}_{xy}$ and is given by,

$$\overline{A}_{xy}\ =\ E(v^{T(xy)})\ =\ \int_0^\infty v^t\ {}_tp_{xy}\mu_{xy}(t)dt\ =\ \int_0^\infty v^t\ {}_tp_x\ {}_tp_y(\mu_{x+t} + \mu_{y+t})dt$$

With the knowledge of v, ${}_tp_x$, ${}_tp_y$, μ_{x+t} and μ_{y+t}, $\overline{A}_{xy}$ can be evaluated. If the benefit of a unit amount is payable at the end of the year in which the joint life status fails, the actuarial present value of the benefit payment is given by,

$$A_{xy}\ =\ \sum_{k=0}^\infty v^{k+1}\{\ {}_kp_x\ {}_kp_y(q_{x+k} + q_{y+k} - q_{x+k}q_{y+k})\}$$

It has been shown in Section 9.8 of Bowers et al [1] that under the uniformity assumption in unit age intervals for both the ages,

$$\overline{A}_{xy}\ \cong\ (i/\delta)A_{xy}$$

The actuarial present value of the benefit in other insurance products are defined on similar lines. Thus, the actuarial present value of unit amount payable immediately on the failure of the joint life status in n-year term insurance, is denoted by $\overline{A}\,{}^1_{xy:\overline{n}|}$ and is given by,

$$\overline{A}^1_{xy:\overline{n}|}\ =\ \int_0^n v^t\ {}_tp_{xy}\mu_{xy}(t)dt$$

If the benefit of a unit amount is payable at the end of the year in which the joint life status fails in n-year term insurance, then the actuarial present value of the benefit payment is given by,

$$A^1_{xy:\overline{n}|} = \sum_{k=0}^{n-1} v^{k+1}\{\, _kp_x\ _kp_y(q_{x+k} + q_{y+k} - q_{x+k}q_{y+k})\}$$

For an n-year endowment insurance, the corresponding formulae are as follows.

$$\overline{A}_{xy:\overline{n}|} = \overline{A}^1_{xy:\overline{n}|} + v^n\ _np_{xy} \quad \& \quad A_{xy:\overline{n}|} = A^1_{xy:\overline{n}|} + v^n\ _np_{xy}$$

Various types of annuities and their actuarial present values as discussed in Chapter 6 are easily extendable to the group contracts. The notations are very similar to single life cases, we just need to replace x by xy. The relations between Z and Y as given by,

$$Y = (1 - Z)/d \quad \text{or} \quad Y = (1 - Z)/\delta$$

in Chapter 6, are also applicable in this setup. This relation is also useful to find the variance of the present value random variable of annuity. The following are the formulae for the actuarial present value of some annuities. $\overline{a}_{xy}$ denotes the actuarial present value of an annuity of amount 1 unit per annum, payable continuously as long as both (x) and (y) are alive. It is given by

$$\overline{a}_{xy} = E(\overline{a}_{T(xy)}) = \int_0^\infty \overline{a}_{\overline{t}|}\ _tp_{xy}\mu_{x+t\ y+t}dt$$
$$= \int_0^\infty v^t\ _tp_{xy}dt = (1 - \overline{A}_{xy})/\delta$$

The actuarial present value of an annuity of amount 1 unit, payable at the beginning of each year so long as both (x) and (y) are alive, is denoted by $\ddot{a}_{xy}$ and is given by,

$$\ddot{a}_{xy} = \sum_{t=0}^\infty v^t\ _tp_{xy} = (1 - A_{xy})/d$$

The actuarial present value of an annuity of amount 1 unit payable at the end of each year so long as both (x) and (y) are alive is denoted by a_{xy} and is given by,

$$a_{xy} = \sum_{t=1}^\infty v^t\ _tp_{xy} = \ddot{a}_{xy} - 1$$

The actuarial present value of an annuity of amount 1 in n-year temporary annuity due, in a joint life status, is denoted by $\ddot{a}_{xy:\overline{n}|}$ and is defined as

$$\ddot{a}_{xy:\overline{n}|} = \sum_{t=0}^{n-1} v^t\ _tp_{xy} = (1 - A_{xy:\overline{n}|})/d = \ddot{a}_{xy} - \ _np_{xy}\, v^n\, \ddot{a}_{x+n\ y+n}$$

Other annuity functions are defined analogously.

With the extension of two components of premium calculation to multiple lives, premiums for multiple life contracts can be defined on similar lines as those for single life studied in

Chapter 7. As an illustration, suppose a whole life insurance of unit benefit, is purchased by (xy) with annual premiums payable as whole life annuity due. Thus, the benefit is payable at the moment of the first death and premium payments will cease when the first death occurs, since the joint life status terminates and the policy becomes inactive at the first death. Using the equivalence principle, the annual premium P is given by,

$$P = \overline{A}_{xy}/\ddot{a}_{xy}$$

If the premiums are payable as n-year temporary annuity due, the premium P is given by

$$P = \overline{A}_{xy}/\ddot{a}_{xy:\overline{n}|}$$

Premiums for other insurance and annuity products can be computed on similar lines.

In the next example, we derive bounds on $\overline{A}^1_{xy:\overline{n}|}$ and on $\overline{A}_{xy}$.

Example 9.2.1. Prove that

$$_np_y\,\overline{A}^1_{x:\overline{n}|} + {}_np_x\overline{A}^1_{y:\overline{n}|} < \overline{A}^1_{xy:\overline{n}|} < \overline{A}^1_{x:\overline{n}|} + \overline{A}^1_{y:\overline{n}|}$$
$$\text{and} \quad \overline{A}_{xy} < \overline{A}_x + \overline{A}_y$$

Solution: From Eq. 9.2.2, the probability density function of $T(xy)$ is given by,

$$f_T(t) = {}_tp_x\,{}_tp_y(\mu_{x+t} + \mu_{y+t})$$

Thus, using the formula of $\overline{A}^1_{xy:\overline{n}|}$ we get,

$$\overline{A}^1_{xy:\overline{n}|} = \int_0^n v^t\,{}_tp_x\,{}_tp_y(\mu_{x+t} + \mu_{y+t})$$
$$\leq \int_0^n v^t\,{}_tp_x\,\mu_{x+t}\,dt + \int_0^n v^t\,{}_tp_y\,\mu_{y+t}\,dt$$
$$= \overline{A}^1_{x:\overline{n}|} + \overline{A}^{\,1}_{y:\overline{n}|}$$

The second step follows from the fact that being probability, the values of both $_tp_x$ and $_tp_y$ are less than or equal to 1. If both $_tp_x$ and $_tp_y$ are 1, then in this case, both μ_{x+t} and μ_{y+t} are 0. As a consequence, the probability density function of $T(xy)$, $T(x)$ and $T(y)$ at t is 0. If it is true for all $t \leq n$, then $\overline{A}^1_{xy:\overline{n}|}$, $\overline{A}^1_{x:\overline{n}|}$ and $\overline{A}^1_{xy:\overline{n}|}$ are all 0. To prove the first part of the inequality, we use the result that being survival function, $_tp_x$ is a decreasing function of t. Thus,

$$t < n \implies {}_tp_x > {}_np_x \ \& \ {}_tp_y > {}_np_y$$
$$\implies \overline{A}^1_{xy:\overline{n}|} = \int_0^n v^t\,{}_tp_x\,{}_tp_y(\mu_{x+t} + \mu_{y+t})$$
$$= \int_0^n v^t\,{}_tp_x\,\mu_{x+t}\,{}_tp_y dt + \int_0^n v^t\,{}_tp_y\,\mu_{y+t}\,{}_tp_x dt$$
$$> {}_np_y\int_0^n v^t\,{}_tp_x\,\mu_{x+t}\,dt + {}_np_x\int_0^n v^t\,{}_tp_y\,\mu_{y+t}\,dt$$
$$= {}_np_y\overline{A}^1_{x:\overline{n}|} + {}_np_x\overline{A}^1_{y:\overline{n}|}$$

Inequality for whole life insurance follows on similar lines as in the proof of $\overline{A}^{1}_{xy:\overline{n}|} < \overline{A}^{1}_{x:\overline{n}|} + \overline{A}^{1}_{y:\overline{n}|}$. It can also be obtained by allowing $n \to \infty$ in the inequality $\overline{A}^{1}_{xy:\overline{n}|} < \overline{A}^{1}_{x:\overline{n}|} + \overline{A}^{1}_{y:\overline{n}|}$. In Example 9.3.5 we verify these results when the underlying mortality pattern is Gompertz' law. ∎

In the next section we derive the results for the last survivor status. These are derived on similar lines as those for the joint life status. We also establish a link among the random variables $T(xy)$, $T(\overline{xy})$, $T(x)$ and $T(y)$. This link is heavily used to derive formulae for the actuarial present values in insurance and annuity products in the setup of last survivor status.

9.3   Last Survivor Status

In the last survivor status the distribution function of $T(\overline{xy})$, the time to failure of the last survivor status of a group consisting of two lives, is given by,

$$
\begin{aligned}
{}_{t}q_{\overline{xy}} &= P[T(\overline{xy}) \leq t] = P[\max\{T(x), T(y)\} \leq t] \\
&= P[T(x) \leq t]P[T(y) \leq t] = {}_{t}q_{x}\, {}_{t}q_{y} \qquad (9.3.1) \\
\Rightarrow \quad {}_{t}p_{\overline{xy}} &= 1 - (1 - {}_{t}p_{x})(1 - {}_{t}p_{y}) = {}_{t}p_{x} + {}_{t}p_{y} - {}_{t}p_{x}\, {}_{t}p_{y} \qquad (9.3.2)
\end{aligned}
$$

Further, the probability density function is given by,

$$
\begin{aligned}
f_{T(\overline{xy})}(t) &= \frac{d}{dt}\{(1 - {}_{t}p_{x})(1 - {}_{t}p_{y})\} \\
&= (1 - {}_{t}p_{x})(\mu_{y+t}\, {}_{t}p_{y}) + (1 - {}_{t}p_{y})({}_{t}p_{x}\mu_{x+t}) \\
&= {}_{t}p_{x}\, \mu_{x+t} + {}_{t}p_{y}\mu_{y+t} - {}_{t}p_{x}\, {}_{t}p_{y}(\mu_{x+t} + \mu_{y+t}) \qquad (9.3.3) \\
&= f_{T(x)} + f_{T(y)} - {}_{t}p_{xy}\mu_{xy}(t) \\
&= f_{T(x)}(t) + f_{T(y)}(t) - f_{T(xy)}(t) \qquad (9.3.4)
\end{aligned}
$$

The probability density function of $T(\overline{xy})$ is expressed in terms of single life functions in Eq. 9.3.3, while Eq. 9.3.4 establishes a relation among the probability density functions of $T(x)$, $T(y)$ and $T(xy)$. The force of mortality corresponding to this status is obtained as,

$$
\mu_{\overline{xy}}(t) = \frac{f_{T(\overline{xy})}(t)}{(1 - F_{T(\overline{xy})}(t))} = \frac{\{ {}_{t}p_{x}\mu_{x+t} + {}_{t}p_{y}\mu_{y+t} - {}_{t}p_{xy}\mu_{xy}(t)\}}{\{ {}_{t}p_{x} + {}_{t}p_{y} - {}_{t}p_{xy}\}}
$$

Remark 9.3.1. When x and y are specified numbers, to avoid confusion we denote xy and $\overline{xy}$ as $x:y$ and $\overline{x:y}$. For example, if $x = 30$ and $y = 40$, then we denote ${}_{t}q_{\overline{xy}}$ by ${}_{t}q_{\overline{30:40}}$.

Example 9.3.1. It is given that (i) ${}_{5}p_{50} = 0.9$, (ii) ${}_{5}p_{60} = 0.8$, (iii) $q_{55} = 0.03$, (iv) $q_{65} = 0.05$ and (v) $T(50)$ and $T(60)$ are independent. Calculate ${}_{5|}q_{\overline{50:60}}$.

Solution: Under the assumption of independence, ${}_{t}q_{\overline{xy}} = {}_{t}q_{x}\, {}_{t}q_{y}$ for the last-survivor status. Hence,

$$
{}_{5|}q_{\overline{50:60}} = {}_{6}q_{\overline{50:60}} - {}_{5}q_{\overline{50:60}} = {}_{6}q_{50}\, {}_{6}q_{60} - {}_{5}q_{50}\, {}_{5}q_{60}
$$

From the given information,

$$\begin{aligned}
{}_{6}p_{50} &= {}_{5}p_{50}\, p_{55} = (0.9)(0.97) = 0.873 \\
\text{and } {}_{6}p_{60} &= {}_{5}p_{60}\, p_{65} = (0.8)(0.95) = 0.76 \\
\Rightarrow \quad {}_{5|}q_{\overline{50:60}} &= (0.127)(0.24) - (0.1)(0.2) = 0.01048
\end{aligned}$$

$\blacksquare$

The probability mass function of $K(\overline{xy})$ is derived as follows. For a non-negative integer k, applying the basic probability laws and the assumption of independence of $K(x)$ and $K(y)$ we have,

$$\begin{aligned}
P[K(\overline{xy}) = k] \;=\;& P[K(x) = k, K(y) < k] \;+\; P[K(y) = k, K(x) < k] \\
+\;& P[K(x) = k, K(y) = k] \\
=\;& (1 - {}_{k}p_{y})\, {}_{k}p_{x} q_{x+k} + (1 - {}_{k}p_{x})\, {}_{k}p_{y} q_{y+k} + {}_{k}p_{x}\, {}_{k}p_{y}\, q_{x+k} q_{y+k} \\
=\;& {}_{k}p_{x} q_{x+k} + {}_{k}p_{y} q_{y+k} - {}_{k}p_{x}\, {}_{k}p_{y}(q_{x+k} + q_{y+k} - q_{x+k} q_{y+k}) \\
=\;& P[K(x) = k] + P[K(y) = k] - P[K(xy) = k]
\end{aligned}$$

Note the relation between $T(xy)$ and $K(xy)$ in Section 2. We have similar relation between $T(\overline{xy})$ and $K(\overline{xy})$. Observe that If $T(\overline{xy}) = 7.8$, then one of (x) or (y) survive 7 complete years and the other survives less than 7 complete years, or both survive 7 complete years. Hence, the maximum of $K(x)$ or $K(y)$ is 7, thus $K(\overline{xy}) = 7$. If $T(\overline{xy}) = 0.8$, then both $K(x)$ and $K(y)$ are 0 and hence $K(\overline{xy}) = 0$. Similarly if $T(\overline{xy}) = 10$, then the maximum of $K(x)$ and $K(y)$ is 9 and hence $K(\overline{xy}) = 9$. Thus, $K(\overline{xy})$ is the largest integer strictly smaller than $T(\overline{xy})$.

Equation 9.3.4 exhibits a relationship among the distributions of $T(xy)$, $T(\overline{xy})$, $T(x)$ and $T(y)$ under the assumption of independence of $T(x)$ and $T(y)$. We have a similar relationship among the distributions of $K(x)$, $K(y)$ and $K(xy)$, under the assumption of independence of $T(x)$ and $T(y)$. Below, we show that such a relationship exists among $T(xy), T(\overline{xy}), T(x)$ and $T(y)$ and also among $K(x)$, $K(y)$ and $K(xy)$, irrespective of the assumption of independence. It is to be noted that,

$$T(xy) = \begin{cases} T(x) & \text{if } (x) \text{ dies before } (y) \\ T(y) & \text{if } (y) \text{ dies before } (x) \end{cases}$$

and

$$T(\overline{xy}) = \begin{cases} T(y) & \text{if } (x) \text{ dies before } (y) \\ T(x) & \text{if } (y) \text{ dies before } (x) \end{cases}$$

So, even if $T(x)$ and $T(y)$ are not independent,

$$T(xy) + T(\overline{xy}) \;=\; T(x) + T(y) \tag{9.3.5}$$

Equation 9.3.5 essentially follows from the following mathematical identity.

$$\min\{a, b\} \;+\; \max\{a, b\} \;=\; a + b \;\; \forall \;\; a, b \in \mathbb{R}$$

Further,

$$F_{T(xy)}(t) + F_{T(\overline{xy})}(t) = F_{T(x)}(t) + F_{T(y)}(t) \quad \text{if } (x) \text{ dies before } (y)$$
$$= F_{T(y)}(t) + F_{T(x)}(t) \quad \text{if } (y) \text{ dies before } (x)$$

Either (x) dies before (y) or (y) dies before (x), hence

$$F_{T(xy)}(t) + F_{T(\overline{xy})}(t) = F_{T(x)}(t) + F_{T(y)}(t) \tag{9.3.6}$$
$$\Rightarrow f_{T(xy)}(t) + f_{T(\overline{xy})}(t) = f_{T(x)}(t) + f_{T(y)}(t) \tag{9.3.7}$$

by taking derivatives with respect to t. This equation has been derived in Eq. 9.3.4 under the assumption of independence of $T(x)$ and $T(y)$. From Eq. 9.3.6,

$$F_{T(\overline{xy})}(t) = F_{T(x)}(t) + F_{T(y)}(t) - F_{T(xy)}(t)$$
$$= 1 - {}_tp_x + 1 - {}_tp_y - 1 + {}_tp_{xy}$$
$$\Rightarrow {}_tp_{\overline{xy}} = {}_tp_x + {}_tp_y - {}_tp_{xy}$$

in international actuarial notation. From Eq. 9.3.7,

$$f_{T(\overline{xy})}(t) = f_{T(x)}(t) + f_{T(y)}(t) - f_{T(xy)}(t)$$
$$= {}_tp_x\mu_{x+t} + {}_tp_y\mu_{y+t} - {}_tp_{xy}\mu_{xy}(t)$$
$$= {}_tp_x\mu_{x+t} + {}_tp_y\mu_{y+t} - {}_tp_x\,{}_tp_y(\mu_{x+t} + \mu_{y+t})$$

The last equality follows under the assumption of independence of $T(x)$ and $T(y)$. A relationship similar to that among $T(xy), T(\overline{xy}), T(x)$ and $T(y)$ exists among $K(xy), K(\overline{xy}), K(x)$ and $K(y)$. Thus, using the arguments similar to those for $T(xy), T(\overline{xy})$ we get,

$$K(xy) + K(\overline{xy}) = K(x) + K(y) \tag{9.3.8}$$

For example, suppose $T(x) = 3.7$ and $T(y) = 3.9$, then $K(x) = K(y) = 3$, $T(xy) = 3.7$, $T(\overline{xy}) = 3.9$, $K(xy) = 3$ and $K(\overline{xy}) = 3$. It is easy to check that $T(xy) + T(\overline{xy}) = T(x) + T(y)$ and $K(xy) + K(\overline{xy}) = K(x) + K(y)$.

Relations developed in Eq. 9.3.7 and in Eq. 9.3.8 can be exploited to obtain expectations, variances and the covariances of the joint and last survivor future lifetimes.

Suppose expectations of $T(xy)$ and $T(\overline{xy})$ are denoted by e^0_{xy} and $e^0_{\overline{xy}}$ respectively. Then

$$e^0_{xy} = \int_0^\infty {}_tp_{xy}\,dt \quad \text{and} \quad e^0_{\overline{xy}} = \int_0^\infty {}_tp_{\overline{xy}}\,dt$$

Since ${}_tp_{\overline{xy}} = {}_tp_x + {}_tp_y - {}_tp_{xy}$, we get a relation

$$e^0_{\overline{xy}} = e^0_x + e^0_y - e^0_{xy}$$

It also follows from Eq. 9.3.5. Similarly for curtate future life time, the expectation of $K(xy)$ is given by, $e_{xy} = \sum {}_kp_{xy}$ and that of $K(\overline{xy})$ is given by, $e_{\overline{xy}} = \sum {}_kp_{\overline{xy}}$. Further, we get a relation,

$$e_{\overline{xy}} = e_x + e_y - e_{xy}$$

Variances of $T(xy)$ and $T(\overline{xy})$ are

$$V(T(xy)) \;=\; 2\int_0^\infty t \; {}_tp_{xy}dt - (e^0_{xy})^2, \quad V(T(\overline{xy})) \;=\; 2\int_0^\infty t \; {}_tp_{\overline{xy}}dt - (e^0_{\overline{xy}})^2$$

To find $Cov(T(xy), T(\overline{xy}))$, again it is noted that,

$$T(xy)\; T(\overline{xy}) = T(x)T(y)$$

hence, under the assumption of independence,

$$
\begin{aligned}
Cov(T(xy), T(\overline{xy})) &= E(T(xy)T(\overline{xy})) - e^0_{xy}e^0_{\overline{xy}} \;=\; E(T(x)T(y)) - e^0_{xy}e^0_{\overline{xy}} \\
&= E(T(x))E(T(y)) - e^0_{xy}e^0_{\overline{xy}} \;=\; e^0_x e^0_y - e^0_{xy}e^0_{\overline{xy}} \\
&= e^0_x e^0_y - e^0_{xy}\,(e^0_x + e^0_y - e^0_{xy}) \\
&= e^0_x e^0_y - e^0_{xy}\, e^0_x - e^0_{xy}e^0_y + e^0_{xy}e^0_{xy} \\
&= e^0_x(e^0_y - e^0_{xy}) - e^0_{xy}(e^0_y - e^0_{xy}) \;=\; (e^0_x - e^0_{xy})(e^0_y - e^0_{xy})
\end{aligned}
$$

Note that $e^0_x \geq e^0_{xy}$ and $e^0_y \geq e^0_{xy}$, consequently, $Cov(T(xy), T(\overline{xy})) \geq 0$. Thus, $T(xy)$ and $T(\overline{xy})$ are positively correlated except in trivial cases where e^0_x or e^0_y equals e^0_{xy}. The following examples illustrate the relations derived above.

Example 9.3.2. It is given that (i) $T(x)$ and $T(y)$ are independent,
(ii) $E(T(x)) = E(T(y)) = 4.0$, (iii) $Cov(T(xy), T(\overline{xy})) = 0.09$. Calculate $E(T(xy))$.

Solution: If $T(x)$ and $T(y)$ are independent,

$$
\begin{aligned}
Cov(T(xy), T(\overline{xy})) &= (e^0_x - e^0_{xy})(e^0_y - e^0_{xy}) \\
\Rightarrow \;\; 0.09 &= (4 - e^0_{xy})(4 - e^0_{xy}) = (4 - e^0_{xy})^2 \\
\Rightarrow \;\; 4 - e^0_{xy} &= \pm 0.3 \;\; \Rightarrow \;\; e^0_{xy} = 3.7 \text{ or } 4.3
\end{aligned}
$$

Since $T(xy) = \min\{T(x), T(y)\}$, it follows that $e^0_{xy} \leq e^0_x$ and therefore, we choose the smaller root. Hence, $e^0_{xy} = 3.7$. Note that $e^0_{\overline{xy}} = 4.3$. ∎

Example 9.3.3. Show that the probability of two lives (x) and (y) dying in the same year can be expressed as

$$1 + e_{xy} - p_x(1 + e_{(x+1):y}) - p_y(1 + e_{x:(y+1)}) + p_{xy}(1 + e_{(x+1):(y+1)})$$

Assume that $T(x)$ and $T(y)$ are independent and identically distributed.

Solution: Under the assumption of independence of $T(x)$ and $T(y)$, and using the formula $e_{xy} = \sum_{k=1}^{\infty} {}_k p_{xy}$, the probability that both die in the same year is derived as follows.

$$\sum_{t=1}^{\infty} P[t-1 < T(x) < t,\ \ t-1 < T(y) < t]$$

$$= \sum_{t=1}^{\infty} ({}_{t-1}p_x - {}_t p_x)({}_{t-1}p_y - {}_t p_y)$$

$$= \sum_{t=1}^{\infty} \left\{ {}_{t-1}p_{xy} - p_y\,{}_{t-1}p_{x:(y+1)} - p_x\,{}_{t-1}p_{(x+1):y} + {}_t p_{xy} \right\}$$

$$= \sum_{t=1}^{\infty} \left\{ {}_{t-1}p_{xy} - p_y\,{}_{t-1}p_{x:(y+1)} - p_x\,{}_{t-1}p_{(x+1):y} + p_x p_y\,{}_{t-1}p_{(x+1):(y+1)} \right\}$$

$$= 1 + e_{xy} - p_x(1 + e_{(x+1):y}) - p_y(1 + e_{x:(y+1)}) + p_{xy}(1 + e_{(x+1):(y+1)})$$

∎

Example 9.3.4. Show that the probability of two lives (30) and (40) dying at the same age at last birthday can be expressed as

$$ {}_{10}p_{30}(1 + e_{40:40}) - 2\,{}_{11}p_{30}(1 + e_{40:41}) + p_{40}\,{}_{11}p_{30}(1 + e_{41:41}) $$

Assume that $T(30)$ and $T(40)$ are independent and identically distributed.

Solution: Under the assumption of independence of $T(30)$ and $T(40)$, and using the formula $e_{xy} = \sum_{k=1}^{\infty} {}_k p_{xy}$, the probability that both die in the same year can be obtained as follows.

$$\sum_{t=0}^{\infty} P[10 + t < T(30) < 11 + t,\ \ t < T(40) < t+1]$$

$$= \sum_{t=0}^{\infty} ({}_{10+t}p_{30} - {}_{11+t}p_{30})({}_t p_{40} - {}_{t+1}p_{40})$$

$$= \sum_{t=0}^{\infty} \left\{ {}_{10}p_{30}\,{}_t p_{40}\,{}_t p_{40} - {}_{10}p_{30}\,p_{40}\,{}_t p_{40} p_{41} - {}_{11}p_{30}\,{}_t p_{41}\,{}_t p_{40} \right\}$$

$$+ \sum_{t=0}^{\infty} {}_{11}p_{30}\,{}_t p_{41} p_{40}\,{}_t p_{41}$$

$$= \sum_{t=0}^{\infty} \left\{ {}_{10}p_{30}\,{}_t p_{40:40} - {}_{11}p_{30}\,{}_t p_{40:41} - {}_{11}p_{30}\,{}_t p_{40:41} + {}_{11}p_{30}\,p_{40}\,{}_t p_{41:41} \right\}$$

$$= {}_{10}p_{30}(1 + e_{40:40}) - 2\,{}_{11}p_{30}(1 + e_{40:41}) + p_{40}\,{}_{11}p_{30}(1 + e_{41:41})$$

∎

The actuarial present values of benefit in insurance products and of annuity payments in annuity products for the last survivor status are defined exactly on similar lines as those for the joint life status. However, instead of computing these separately, the general relationship among $T(xy), T(\overline{xy}), T(x)$ and $T(y)$ and also among $K(xy), K(\overline{xy}), K(x)$ and $K(y)$, is useful to establish relations among the actuarial present values of benefit in insurance products and of annuity payments in annuity products. Thus from Eqs 9.3.5 and 9.3.8, we get relationships among the actuarial present values. Some of these are listed below.

$$\overline{A}_{\overline{xy}} + \overline{A}_{xy} = \overline{A}_x + \overline{A}_y \qquad \text{and} \qquad A_{\overline{xy}} + A_{xy} = A_x + A_y$$

$$\overline{A}^1_{\overline{xy}:\overline{n}|} + \overline{A}^1_{xy:\overline{n}|} = \overline{A}^1_{x:\overline{n}|} + \overline{A}^1_{y:\overline{n}|}$$

$$A^1_{\overline{xy}:\overline{n}|} + A^1_{xy:\overline{n}|} = A^1_{x:\overline{n}|} + A^1_{y:\overline{n}|}$$

$$_n E_{\overline{xy}} +_n E_{xy} = {}_n E_x +_n E_y$$

$$\overline{A}_{\overline{xy}:\overline{n}|} + \overline{A}_{xy:\overline{n}|} = \overline{A}_{x:\overline{n}|} + \overline{A}_{y:\overline{n}|}$$

$$A_{\overline{xy}:\overline{n}|} + A_{xy:\overline{n}|} = A_{x:\overline{n}|} + A_{y:\overline{n}|}$$

$$\overline{a}_{\overline{xy}} + \overline{a}_{xy} = \overline{a}_x + \overline{a}_y \qquad \text{and} \qquad \ddot{a}_{\overline{xy}} + \ddot{a}_{xy} = \ddot{a}_x + \ddot{a}_y$$

$$\overline{a}_{\overline{xy}:\overline{n}|} + \overline{a}_{xy:\overline{n}|} = \overline{a}_{x:\overline{n}|} + \overline{a}_{y:\overline{n}|}$$

$$\ddot{a}_{\overline{xy}:\overline{n}|} + \ddot{a}_{xy:\overline{n}|} = \ddot{a}_{x:\overline{n}|} + \ddot{a}_{y:\overline{n}|} \tag{9.3.9}$$

These formulae are useful to express the actuarial present values of annuities and insurances corresponding to the last survivor status, in terms of those for the individual lives and the joint life status. Computation of premiums is similar to that in joint life status. The following examples illustrate the computations.

Example 9.3.5. Suppose the future lifetimes $T(x)$ and $T(y)$ are independent and each has the distribution defined by Gompertz' law with the force of mortality, $\mu_x = BC^x$ and $\mu_y = BC^y$ with $B = 0.0001151$ and $C = 1.096$. It is given that $\delta = 0.05$, $x = 25$ and $y = 30$. (i) Determine the survival function, the force of mortality and the probability density function of $T(xy)$ for the joint life status. (ii) Calculate $\overline{A}^1_{xy:\overline{n}|}$ and $\overline{A}^1_{\overline{xy}:\overline{n}|}$ for $n = 1$ to 10. Verify the result proved in Example 9.2.1. (iii) Calculate $\overline{A}_{xy:\overline{n}|}$ and $\overline{A}_{\overline{xy}:\overline{n}|}$ for $n = 1$ to 10. (iv) Calculate the annual premium, payable as n-year temporary continuous life annuity, for the benefit of Rs 1000, payable at the moment of first death in an n-year term insurance and in an n-year endowment insurance, for $n = 1$ to 10. Calculate the annual premium, payable as an n-year temporary continuous life annuity, for a benefit of Rs 1000, payable at the moment of last death in an n-year term insurance and in an n-year endowment insurance, for $n = 1$ to 10. (v) Calculate $\overline{A}_{xy}$ and $\overline{A}_{\overline{xy}}$. Calculate the annual premium, payable as whole life continuous annuity, for a benefit of Rs 1000, payable at the moment of first death in whole life insurance issued to (x, y). Calculate the annual premium, payable as whole life continuous annuity, for a benefit of Rs 1000, payable at the moment of last death in whole life insurance issued to (x, y).

Solution: From Example 4.2.11, the survival function of $T(x)$ for Gompertz' law is given by, $_t p_x = \exp[-mC^x(C^t - 1)]$, $t > 0$. Hence, in joint life status, the survival function of $T(xy)$ is given by,

$$_t p_{xy} = {}_t p_x \, {}_t p_y = \exp[-m(C^x + C^y)(C^t - 1)]$$

By the definition of force of mortality, $\mu_{x+t} = BC^{x+t}$. Thus, from Eq. 9.2.3,

$$\mu_{xy}(t) = \mu_{x+t} + \mu_{y+t} = BC^t(C^x + C^y)$$

Hence, the probability density function of $T(xy)$ is given by,

$$g_{T(xy)}(t) = {}_tp_{xy}\,\mu_{xy}(t) = e^{-m(C^t-1)(C^x+C^y)}\,BC^t(C^x+C^y), \quad t>0$$

Suppose $m(C^x+C^y)$ is denoted by α_{xy}. Then $g_{T(xy)}(t)$ is expressible as,

$$g_{T(xy)}(t) = \alpha_{xy}\,\log C\,e^{\alpha_{xy}}\,e^{-\alpha_{xy}C^t}\,C^t, \quad t>0$$

In Example 5.3.7, we have derived the probability density function of $T(x)$ and it is given by,

$$g_{T(x)}(t) = \alpha_x\,\log C\,e^{\alpha_x}\,e^{-\alpha_x C^t}\,C^t, \quad \text{for } t>0, \text{ with } \alpha_x = mC^x$$

The two densities are exactly of the same form with α_{xy} replaced by α_x. This observation is useful to find $\overline{A}^{\,1}_{xy:\overline{n}|}$, $_nE_{xy}$, $\overline{A}_{xy:\overline{n}|}$ and $\overline{A}_{xy}$ using the same approach of computing probabilities of certain intervals in gamma distribution with appropriate parameters, as adopted in Example 5.3.7.

(ii) From (i) we have,

$$\overline{A}^{\,1}_{xy:\overline{n}|} = \alpha_{xy}\,e^{\alpha_{xy}}\Gamma(\lambda)\,\alpha_{xy}^{-\lambda}\,P[1 \le W \le C^n]$$

where W follows a gamma distribution with a shape parameter λ and scale parameter α_{xy}. We find the values of $1000\overline{A}^{\,1}_{xy:\overline{n}|}$ for $x=25$ and $y=30$ using the formulae derived above and Part III of Code 9.4.1. These are presented in the fifth column of Table 9.1. Using the relation $\overline{A}^{\,1}_{\overline{xy}:\overline{n}|} + \overline{A}^{\,1}_{xy:\overline{n}|} = \overline{A}^{\,1}_{x:\overline{n}|} + \overline{A}^{\,1}_{y:\overline{n}|}$, we find the values of $1000\overline{A}^{\,1}_{\overline{xy}:\overline{n}|}$. These are reported in the last column of Table 9.1. The fourth and sixth columns report the lower limit $L = 1000(\,_np_y\,\overline{A}^{\,1}_{x:\overline{n}|} + \,_np_x\overline{A}^{\,1}_{y:\overline{n}|})$ and the upper limit $U = 1000\overline{A}^{\,1}_{x:\overline{n}|} + 1000\overline{A}^{\,1}_{y:\overline{n}|}$ respectively, as derived in Example 9.2.1. Comparing the values in columns 4, 5 and 6, we note that the result proved in Example 9.2.1 is verified.

n	$1000\overline{A}^{\,1}_{x:\overline{n}|}$	$1000\overline{A}^{\,1}_{y:\overline{n}|}$	L	$1000\overline{A}^{\,1}_{xy:\overline{n}|}$	U	$1000\overline{A}^{\,1}_{\overline{xy}:\overline{n}|}$
1	1.16	1.84	3.00	3.00	3.00	0.00
2	2.37	3.75	6.10	6.11	6.12	0.01
3	3.63	5.74	9.32	9.35	9.37	0.02
4	4.94	7.80	12.65	12.71	12.74	0.03
5	6.31	9.95	16.12	16.19	16.26	0.07
6	7.73	12.19	19.70	19.81	19.92	0.11
7	9.21	14.51	23.40	23.57	23.72	0.15
8	10.74	16.93	27.22	27.46	27.67	0.21
9	12.34	19.43	31.16	31.49	31.77	0.28
10	14.00	22.04	35.23	35.67	36.04	0.37

Table 9.1 Gompertz' law: $1000\overline{A}^{\,1}_{xy:\overline{n}|}$ and $1000\overline{A}^{\,1}_{\overline{xy}:\overline{n}|}$

All the values in the Table 9.1 are obtained using Part II and III of Code 9.4.1. From the values reported in the table, we note that for the underlying mortality pattern, values of $1000\overline{A}^{\,1}_{\overline{xy}:\overline{n}|}$

are very low $\forall\ n$, in view of the fact that the chance of death of both (25) and (30) within the term of 10 years is very less. It is to be noted that

$$\overline{A}^{\,1}_{\overline{xy}:\overline{n}|} < \overline{A}^{\,1}_{x:\overline{n}|} \quad \text{and} \quad \overline{A}^{\,1}_{\overline{xy}:\overline{n}|} < \overline{A}^{\,1}_{y:\overline{n}|}$$

It is logically appealing since the random time period for which the last survivor status is alive is larger than the future life time random variable of an individual member in the group.

(iii) and (iv): In (ii) we have obtained the values of $\overline{A}^{\,1}_{\overline{xy}:\overline{n}|}$ and $\overline{A}^{\,1}_{\overline{xy}:\overline{n}|}$. To find the values of $\overline{A}_{xy:\overline{n}|}$ and $\overline{A}_{\overline{xy}:\overline{n}|}$ and the premiums for term and endowment insurances, we use the following formulae.

$$
\begin{aligned}
{}_nE_x &= v^n\ {}_tp_x = \exp[-\delta n - mC^x(C^t - 1)] \\
{}_nE_y &= v^n\ {}_tp_y = \exp[-\delta n - mC^y(C^t - 1)] \\
{}_nE_{xy} &= v^n\ {}_tp_{xy} = \exp[-\delta n - m(C^x + C^y)(C^t - 1)] \\
{}_nE_{\overline{xy}} &= {}_nE_x + {}_nE_y - {}_nE_{xy}, \quad \text{by Eq. 9.3.9.}
\end{aligned}
$$

$$
\begin{aligned}
\overline{A}_{xy:\overline{n}|} &= \overline{A}^{\,1}_{xy:\overline{n}|} + {}_nE_{xy}, \quad \overline{A}_{\overline{xy}:\overline{n}|} = \overline{A}^{\,1}_{\overline{xy}:\overline{n}|} + {}_nE_{\overline{xy}} \\
\overline{a}_{xy:\overline{n}|} &= (1 - \overline{A}_{xy:\overline{n}|})/\delta \quad \text{and} \quad \overline{a}_{\overline{xy}:\overline{n}|} = (1 - \overline{A}_{\overline{xy}:\overline{n}|})/\delta
\end{aligned}
$$

Using the above formulae, we obtain the values of net single premiums in n-year endowment insurance issued to single life $x = (25)$ and $y = (30)$ and in joint life and last survivor status. We use Part II and III of Code 9.4.1 to compute these functions. The values are reported in Table 9.2.

n	$1000\overline{A}_{x:\overline{n}|}$	$1000\overline{A}_{y:\overline{n}|}$	$1000\overline{A}_{xy:\overline{n}|}$	$1000\overline{A}_{\overline{xy}:\overline{n}|}$
1	951.26	951.27	951.30	951.23
2	904.95	905.02	905.13	904.84
3	860.96	861.11	861.36	860.71
4	819.18	819.44	819.89	818.73
5	779.50	779.91	780.61	778.81
6	741.83	742.41	743.42	740.83
7	706.06	706.86	708.22	704.70
8	672.11	673.15	674.92	670.34
9	639.90	641.21	643.44	637.66
10	609.33	610.95	613.70	606.58

Table 9.2 Gompertz' law: Endowment insurance in joint and last survivor status

We now proceed to computation of premiums. In term insurance, the annual premium payable as n-year temporary annuity, which ceases at the first death in the joint life status, if it occurs before the end of the term, are obtained using the formula $1000\overline{A}^{\,1}_{xy:\overline{n}|}/\overline{a}_{xy:\overline{n}|}$. These are reported in the second column of Table 9.3.

n	Term joint life	Term last survivor	Endowment joint life	Endowment last survivor
1	3.08	0.00	976.75	975.21
2	3.22	0.00	477.03	475.42
3	3.37	0.01	310.65	308.96
4	3.53	0.01	227.61	225.84
5	3.69	0.02	177.90	176.05
6	3.86	0.02	144.87	142.92
7	4.04	0.03	121.36	119.32
8	4.22	0.03	103.81	101.67
9	4.42	0.04	90.23	87.99
10	4.62	0.05	79.43	77.09

Table 9.3 Gompertz' law: Premiums for term and endowment insurance in joint and last survivor status

For the setup of last survivor status, annual premium payable as n-year temporary annuity, ceases at the last death. It is obtained using the formula $1000\overline{A}^{1}_{\overline{xy}:\overline{n}|}/\overline{a}_{\overline{xy}:\overline{n}|}$. These are reported in the third column of Table 9.3. Premiums for n-year endowment insurance are obtained as $1000\overline{A}_{xy:\overline{n}|}/\overline{a}_{xy:\overline{n}|}$ and $1000\overline{A}_{\overline{xy}:\overline{n}|}/\overline{a}_{\overline{xy}:\overline{n}|}$ for joint life and last survivor status respectively. These are reported in the fourth and fifth columns of Table 9.3. We use Part IV of Code 9.4.1 to compute these values. From the values reported in the table, we note that for the underlying mortality pattern, values of premium for last survivor status are very low for all n, for the same reason as stated above.

(v) To find $\overline{A}_{xy}$, we again use the same approach, as adopted in Example 5.3.7. Thus, $\overline{A}_{xy}$ is given by,

$$\overline{A}_{xy} = \alpha_{xy}\, e^{\alpha_{xy}}\Gamma(\lambda)\, \alpha_{xy}^{-\lambda}\, P[W \geq 1]$$

where W follows a gamma distribution with a shape parameter λ and scale parameter α_{xy}. For the given mortality and interest pattern we have calculated $\overline{A}_x$ for $x = 25$ and 30 in Example 5.3.7. Hence we get the value of $\overline{A}_{\overline{xy}}$. These are as given below.

$$1000\overline{A}_{xy} = 235.98, \quad 1000\overline{A}_x = 152.54, \quad 1000\overline{A}_y = 189.12, \quad 1000\overline{A}_{\overline{xy}} = 105.68$$

It is to be noted that,

$$\overline{A}_{\overline{xy}} < \overline{A}_x < \overline{A}_y < \overline{A}_{xy}, \quad \text{for} \quad x < y$$

Annual Premium P(Joint) for whole life insurance in joint life status and P(Last) for whole life insurance in last survivor status are given by,

$$\text{P(Joint)} = 1000\delta\overline{A}_{xy}/(1 - \overline{A}_{xy}) = 15.44 \text{ and } \text{P(Last)} = 1000\delta\overline{A}_{\overline{xy}}/(1 - \overline{A}_{\overline{xy}}) = 5.91$$

Note that the premium in the joint life status is higher than that in the last survivor status. All these values are obtained using Part V of Code 9.4.1. ∎

In the next example we discuss similar computations for fully discrete policies.

Example 9.3.6. Suppose the curtate future lifetimes $K(x)$ and $K(y)$ are independent and each has the distribution defined by Gompertz' law with the force of mortality, $\mu_x = BC^x$ and $\mu_y = BC^y$ with $B = 0.0001151$ and $C = 1.096$. It is given that $\delta = 0.05$, $x = 25$ and $y = 30$. (i) Determine the probability mass function of $K(xy)$. Calculate e_{xy}. (ii) Calculate $1000A^1_{x:\overline{n}|}$, $1000A^1_{y:\overline{n}|}$, $1000A^1_{xy:\overline{n}|}$ and $1000A^1_{\overline{xy}:\overline{n}|}$ for $n = 1$ to 10. Calculate $1000A_{x:\overline{n}|}$, $1000A_{y:\overline{n}|}$, $1000A_{xy:\overline{n}|}$ and $1000A_{\overline{xy}:\overline{n}|}$ for $n = 1$ to 10. (iii) Calculate the premiums for insurance products specified in (ii). (iv) Calculate $1000A_{xy}$ and $1000A_{\overline{xy}}$. Calculate the annual premium, payable as whole life annuity due, payable till the first death, for the benefit of 1000, payable at the end of year of first death in whole life insurance issued to (x, y). Calculate the annual premium, payable as whole life annuity due, payable till the last death, for the benefit of 1000, payable at the end of year of last death in whole life insurance issued to (x, y).

Solution: To find the required actuarial present values, we need to find the probability mass function of $K(25:30)$ corresponding to Gompertz' law. The probability mass function of $K(xy)$ is given by,

$$P[K(xy) = k] \;=\; {}_kp_{xy} - {}_{(k+1)}p_{xy}$$

For Gompertz' law, in Example 9.3.5 we have obtained,

$$_kp_{xy} \;=\; {}_kp_x\, {}_kp_y \;=\; \exp[-m(C^x + C^y)(C^k - 1)] \;=\; e^{-\alpha_{xy}(C^k - 1)}$$

where $\alpha_{xy} = m(C^x + C^y)$ is as defined in Example 9.3.5. Further
$q_{x+k} = 1 - e^{-m\,C^{x+k}(C-1)}$. Substituting these expressions in the expression for the probability mass function of $K(xy)$ and after some simplification we get,

$$P[K(xy) = k] \;=\; e^{\alpha_{xy}}\left(e^{-\alpha_{xy}C^k} - e^{-\alpha_{xy}C^{k+1}}\right)$$

In Example 4.3.4, we have obtained the probability mass function of $K(x)$ and it is given by,

$$P[K(x) = k] \;=\; e^{\alpha_x}\left(e^{-\alpha_x C^k} - e^{-\alpha_x C^{k+1}}\right)$$

where $\alpha_x = mC^x$. It is to be noted that the probability mass function of $K(xy)$ and that of $K(x)$ are exactly of the same form, with the only change of α_{xy} and α_x. We use Part III of Code 9.4.2 to compute the probability mass function of $K(xy)$. Table 9.4 displays the probabilities for $k = 0$ to $k = 71$. The probabilities for $k = 65$ to $k = 71$ are 0 up to 5 decimal places. Beyond 71 these are almost 0 and are not reported in the table. $P[K(25:30) = 0] = 0.00307$ implies that the chance of death of one of (25) and (30) in the first year after signing the contract is 0.00307. From Table 4.2, $P[K(30) = 0] = 0.00188$. In fact note that, $P[K(x) = k] < P[K(xy) = k]\ \forall\ k$. Thus, in a group if two lives are involved then the chance of death of one of the two in any year is higher than the chance of death of the single individual. From the distribution of $K(25:30)$ we get, $e_{25:30} = E(K(25:30)) = 32.11$. From the probability distribution of $K(25:30)$, we note that the mode of the distribution is at 37. Figure 9.1 shows the graph of probability mass function of $K(25:30)$. It is slightly negatively skewed but close to a bell shaped curve, with probabilities approaching to 0 after 64, but comparatively heavy tail at the left side.

For n-year term insurance $1000A^1_{xy:\overline{n}|}$ for $n = 1$ to 10 are obtained using the following formula and the probability mass function of $K(xy)$ derived above.

$$A^1_{xy:\overline{n}|} \;=\; \sum_{k=0}^{n-1} v^{k+1}\left\{ e^{\alpha_{xy}}\left(e^{-\alpha_{xy}C^k} - e^{-\alpha_{xy}C^{k+1}}\right)\right\}$$

k	Probability	k	Probability	k	Probability
0	0.00307	24	0.02118	48	0.01681
1	0.00336	25	0.02255	49	0.01418
2	0.00367	26	0.02394	50	0.01166
3	0.00400	27	0.02533	51	0.00932
4	0.00437	28	0.02672	52	0.00724
5	0.00477	29	0.02808	53	0.00544
6	0.00520	30	0.02939	54	0.00394
7	0.00566	31	0.03063	55	0.00274
8	0.00617	32	0.03177	56	0.00183
9	0.00672	33	0.03277	57	0.00116
10	0.00731	34	0.03361	58	0.00070
11	0.00795	35	0.03425	59	0.00040
12	0.00863	36	0.03466	60	0.00021
13	0.00937	37	0.03481	61	0.00011
14	0.01016	38	0.03467	62	0.00005
15	0.01101	39	0.03421	63	0.00002
16	0.01191	40	0.03342	64	0.00001
17	0.01287	41	0.03229	65	0.00000
18	0.01389	42	0.03082	66	0.00000
19	0.01497	43	0.02903	67	0.00000
20	0.01611	44	0.02696	68	0.00000
21	0.01731	45	0.02463	69	0.00000
22	0.01855	46	0.02212	70	0.00000
23	0.01985	47	0.01949	71	0.00000

Table 9.4 Gompertz' law: Probability distribution of $K(25:30)$

To compute $A^1_{\overline{xy}:\overline{n}|}$, we compute $1000A^1_{x:\overline{n}|}$ and $1000A^1_{y:\overline{n}|}$ from the probability mass function of $K(x)$ and $K(y)$ respectively. Parts II, III and IV of Code 9.4.2 are used to compute these values. These are presented in Table 9.5. From the table we note that the net single premiums for the term insurance are small for all n, as usual. However, in the last survivor setup these are very small. When these values are compared to those of $\overline{A}^1_{\overline{xy}:\overline{n}|}$ reported in Table 9.1, we note that $A^1_{\overline{xy}:\overline{n}|} < \overline{A}^1_{\overline{xy}:\overline{n}|}$, as expected. To find the corresponding values for n-year endowment insurance we proceed as follows. Using the survival function of $K(xy)$, we obtain $_nE_{xy} = v^n e^{-\alpha_{xy}(C^n-1)}$. For n-year endowment insurance $A_{xy:\overline{n}|}$ for $n=1$ to 10 are then given by $A_{xy:\overline{n}|} = A^1_{xy:\overline{n}|} + {}_nE_{xy}$. To find the values of $A_{\overline{xy}:\overline{n}|}$, we compute $_nE_x$, $_nE_y$ and hence $A_{x:\overline{n}|}$ and $A_{y:\overline{n}|}$. The value of $A_{\overline{xy}:\overline{n}|}$ is then obtained by the formula given in Eq. 9.3.9, which states that

$$A_{\overline{xy}:\overline{n}|} = A_{x:\overline{n}|} + A_{y:\overline{n}|} - A_{xy:\overline{n}|}$$

Parts II, III and IV of Code 9.4.2 are used to compute these values. These are presented in Table 9.6. Using the actuarial present values as displayed in Table 9.5 and Table 9.6, we find the premiums for all n-year term and endowment insurances in joint life status and the last

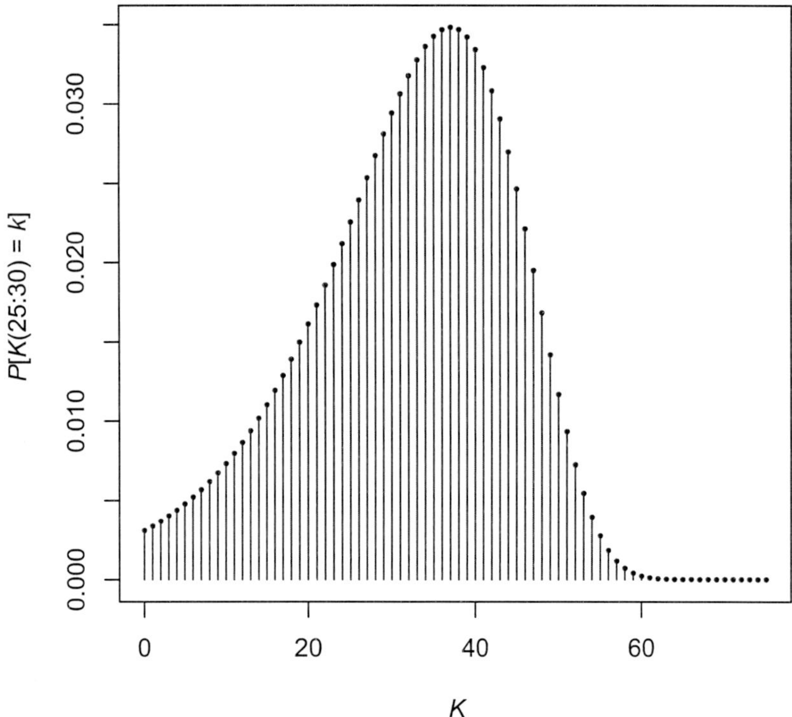

Figure 9.1 Distribution of $K(25:30)$

n	$1000A^1_{x:\overline{n}|}$	$1000A^1_{y:\overline{n}|}$	$1000A^1_{xy:\overline{n}|}$	$1000A^1_{\overline{xy}:\overline{n}|}$
1	1.13	1.79	2.92	0.00
2	2.31	3.66	5.96	0.01
3	3.54	5.60	9.12	0.02
4	4.82	7.61	12.39	0.04
5	6.15	9.71	15.80	0.07
6	7.54	11.89	19.33	0.10
7	8.98	14.16	22.99	0.15
8	10.48	16.51	26.79	0.21
9	12.04	18.96	30.72	0.28
10	13.66	21.50	34.80	0.36

Table 9.5 Gompertz' law: Term insurance in joint and last survivor status

survivor status. The general formula for finding the premium is,

$$\text{Premium} = \frac{d\,1000\,\text{Actuarial present value of benefit}}{(1 - \text{Actuarial present value of benefit in endowment insurance})}$$

n	$1000A_{x:\overline{n}|}$	$1000A_{y:\overline{n}|}$	$1000A_{xy:\overline{n}|}$	$1000A_{\overline{xy}:\overline{n}|}$
1	951.23	951.23	951.23	951.23
2	904.89	904.92	904.98	904.84
3	860.87	860.97	861.13	860.71
4	819.06	819.25	819.58	818.73
5	779.35	779.67	780.21	778.80
6	741.64	742.12	742.93	740.82
7	705.84	706.50	707.64	704.70
8	671.85	672.74	674.25	670.34
9	639.60	640.73	642.67	637.66
10	608.99	610.41	612.83	606.57

Table 9.6 Gompertz' law: Endowment insurance in joint and last survivor status for fully discrete policies

where d is the discount factor. We use Part V of Code 9.4.2 to compute premiums. These are reported in Table 9.7.

n	Term joint life	Term last survivor	Endowment joint life	Endowment last survivor
1	2.92	0.0021	951.23	951.23
2	3.06	0.0046	464.50	463.73
3	3.20	0.0075	302.44	301.36
4	3.35	0.0109	221.55	220.28
5	3.51	0.0148	173.13	171.72
6	3.67	0.0193	140.95	139.41
7	3.84	0.0245	118.05	116.39
8	4.01	0.0305	100.95	99.17
9	4.19	0.0374	87.72	85.83
10	4.38	0.0452	77.20	75.19

Table 9.7 Premiums for benefit of 1000 in joint life and last survivor status

The actuarial present values of benefit in whole life insurance under both the setup are obtained using the appropriate formulae and the probability mass function of $K(xy)$. These are as given below.

$$1000A_{xy} = 230.17, \quad 1000A_x = 148.78, \quad 1000A_y = 184.47, \quad 1000A_{\overline{xy}} = 103.08$$

Here also $A_{\overline{xy}} < A_x < A_y < A_{xy}$. Annual premium P(Joint) for whole life insurance in joint life status and P(Last) for whole life insurance in last survivor status are given by,

$$\text{P(Joint)} = 1000dA_{xy}/(1 - A_{xy}) = 14.58 \quad \text{and} \quad \text{P(Last)} = 1000dA_{\overline{xy}}/(1 - A_{\overline{xy}}) = 5.61$$

We use Code 9.4.2 to compute the premiums. Again it is to be noted that the premium in the joint life status is higher than that in the last survivor status. ∎

As stated in Section 4.7, the mortality pattern is usually specified by a set of q_x values for integer values of x starting from 0 to some limiting age w. We have used Code 4.7.2 to compute the probability distribution of the curtate future life time random variable $K(25)$ when values of q_x are as specified in Table 4.8. In the following example we assume that the mortality law of both $x = (25)$ and $y = (30)$ is as depicted by q_x in Table 4.8. We find the actuarial present values of the benefit involved in term insurance, endowment insurance, and whole life insurance issued to a group of two individuals (25) and (30), in both the status. We also compute the premiums. Given a set of q_x values, we cannot find the exact values of the actuarial present values of the benefit to be paid at the moment of termination of the status. We find these values from the discrete setup, under the assumption of uniformity of deaths in a unit age interval.

Example 9.3.7. Suppose the mortality law of both $x = (25)$ and $y = (30)$ is specified by values of q_x as given in Table 4.8. Compute the actuarial present values of the benefit of 1000 and the fully discrete premiums in the term insurance, the endowment insurance and whole life insurance issued to a group of two individuals (25) and (30), in both the status. Find fully continuous premiums under the assumption of uniformity of deaths in a unit age interval for the whole life insurance in both the status. Suppose the effective rate of interest is $i = 0.06$.

Solution: As in Example 9.3.6, to find the required monetary functions, we find the probability mass functions of $K(25)$ and $K(30)$ from the q_x as specified in Table 4.8. We use these to find the actuarial present values of benefits, annuities and premiums for all the insurance products involving single life. We then find the probability mass function of $K(25 : 30)$ and use it to find the actuarial present values of benefits, annuities and premiums for all the insurance products issued to a group of (25) and (30) in the joint life status. We use the formulae in Eq. 9.3.9 to compute the corresponding monetary functions in the last survivor status. All these computations are done using Code 9.4.3. The probability mass function of $K(xy)$ is given by,

$$P[K(xy) = k] \quad = \quad {}_kp_x\, {}_kp_y(q_{x+k} + q_{y+k} - q_{x+k}q_{y+k}) , \qquad k = 0, 1, \cdots$$

For an n-year term insurance $1000A^1_{xy:\overline{n}|}$ for $n = 1$ to 10 are obtained using the following formula and the probability mass function of $K(xy)$.

$$A^1_{xy:\overline{n}|} \quad = \quad \sum_{k=0}^{n-1} v^{k+1}\{ {}_kp_x\, {}_kp_y(q_{x+k} + q_{y+k} - q_{x+k}q_{y+k})\}$$

To compute $A^1_{\overline{xy}:\overline{n}|}$, we compute $1000A^1_{x:\overline{n}|}$ and $1000A^1_{y:\overline{n}|}$ from the probability mass function of $K(x)$ and $K(y)$ respectively. We use the formulae in Eq. 9.3.9 to compute $A^1_{\overline{xy}:\overline{n}|}$. Parts I, II, III and IV of Code 9.4.3 are used to compute these values. These are presented in Table 9.8. From the table we note that the net single premiums for the term insurance are small for all n, as usual. However, in the last survivor setup these are very small. To find the corresponding values for n-year endowment insurance we proceed as follows. We obtain ${}_nE_{xy} = v^n\, {}_np_{xy}$. For an n-year endowment insurance $A_{xy:\overline{n}|}$ for $n = 1$ to 10 are then given by $A_{xy:\overline{n}|} = A^1_{xy:\overline{n}|} + {}_nE_{xy}$. To find the values of $A_{\overline{xy}:\overline{n}|}$, we compute ${}_nE_x$, ${}_nE_y$ and hence $A_{x:\overline{n}|}$ and $A_{y:\overline{n}|}$. The value of $A_{\overline{xy}:\overline{n}|}$ is then obtained by the formula given in Eq. 9.3.9, which states that

$$A_{\overline{xy}:\overline{n}|} = A_{x:\overline{n}|} + A_{y:\overline{n}|} - A_{xy:\overline{n}|}$$

n	$1000A^1_{x:\overline{n}|}$	$1000A^1_{y:\overline{n}|}$	$1000A^1_{xy:\overline{n}|}$	$1000A^1_{\overline{xy}:\overline{n}|}$
1	1.10	1.32	2.42	0.00
2	2.16	2.63	4.78	0.01
3	3.18	3.92	7.08	0.01
4	4.17	5.19	9.34	0.02
5	5.16	6.46	11.58	0.04
6	6.14	7.70	13.79	0.05
7	7.11	8.94	15.98	0.08
8	8.07	10.17	18.14	0.10
9	9.02	11.38	20.27	0.13
10	9.95	12.57	22.37	0.16

Table 9.8 Term insurance: Joint and last survivor status

n	$1000A_{x:\overline{n}|}$	$1000A_{y:\overline{n}|}$	$1000A_{xy:\overline{n}|}$	$1000A_{\overline{xy}:\overline{n}|}$
1	943.40	943.40	945.82	940.97
2	890.06	890.07	892.49	887.64
3	839.80	839.84	842.31	837.32
4	792.44	792.52	795.13	789.83
5	747.83	747.96	750.76	745.02
6	705.79	705.98	709.02	702.76
7	666.18	666.46	669.77	662.88
8	628.88	629.24	632.86	625.26
9	593.74	594.20	598.15	589.78
10	560.64	561.21	565.53	556.32

Table 9.9 Endowment insurance: Joint and last survivor status

Parts I, II, III and IV of Code 9.4.3 are used to compute these values. These are presented in Table 9.9.

Using the actuarial present values as displayed in Table 9.8 and Table 9.9, we find the premiums for n-year term and endowment insurances in joint life status and the last survivor status. The general formula for finding the premium is,

$$\text{Premium} = \frac{d\ 1000\ \text{Actuarial present value of benefit}}{(1 - \text{Actuarial present value of benefit in endowment insurance})}$$

The premiums for an n-year term insurance for single lives and in the joint and the last survivor status are reported in Table 9.10. The premiums for an n-year endowment insurance for single lives and in the joint and the last survivor status are reported in Table 9.11.

The actuarial present values of the benefit of 1000, when paid at the end of the year of the termination of the status, in whole life insurance under both the status are obtained using the appropriate formulae and the probability mass function of $K(xy)$. These are as given below.

$$1000A_x = 73.88, \quad 1000A_y = 92.54, \quad 1000A_{xy} = 125.86 \ \ \& \ \ 1000A_{\overline{xy}} = 40.57$$

n	$1000P^1_{x:\overline{n}\rvert}$	$1000P^1_{y:\overline{n}\rvert}$	$1000P^1_{xy:\overline{n}\rvert}$	$1000P^1_{\overline{xy}:\overline{n}\rvert}$
1	1.1038	1.3208	2.5313	0.0015
2	1.1129	1.3528	2.5182	0.0031
3	1.1219	1.3842	2.5408	0.0047
4	1.1371	1.4170	2.5805	0.0065
5	1.1579	1.4501	2.6293	0.0084
6	1.1812	1.4828	2.6819	0.0104
7	1.2056	1.5172	2.7382	0.0126
8	1.2304	1.5524	2.7963	0.0149
9	1.2562	1.5870	2.8549	0.0174
10	1.2824	1.6216	2.9141	0.0200

Table 9.10 Premiums in term insurance: Joint life and last survivor status

n	$1000P_{x:\overline{n}\rvert}$	$1000P_{y:\overline{n}\rvert}$	$1000P_{xy:\overline{n}\rvert}$	$1000P_{\overline{xy}:\overline{n}\rvert}$
1	943.40	943.40	988.12	902.35
2	458.25	458.31	469.91	447.15
3	296.73	296.81	302.36	291.35
4	216.11	216.21	219.69	212.72
5	167.86	167.98	170.50	165.39
6	135.79	135.92	137.92	133.82
7	112.96	113.10	114.80	111.30
8	95.92	96.07	97.57	94.44
9	82.72	82.88	84.26	81.38
10	72.23	72.40	73.68	70.97

Table 9.11 Premiums in endowment insurance: Joint life and last survivor status

The annual premiums for benefit of 1000 when paid at the end of the year of the termination of the status, in whole life insurance the for joint life status and for the last survivor status are given by,

$$1000P_x = 4.52, \quad 1000P_y = 5.77, \quad 1000P_{xy} = 8.15 \ \ \& \ \ 1000P_{\overline{xy}} = 2.39$$

Under the assumption of uniformity in unit age interval, we find the actuarial present values of benefit of 1000 in whole life insurance, when paid at the termination of the status. These are as given below.

$$1000\overline{A}_x = 76.08, \quad 1000\overline{A}_y = 95.29, \quad 1000\overline{A}_{xy} = 129.60 \ \ \& \ \ 1000\overline{A}_{\overline{xy}} = 41.77$$

slightly higher than the corresponding monetary functions when the benefit is paid at the end of the year of the termination of the status. The annual premiums for benefit of 1000 in whole life insurance for the joint life status and for the last survivor status, when the benefit paid at the termination of the status are given by,

$$1000P_x = 4.80, \quad 1000P_y = 6.14, \quad 1000P_{xy} = 8.68 \ \ \& \ \ 1000P_{\overline{xy}} = 2.54$$

Again it is to be noted that the premium in the joint life status is higher than that in last survivor status. ∎

The following are some more examples to illustrate the concepts discussed in this and previous sections.

Example 9.3.8. Z is a present-value random variable for a discrete whole life insurance of 1 issued to (x) and (y) which pays 1 at the first death and 1 at the second death. It is given that (i) $a_x = 9$, (ii) $a_y = 13$, (iii) $i = 0.04$. Calculate $E(Z)$.

Solution: From the given information we have,

$$\begin{aligned} E(Z) &= A_{xy} + A_{\overline{xy}} = A_x + A_y \\ &= 1 - d\ddot{a}_x + 1 - d\ddot{a}_y = 1 - d(a_x + 1) + 1 - d(a_y + 1) \\ &= 2 - (.04/1.04)(9 + 1 + 13 + 1) = 1.08 \end{aligned}$$

∎

Example 9.3.9. Y is a present-value random variable for an whole life annuity-due of 1 issued to the group of lives (x) and (y). For the first 15 years, a payment is made if at least one of (x) and (y) is alive. Thereafter, a payment is made only if exactly one of (x) and (y) is alive. It is given that (i) $\ddot{a}_{xy} = 7.6$,
(ii) $\ddot{a}_x = 9.8$, (iii) $\ddot{a}_y = 11.6$ and (iv) $_{15|}\ddot{a}_{xy} = 3.7$. Calculate $E(Y)$.

Solution: For the first 15 years, this is a 15 year temporary last-survivor annuity-due with the actuarial present value,

$$\ddot{a}_{\overline{xy}:\overline{15|}} = \ddot{a}_{x:\overline{15|}} + \ddot{a}_{y:\overline{15|}} - \ddot{a}_{xy:\overline{15|}}$$

After 15 years, payment is made if exactly one is alive. The deferred annuity value is,

$$1(\,_{15|}\ddot{a}_x - \,_{15|}\ddot{a}_{xy}) + 1(\,_{15|}\ddot{a}_y - \,_{15|}\ddot{a}_{xy}) = \,_{15|}\ddot{a}_x - \,_{15|}\ddot{a}_{xy} + \,_{15|}\ddot{a}_y - \,_{15|}\ddot{a}_{xy}$$

The actuarial present value of the total annuity a is then given by,

$$\begin{aligned} a &= \ddot{a}_{x:\overline{15|}} + \ddot{a}_{y:\overline{15|}} - \ddot{a}_{xy:\overline{15|}} + \,_{15|}\ddot{a}_x - \,_{15|}\ddot{a}_{xy} + \,_{15|}\ddot{a}_y - \,_{15|}\ddot{a}_{xy} \\ &= \ddot{a}_x + \ddot{a}_y - \ddot{a}_{xy} - \,_{15|}\ddot{a}_{xy} = 10.1 \end{aligned}$$

∎

Example 9.3.10. Suppose the mortality for each life in a group follows Gompertz' law, given by, $\mu_x = BC^x$ and $\mu_y = BC^y$. Show that there exists u such that $_tp_u = \,_tp_{xy}$.

Solution: For the joint life status, under the assumption of independence,

$$\mu_{xy}(t) = \mu_{x+t} + \mu_{y+t} = BC^{x+t} + BC^{y+t} = BC^t(C^x + C^y)$$

Suppose u is such that for $s \geq 0$,

$$\mu_{xy}(s) = \mu_{u+s} \iff BC^{x+s} + BC^{y+s} = BC^{u+s} \iff C^x + C^y = C^u$$

For each specified x and y, it is possible to find $u = \log_c(C^x + C^y)$. For u obtained in this way,

$$_tp_u = \exp\left\{-\int_0^t \mu_{u+s}ds\right\} = \exp\left\{-\int_0^t \mu_{xy}(s)ds\right\} = {}_tp_{xy}$$

Note that to find $\mu_{xy}(t)$, for fixed B and C, we need to prepare a bivariate table for each fixed x and y for t varying over $(0, \infty)$. However, for u obtained in this way, all probabilities, expected values, variances for the joint life status (xy) equal those for the single life u. Thus, a two-dimensional array is no more required and one-dimensional array is enough to obtain survival probabilities. However, in most of the cases u will be non-integral and therefore, the determination of its value will require interpolation in the single array. ∎

Example 9.3.11. Suppose each life follows Makeham's mortality law with the force of mortality $\mu_{x+s} = A + BC^{x+s}$ and $\mu_{y+s} = A + BC^{y+s}$. Show that there exists u such that $({}_tp_u)^2 = {}_tp_{xy}$.

Solution: For the joint life status, under the assumption of independence, for Makeham's law,

$$\mu_{xy}(s) = \mu_{x+s} + \mu_{y+s} = 2A + BC^{x+s} + BC^{y+s}$$

Due to the term $2A$, as in Gompertz' law, it is not possible to find a single u corresponding to joint life status xy so that $\mu_{xy}(s) = \mu_{u+s}$. Suppose there exists u such that for joint life status uu, $\mu_{uu}(s) = \mu_{xy}(s)$. Thus,

$$
\begin{aligned}
\mu_{xy}(s) &= \mu_{uu}(s) \\
&\Longleftrightarrow \quad 2A + BC^s\, C^x + BC^y\, C^s = 2A + 2BC^s\, C^u \\
&\Longleftrightarrow \quad C^x + C^y = 2C^u \quad \Longleftrightarrow \quad u = \log_c((C^x + C^y)/2)
\end{aligned}
$$

Thus, for specified x and y, u can be obtained such that $\mu_{xy}(s) = \mu_{uu}(s) = 2\mu_{u+s}$. Consequently, the survival function is given by,

$$
\begin{aligned}
_tp_u &= \exp\left(-\int_0^t \mu_{u+s}ds\right) = \exp\left(-\frac{1}{2}\int_0^t \mu_{xy}(s)ds\right) \\
&= \left(\exp - \int_0^t \mu_{xy}(s)ds\right)^{1/2} = {}_tp_{xy}^{1/2} \Rightarrow ({}_tp_u)^2 = {}_tp_{xy}
\end{aligned}
$$

∎

Example 9.3.12. For independent lives (40) and (60), the following data are available. (i) $a_{40:60} = a_{61.6}$ for a Gompertz' law with $\log_{10} C = 0.04$. (ii) $a_{40:60} = a_{xx}$ for Makeham's law with the same value of C. (iii) $\log_{10} 2 = 0.3$. Calculate x.

Solution: Under Gompertz' law, $C^{40} + C^{60} = C^{61.6}$ and under Makeham's law, $C^{40} + C^{60} = 2C^x$. It then follows that,

$$
\begin{aligned}
2C^x &= C^{61.6} \Rightarrow \log_{10} 2 + x\log_{10} C = 61.6\log_{10} C \\
\Rightarrow \quad 0.3 + 0.04x &= (61.6)(0.04) \quad \Rightarrow \quad x = 54
\end{aligned}
$$

∎

Example 9.3.13. For a joint life status (xy) for a pair of independent lives, the following information is given. (i) If mortality follows Gompertz' law with $C = 2^{0.2}$, then $\ddot{a}_{xy} = \ddot{a}_{56}$. (ii) If mortality follows Makeham's law with $C = 2^{0.2}$, then $\ddot{a}_{xy} = \ddot{a}_{zz}$. Find z.

Solution: By Gompertz' law, $2^{0.2x} + 2^{0.2y} = 2^{0.2(56)}$. By Makeham's law, $2^{0.2x} + 2^{0.2y} = 2(2^{0.2})^z$. Hence,

$$0.2z + 1 = 11.2 \quad \Rightarrow \quad z = 51$$

$\blacksquare$

Example 9.3.14. Derive the probability density function of $T(xy)$ and $T(\overline{xy})$ under the assumption of uniform distribution of deaths in each year of age for each individual in the joint life status.

Solution: Under the assumption of uniform distribution of deaths for each year of age, we know that,

$$_tp_x = 1 - tq_x \quad \text{and} \quad {}_tp_x\mu_{x+t} = \frac{d}{dt}(1 - {}_tp_x) = \frac{d}{dt}(tq_x) = q_x$$

For joint life status (xy), with independent $T(x)$ and $T(y)$, we have for $t \in [0, 1]$.

$$\begin{aligned}
_tp_{xy}\mu_{xy}(t) &= {}_tp_x\,{}_tp_y(\mu_{x+t} + \mu_{y+t}) = {}_tp_y({}_tp_x\mu_{x+t}) + {}_tp_x({}_tp_y\mu_{y+t}) \\
&= (1 - tq_y)q_x + (1 - tq_x)q_y = q_x + q_y - 2tq_xq_y \\
&= q_x + q_y - q_xq_y + (1 - 2t)q_xq_y = q_{xy} + (1 - 2t)q_xq_y
\end{aligned}$$

With the distributions of $T(x), T(y)$ and $T(xy)$, the distribution of $T(\overline{xy})$ can be obtained easily from the relation given in Eq. 9.3.5. $\blacksquare$

In sections 2 and 3 we have noted how the entire theory developed in Chapters $4, 5, 6$ and 7 can be easily generalized to obtain the premiums for group insurance and group annuities. Another direction of generalization of these chapters is in multiple decrement models. The theory developed in all the previous chapters is subject to a single contingency of death. In many situations, a single life or a group, that is, multiple lives, are subject to multiple contingencies. For example, a death of an individual may be due to cardiovascular disease or neoplasmic disease or accident or other causes. In these cases, the benefit structure and consequently the premium structure may depend on the cause of the death. Many insurance products provide payment of some benefit even if premiums stop before the end of the specified premium payment term. In such situations, time until termination of premium payment and the cause of termination are the two random variables of interest to decide the benefit to be paid.

Similar situations arise in the industrial field also. For example, failure of a metal strip under test conditions can occur in a number of different ways—by cracking, bucking, shearing and so on. Thus, the time to failure and cause of failure are two variables of interest. Here we use the terminology of mortality and survival function, because this is our primary interest. In actuarial science, the termination from a given status is known as decrement. Models which take into account various causes of decrement are known as multiple decrement models in the insurance business. In Biostatistics, related theory is known as the theory of competing risks.

The theory developed so far can be extended to a general theory of multiple decrement models involving the effect of several causes of decrement on a group of individuals. Multiple decrement models consider a large number of lives subject to several causes of decrement. The multiple decrement table is very similar to the life-table discussed in Chapter 4, except that the l_x column is reduced by several d_x's rather than one d_x. These several d_x columns correspond to several causes of decrement. The life table studied in Chapter 4, is labelled as a single decrement table. The basic underlying random variable in a single decrement table is the future life time random variable. In the theory of multiple decrement models, one more random variable $J(x)$, specifying the cause of termination comes in a picture and it becomes necessary to study the joint distribution of $J(x)$ and the random variable $T(x)$. We will not discuss all these details in this book. Interested readers may refer to Chapter 10 in Bowers et al [1] or Deshmukh [4].

9.4   R Codes

Computations of monetary functions in multiple life contracts is parallel to those for single life contracts. We have discussed two types of status in this chapter when the contract involves more than a single life. Thus in the computations of 'A' or 'a' functions we have to use the probability density function or probability mass function of the time to failure of the underlying status. We make use of the relations between single life functions and multiple life functions, as stated in Eq. 9.3.9, to compute the monetary functions corresponding to the last survivor status when functions corresponding to single life functions and joint life status are computed. Once these monetary functions are computed, premiums and reserves can be computed easily.

Code 9.4.1. In Example 9.3.5, we have obtained net single and annual premiums for fully continuous n-year term and endowment insurance, also for whole life insurance, in both the joint life and the last survivor status. The mortality law for both $x = 25$ and $y = 30$ is Gompertz' law, with the same parameters. We use this code to compute the required values in this example. It consists of five parts. The first part specifies the parameters of the morality law and interest parameters. The second part computes the net single premiums for term and endowment insurance for the setup of singe life. The next part computes these for multiple lives. Part IV is concerned with the computation of annual premiums in term and endowment insurance while the last part deals with the whole life insurance. The values are reported in Table 9.1, Table 9.2 and Table 9.3.

```
# Part I: Parameters of Gompertz' law and interest pattern
B=0.0001151; C=1.096; m=B/log(C); del=0.05; v=exp(-del)
la=(-del/log(C))+1; x=c(25,30); alx=m*C^x; n=1:10
# Part II: Single life functions, term and endowment insurance
T=P=S=matrix(nrow=length(n),ncol=length(x))
for(i in 1:length(n))
{
```

```
for(j in 1:length(x))
{
S[i,j]=pgamma(C^n[i],la,alx[j])-pgamma(1,la,alx[j])
T[i,j]=exp(alx[j])*gamma(la)*alx[j]^(1-la)*S[i,j]
P[i,j]=exp(-del*n[i]-alx[j]*(C^n[i]-1))
}
}
E=T+P # Endowment insurance
# Part III: Two life functions, term and endowment insurance
alJ=m*(C^x[1]+C^x[2])
TJ=PJ=SJ=c()
for(i in 1:length(n))
{
SJ[i]=pgamma(C^n[i],la,alJ)-pgamma(1,la,alJ)
TJ[i]=exp(alJ)*gamma(la)*alJ^(1-la)*SJ[i]
PJ[i]=exp(-del*n[i]-alJ*(C^n[i]-1))
}
EJ=TJ+PJ; TL=T[,1]+T[,2]-TJ; TL
EL=E[,1]+E[,2]-EJ; EL
Llimit=(P[,2]*T[,1]+P[,1]*T[,2])*exp(del*n); Llimit
d=round(1000*data.frame(T[,1],T[,2],Llimit,TJ,T[,1]+T[,2],TL),2)
d1=data.frame(n,d); d1 # Table 9.1
En=round(1000*data.frame(E,EJ,EL),2)
d2=data.frame(n,En); d2 # Table 9.2
# Part IV: Premiums for term and endowment insurance, paid
# as n-year temporary continuous life annuity
abarJ=(1-EJ)/del; abarL=(1-EL)/del
PrTJ=TJ/abarJ; PrTL=TL/abarL; PrEJ=EJ/abarJ; PrEL=EL/abarL
d3=round(1000*data.frame(PrTJ,PrTL,PrEJ,PrEL),2)
d4=data.frame(n,d3); d4 # Table 9.3
# Part V: Whole life insurance, premiums as whole life annuity
w=c()
for(j in 1:length(x))
{
w[j]=exp(alx[j])*gamma(la)*alx[j]^(1-la)*(1-pgamma(1,la,alx[j]))
}
wJ=exp(alJ)*gamma(la)*alJ^(1-la)*(1-pgamma(1,la,alJ)); wJ
wL=w[1]+w[2]-wJ; wL
wh=round(1000*c(w,wJ,wL),2); wh
```

```
abarwJ=(1-wJ)/del; abarwL=(1-wL)/del
PbarWJ=round(1000*wJ/abarwJ,2); PbarWJ
PbarWL=round(1000*wL/abarwL,2); PbarWL
```

∎

Code 9.4.2. In Example 9.3.6, we have obtained net single and annual premiums for fully discrete n-year term and endowment insurance, also for whole life insurance, in both the joint life and the last survivor status. The mortality law for both $x = 25$ and $y = 30$ is Gompertz' law, with the same parameters. We use this code to compute the required actuarial present values and premiums in this example. Table 9.4 displays the probability distribution of $K(xy)$. Values of $1000A^1_{x:\overline{n}|}$, $1000A^1_{y:\overline{n}|}$, $1000A^1_{xy:\overline{n}|}$ and $1000A^1_{\overline{xy}:\overline{n}|}$ are presented in Table 9.5. Values of $1000A_{x:\overline{n}|}$, $1000A_{y:\overline{n}|}$, $1000A_{xy:\overline{n}|}$ and $1000A_{\overline{xy}:\overline{n}|}$ are presented in Table 9.6. Values of premiums are reported in Table 9.7.

```
# Part I: Gompertz' Mortality law and interest rate
B=0.0001151; C=1.096; m=B/log(C); del=0.05; v=exp(-del)
d=1-v; int=1/v-1; x=c(25,30); k=0:75
y=t=matrix(nrow=length(k),ncol=length(x))# pmf of K(25),k(30)
for(i in 0:length(k))
{
for(j in 1:length(x))
{
y[i,j]=exp(-m*C^x[j]*(C^k[i]-1))-exp(-m*C^x[j]*(C^(k[i]+1)-1))
}
}
apply(y,2,sum)
b=v^(k+1);w=c()
for(j in 1:length(x))
{
w[j]=sum(y[,j]*b)
t[,j]=cumsum(y[,j]*b)
}
n=1:10; T=t[1:length(n),] # Term insurance
P=matrix(nrow=length(n),ncol=length(x))
for(i in 1:length(n))
{
for(j in 1:length(x))
{
P[i,j]=v^(n[i])*exp(-m*C^x[j]*(C^n[i]-1))
}
}
```

```
E=T+P; # Endowment insurance
# Part III: Multiple lives, pmf of K(xy)
pxy=c(); alxy=m*(C^x[1]+C^x[2])
for(i in 0:length(k))
{
pxy[i]=exp(alxy)*(exp(-alxy*C^k[i])-exp(-alxy*C^(k[i]+1)))
}
sum(pxy)
exy=sum(k*pxy); exy
d1=round(data.frame(k,pxy),5);d1 # Table 9.4
plot(k,pxy,"h",xlab="k",ylab="P[K(25:30)=k]",
col="dark blue",lty=1)
points(k,pxy,pch=16,cex=.5,col=" blue")
# Part IV: APV, whole life,term and endowment
# in joint life status
wJ=sum(pxy*b); wL=w[1]+w[2]-wJ
Wh=round(1000*c(w[1],w[2],wJ,wL),2);Wh
tJ=cumsum(pxy*b)
n=1:10; TJ=tJ[1:length(n)] # Term insurance
TL=T[,1]+T[,2]-TJ
d2=round(1000*data.frame(T,TJ,TL),2)
d3=data.frame(n,d2); d3 # Table 9.5
PuJ=v^n*exp(-alxy*(C^n-1)); EJ=TJ+PuJ
EL=E[,1]+E[,2]-EJ
d4=round(1000*data.frame(E,EJ,EL),2)
d5=data.frame(n,d4); d5 # Table 9.6
# Part V: Premiums, whole life,term and endowment
# in joint life status
PrWJ=round(1000*d*wJ/(1-wJ),2); PrWJ
PrWL=round(1000*d*wL/(1-wL),2); PrWL
PrTJ=round(1000*d*TJ/(1-EJ),2)
PrTL=round(1000*d*TL/(1-EL),4)
PrEJ=round(1000*d*EJ/(1-EJ),2)
PrEL=round(1000*d*EL/(1-EL),2)
d6=data.frame(n,PrTJ,PrTL,PrEJ,PrEL); d6 # Table 9.7
```

■

Code 9.4.3. In Example 9.3.7, the mortality law of both $x = (25)$ and $y = (30)$ is specified by values of q_x as given in Table 4.8. We use this code to compute the actuarial present values of the benefit of 1000 and the fully discrete premiums in the term insurance, the endowment insurance and whole life insurance issued to a group of two individuals (25) and (30), in both the status. Values of $1000A^1_{x:\overline{n}|}$, $1000A^1_{y:\overline{n}|}$, $1000A^1_{xy:\overline{n}|}$ and $1000A^1_{\overline{xy}:\overline{n}|}$ are presented in Table 9.8. Values of $1000A_{x:\overline{n}|}$, $1000A_{y:\overline{n}|}$, $1000A_{xy:\overline{n}|}$ and $1000A_{\overline{xy}:\overline{n}|}$ are presented in Table 9.9. Premiums for term insurance and for endowment insurance under both the status are reported in Table 9.10 and Table 9.11 respectively.

```r
z=read.table("F://qx.txt", header=T)
x=z[,1]; q=z[,2]; p=1-q; w=length(p); w; age=c(25,30)
# Part I: Pmf of K(25)
p25=p[(age[1]+1):w]
p125=c(p25[1],2:(w-age[1]-1))
for (i in 2:(w-age[1]-1))
{
p125[i]=p125[i-1]*p25[i]
}
q25=1-p25
k25=c(1,p125)*q25; sum(k25) # pmf of K(25)
length(k25)
#Part II: Term,endowment and whole life insurance
# APVs of benefits
int=0.06; v=(1 + int)^(-1); del=log((1+int));d=1-v
k=0:85; n=1:10; b=v^(k+1); v1=v^n
w25=cumsum(k25*b); t25=w25[1:10] # Term insurance
tbar25=(int/del)*t25; pu25=p125[1:10]*v1
e25=t25+pu25; ebar25=tbar25+pu25
wh25=sum(k25*b);whbar25=(int/del)*wh25
# Part III: APVs of temporary annuities
abar25=(1-ebar25)/del; ad25=(1-e25)/d; wa25=(1-wh25)/d
wabar25=(1-whbar25)/del
# Part IV: Annual premiums for n-year term and endowment insurance
Pt25=t25/ad25; cPt=tbar25/abar25; PEn25=e25/ad25
cPEn=ebar25/abar25; Pwh25=wh25/wa25
Pwhbar25=whbar25/wabar25
# Part IV: Pmf of K(30)
p30=p[(age[2]+1):w]
p130=c(p30[1],2:(w-age[2]-1))
for (i in 2:(w-age[2]-1))
{
p130[i]=p130[i-1]*p30[i]
}
```

```
q30=1-p30
k30=c(1,p130)*q30; sum(k30) # pmf of K(30)
length(k30)
# Part V: Term,endowment and whole life insurance
# APVs of benefits
k=0:80; b=v^(k+1); v1=v^n
w30=cumsum(k30*b); t30=w30[1:10] # Term insurance
tbar30=(int/del)*t30; pu30=p130[1:10]*v1
e30=t30+pu30; ebar30=tbar30+pu30
wh30=sum(k30*b);whbar30=(int/del)*wh30
# Part VI: APVs of temporary annuities
abar30=(1-ebar30)/del; ad30=(1-e30)/d; wa30=(1-wh30)/d
wabar30=(1-whbar30)/del
# Part VII: Annual premiums for n-year term and endowment insurance
Pt30=t30/ad30; PEn30=e30/ad30; Pwh30=wh30/wa30
Pwhbar30=whbar30/wabar30
# Part VIII: Pmf of K(25:30)
q30J<-q30[1:80]; p1<-c(1,p125[1:79])
p2<-c(1,p130[1:79]); q25J<-q25[1:80]
kxy<-p1*p2*(q25J+q30J-q25J*q30J); sum(kxy) # pmf of K(25:30)
length(kxy)
# Part IX: Term,endowment and whole life insurance
# APVs of benefits
k<-0:79; n=1:10; b=v^(k+1); v1=v^n
wJ=cumsum(kxy*b); tJ=wJ[1:10] # Term insurance
tbarJ=(int/del)*tJ; pJ=p1*p2; puJ=pJ[1:10]*v1
eJ=tJ+puJ; ebarJ=tbarJ+puJ
whJ=sum(kxy*b);whbarJ=(int/del)*whJ
whJ; whbarJ
# Part X: APVs of temporary annuities
abarJ=(1-ebarJ)/del; adJ=(1-eJ)/d; waJ=(1-whJ)/d
wabarJ=(1-whbarJ)/del
# Part XI: Annual premiums for n-year term and endowment insurance
PtJ=tJ/adJ; PEnJ=eJ/adJ; PwhJ=whJ/waJ; PwhbarJ=whbarJ/wabarJ
PwhJ
# Part XII: Last survivor status
tL=t25+t30-tJ; tbarL=tbar25+tbar30-tbarJ
d1=round(1000*data.frame(t25,t30,tJ,tL),2)
d2=data.frame(n,d1);d2 # Table 9.8
```

```
eL=e25+e30-eJ; ebarL=ebar25+ebar30-ebarJ
d3=round(1000*data.frame(e25,e30,eJ,eL),2)
d4=data.frame(n,d3);d4 # Table 9.9
whL=wh25+wh30-whJ
whbarL=whbar25+whbar30-whbarJ
# Part XIII: APVs of temporary annuities
abarL=(1-ebarL)/del; adL=(1-eL)/d; waL=(1-whL)/d
wabarL=(1-whbarL)/del
# Part XIV: Annual premiums for n-year term and endowment insurance
PtJ=tJ/adJ; PEnJ=eJ/adJ; PwhJ=whJ/waJ; PwhbarJ=whbarJ/wabarJ
PtL=tL/adL; PEnL=eL/adL; PwhL=whL/waL; PwhbarL=whbarL/wabarL
d5=round(1000*data.frame(Pt25,Pt30,PtJ,PtL),4)
d6=data.frame(n,d5);d6 # Table 9.10
d7=round(1000*data.frame(PEn25,PEn30,PEnJ,PEnL),2)
d8=data.frame(n,d7);d8 # Table 9.11
wh=round(1000*c(wh25,wh30,whJ,whL),2);wh
whbar=round(1000*c(whbar25,whbar30,whbarJ,whbarL),2); whbar
Pwh=round(1000*c(Pwh25,Pwh30,PwhJ,PwhL),2); Pwh
Pwhbar=round(1000*c(Pwhbar25,Pwhbar30,PwhbarJ,PwhbarL),2); Pwhbar
```

■

9.5   Conceptual Exercises

9.5.1 It is given that $q_{25} = 0.145, q_{26} = 0.151, q_{27} = 0.156, q_{28} = 0.174, q_{29} = 0.185, q_{30} = 0.194$. Find (a) $_3q_{25:27}$, (b) $_3q_{\overline{25:27}}$, (c) $_2p_{26:28}$, (d) $_2q_{\overline{26:28}}$ and (e) $_{2|}q_{\overline{25:27}}$.

9.5.2 Suppose the mortality pattern of (25) and (27) is governed by a law for which q_x values for ages 25 to 30 are as given in Exercise 9.5.1. Suppose $i = 0.06$. Find the actuarial present value of 3-year temporary annuity due and immediate, at the rate of 10000 per annum, issued to a group of two individuals (25) and (27), when the annuity payments cease (a) at the first death and (b) at the second death.

9.5.3 For the set up of Exercise 9.5.2, (a) find the actuarial present value of benefit of 10000, payable at the end of year of first death, in 3-year endowment insurance and 3-year term insurance. (b) Find the actuarial present value of benefit of 10000, payable at the end of year of the second death. (c) Find the annual premium for a benefit of 10000, under both joint life status and last survivor status, for a 3-year endowment insurance and 3-year term insurance, payable as 3-year temporary annuity due.

9.5.4 Suppose the curtate future lifetimes $K(x)$ and $K(y)$ are independent and each has the distribution defined by Makeham's law with the force of mortality, $\mu_x = A + BC^x$ and

$\mu_y = A + BC^y$ with $A = 0.0007$, $B = 0.0001151$ and $C = 1.096$. It is given that $\delta = 0.05$, $x = 35$ and $y = 40$. (a) Determine the probability mass function of $K(xy)$. Calculate e_{xy} and $e_{\overline{xy}}$. (b) Calculate $1000A_{xy}$ and $1000A_{\overline{xy}}$. (c) Calculate the annual premium, payable as whole life annuity due, payable till the first death, for the benefit of 1000, payable at the end of the year of first death in the whole life insurance issued to (x, y). (d) Calculate the annual premium, payable as whole life annuity due, payable till the last death, for the benefit of 1000, payable at the end of year of last death in the whole life insurance issued to (x, y). (e) Calculate $1000A\,{}^{1}_{xy:\overline{n}|}$, $1000A\,{}_{xy:\overline{n}|}^{\ 1}$ and $1000A_{xy:\overline{n}|}$ for $n = 1$ to 10, for joint life status and similar functions for the last survivor status. (f) Calculate premiums for the insurance products specified in (e).

9.5.5 Two independent lives, both of age x, are subject to the same mortality table. Calculate the maximum possible value of ${}_tp_{\overline{xx}} - {}_tp_x$.

9.5.6 It is given that ${}_tp_x = 1 - t^2 q_x$, ${}_tp_y = 1 - t^2 q_y$, $0 \le t \le 1, q_x = 0.08$, $q_y = .004$ and $T(x)$ and $T(y)$ are independent. Derive an expression for ${}_tp_{xy}\mu_{xy}(t)$ in terms of t.

9.5.7 For a fully continuous whole life insurance issued to (x) and (y), it is given that (i) the death benefit of 100 is payable at the second death, (ii) premiums are payable until the first death, (iii) $\delta = 0.05$, (iv) for $t > 0$ ${}_tp_x = \exp(-0.01t)$ and ${}_tp_y = \exp(-0.02t)$, ${}_tp_{xy} = 0.25\exp(-0.01t) + 0.75\exp(-0.03t)$, (iii) the future lifetimes of (x) and (y) are dependent. Calculate the annual benefit premium rate for this insurance.

9.5.8 In a fully continuous insurance, 1 unit benefit is paid when the last survivor status of (x) and (y) terminates. Calculate the annual benefit premium payable until the first of (x) and (y) dies, given the following information: (i) Future life times of (x) and (y) are independent, (ii) $\mu_{x+t} = 0.09$, $\mu_{y+t} = 0.06$, $t > 0$ and (iii) $\delta = 0.05$.

9.5.9 Find $\overset{0}{e}_{\overline{50:55}}$ when $l_x = 100(100 - x)$, $0 \le x \le 100$ and when future lifetimes of (50) and (55) are independent.

9.5.10 Prove that $\ddot{a}_{xy} = 1 + p_{xy}\,\ddot{a}_{(x+1):(y+1)}$. Hence show that
$A_{xy} = vq_{xy} + vp_{xy}A_{(x+1):(y+1)}$.

9.6   Computational Exercises

Note: For the following two questions, use the mortality pattern and interest pattern that you have adopted in the computational exercise 5.8.2.

9.6.1 Suppose the future lifetimes $T(x)$ and $T(y)$ are independent and each has the distribution defined by Makeham's law, where $x = 30$ and $y = 35$.
(i) Compute $\overline{A}\,{}^{1}_{xy:\overline{n}|}$ and $\overline{A}\,{}_{xy:\overline{n}|}^{\ 1}$ for $n = 1$ to 10. Verify the result proved in Example 9.2.1. (ii) Compute $\overline{A}_{xy:\overline{n}|}$ and $\overline{A}_{\overline{xy}:\overline{n}|}$ for $n = 1$ to 10.
(iv) Compute the annual premium, payable as n-year temporary continuous life annuity, for the benefit of 1000, payable at the moment of first death in n-year term insurance

and in n-year endowment insurance, for $n = 1$ to 10. Compute the annual premium, payable as n-year temporary continuous life annuity, for the benefit of 1000, payable at the moment of last death in n-year term insurance and in n-year endowment insurance, for $n = 1$ to 10. (v) Compute $\overline{A}_{xy}$ and $\overline{A}_{\overline{xy}}$. Compute the annual premium, payable as whole life continuous annuity, for the benefit of 1000, payable at the moment of first death in whole life insurance issued to (x, y). Compute the annual premium, payable as whole life continuous annuity, for the benefit of 1000, payable at the moment of last death in whole life insurance issued to (x, y).

9.6.2 Suppose the curtate future lifetimes $K(x)$ and $K(y)$ are independent and each has the distribution defined by Makeham's law, where $x = 30$ and $y = 35$. (i) Find the probability mass function of $K(xy)$. Compute e_{xy}. (ii) Compute $1000A\,{}^{1}_{xy:\overline{n}|}$ and $1000A\,{}^{1}_{\overline{xy}:\overline{n}|}$ for $n = 1$ to 10. Compute $1000A_{xy:\overline{n}|}$ and $1000A_{\overline{xy}:\overline{n}|}$ for $n = 1$ to 10. (iii) Compute premiums for insurance products specified in (ii). (iv) Compute $1000A_{xy}$ and $1000A_{\overline{xy}}$. Compute the annual premium, payable as whole life annuity due, payable till the first death, for the benefit of 1000, payable at the end of year of first death in whole life insurance issued to (x, y). Compute the annual premium, payable as whole life annuity due, payable till the last death, for the benefit of 1000, payable at the end of year of last death in whole life insurance issued to (x, y).

9.6.3 Suppose the mortality law of both (30) and (35) is specified by values of q_x as specified in Table 4.21. Compute the actuarial present values of the benefit of 1000 and the fully discrete premiums in the term insurance, the endowment insurance and whole life insurance issued to a group of two individuals (25) and (30), in both the status. Find fully continuous premiums under the assumption of uniformity in unit age interval for the whole life insurance in both the status. Suppose the effective rate of interest is $i = 0.05$. Compute all these monetary functions when the mortality law of both (30) and (35) is specified by values of q_x as specified in Table 4.22. Compare the premiums.

9.7   Multiple Choice Questions

Note: Unless specified otherwise, you have to identify which of the options is correct. Answers are given in the solutions of conceptual exercises.

9.7.1 Following are three statements.
   (I) $P[T(70:75) \leq 1] = P[K(70:75) = 1]$.
   (II) $P[T(70:75) \leq 1] = P[K(70:75) = 0]$.
   (III) $P[T(70:75) \leq 1] = P[K(70:75) = 0] + P[K(70:75) = 1]$.

   (a) (I) is true
   (b) (II) is true
   (c) (III) is true
   (d) All are false

9.7.2 Suppose $q_{70} = 0.14, q_{71} = 0.15, q_{75} = 0.18$, and $q_{76} = 0.19$. Then $P[T(70:75) \leq 2]$ is

(a) 0.2948
(b) 0.9993
(c) 0.5145
(d) 0.0007

9.7.3 Suppose $q_{70} = 0.14, q_{71} = 0.15, q_{75} = 0.18$, and $q_{76} = 0.19$. Then $P[K(70:75) \leq 1]$ is

(a) 0.2948
(b) 0.9993
(c) 0.5145
(d) 0.0007

9.7.4 Suppose the future life times of (70) and (75) are independent with $p_{70} = 0.87, p_{71} = 0.86, p_{72} = 0.85, p_{75} = 0.81, p_{76} = 0.80, p_{77} = 0.79$. The probability that the joint life status terminates in the second year is

(a) 0.3764
(b) 0.5037
(c) 0.2199
(d) 0.5152

9.7.5 Suppose the future life times of (70) and (75) are independent with $p_{70} = 0.87, p_{71} = 0.86, p_{72} = 0.85, p_{75} = 0.81, p_{76} = 0.80, p_{77} = 0.79$. The probability that the last survivor status terminates in the second year is

(a) 0.5744
(b) 0.0639
(c) 0.8432
(d) 0.2199

9.7.6 The distribution function of time $T(xyz)$ to failure of a joint life status of a group of three individuals of ages x, y and z, under the assumption of independence of $T(x), T(y)$ and $T(z)$ is given by

(a) $_tq_x \, _tq_y \, _tq_z$
(b) $_tp_x \, _tp_y \, _tp_z$
(c) $1 - \, _tq_x \, _tq_y \, _tq_z$
(d) $1 - \, _tp_x \, _tp_y \, _tp_z$

9.7.7 The survival function of time $T(xyz)$ to failure of a joint life status of a group of three individuals of ages x, y and z, under the assumption of independence of $T(x), T(y)$ and $T(z)$ is given by

(a) $_tq_x \, _tq_y \, _tq_z$
(b) $_tp_x \, _tp_y \, _tp_z$
(c) $1 - \, _tq_x \, _tq_y \, _tq_z$
(d) $1 - \, _tp_x \, _tp_y \, _tp_z$

9.7.8 The distribution function of time $T(\overline{xyz})$ to failure of a last survivor status of a group of three individuals of ages x, y and z, under the assumption of independence of $T(x), T(y)$ and $T(z)$ is given by

(a) $_tq_x \; _tq_y \; _tq_z$
(b) $_tp_x \; _tp_y \; _tp_z$
(c) $1 - \; _tq_x \; _tq_y \; _tq_z$
(d) $1 - \; _tp_x \; _tp_y \; _tp_z$

9.7.9 The survival function of time $T(\overline{xyz})$ to failure of a last survivor status of a group of three individuals of ages x, y and z, under the assumption of independence of $T(x), T(y)$ and $T(z)$ is given by

(a) $_tq_x \; _tq_y \; _tq_z$
(b) $_tp_x \; _tp_y \; _tp_z$
(c) $1 - \; _tq_x \; _tq_y \; _tq_z$
(d) $1 - \; _tp_x \; _tp_y \; _tp_z$

9.7.10 The survival function of time $T(xy)$ to failure of a joint life status of a group of two individuals of ages x and y, under the assumption of independence of $T(x)$ and $T(y)$ is given by (I) $1 - \; _tq_x \; _tq_y$. (II) $_tp_x + \; _tp_y - \; _tp_x \; _tp_y$.

(a) (I) is true but (II) is false
(b) (I) is false but (II) is true
(c) Both (I) and (II) are true
(d) Both (I) and (II) are false

9.7.11 The following are two statements. (I) The force of failure of the joint life status at time t is $\mu_{x+t} + \mu_{y+t}$. (II) The force of failure of the last survivor status at time t is $\mu_{x+t} + \mu_{y+t}$.

(a) (I) is true but (II) is false
(b) (I) is false but (II) is true
(c) Both (I) and (II) are true
(d) Both (I) and (II) are false

9.7.12 It is given that $p_{70} = 0.8$ and $p_{80} = 0.7$. Suppose pr_1 is the probability that the joint life status of (70) and (80) is not terminated for the next half year and pr_2 is the probability that the last survivor status of (70) and (80) is not terminated for next half year. Under the assumption of constant force of mortality in each unit interval $(x, x + 1)$ for integer x,

(a) $pr_1 = 0.7517, pr_2 = 0.7483$
(b) $pr_1 = 0.5600, pr_2 = 0.4400$
(c) $pr_1 = 0.7483, pr_2 = 0.9828$
(d) pr_1 and pr_2 cannot be computed in view of insufficient information

9.7.13 For $x = 25$ and $y = 27$, it is given that $1000A_{x:\overline{n}|} = 743.40$, $1000A_{y:\overline{n}|} = 745.60$ and $1000A_{xy:\overline{n}|} = 754.82$. Then $1000A_{\overline{xy}:\overline{n}|}$

(a) is 754.82
(b) is 734.18

(c) is > 743.40

(d) cannot be computed in view of insufficient information

9.7.14 The following are four statements. (I) $\bar{a}_{xy} = (1 - \overline{A}_{xy})/\delta$.
(II) $\bar{a}_{xy} = (1 - \overline{A}_{xy})/d$. (III) $\ddot{a}_{xy} = (1 - A_{xy})/\delta$. (IV) $\ddot{a}_{xy} = (1 - A_{xy})/d$.

(a) Only (I) is true

(b) Only (I) and (III) are true

(c) Only (II) and (III) are true

(d) Only (I) and (IV) are true

9.7.15 The following are three statements.   (I) $\overline{A}_{xy} < \overline{A}_x + \overline{A}_y$.   (II) $\overline{A}_{xy} > \overline{A}_x + \overline{A}_y$.
(III) $\overline{A}_{xy} + \overline{A}_{\overline{xy}} = \overline{A}_x + \overline{A}_y$.

(a) Only (I) is true

(b) Only (II) is true

(c) Only (I) and (III) are true

(d) Only (II) and (III) are true

9.7.16 It is given that $e^0_{30} = 58$, $e^0_{40} = 54$ and $Cov(T(30:35), T(\overline{30:35})) = 32$. Then $e^0_{\overline{30:40}}$

(a) is < 54

(b) is 50

(c) is 62

(d) cannot be computed in view of insufficient information

9.7.17 It is given that $e_{30} = 56$, $e_{40} = 54$ and $e_{30:40} = 52$. Then $e_{\overline{30:40}}$

(a) is < 52

(b) is 58

(c) is > 58

(d) cannot be computed in view of insufficient information

9.7.18 In a fully continuous insurance, Rs 3 lakh benefit is paid when the last survivor status of (30) and (35) terminates. It is given that future life times of (30) and (35) are independent with $\mu_{30+t} = \mu_{35+t} = 0.01$, $t > 0$ and $\delta = 0.05$. The annual benefit premium payable until the first of (30) and (35) dies

(a) is 3000

(b) is 1000

(c) is > 1000

(d) cannot be computed in view of insufficient information

9.7.19 It is given that future life times of (30) and (35) are independent with $\mu_{30+t} = \mu_{35+t} = 0.02$, $t > 0$. Then $e^0_{\overline{30:35}}$

(a) is 25

(b) is 50

(c) is 75

(d) cannot be computed in view of insufficient information

9.7.20 For a fully continuous whole life insurance issued on (x) and (y), it is given that (i) the death benefit of 1000 is payable at the first death, (ii) premiums are payable until the first death, (iii) $\delta = 0.05$ and $_tp_{xy} = 0.25\exp(-0.01t) + 0.75\exp(-0.02t)$. Then the annual benefit premium for this insurance with benefit of Rs 1000

(a) is 0.0172
(b) is 1.720
(c) is 17.20
(d) cannot be computed in view of insufficient information

9.7.21 Two independent lives, both of age x, are subject to the same mortality pattern. The maximum possible value of $_tp_{\overline{xx}} - {}_tp_x$

(a) is 0.25
(b) depends on x
(c) depends on t
(d) cannot be computed in view of insufficient information

9.7.22 It is given that $e^0_{30} = 58$, $e^0_{40} = 54$ and $Cov(T(30:35), T(\overline{30:35})) = 32$. Then $e^0_{30:40}$

(a) is > 54
(b) is 50
(c) is 62
(d) cannot be computed in view of insufficient information

Answers to Conceptual Exercises

Chapter 3

3.5.1 (a) $U'(w) > 0$, for all w and $U''(w) < 0$ for w in a set
$A = \{w | 4.2928 < w < 5, w > 5.7071\}$. Hence $U(w)$ is a utility function of a risk averse
individual for $w \in A$.

3.5.2 Upper limit $= 208.33$, $H = 2.31$

3.5.3 $G = 52.68$, $H = 51.29$. The insurance contract is feasible as $G > H$.

3.5.4 $E(X) = 25$, $G = 27.89$, $H = 25.25$. The insurance contract is feasible as $G > H$.

3.5.5 $E(X) = 10$, $G = 11.16$, $H = 10.10$. The insurance contract is feasible as $G > H$. In
Exercise 3.5.4, $E(X) = 25$ and in this exercise $E(X) = 10$. Hence, the values of G and
H are smaller than those in Exercise 3.5.4.

3.5.6 $E(X) = 60$, $G = 78$, $H = 60.72$. The insurance contract is feasible as $G > H$.

3.5.7 A and B.

3.5.8 G for A, B, C, D is $45.5148, 43.8175, 43.0933, 44.5588$ respectively. Since $G > 45$ for A,
only A will purchase the insurance.

3.5.9 Rs 1,85,493.75

Answers to the multiple choice questions, based on Chapter 3, are given in Table 1.

Q.No.	1	2	3	4	5	6
Ans	b	c	b	b	a	c

Table 1 Answer Key to MCQs in Chapter 3

Chapter 4

4.8.1 (a) Since $S(0) = 0$, it is not a survival function. (b) $S(x)$ is a survival function for any
value of A. $\mu_x = A/(1 + Ax)$, $f(x) = A/(1 + Ax)^2$,
$F(x) = 1 - 1/(1 + Ax)$, for $x \geq 0$.

4.8.2 For a set $A = \{n | n \leq 1\}$, μ_x is a mortality function as it is non-negative and
$\int_0^\infty \mu_x dx = \infty$. For these values of n, survival function is
$S(x) = \exp\{-1/(n-1)(1 - (1+x)^{(-n+1)})\}, x \geq 0$ and probability density function is
$f_X(x) = (1+x)^{(-n)} \exp\{-1/(n-1)(1 - (1+x)^{(-n+1)})\}, x \geq 0$.

4.8.3 $S'(1.5) > 0$ implies that there exists a neighbourhood of 1.5 where S' has the same sign, that is, in that neighbourhood $S(x)$ is an increasing function. Hence it is not a survival function.

4.8.4 (a) 0.0208   (b) 0.0052   (c) 0.0401 (d) 0.0874

4.8.5 (a) 0.9608   (b) 0.0769   (c) 0.0724

4.8.6 0.0102

4.8.7 0.77698

4.8.8 (a) $l_{65}/l_{25} = 0.8142$,   (b) $(l_{85} - l_{95})/l_{25} = 0.26097$
(c) $l_{40}/l_{30} = 0.9825$,   (d) $l_{65}/l_{40} = 0.8185$
(e) $(l_{50} - l_{60})/l_{50} = 0.07398$,   (f) $(l_{50} - l_{70})/l_{50} = 0.2275$
(g) $(l_{80} - l_{85})/l_{60} = 0.1790$,   (h) $(l_{65} - l_{70})/l_{60} = 0.09541$

4.8.9 (a) $l_{40}/l_{30} = 0.9825$,   (b) $(l_{30} - l_{50})/l_{30} = 0.04841$
(c) $d_{49}/l_{30} = 0.0043$,   (d) $(l_{40} - l_{50})/l_{30} = 0.03094$
(e) $((l_{35} - l_{45}) + (l_{70} - l_{80})/l_{30} = 0.2724$

4.8.10 (a) (i) $\exp(-\int_0^5 0.005dt) = 0.97531$, (ii) 0.8661, (b) 0.004699

4.8.11 (a) $e_x^0 = \int_0^\infty {}_x p_0 \, dx = \int_0^\infty (l_x/l_0)dx = 1$
(b) $\mu_x = -l'_x/l_x = 2/(1+x) \;\Rightarrow\; \mu_1 = 1$
(c) ${}_{1|}q_0 = (l_1 - l_2)/l_0 = 0.13889$

4.8.12 Table 2 gives the values of l_x and d_x, which are rounded to integers.

x	l_x	d_x
30	75000.00	750.00
31	74250.00	742.50
32	73507.50	1470.15
33	72037.35	2161.12
34	69876.23	2795.05
35	67081.18	3354.06

Table 2 Values of l_x and d_x

4.8.13 (a) l_{35}/l_{31} (b) d_{34}/l_{32} (c) $3000(l_{31} - l_{35})/l_{30}$
(d) $\frac{l_{31}-l_{51})}{l_{31}} \times \left(\frac{l_{33}-l_{53}}{l_{33}}\right)$,   (e) $\left(\frac{l_{20}}{l_0}\right)\left(\frac{l_{31}-l_{51}}{l_{31}}\right)\left(\frac{l_{33}-l_{53}}{l_{33}}\right)$
(f) $\frac{T_0-T_{15}}{T_0} \times 100$ (g) $\frac{T_{65}}{T_0} \times 100$ (h) $20 + e_{20}$

4.8.14 2081.61

4.8.15 $f_X(x) = -\frac{d}{dx}(l_x/l_0) = (1/15)(64 - 0.8x)^{-2/3}$, $0 \le x < 80$

4.8.16 $\mu_x = (-\frac{d}{dx}l_x)/l_x = (1/2)(100 - x)^{-1}$. Hence the exact value of $\mu_{36+1/4}$ is 0.0078431. Under the assumption of uniformity, $\mu_{36+1/4} = 0.007858$. Under the assumption of uniformity, $\mu_{36+1/4}$ is slightly higher than the exact value.

4.8.17 All the three relationships are correct.

4.8.18 25.1 years

4.8.19 (a) 0.2444   (b) 0.2589

4.8.20 0.03959

4.8.21 0.9505

4.8.22 (a) 0.9139   (b) 0.006091

4.8.23 0.0782

4.8.24 19/27

4.8.25 (a) $1 - e^{-0.004}$ (b) $1 - e^{-0.001}$ (c) $1 - e^{-0.0005}$

4.8.26 0.3

4.8.27 1/8

4.8.28 (a) 0.81   (b) 0.242   (c) 0.924

4.8.29 (a) 287   (b) 194

4.8.30 0.0036

4.8.31 (a) $_{10}p_{[35]}$ (b) $_{10|5}q_{[35]}$

4.8.32 (a) 0.01296   (b) 0.006923

4.8.33 4590.571

4.8.34 (i) is true and (ii) and (iii) are false

4.8.35 41.56

4.8.36 8224.6246

4.8.37 Table 3 presents all the values

$[x]$	$l_{[x]}$	$l_{[x]+1}$	l_{x+2}	$x + 2$
45	10000	9908	9823.78	47
46	9901.01	9817.84	9752.07	48

Table 3 2 Year select and ultimate life table

Answers to the multiple choice questions, based on Chapter 4, are given in Table 4.

Q.No.	1	2	3	4	5	6	7	8	9
Ans	b	a	c	b	d	c	c	c	a

Q.No.	10	11	12	13	14	15	16	17	18
Ans	b	d	b	c	a	c	a	c	b

Q.No.	19	20	21	22	23	24	25		
Ans	b	c	b	b	d	d	c		

Table 4 Answer Key to MCQs in Chapter 4

Chapter 5

5.7.1 (a) $d = 0.03846154$, $v = 0.9615385$, $i^{(2)} = 0.03960781$
$i^{(4)} = 0.03941363$, $i^{(12)} = 0.03928488$, $i^{(365)} = 0.03922282$ and
$\delta = 0.03922071$
(b) $i = 0.06183655$, $d = 0.05823547$, $v = 0.9417645$, $i^{(2)} = 0.06090907$
$i^{(4)} = 0.06045226$, $i^{(12)} = 0.06015025$, $i^{(365)} = 0.06000493$

5.7.2 (a) 50005.48   (b) 50187.50   (c) 52500

5.7.3 26789.76

5.7.4 6232.94

5.7.5 Seven days: 0.06003463   one month: 0.06015025,   six months: 0.06090907

5.7.6 (a) 40204.36,   (b) 41357.32,   (c) 47888.69

5.7.7 (a) $(\beta/(\delta + \beta))^\alpha\ P[W < 25]$, where W follows gamma distribution with scale parameter
$\delta + \beta$ and shape parameter α.
(b) $P[W < 25]$, where W follows gamma distribution with scale parameter β and shape
parameter α.
(c) $\overline{A}_{30} = (\beta/(\delta + \beta))^\alpha$, $\overline{IA}_{30} = \beta^\alpha \Gamma(\alpha + 1)/(\beta + \delta)^{\alpha+1}\Gamma\alpha$

5.7.8 a and b are true

5.7.9 (a) 0.5   (b) 0.05

5.7.10 0.0526

5.7.11 0.018

5.7.12 0.1797

5.7.13 0.065

5.7.14 $A_{61} = 0.57905$, $A_{60} = 0.55749$

5.7.15 62,000

5.7.16 $A^1_{25:\overline{6}|} = 553.57,\quad \overline{A}^1_{25:\overline{6}|} = 570.02,\quad A_{25:\frac{1}{\overline{6}|}} = 237.10$

$A_{25:\overline{6}|} = 790.67,\quad \overline{A}_{25:\overline{6}|} = 807.12\ \ {}_{2|}A^1_{25:\overline{2}|} = 181.63$

${}_{2|}A_{25:\overline{2}|} = 587.19$

5.7.17 Table 5 displays the values required in (a) and (b).

Age x	1000 A_x	1000 $\overline{A}_x$
25	156.90	160.89
30	191.77	196.64
35	233.39	239.33
40	282.24	289.42

Table 5 $1000A_x$ and $1000\overline{A}_x$

(c) Table 6, Table 7 and Table 8 present the required values.

n	25	30	35	40
1	1.80	2.46	3.50	5.14
2	3.61	4.95	7.07	10.40
3	5.44	7.48	10.72	15.80
4	7.28	10.06	14.45	21.34
5	9.15	12.69	18.27	27.02
6	11.04	15.38	22.19	32.85
7	12.96	18.13	26.20	38.83
8	14.91	20.92	30.32	44.97
9	16.90	23.78	34.54	51.26
10	18.93	26.71	38.87	57.71

Table 6 Makeham's law: $1000A^1_{x:\overline{n}|}$

n	25	30	35	40
1	949.43	948.77	947.73	946.09
2	901.32	900.01	897.94	894.68
3	855.53	853.58	850.50	845.65
4	811.96	809.37	805.29	798.87
5	770.50	767.27	762.19	754.23
6	731.03	727.16	721.10	711.61
7	693.45	688.96	681.92	670.93
8	657.68	652.56	644.55	632.07
9	623.62	617.87	608.89	594.95
10	591.18	584.80	574.86	559.49

Table 7 Makeham's law: $1000A_{x:\frac{1}{\overline{n}|}}$

n	25	30	35	40
1	951.23	951.23	951.23	951.23
2	904.93	904.96	905.01	905.09
3	860.97	861.06	861.21	861.45
4	819.24	819.43	819.74	820.21
5	779.64	779.96	780.46	781.25
6	742.07	742.54	743.29	744.47
7	706.41	707.08	708.12	709.76
8	672.59	673.48	674.86	677.04
9	640.52	641.65	643.43	646.21
10	610.10	611.52	613.73	617.20

Table 8 Makeham's law: $1000A_{x:\overline{n}|}$

5.7.18 (a) $30,000\ A_{30:\overline{20}|}$ (b) $10^5\ _{10|}\overline{A}_{40}$, $30,000\ _{15|}\overline{A}^1_{40:\overline{10}|}$

5.7.19 0.04535

5.7.20 0.6529

Answers to the multiple choice questions, based on Chapter 5, are given in Table 9.

Q.No.	1	2	3	4	5	6	7	8	9	10
Ans	d	a	d	d	c	c	d	d	c	d
Q.No.	11	12	13	14	15	16	17	18	19	20
Ans	b	a	c	b	d	c	c	c	d	d
Q.No.	21	22	23	24	25	26				
Ans	c	c	b	c	b	c				

Table 9 Answer Key to MCQs in Chapter 5

Chapter 6

6.7.1 $15000\ a_{15} = 166775.80$, $15000\ S_{15} = 300353.80$
$15000\ \ddot{a}_{15} = 173446.80$, $15000\ \ddot{S}_{15} = 312368.00$

6.7.2 $10000\ \ddot{a}_{10} = 84353.32$

6.7.3 Monthly installment: 26273.84 for 10 years,   19545.89 for 15 years,   16279.59 for 20 years

6.7.4 Purchase price $= 395115.10$,   Accumulated value $= 584866.90$

6.7.5 Purchase price $= 122426.20$,   Accumulated value $= 161104.60$

6.7.6 Purchase price $= 133406.50$

6.7.7 4753.51

6.7.8 (a) 67863.73   (b) 64632.13   (c) 66369.53
(d) 66100.23   (e) 66234.79

6.7.9 Table 10 and Table 11 report the present values and the accumulated values of the required annuities.

n	$10^3 a_{\overline{n}\rvert}$	$10^3 a_{\overline{n}\rvert}^{(4)}$	$10^3 \overline{a}_{\overline{n}\rvert}$	$10^3 \ddot{a}_{\overline{n}\rvert}^{(4)}$	$10^3 \ddot{a}_{\overline{n}\rvert}$
1	943.40	964.36	971.42	978.52	1000.00
2	1833.39	1874.14	1887.86	1901.64	1943.40
3	2673.01	2732.42	2752.42	2772.52	2833.39
4	3465.11	3542.12	3568.05	3594.10	3673.01
5	4212.36	4305.99	4337.51	4369.18	4465.11
6	4917.32	5026.62	5063.41	5100.38	5212.36
7	5582.38	5706.46	5748.23	5790.20	5917.32
8	6209.79	6347.82	6394.28	6440.97	6582.38
9	6801.69	6952.87	7003.76	7054.90	7209.79
10	7360.09	7523.68	7578.75	7634.08	7801.69

Table 10 Present values of annuities

n	$10^3 S_{\overline{n}\rvert}$	$10^3 S_{\overline{n}\rvert}^{(4)}$	$10^3 \overline{S}_{\overline{n}\rvert}$	$10^3 \ddot{S}_{\overline{n}\rvert}^{(4)}$	$10^3 \ddot{S}_{\overline{n}\rvert}$
1	1000.00	1022.23	1029.71	1037.23	1060.00
2	2060.00	2105.79	2121.20	2136.69	2183.60
3	3183.60	3254.36	3278.18	3302.12	3374.62
4	4374.62	4471.85	4504.58	4537.47	4637.09
5	5637.09	5762.39	5804.56	5846.94	5975.32
6	6975.32	7130.36	7182.55	7234.99	7393.84
7	8393.84	8580.41	8643.21	8706.31	8897.47
8	9897.47	10117.46	10191.51	10265.92	10491.32
9	11491.32	11746.73	11832.71	11919.10	12180.79
10	13180.79	13473.76	13572.38	13671.47	13971.64

Table 11 Accumulated values of annuities

6.7.10 $8000\, \overline{a}_{32:\overline{5}\rvert} = 26374.67$ Standard deviation $= 11227.90$

6.7.11 $8000\, \overline{a}_{32} = 69824.00$ Standard deviation $= 13077.04$

6.7.12 $8000\, _{5\rvert}\overline{a}_{32} = 43449.33$ $8000\, \overline{a}_{\overline{32:5}\rvert} = 78006.90$

6.7.13 (a) 13.8484   (b) 3.2934

6.7.14 $10000\ddot{a}_{25:\overline{6}\rvert} = 36981.01$, $10000 a_{25:\overline{6}\rvert} = 29352.05$, $10000\overline{a}_{25:\overline{6}\rvert} = 33101.81$
$10000\, \ddot{a}_{25:\overline{6}\rvert}\ >\ 10000\, \overline{a}_{25:\overline{6}\rvert}\ >\ 10000\, a_{25:\overline{6}\rvert}$

6.7.15 (a) $1000\ _{5|}\ddot{a}_{40} = 10006.48$ $1000\ _{5|}a_{40} = 9277.37$

   (b) $1000\ \ddot{a}_{\overline{40:\overline{5}|}} = 14457.06$ $1000\ a_{\overline{40:\overline{5}|}} = 13468.77$

   (c) Table 12 reports the actuarial present values of the quarterly annuities.

Age	$1000\ \ddot{a}_{40}^{(4)}$	$1000\ a_{40}^{(4)}$
40	14052.22	13802.22
41	13921.43	13671.43
42	13785.02	13535.02
43	13643.97	13393.97
44	13496.50	13246.50
45	13343.52	13093.52
46	13183.19	12933.19
47	13017.66	12767.66
48	12845.13	12595.13
49	12666.32	12416.32

Table 12 Actuarial present values of annuities with quarterly payment

6.7.16 $10000\ddot{a}_{25:\overline{6}|}^{(12)} = 33420.13$ $10000a_{25:\overline{6}|}^{(12)} = 32784.39$

6.7.17 (c) is the correct expression.

6.7.18 (b) $_nE_x = 1/4$, $\ddot{a}_{x:\overline{n}|} = 18.75$

6.7.19 $a_x = vp_x(1 + a_{x+1})$

6.7.20 0.5217

6.7.21 (a) $5000\ a_{\overline{15}|}$ (b) $10000\ _{10|}\ddot{a}_{\overline{20}|}$ (c) $12000\ \ddot{a}_{\overline{10}|}^{(12)}$

   (d) $20000\ a_{50:\overline{20}|}$ (e) $60000\ \ddot{a}_{40}^{(12)}$ (f) $8000\ _{10|}\overline{a}_{50:\overline{20}|}$, (g) $5000\ \ddot{a}_{\overline{55:\overline{5}|}}$

6.7.22 $\ddot{a}_{74}^{(2)} = 8.59$

6.7.23 $\overline{a}_x = 11.2058$

6.7.24 $Var(\overline{a}_{\overline{T}|}) = 52.0833$

Answers to the multiple choice questions, based on Chapter 6, are given in Table 13.

Chapter 7

7.8.1   2500

7.8.2   0.2506

Q.No.	1	2	3	4	5	6	7	8	9	10
Ans	b	c	b	d	a	a	c	d	c	a
Q.No.	11	12	13	14	15	16	17	18	19	20
Ans	a	d	c	b	a	b	b	d	b	c
Q.No.	21	22	23	24	25	26	27	28	29	30
Ans	b	a	c	d	d	d	d	b	c	d
Q.No.	31	32	33	34	35	36	37	38		
Ans	c	c	d	c	d	c	d	d		

Table 13 Answer Key to MCQs in Chapter 6

7.8.3 (a) $L(K) = v^{K+1} - P\ddot{a}_{\overline{K+1|}} = (1 + (P/d))\, v^{K+1} - (P/d)$

(b) (i), (c) (iii), (d) $E(L) = -0.2,\ \ Var(L) = 0.16,\ $ (e) 0.023

7.8.4 (a)$L(T) = 1000(v^T - P\,\overline{a}_{\overline{T}|}),\ \ $ if $T \leq 5$ and $1000(v^5 - P\,\overline{a}_{\overline{5}|}),\ $ if $T > 5$

$$Var(L(T)) = (1 + P/\delta)^2\, (\,{}^{2}\overline{A}_{32:\overline{5}|} - \overline{A}^{2}_{32:\overline{5}|})\, \times 10^6$$

(b)$1000\,\overline{P}(\overline{A}_{32:\overline{5}|})$=243.32,   variance $= 181227.2$
(c) Expense loaded premium $= 285.12,$ loading for expenses $= 41.80$

$$(d)L(T) = 1000(v^T\, -\, P\,\overline{a}_{\overline{T}|}),\ \ Var(L(T)) = (1 + P/\delta)^2\, (\,{}^{2}\overline{A}_{32} - \overline{A}^{2}_{32})\, \times 10^6$$

$1000\,\overline{P}(\overline{A}_{32}) = 54.57,$ variance $= 35075.96$
Expense loaded premium $= 64.92,$ loading for expenses $= 10.35$
(e) $1000\,\overline{P}({}_{5|}\overline{a}_{32}) = 1257.30,$ $1000\,\overline{P}(\overline{a}_{\overline{32:\overline{5}|}}) = 2257.30$

7.8.5 (a) $1000\,P^{\,1}_{25:\overline{6}|} = 149.69,$ $1000\,P_{25:\overline{6}|}^{\ \ 1} = 64.12,$ $1000\,P_{25:\overline{6}|} = 213.81$
(b) $1000\,P(\overline{A}_{25:\overline{6}|}) = 218.25,$ under the assumption of uniformity
(c) $1000\,{}_{3}P^{\,1}_{25:\overline{6}|} = 225.00,$ $1000\,{}_{3}P_{25:\overline{6}|}^{\ \ 1} = 96.37,$ $1000\,{}_{3}P_{25:\overline{6}|} = 321.38$
(d) Table 14 gives the expense loaded premiums, net premiums and the loading for expenses for insurance products mentioned in (a) to (c).

7.8.6 (a)$1000P^{(12)}_{30}/12 = 0.95,$ (b) $1000P^{(12)}(\overline{A}_{30})/12 = 0.97$
(c) $1000\,{}_{6}P^{(4)}_{30}/4 = 8.89,$ $1000\,{}_{6}P^{(4)}(\overline{A}_{30})/4 = 9.12$
(d) $1000\,{}_{6}P^{(2)}(\ddot{a}_{30})/2 = 1601.91$
(e) $1000P^{(12)}_{30:\overline{6}|}/12 = 11.978,$ (f) $1000P^{1(12)}_{30:\overline{6}|}/12 = 0.192$
(g) $1000P^{(12)}(\overline{A}_{30:\overline{6}|})/12 = 11.983,$ $1000P^{(12)}(\overline{A}^{\,1}_{30:\overline{6}|})/12 = 0.197$

7.8.7 0.0413

7.8.8 $\overline{P}(\overline{A}_{40:\overline{25}|}) > P^{(2)}(\overline{A}_{40:\overline{25}|}) > P(\overline{A}_{40:\overline{25}|})$

7.8.9 0.008

Insurance product	Gross premium	Net premium	Loading
6-Year term	189.74	149.69	40.05
6-Year pure endowment	86.15	64.12	22.03
6-Year endowment	267.36	213.81	53.55
6-Year term, 3-payment years	285.21	225.00	60.21
6-Year pure endowment, 3-payment years	129.50	96.37	33.13
6-Year endowment, 3-payment years	401.87	321.38	80.49
6-Year endowment, semi-continuous	272.74	218.25	54.49

Table 14 Net and gross premiums for benefit of 1000

7.8.10 (a) $20000\, A_{40:\overline{20|}}^{\ 1}/\ddot{a}_{40:\overline{20|}}$ (b) $(12 \times 8000\,_{30|}a_{30})/(12\, \ddot{a}_{30:\overline{30|}}^{(12)})$

(c) $25000\, \ddot{a}_{\overline{20|}}/\ddot{a}_{40:\overline{20|}}$ (d) $(20000\,_{10|}\overline{a}_{50})/(12\, \ddot{a}_{50:\overline{5|}}^{(12)})$

(e) $(10^5\, A_{25:\overline{25|}})/(4\, \ddot{a}_{25:\overline{10|}}^{(4)})$

(f) Suppose G denotes the monthly premium, then G is a solution of following equation.

$$12G\ddot{a}_{40:\overline{20|}}^{(12)} = 12 \times 5000\,_{20|}\ddot{a}_{40}^{(12)} + 0.2 \times 5000 + 12 \times 0.05G\ddot{a}_{40:\overline{20|}}^{(12)} - 0.05G$$

Answers to the multiple choice questions, based on Chapter 7, are given in Table 15.

Q.No.	1	2	3	4	5	6	7	8
Ans	c	c	a	b	d	c	b	d
Q.No.	9	10	11	12	13	14	15	16
Ans	a	b	c	c	b	d	a	d
Q.No.	17	18	19	20	21	22	23	24
Ans	a	d	a	d	c	b	d	c

Table 15 Answer Key to MCQs in Chapter 7

Chapter 8

8.5.1 Table 16 presents the actuarial present values required to find prospective reserves. Premiums for term and endowment insurance for a unit benefit are given by 0.1466 and 0.2387 respectively. Further, $\ddot{a}_{25:\overline{2}|} = 1.8066$, $_2P^1_{25:\overline{5}|} = 0.2748$ and $_2P_{25:\overline{5}|} = 0.4474$. Table 17 and Table 18 display the values of prospective and retrospective reserves respectively for a fully discrete policy. Both the approaches give the same values of reserve for all the products.

k	$A^1_{25+k:\overline{5-k}|}$	$A_{25+k:\overline{5-k}|}^{1}$	$A_{25+k:\overline{5-k}|}$	$\ddot{a}_{25+k:\overline{5-k}|}$
0	0.4965	0.3118	0.8083	3.3863
1	0.4460	0.3866	0.8325	2.9584
2	0.3862	0.4770	0.8632	2.4167
3	0.3002	0.5991	0.8993	1.7792
4	0.1745	0.7689	0.9434	1.0000
5	0.0000	1.0000	1.0000	0.0000

Table 16 Actuarial present values to compute prospective reserve

k	Term	Pure endowment	Endowment	2-Payments term	2-Payments endowment
1	12.19	114.16	126.35	171.13	385.11
2	31.82	254.51	286.33	386.16	863.21
3	39.28	435.30	474.57	300.15	899.29
4	27.91	676.78	704.69	174.53	943.40
5	0.00	1000.00	1000.00	0.00	1000.00

Table 17 Reserve by prospective approach: Fully discrete policy

k	Term	Pure endowment	Endowment	2-Payments term	2-Payments endowment
1	12.19	114.16	126.35	171.13	385.11
2	31.82	254.51	286.33	386.16	863.21
3	39.28	435.30	474.57	300.15	899.29
4	27.91	676.78	704.69	174.53	943.40
5	0.00	1000.00	1000.00	0.00	1000.00

Table 18 Reserve by retrospective approach: Fully discrete policy

8.5.2 The values of premiums are as follows. $P(\overline{A}^{\,1}_{x:\overline{n}|}) = 0.1510$, $P(\overline{A}_{x:\overline{n}|}^{\,\,1}) = 0.09208$, $P(\overline{A}_{x:\overline{n}|}) = 0.2431$, $_2P(\overline{A}^{\,1}_{x:\overline{n}|}) = 0.2830$, $_2P(\overline{A}_{x:\overline{n}|}) = 0.4556$. To obtain the reserve values for a semi-continuous policy, we need to assume the uniformity of deaths in a unit interval. Table 19 and Table 20 display the values of prospective and retrospective reserves respectively for a semi-continuous policy. Both the approaches give the same values of reserve for all the products. We note that the reserve values for semi-continuous policies are slightly higher than those for fully discrete policies.

k	Term	Pure endowment	Endowment	2-Payments term	2-Payments endowment
1	12.55	114.16	126.71	176.21	390.20
2	32.77	254.51	287.28	397.63	874.68
3	40.44	435.30	475.74	309.07	908.21
4	28.74	676.78	705.52	179.71	948.58
5	0.00	1000.00	1000.00	0.00	1000.00

Table 19 Reserve by prospective approach: Semi-continuous policy

k	Term	Pure endowment	Endowment	2-Payments term	2-Payments endowment
1	12.55	114.16	126.71	176.21	390.20
2	32.77	254.51	287.28	397.63	874.68
3	40.44	435.30	475.74	309.07	908.21
4	28.74	676.78	705.52	179.71	948.58
5	0.00	1000.00	1000.00	0.00	1000.00

Table 20 Reserve by retrospective approach: Semi-continuous policy

8.5.3 Table 21 displays the values of the prospective reserve when the premiums are paid quarterly. Using the retrospective approach we get exactly the same values as obtained with the prospective approach. To obtain the reserve values when the premiums are paid quarterly, we need to assume the uniformity of deaths in a unit interval.

8.5.4 Table 22 presents the formulae for variance

8.5.5 $_nV_x = 1/3, \qquad _nV_{x+n} = 1/2$

8.5.6 By retrospective method, $_1V_{40} = 0.01264$

8.5.7 0.005

8.5.8 0.240

8.5.9 (b) and (c)

k	Term	Pure endowment	Endowment	2-Payments term	2-Payments endowment
1	12.93	114.63	127.56	170.89	384.73
2	33.77	255.73	289.50	386.16	863.21
3	41.68	436.80	478.48	300.15	899.29
4	29.61	677.85	707.47	174.53	943.40
5	0.00	1000.00	1000.00	0.00	1000.00

Table 21 Reserve by prospective approach: Quarterly premiums

Product	Variance of $_tL$
n-Year term	$(1 + \overline{P}(\overline{A}^1_{x:\overline{n}|})/\delta)^2\ (^2\overline{A}^1_{x+t:\overline{n-t}|} - (\overline{A}^1_{x+t:\overline{n-t}|})^2)$
n-Year endowment	$(1 + \overline{P}(\overline{A}_{x:\overline{n}|})/\delta)^2\ (^2\overline{A}_{x+t:\overline{n-t}|} - (\overline{A}_{x+t:\overline{n-t}|})^2)$

Table 22 Variance of prospective loss random variable: Fully continuous policy

8.5.10 36497.16

8.5.11 0.0099

8.5.12 0.8514

Answers to the multiple choice questions, based on Chapter 8, are given in Table 23.

Q.No.	1	2	3	4	5	6	7	8
Ans	a	d	b	d	b	d	a	b
Q.No.	9	10	11	12	13	14	15	16
Ans	b	d	b	b	b	c	b	a
Q.No.	17	18	19	20				
Ans	d	b	c	b				

Table 23 Answer Key to MCQs in Chapter 8

Chapter 9

9.5.1 (a) 0.6519, (b) 0.1673 (c) 0.4824 (d) 0.09263 (e) 0.08425

9.5.2

$$(a)\,10000\,\ddot{a}_{25:27:\overline{3}|} = 21311.59$$

$$10000\,a_{25:27:\overline{3}|} = 10000\,\{\ddot{a}_{25:27:\overline{3}|} + {}_3E_{25\ 27} - 1\}$$

$$= 10000\,\{2.131159 + 0.2922663 - 1\} = 14234.26$$

$$(b)\,\ddot{a}_{\overline{25:27}:\overline{3}|} = \{\ddot{a}_{25:\overline{3}|} + \ddot{a}_{27:\overline{3}|} - \ddot{a}_{25:27:\overline{3}|}\}$$

$$= \{2.452648 + 2.416682 - 2.131159\} = 2.738171$$

$$\Rightarrow 10000\,\ddot{a}_{\overline{25:27}:\overline{3}|} = 27381.71$$

$$a_{\overline{25:27}:\overline{3}|} = \{a_{25:\overline{3}|} + a_{27:\overline{3}|} - a_{25:27:\overline{3}|}\}$$

$$= \{1.967045 + 1.893731 - 1.423426\} = 2.43735$$

$$\Rightarrow 10000\,a_{\overline{25:27}:\overline{3}|} = 24373.50$$

9.5.3 (a) $10000A_{25:27:\overline{3}|} = 8793.68,\qquad 10000A_{25:27:\overline{3}|}^{\ 1} = 5871.02$

(b) $10000A_{\overline{25:27}:\overline{3}|} = 8450.09,\qquad 10000A_{\overline{25:27}:\overline{3}|}^{\ \ 1} = 1458.30$

(c) Joint life status: endowment insurance: 4126.25,
term insurance: 2754.85
Last survivor status: endowment insurance: 3086.04,
term insurance: 532.58

9.5.4 (a) Table 24 displays the probability distribution of $K(35:40)$. The expected values

k	Probability	k	Probability	k	Probability	k	Probability
0	0.00906	15	0.02422	30	0.03399	45	0.00271
1	0.00970	16	0.02557	31	0.03276	46	0.00180
2	0.01039	17	0.02693	32	0.03120	47	0.00114
3	0.01113	18	0.02828	33	0.02933	48	0.00069
4	0.01193	19	0.02959	34	0.02717	49	0.00039
5	0.01278	20	0.03084	35	0.02477	50	0.00021
6	0.01368	21	0.03202	36	0.02220	51	0.00010
7	0.01464	22	0.03308	37	0.01952	52	0.00005
8	0.01566	23	0.03401	38	0.01681	53	0.00002
9	0.01674	24	0.03476	39	0.01415	54	0.00001
10	0.01787	25	0.03532	40	0.01161	55	0.00000
11	0.01905	26	0.03563	41	0.00927	56	0.00000
12	0.02029	27	0.03568	42	0.00718	57	0.00000
13	0.02156	28	0.03544	43	0.00539	58	0.00000
14	0.02288	29	0.03488	44	0.00389	59	0.00000

Table 24 Makeham's law: Probability distribution of $K(35:40)$

are as follows. $e_{35:40} = 22.96,\quad e_{35} = 31.996,\quad e_{40} = 27.65$ and $e_{\overline{35:40}} = 36.69$.
(b) $1000A_{35:40} = 347.00,\quad 1000A_{35} = 233.39,\quad 1000A_{40} = 282.24$ and

$1000A_{\overline{35:40}} = 168.63.$

(c) Premium for whole life insurance in joint life status = 25.92.

(d) Premium for whole life insurance in last survivor status = 9.89.

(e) Table 25 and Table 26 display the required values.

(f) Table 27 and Table 28 display the required values.

n	Term	Pure endowment	Endowment
1	8.61	942.61	951.23
2	17.39	887.87	905.26
3	26.34	835.62	861.96
4	35.45	785.75	821.20
5	44.74	738.14	782.88
6	54.21	692.67	746.88
7	63.85	649.25	713.10
8	73.67	607.77	681.43
9	83.65	568.14	651.79
10	93.81	530.28	624.09

Table 25 Actuarial present values of benefit of 1000 in joint life status

n	Term	Pure endowment	Endowment
1	0.02	951.21	951.23
2	0.08	904.76	904.84
3	0.18	860.53	860.71
4	0.34	818.40	818.74
5	0.55	778.28	778.83
6	0.83	740.04	740.87
7	1.18	703.60	704.78
8	1.62	668.85	670.47
9	2.14	635.70	637.85
10	2.77	604.07	606.84

Table 26 Actuarial present values of benefit of 1000 in last survivor status

9.5.5 0.25

9.5.6 $0.168t - 0.00128t^3$

9.5.7 0.9563

9.5.8 0.08766

9.5.9 31.75

Answers to the multiple choice questions, based on Chapter 9, are given in Table 29.

n	Term	Pure endowment	Endowment
1	8.61	942.61	951.23
2	8.95	457.05	466.00
3	9.30	295.22	304.53
4	9.67	214.33	224.00
5	10.05	165.81	175.86
6	10.44	133.46	143.91
7	10.85	110.37	121.22
8	11.28	93.05	104.32
9	11.72	79.57	91.29
10	12.17	68.80	80.97

Table 27 Premiums for benefit of 1000 in joint life status

n	Term	Pure endowment	Endowment
1	0.02	951.21	951.23
2	0.04	463.69	463.73
3	0.06	301.31	301.37
4	0.09	220.21	220.30
5	0.12	171.62	171.74
6	0.16	139.28	139.44
7	0.20	116.24	116.43
8	0.24	98.99	99.23
9	0.29	85.61	85.90
10	0.34	74.93	75.28

Table 28 Premiums for benefit of 1000 in last survivor status

Q.No.	1	2	3	4	5	6	7	8
Ans	b	c	c	c	b	d	b	a

Q.No.	9	10	11	12	13	14	15	16
Ans	c	c	a	c	b	d	c	c

Q.No.	17	18	19	20	21	22		
Ans	b	b	c	c	a	b		

Table 29 Answer Key to MCQs in Chapter 9

Bibliography

[1] **Bowers Jr N L, Gerber H U, Hickman J C, Jones D A and Nesbitt C J** (1997). *Actuarial Mathematics*, Second Edition, *The Society of Actuaries*, Sahaumburg, Illinois.

[2] **Crawley M J** (2007). *The R Book*, John Wiley, London.

[3] **Dalgaard P** (2008). *Introductory Statistics with R*, Second Edition, Springer, New York.

[4] **Deshmukh Shailaja** (2012). *Multiple Decrement Models in Insurance: An Introduction Using with R*, Springer, New Delhi.

[5] **Douglas Bates and Martin Maechler** (2019). Matrix: Sparse and Dense Matrix Classes and Methods. R package version 1.2-17. https://CRAN.R-project.org/package=Matrix

[6] **Frederick Novomestky** (2012). matrixcalc: Collection of functions for matrix calculations. R package version 1.0-3. https://CRAN.R-project.org/package=matrixcalc

[7] **Gregory R Warnes, Ben Bolker and Thomas Lumley** (2018). gtools: Various R Programming Tools. R package version 3.8.1. https://CRAN.R-project.org/package=gtools

[8] **Hans Werner Borchers** (2018). numbers: Number-Theoretic Functions. R package version 0.7-1. https://CRAN.R-project.org/package=numbers

[9] **Johnson N L, Kotz S and Balakrishnan N** (1995). *Discrete Univariate Distributions*, John Wiley, New York.

[10] **Johnson N L, Kotz S and Balakrishnan N** (1995). *Continuous Univariate Distributions - I*, John Wiley, New York.

[11] **Johnson N L, Kotz S and Balakrishnan N** (1995). *Continuous Univariate Distributions*, Vol II, Second Edition, John Wiley, New York.

[12] **Palande P S, Shah R S and Lunawat M L** (2003). *Insurance in India - Changing Policies and Emerging Opportunities*, Response Books, New Delhi.

[13] **Pollard A H, Yusuf F and Pollard G N** (1981). *Demographic Techniques*, Pergamon Press, Sydney.

[14] **Purohit S G, Gore S D and Deshmukh S R** (2008). *Statistics using R*, second edition. Narosa Publishing House, New Delhi.

[15] **R Core Team** (2019). *R: A language and environment for statistical computing*, R Foundation for Statistical Computing, Vienna, Austria. URL- https://www.R-project.org/.

[16] **Rohatgi V K and Saleh A K MD E** (2002). *An Introduction to Probability and Statistics*, John Wiley, New York.

[17] **Ross S M** (2014) *Introduction to Probability Models*, 11th edition, Academic Press, New York.

[18] **Siegel J S and David A S** (2004). *The Methods and Materials of Demography*, Second Edition, Elsevier Academic Press, San Diego.

[19] **signal developers** (2013). signal:Signal processing. URL: http://r-forge.r-project.org/projects/signal/.

[20] **Venables W N and Ripley B D** (2002). *Modern Applied Statistics with S*, fourth edition, Springer, New York, URL- http://www.stats.ox.ac.uk/pub/MASS4

[21] **Verzani J** (2005). *Using R for Introductory Statistics*, Chapman and Hall/CRC Press, New York.

[22] **Vincent Goulet, Christophe Dutang, Martin Maechler, David Firth, Marina Shapira and Michael Stadelmann** (2019). expm: Matrix Exponential, Log, 'etc'. R package version 0.999-4. https://CRAN.R-project.org/package=expm

Index